Elevator Traffic Handbook

This second edition of this well-respected book covers all aspects of the traffic design and control of vertical transportation systems in buildings, making it an essential reference for vertical transportation engineers, other members of the design team, and researchers. The book introduces the basic principles of circulation, outlines traffic design methods and examines and analyses traffic control using worked examples and case studies to illustrate key points. The latest analysis techniques are set out, and the book is up-to-date with current technology. A unique and well-established book, this much-needed new edition features extensive updates to technology and practice, drawing on the latest international research.

Gina Barney is principal of Gina Barney Associates. Through her 50 years of experience in practice, research and lecturing in vertical transportation and control systems, she has had a profound influence on this field. Her involvement continues with her activities with the BSI Committees on Lifts, Escalators & Moving Walks, the CIBSE Lifts Group Executive Committee and as English Editor of *Elevatori* (the European lift magazine).

Lutfi Al-Sharif is an Associate Professor at The University of Jordan. He has 30 years of experience in the area of lifts and escalators, including working for a manufacturer (Jordan Lift & Crane), for a client (London Underground), for a consultancy (WSP and Al-Sharif VTC) and in academia (The University of Jordan). He has had over 20 papers published in peer-reviewed journals, is co-author of three patents and is a member of the scientific committee of the Lift Symposium held annually at the University of Northampton.

Elevator Traffic Handbook

Theory and practice

Second edition

Gina Barney and Lutfi Al-Sharif

Routledge
Taylor & Francis Group

LONDON AND NEW YORK

First published 2002 by Routledge
Second edition published 2016
by Routledge

2 Park Square, Milton Park, Abingdon, Oxon OX14 4RN
605 Third Avenue, New York, NY 10017

Routledge is an imprint of the Taylor & Francis Group, an informa business

First issued in paperback 2021

Publisher's Note

The publisher has gone to great lengths to ensure the quality of this reprint but points out that some imperfections in the original copies may be apparent.

British Library Cataloguing-in-Publication Data
A catalogue record for this book is available from the British Library

Library of Congress Cataloging-in-Publication Data
Barney, G. C. (Gina Carol), 1935–
    Elevator traffic handbook: theory and practice / Gina Barney and Lutfi Al-Sharif. — 2nd edition.
        pages cm
    Includes bibliographical references and index.
    1. Elevators.   I. Al-Sharif, Lutfi.   II. Title.
    TJ1374.B37 2016
    621.8'77—dc23
    2015005139

ISBN: 978-1-138-85232-7 (hbk)
ISBN: 978-1-03-217965-0 (pbk)
DOI: 10.4324/9781315723600

Typeset in Times New Roman
by Apex CoVantage, LLC

To Josie (1939–2008)
without whose love, friendship and support
many things would not have been
and still would not be possible.

To my wife, Mona,
my daughter Shaima and my sons Khaled, Omar and little Rawhi,
. . . especially Rawhi, without whom I could have completed the draft of
this book much earlier!

Contents

5   Evaluating the round trip time equation   87

TCS Traffic case studies 213

PART C
Traffic control **233**

10 Legacy traffic control 235

11 Computer-based traffic control 261

Figures

Tables

Foreword

To the second edition

When first asked by the authors to be a luminary and writer of the foreword to this latest edition I thought it was fitting, but not for the reasons envisaged by the authors.

During my career within the lifts industry one of the 'dark arts', for me at least, has always been that of traffic analysis and the prediction of people's behaviour in planning the best possible usage of lifts within buildings.

Certainly, I have sat in many lectures on this subject, being beguiled as the lecturer demonstrates their mathematic skill with flawless presentations, but only to come away with more questions than answers.

And so it was that I arrived at the first edition of this book, as an acolyte looking for illumination into the mysterious world of lift traffic analysis.

The first edition helped to open my eyes to the fact that, although without doubt a complicated subject, it was possible to become at least proficient, and in working through the book, to come to an understanding that, while I will never be considered an expert in the field, I can at least say that my curiosity has been sated, the blindfold removed, and my knowledge expanded to the degree that enlightenment has been achieved.

This new edition builds on that achievement, and the collaboration between Dr Barney and Dr Al-Sharif in authoring this revision has resulted in the shedding of new light on this subject.

'Though much is taken, much abides' holds true to this re-crafted work. That text which was part of the first edition has been extensively revised and reordered to help the reader to navigate the subject matter in a more orderly way, and to assist in the formulation of the worked examples. The result is a more user-friendly experience when trying to assimilate the great wealth of knowledge placed within this tome by these two world-renowned experts in the field.

As with all things, the study of traffic has not stood still in the years since first publication. With new experience of differing building types and emerging technology in lift control systems, the authors have elaborated new methodologies in this edition. New chapters regarding the use of queuing theory and chaotic system modelling, among a general discussion on new advanced techniques, show an ever-present will to create the perfect model simulation. Certainly, the current simulation of traffic patterns has far exceeded those of old, but as experience is gained in economic areas outside of the traditional Europe and North American boundaries, then models and simulations have had to be adapted to meet these new market needs.

Particularly in the last few years, there has been increasing debate among lift professionals regarding all aspects of traffic handling. This has had an invigorating effect on standards writers at all levels, but particularly in ISO where new work has commenced to aid the common terminology, understanding and calculation of traffic studies, work in which Dr Barney and Dr Al-Sharif are actively participating. There is no doubt that their redrafting of this book will contribute towards those new standards to the benefit of all.

Some challenges to be overcome at ISO level will be the continued reliance on standards for capacity based on a given passenger mass and then its translation into a given capacity for persons.

Even the latest CEN/TR81–12 has acknowledged that each economic area has its own idea of what the average weight of a person is, predictably showing a smaller mass for those in the Far East to those in the West. This results in the disparity of market solutions where a 1,600 kg lift can have different capacities depending on where in the world it is installed, and therefore subsequent effects on traffic study.

The help of works such as this in bringing these problems into the light, so that they can be closely studied and resolved within our prescriptive standards committees, will lead to further global unification of our trade.

Finally, the new edition looks towards the future. Control systems that were previously only imagined in the past have become common place, such as destination control on entry to the building. Add this to the increased awareness of the needs of disabled persons and it is easy to see that there are still areas of traffic study yet to be conquered.

Who knows what the future may bring and how far away we are in terms of face recognition and a cheery greeting from the control system of 'Good morning Mr Jones, your lift awaits you'!

Ian G. Jones
Chairman – MHE/4, British Standards Institution
Committee for Lifts and Escalators
Convenor – CEN/TC10/WG1, European Committee for
Standardisation of Passenger and Goods Lifts

Foreword
To the first edition

People who are familiar with Dr Barney's book, *Elevator Traffic Analysis Design and Control* (written with S. M. Dos Santos), will be quite surprised when they review Dr Barney's latest – this volume. Where the initial book left off, this one picks up, repeats some of the previous information and goes way beyond.

I was impressed that the title of the book is 'Elevator Traffic Handbook' yet most of the text refers to 'lifts'. Those familiar with the current lift (one must get used to the word 'lift' instead of 'elevator' when reading this book) and escalator literature can recognise some information and will appreciate its association in Dr Barney's context.

Dr Barney is a good student and an astute observer. She has taken the appropriate information about pedestrian handling and movement and consolidated it into this handbook. It is quite interesting that Dr Barney starts off with pedestrian circulation in her initial chapters. This is one area which few elevator (lift) consultants consider and its proper provision can be an asset or detriment to any vertical transportation scheme. This is particularly true of escalators, which Dr Barney emphasises.

An interesting feature is added after the initial chapters in a series of case studies where various traffic situations are presented and solutions offered. I am certain many readers will relate to some of these cases and could add others from their own experience.

Beyond that, the book settles into the fundamentals of elevators, people handling, theory and design. How elevators serve people, how people react to elevators, the many time factors, the estimation of demand, various traffic patterns – the myriad aspects that constitute the entire art of elevator traffic design are covered in these ensuing chapters. One may or may not agree with what is presented but the debate goes on as long as anyone involved in the process has any judgement or opinion.

One of the most controversial aspects of vertical transportation is covered in Chapter 6, which is the determination of passenger demand. This relates to the population and the use of various buildings, which is probably the most 'blue sky' aspect of the whole design process. Who can predict what will happen in the years or more between the initiation and the completion of any building? As an observation, some consultants have either over populated or over elevated a building and have been overly pessimistic about the elevator's ability to handle the projected population and have ended up 'playing safe'.

Leaving the theoretical behind, the book proceeds into the various aspects of elevator operation and traffic situations. The mathematics becomes more extensive, and various individuals who have had input to the process are quoted and discussed. I was pleased to see the reference to loading based on volume rather than weight, which those familiar with my work will recognise as one of my pleas.

Dr Barney does an extensive job in this middle third of the book. Special elevator situations such as sky lobbies, shuttles and double decks are investigated, as well as the more esoteric aspects of elevator shape, door design, lobby size and so on. Tables and case studies are provided to illustrate and consolidate information. If you are looking for something, you will find it, and the extensive table of contents and comprehensive index will help.

The ensuing chapters are devoted to various operating systems and approaches to traffic handling. Included are discussions of call allocation and fixed and dynamic sectoring. Common and unusual traffic situations are discussed, as well as some of the aspects of traffic handling during peak periods.

Modern elevator literature would not be complete without investigation of computer control, which Dr Barney does in depth. I have to confess that much of it goes over my head, being a veteran of the transition from relay control to solid state. I trust that the modern elevator person will have little trouble understanding what Dr Barney presents. Reinforcing the chapters are a number of case studies which enhance the related information.

The final chapters are an in-depth study of elevator traffic handling, which is an expansion of the studies presented in the earlier book. Helping the discussion is a study of simulation which, of course, was a difficult approach when the original volume was published in 1977 and revised in 1985. Modern computers have made this a viable approach and a tool for current elevator engineers.

All in all, Dr Barney has added a substantial volume to the growing library of literature dealing with vertical transportation. The book contributes to the final product of our vertical transportation efforts. We are not dealing with the hardware of elevators and escalators but rather with their use and purpose – that is – moving people safely and swiftly from place to place. The information found in this book will help in accomplishing that goal.

George Strakosch
Elevator Consultant *Emeritus*
c/o Elevator World, Mobile, USA
21 June 2002

Preface

The first author's work in the lift industry started in early 1968, when she was an academic at the University of Manchester Institute of Science and Technology (UMIST). At that time, *Encyclopaedia Britannica* listed four books on lift technology. Of these, the most important to her was *Elevators and Escalators* by George Strakosch (1967). This book, now in its fourth edition (2010), provided her with an excellent start to what has become her life's work. Strakosch brought together, in one place, a wide wealth of knowledge, much of which still stands today and readers who do not possess a copy would be advised to acquire one.

By 1977, the first author's student, friend and colleague, Sergio dos Santos, past Rector of Minho University, Portugal, and the first author decided to share the results of their research by writing *Lift Traffic Analysis, Design and Control*, which was republished in 1985 as *Elevator Traffic Analysis Design and Control*. That book underscored the practice described in Strakosch with mathematical and computer analysis. In 2003, the first author published *Elevator Traffic Handbook*.

The second author was fascinated by electronic design and spent four years, during and after his first degree in electrical engineering, designing speed and logic lift controllers in Jordan for the Jordan Lift & Crane Manufacturing Company. In 1989, he moved to UMIST in Manchester, England to undertake an M.Sc. and a Ph.D. in the field of lift traffic analysis under the supervision of the first author. He then worked in the area of lifts and escalators for around fourteen years in the role of client (London Underground Ltd.) and then as a consultant (Al-Sharif VTC Ltd.). His research in lift traffic was re-invigorated when he switched to an academic path at the Mechatronics Engineering Department at The University of Jordan, Amman, Jordan in 2006. In 2013, he accepted the kind offer of his former supervisor to act as co-author for this book.

In this book, we have attempted to bring together in one place a complete treatment of lift traffic design and control. In addition to our work in the subject, the book draws on the work of other authors in the field, which are included, commented on, extended, agreed with, or disagreed with. The interpretations are ours, as are the mistakes.

The early chapters concern well-established practice that is generally accepted. The latter chapters contain a more philosophical discussion and present controversial aspects that may require further refinement.

Our careers have led us through industry, research, teaching, consulting, code-making and training. Between us, we have over 75 years of experience in the lift industry. We have written this book for everyone involved in traffic design and the traffic control of lifts. We have based it on theory arising out of practice. Lift traffic design is a practical science.

Readers will see from paragraph two of George Strakosch's Foreword to the first edition that he asks American English speakers to beware of the preference in this book for the word 'lift' over the word 'elevator' ('What's in a word', Bates, 1993). The reason is one of legality. In Europe, British English is used. So we have the Lift Directive and Notified Bodies – Lifts at the European level. At the UK level, we have the Lifts Regulations and Lifting Operations and Lifting Equipment Regulations. Thus the word 'elevator' has no legal standing.

While recognising that the invention of the safety gear which led to a realisable lift (elevator) is due to an American, Elisa Otis in 1852, an early, crude, steam-driven lift was available in England in 1835 called the 'Teagle'. It was developed by the company Frost and Stutt. The lift was belt-driven and used a counterweight for extra power.

At the time of writing, the centre of gravity of the lift world has moved to Europe, with some five million lifts compared to North America with some 600,000 lifts, and will move to Asia where China is said to install that many lifts each year.

In addition, the lift world is becoming more united with the CEN 81–20/50 standards being recognised at ISO level, where they will eventually become ISO 22559-X/Y as worldwide standards.

Dr Gina Barney & Dr Lutfi Al-Sharif
Cumbria, England
1 February 2015

Note to readers

Reader's attention is drawn to the use of end notes to provide additional information. These are shown in the text as superscript indicators, eg: [1] and listed at the end of the respective chapter.

Symbols and abbreviations

Not all symbols and abbreviations are given here if their use is restricted to a specific section.

ACA	*adaptive call allocation*

a	rated acceleration in m/s^2
AJT	passenger average journey time (s)
A_l	area of escalator landing (m^2)
a_{max}	maximum attained acceleration in m/s^2
ASRT	(lift) system response time (s)
ATT	passenger average travel time (s)
AWT	passenger average waiting time (s)
B	number of floors in the lower part of a building (dimensionless)
c	number of servers in system (dimensionless)
C_c	corridor handling capacity (persons/unit time)
CCSTP	car call stops (dimensionless)
C_e	theoretical handling capacity of an escalator (persons per unit time)
CGC	computer group control
CLSTP	common landing and car call stops (dimensionless)
C_s	stairway handling capacity (persons per unit time)
CSTP	total number of car stops (dimensionless)
D	average pedestrian density (persons/m^2)
D	distance terminal floor to terminal floor
d_f	typical height of one floor (m)
d_H	height of building (m)
d_l	depth of the deepest car (mm)
D_l	depth of escalator landing (mm)
D_l	density of occupation on landing (persons/m^2)
DNPAWT	down peak passenger average waiting time (s)
DNPSTPS	down peak average number of stops (dimensionless)
DS	dynamic sectoring
E	expectance
E_H	expectance for H
E_P	expectance for P
E_S	expectance for S
FM	[double deck] lift figure of merit (dimensionless)
FRSTP	algorithmically induced stops (dimensionless)
FS	figure of suitability [nearest car] (dimensionless)
FS4	fixed sectoring priority timed

(*Continued*)

FSO	fixed sectoring common sectors
g	going (m)
g_n	gravity factor (m/s^2)
H	average highest reversal floor (dimensionless)
H_D	average highest reversal floor (dimensionless)
H_M	[basement] average lowest reversal floor (dimensionless)
H_p	average highest reversal floor (Poisson pdf)
IFAWT	interfloor passenger average passenger waiting time (s)
INT	average time between successive lift car arrivals
INT_{act}	[HARint] actual target interval (s)
INT_T	[HARint] target interval (s)
j	rated jerk in m/s^3
j_{max}	maximum attained jerk in m/s^3
k	average density of people (people/escalator step)
k	[HCA] look ahead (dimensionless)
K	number of lifts in the group controlled by hall call allocation (in section 11.6.2.6)
L	number of lifts (dimensionless)
LCSTP	landing call stops (dimensionless)
LJ	likely number of floors jumped (dimensionless)
M	door weight (kg)
M	number of basement floors (dimensionless)
MIDAWT	midday passenger average waiting time (s)
MIDHC	midday handling capacity (persons/5-minutes)
MT	main terminal
n	number of calls registered in time interval T (dimensionless)
N	number of floors above the main terminal (dimensionless)
NIA	net internal area (m^2)
NPER	90% of passengers waiting time (dimensionless)
%POP	percentage population (dimensionless)
P	[lift] average number of passengers (persons)
P	[escalator] pedestrians on escalator landing (persons)
P_{5min}	number passengers arriving in busiest five minutes (passengers/five minutes)
PC	probable car capacity (persons)
pr_0	probability of no calls being registered (dimensionless)
PRBS	pseudo-random binary sequence
PSINT	passenger service interval (s)
Q	number of pedestrians on escalator landing (persons)
r	rise of stair (m)
RL	rated load (kg)
RTT	round trip time (s)
RTT_M	[basement] round trip time (s)
s	number of escalator steps/m (dimensionless)
S	average number of stops (dimensionless)
Sc	[double deck] coincident stops (dimensionless)
S_d	[double deck] expected number of stops (dimensionless)
S_D	down peak average number of stops (dimensionless)
S_M	[basement] average number of stops below main terminal
S_n	[double deck] non-coincident stops (dimensionless)
S_p	average number of stops (Poisson pdf)
S_β	value of S under batch arrivals (dimensionless)
T	performance time (s)
T	number of floors in the upper part of a building (dimensionless)
$t(d_f)$	time taken by the lift to travel one floor starting and finishing at standstill (s)
t_{acc}	time taken to accelerate up to the top speed from standstill (s)
t_c	door closing time (s)
t_{cd}	car call dwell time (s)
t_{cyc}	cycle time (s)

t_d	total door operating time (s)
t_{dec}	time taken to decelerate down from the top speed down to standstill (s)
t_e	express travel time (s)
t_f	time taken to complete a one floor journey in s assuming that the lift attains the top speed v
$t_f(1)$	single floor flight time (s)
$t_f(n)$	multi-floor flight time for a jump of n floors (s)
t_l	passenger loading time (s)
t_{ld}	landing call dwell time (s)
t_o	door opening time (s)
t_p	[one way] passenger transfer time (s)
t_s	stopping time (s)
t_{sm}	[basement] stopping time below main terminal (s)
t_t	transit time for escalator passenger (s)
t_u	passenger unloading time (s)
t_v	single floor transit time (s)
t_v	time taken by the lift to travel one floor while moving at rated speed (s)
t_{vm}	[basement] interfloor transit time below main terminal (s)
U	total building population above main terminal (persons)
U_B	population per upper floors (persons)
U_i	population of floor i. (persons)
UPPAWT	uppeak passenger average waiting time (s)
UPPHC	uppeak handling capacity with 80% car loading (persons/5-minutes)
UPPINT	uppeak interval with 80% car loading (s)
UPPRTT	uppeak round trip time (s)
UPPSTPS	uppeak number of stops (dimensionless)
U_T	population per upper floor (persons)
V	speed along incline (m/s)
v	rated speed in m/s
v_p	average pedestrian walking speed (m/s)
v_{max}	top attained maximum speed in m/s
W	effective corridor width (m)
x	percentage of passengers boarding at basement floors

α	down peak demand (calls/5-minutes)
β	average batch size
β	balanced interfloor 5-minute demand (persons)
γ	balanced interfloor demand per hour (persons)
λ	passenger arrival rate (calls/second)
ρ	system loading (dimensionless)
μ	each lift's passenger processing rate (passengers per second)

Acknowledgements

This book has greatly benefited from the support of our numerous students, associates and colleagues in the lift industry and particularly from many stimulating discussions, corrections and comments from Dr Richard Peters (Peters Research, UK) and Dr Marja-Liisa Siikonen (Kone, Finland).

1 Introduction

1.1 The importance of lifts

Lifts have always been essential to the successful operation of any building and have become especially important in the case of high rise buildings. It is not an exaggeration to state that without lifts, high rise building would not be viable. Lift traffic engineering involves the analysis, design and control of lift systems in order to deal with the passenger traffic flow in buildings.

This chapter sets the scene and presents some preliminary concepts that prepare the reader for the remaining chapters in the book.

1.2 The scope of the book: traffic design vs. engineering design

When discussing the term design in the context of lift systems, it is important to distinguish between two main areas of lift design: engineering design and traffic design.

Engineering design involves the electrical and mechanical design of the various components and systems in the lift.

Traffic design involves the design of the lift system such that it can transport the required number of passengers in a specified period of time under the stipulated performance conditions.

This book is concerned with the traffic design of lift systems, rather than the engineering design. However, it is necessary in certain cases to deal with certain engineering design topics inasmuch as they are closely linked to the traffic design aspects and have a significant impact on decisions made within the traffic design process. An example is the effect of shaft space on the selection of the rated speed of the lift, where the selection of the rated speed of the lift requires minimum clearances in the pit and the head of the shaft, referred to as pit depth and headroom, respectively.

Readers interested in a general view of the engineering of lifts (and escalators and moving walks) might consult CIBSE Guide D: 2015 'Transportation Systems in Buildings'.

1.3 The vertical transportation problem

In any engineering design, it is vital to clearly formulate the problem prior to attempting to solve it. This is no different in the case of the vertical transportation problem.

The vertical transportation problem can be summarised as the requirement to move a specific number of passengers from their origin floors to their respective destination floors with the minimum time for passenger waiting and travelling, using the minimum number of lifts, core space and cost, as well as using the smallest amount of energy.

The aim of lift traffic engineering is to achieve a compromise between cost and performance. A number of parameters have to be optimised such as the average passenger waiting time, the average passenger travelling time and the energy consumed by the lifts. The core space used must also be minimised in the building, in order not to take up valuable net usable area and reduce rental space. The solution of the vertical transportation problem identifies the number of the lifts to be used (as well as their rated speed and rated capacity) in the building in order to achieve the required performance. In effect, the vertical transportation problem is a multiple-constraint-multiple-objective problem that aims to produce a solution that is:

1   Safe.
2   Functional.
3   Reliable.
4   Cost effective.
5   Able to meet the passenger performance requirements (waiting time and travelling time).
6   Able to use the smallest possible core space of the building.
7   Energy efficient.

In order to solve the problem, it is necessary to identify demand and supply. Demand is represented by the arrival of passengers for service. Supply is represented by the number of lifts, their rated speed and rated capacity.

It is possible to think of the lift system as a processor of passengers. Escalators, moving walks, stairs or even corridors and doors can also be thought of as passenger processing devices. As the system is exposed to heavier demand (represented by more passengers arriving for service in a unit of time) the performance of the system changes. The system has handling capacity and is exposed to an arrival rate. The study of lift systems involves understanding how the system processes the passengers and the resulting quality of the performance.

A graphical representation of the vertical transportation design process is shown in Figure 1.1. The user requirements specification (URS) is shown as the input to the system on the left hand side of the figure. The URS usually comprises the quality of service element and the quantity of service element. In some cases, it could also comprise other criteria, such as the energy consumption. The building and passenger parameters are shown in the lower part of the figure.

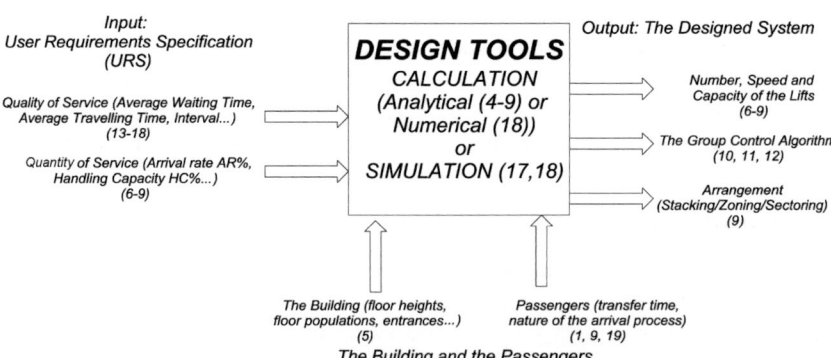

Figure 1.1 Block diagram showing the overall design process

[The numbers shown on the figure represent corresponding chapter numbers]

The building and passenger parameters are combined with the user requirements by the design tool in order to produce a compliant lift traffic design. The design tool could be based on calculation or simulation. The output design is fully described by the following three components:

1 Number, speed and capacity of the lifts in the group.
2 The group control algorithm used.
3 The overall arrangement of the lift groups in the case of high rise buildings. This could include the zoning of the building into separate zones or the use of one or more sky lobbies.

1.4 The core space and the loss of net area

The lift shaft is a vertical space within the core of the building that contains the lift cars, the counterweights, the landing doors and any ancillary equipment. In addition to the core space that each lift takes up, the lift system also requires lobby space to provide an area for passengers to wait for the next lift and to access the lift when it arrives.

This core space (shaft space and lobby space) is a cost to the building developer, due to the loss of net usable area (and rent). This loss in net area is repeated on every floor. One of the main aims of lift system design is to minimise this space by finding the optimum number of lifts required for an optimum design. Zoning the building or using sky lobbies is another tool that reduces the loss in net usable area.

1.5 Peaks in demand for services and products

As is the case with many other services, demand for lift service is usually irregular, peaky, random and time dependent. This presents a problem for the designer of the lift system. Here are some examples of peaks in demand from other industries:

• The demand for fireworks varies annually and peaks in the United Kingdom in November every year (bonfire night).
• The demand for electricity follows a daily cycle, and usually peaks in the evenings just after sunset (especially in the winter) when people get back to their homes and turn the lighting and the electrical heating on.
• The demand for public transportation systems also peaks in the morning rush hour (07:30–09:30) and the evening rush hour (16:30–18:30) when large numbers of passengers require the use of trains, metros and buses.

Peaks in demand are a major challenge for service providers, as they have to provide extra capacity (and the associated infrastructure) in order to meet the peak in demand, which is very costly.

One possible strategy is to try to manage demand by removing the peaks. Suppliers try to find ways of levelling the demand (i.e., move some of the demand from the peaks to the troughs). This can be done, for example, by providing financial incentives for users to use the service during the troughs in demand. For example, electricity providers design tariff structures that offer electricity at lower prices during low demand periods and at higher prices during peak demand periods. Another example is where public transportation providers offer cheaper off-peak train fares, or even different fares for different times of the day or the year.

Another possible strategy to meet the peaks in demand is to employ extra capacity during the peaks, such as a restaurant bringing in extra part-time staff during the peaks.

Nevertheless, peaks in demand will remain and suppliers have to be able to meet these peaks in demand. Designers have to design systems that can meet these expected peaks in demand.

Lift systems are designed based on the peak of the demand. The demand for the lift service is a daily cycle. For office buildings, it usually peaks at the start of the working day (for example, 08:00) when workers arrive for work. A significant peak in demand usually takes place in buildings that have a fixed starting time, and is less dominant in buildings that have a flexible starting time.

An additional complicating factor in the lift traffic design process is the random nature of the process. As will become clearer later in the book, the issue of the randomness of passenger arrivals and the randomness of passenger destinations present a serious complicating factor in the lift traffic design process and require special measures to deal with their effects (Alexandris, 1977).

1.6   Traffic in buildings

It will be noticed throughout the book that the types of traffic in the building have a significant effect on the design process. Thus, it is necessary to define the types of traffic in buildings in this chapter as preparation. In order to define the types of traffic in a building, it is first necessary to classify the types of floors.

1.6.1   *Defining floors*

It is necessary at this point to set the convention for the different types of floors. There are two types of floors in any building: occupant floors and entrance/exit floors. It is assumed that the building is sub-divided into groups of floors: exit/entrance floors and occupant floors.

An exit/entrance floor is a floor that allows passengers to enter or exit the building.

Most buildings have one exit/entrance floor, but some buildings might have multiple entrances. Some reasons for having multiple entrances are the presence of one or more floors of underground car parks, entrances at different street levels due to a sloping ground landscape, adjacent buildings or railway stations. It is usually assumed that no population or traffic demand exists on the entrance/exit floors.

An occupant floor is a floor where passengers spend their time within the building.

These include office floors in an office building or where people live in a residential building. Passengers are not able to leave or enter the building via an occupant floor. Some floors in an office building act as 'magnets', for example, a restaurant floor at the top of a building or a fitness centre in a basement. Strictly speaking, these floors are occupant floors as passengers do not leave the building when they go to the restaurant or gym. However, some designers treat such floors as an entrance/exit floor.

1.6.2   *Defining traffic*

Having defined the types of floors, it is now necessary to define the different types of passenger traffic and their possible journeys in a building. There are four types.

1   Passenger journeys that start at an exit/entrance floor and terminate at an occupant floor are classified as incoming traffic.

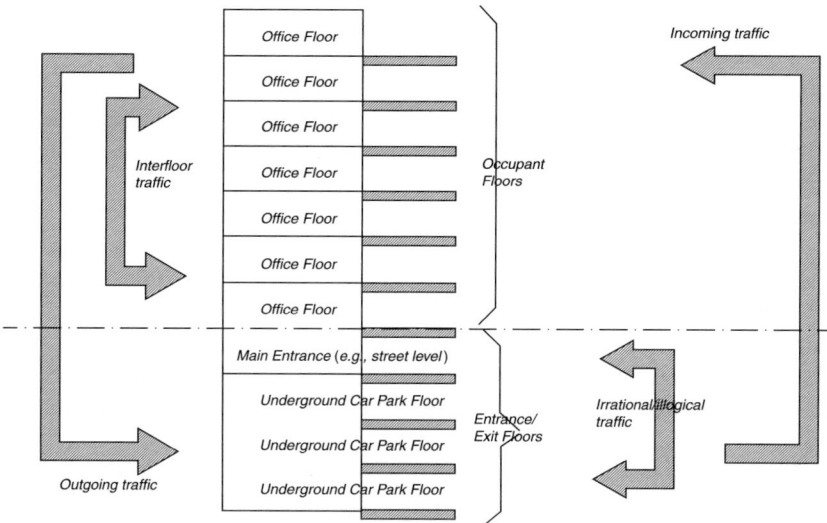

Figure 1.2 The three (possible) types of traffic in an office building

2 Passenger journeys that start at an occupant floor and terminate at an exit/entrance floor are classified as outgoing traffic.

3 Passenger journeys that start and terminate at occupant floors are classified as interfloor traffic.

4 Passenger journeys that start and terminate at entrance/exit floors are illogical/irrational journeys and are thus effectively disallowed and will not be considered any further.

The four types of traffic are further clarified in Figure 1.2.

Surveys have shown that the prevailing traffic during different times of the day in an office building comprises a specific mix of the types of traffic. Examples of traffic mix conditions include: 40%:40%:20% (Barney, 2003a); 45%:45%:10% (CIBSE, 2010); 42%, 42%, 16% (BCO, 2009); 45%, 45%, 10% (BCO, 2014) incoming, outgoing and interfloor traffic, respectively. Such traffic mix ratios are believed to be representative of the lunch-time peak traffic conditions in many modern office buildings.

Classical lift traffic design is based on the assumption of a dominant incoming traffic, usually referred to as uppeak traffic (Chapters 4 to 9). Uppeak traffic is also discussed in detail in Chapter 13. Down peak traffic, where the dominant traffic mode is outgoing traffic, is discussed in detail in Chapter 14. Interfloor traffic is discussed in detail in Chapter 15.

1.7 The seven pillars of lift traffic engineering

Prior to presenting an overview of the various chapters in the book, it is worth listing the seven pillars on which the discipline of lift traffic engineering is built:

1 **Round trip time evaluation**

The evaluation of the round trip time has been, and still is, one of the main tenets of lift traffic engineering. The round trip time is the time that the lift takes in order to complete a full round trip, during which, it picks up passengers from the main entrances, delivers them to their destination and then returns to the main entrance.

2   Design procedure

The design procedure is usually done using calculation and involves using the value of the round trip time in order to decide on the number of lifts in the group, their rated speed and their rated capacity.

3   Performance parameters

The average waiting time and the average travelling time are two of the most important performance parameters of lift traffic design. Barney *et al.* (2005a) and Barney *et al.* (2009)[1] present modern definitions of both.

4   Traffic surveys

Traffic surveys are important tools that allow the designer to assess the traffic performance in a building, and to better understand the nature of the passenger arrival process.

5   Simulation

With the use of more complex building arrangements and more advanced group control algorithms, simulation is usually the tool used to assess the performance of lift traffic systems. 'Elevate' is an example of modern software widely used for simulation (Peters Research 2014).

6   Group control algorithms

The most challenging problem in lift traffic control is the allocation of landing calls to different lifts in the group. A large number of studies have been carried out in this area.

7   The design of high rise buildings

The vertical transportation design for high rise buildings is a much more involved and complicated process than that for low rise buildings. Issues such as zoning, or even the use of sky lobbies, must be considered.

These seven pillars are shown in a diagrammatic format in Figure 1.3. The arrows show the natural progression of complexity in the lift traffic design process, starting with an evaluation of the round trip time and ending with the most advanced process, that of designing lift systems in high rise buildings.

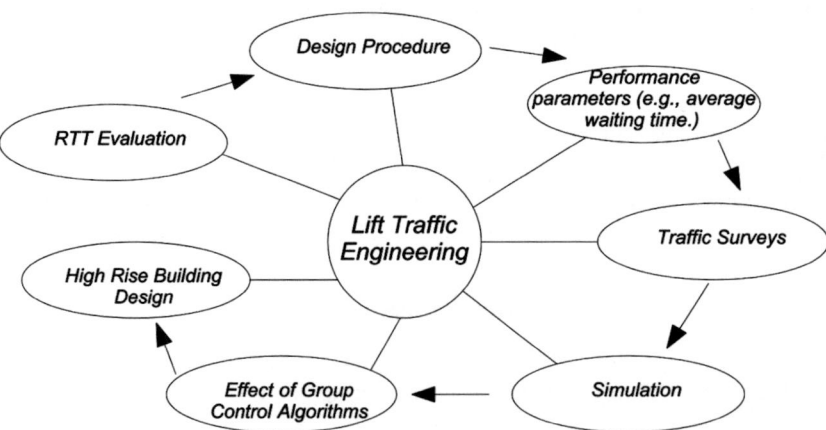

Figure 1.3 The seven pillars of lift traffic engineering

1.8   Overview of the book

The book is divided into five parts.

In Part A, the book starts with two chapters on pedestrian circulation, (Chapters 2 and 3), that present horizontal and quasi-horizontal pedestrian movement in the building as a natural complement to the vertical transportation systems discussed in the rest of the book. The two chapters stand on their own as a self-contained unit on pedestrian flow with supporting case studies.

Part B includes Chapters 4 to 9 and presents an integrated approach to the process of lift traffic design using calculations assuming uppeak traffic. Studying these chapters is an essential starting point for understanding the lift traffic design process. They introduce the process of design starting from the most basic assumptions and adding complexity along the way.

Readers may wish to study Chapter 9 before reading Chapters 4–8 for more elaboration.

Specifically, Chapter 9 presents a set of examples that increase in difficulty, offering the reader a quick-start guide to lift traffic design for the most basic of cases. Chapter 9 also concludes by presenting a case study on the design of a high rise building in three zones. Further case studies complete this part.

Chapters 4 to 9 intentionally address the topic without consideration to other types of traffic in the building and without consideration to the lift group control systems; the latter two parameters are introduced next in Part C. The group control algorithm has a significant effect on the performance of the lift system design. Group control systems are presented in Chapters 10, 11 and 12, whereby Chapter 10 presents the legacy control system and Chapters 11 and 12 present the modern types of systems.

The effect of the traffic type is discussed in detail in Chapters 13, 14, 15 and 16 for uppeak, down peak, interfloor and midday traffic, respectively, in Part D.

Finally, Part E discusses the design of lift traffic systems by simulation (Chapter 17). Chapter 18 presents some of the more advanced techniques that are emerging in the areas of calculation, modelling and simulation of lift traffic systems. A comprehensive treatment of the use of modern techniques in lift traffic systems can be found in Markon *et al.* (2006). Traffic surveys were listed earlier as one of the pillars of lift traffic engineering, and Chapter 19 presents an overview of the different types of traffic surveys that are used to assess the performance of the lift traffic system in a building, to assess the demand or to better model the nature of the passenger arrival process.

As an epilogue, Chapter 20 examines the current trends in lift traffic engineering and attempts to highlight the main drivers that will decide the future of the discipline. It looks at the future developments in the area, and offers a set of pointers to future research priorities.

The book is supported by over 300 references and a bibliography. Readers will find that the list of references at the end of the book indicates where each has been cited in the text.

There are three points worth noting that will become clear to readers as they progress through the book:

1   As with any other field of science and technology, progress in this field has been achieved in a large number of small incremental steps (at times imperceptible) over a long period of time. This started in the thirties of the last century (Cook, 1920[2] and Jones, 1923) with the development of the formula for the expected number of stops in a round trip (driven by the need to assess the number of starts per hour for motor sizing purposes), and has accelerated recently, in pace with the advances in computer processing power (both online

and offline), and with the advances in drive technology (the introduction of the first 2D traffic systems in 2014).

2   Again, as with all other fields of science and technology, the field has changed as it deals with changes that are happening in society. For example, in the last century, the introduction of flexible working hours reduced the demand on lift traffic systems; the lunch-time period is now emerging as the dominant peak in traffic demand.

3   The approach of the progressive introduction of complexity has been adopted throughout the book. This allows the reader to advance as much as they wish in terms of depth, and allows understanding of the basic concepts early on, prior to tackling the more advanced concepts later on.

Notes (shown in text as[1])

1   Page A-73.
2   H.B. Cook offered the first known attempt to develop a round trip time equation in 1920; Gray, L., 2011, *Elevator World*, September 2011.

Part A

Circulation

2 Principles of interior circulation and circulation elements

The question might arise as to why the circulation of people in a building is discussed in a handbook principally concerned with vertical transportation in buildings. The answer is that equipment such as lifts, escalators and moving walks, which provide the mechanical movement of people, must be able to receive and dispatch their passengers efficiently and effectively. Thus, the route taken by pedestrians through the passive elements such as portals (defined here as: entrances, doorways, gates, etc.), corridors and stairs in order to reach and leave the mechanical circulation elements (lifts, escalators, moving walks) is important.[1] A further and important consideration is that efficient and effective circulation is also safe circulation.

Throughout this chapter, the relevance of the circulation elements discussed are illustrated by examples of lift traffic design.

Little is written about the circulation of people in buildings. Tregenza (1971, 1972, 1976) has provided some insight into the UK experience. The classic book on this subject is that of Fruin (1971),[2] and he offers a great deal of information, copiously referenced. Another paper of interest is that of Kavounas (1993d), which presents an analysis of linearly connected vertical modes. Much of the information in this chapter is due to Fruin, extended by the first author's own work.

2.1 General

The circulation of people in the interior of buildings is a complicated activity. It is affected by a number of factors:

- **Mode: horizontal and vertical movement**
  People will generally be walking on the level (except where they are using moving walks). They will change mode from horizontal to vertical movement, and *vice versa*, in order to reach higher or lower building levels. To do this, they will use stairs, moving walks (horizontal and ramped), escalators or lifts.

- **Movement type: natural or mechanically assisted**
  People move naturally when walking, and are mechanically assisted when using moving walks, escalators or lifts.

- **Human behaviour**
  Humankind behaviour is complex. Individuals have: their own concepts of route; their own purpose for travel; their own level of urgency; their own characteristics of age, gender, culture, disabilities, etc. There is unpredictability in human behaviour.

The interior circulation design in a building is not something a lift person has any influence over, but poor design can impact on the traffic design. However, it is sometimes possible to influence the design to ensure an efficient traffic design.

- All circulation routes need careful consideration. These include principal and secondary circulation areas, escape routes, service routes and waiting areas. For example, an evacuation lift should have a clear and protected route to the exterior of the building.

- Clear and obvious circulation routes should be provided. Pedestrians should be able to see the route to take, perhaps assisted by good colour coded signs and open vistas. It is ineffective to place a lift for the use of persons with disabilities in a hidden location.

- Circulation patterns should be rational. An example is the avoidance of pedestrians passing through a lift lobby where other persons are waiting.

- Incompatible types of circulation should not coincide. This would apply to pushing goods trolleys across a pedestrian mall in a shopping centre or sterile/non-sterile movements in a hospital.

- The movement of people and goods should be minimised. This would bring related or associated activities together, e.g., sales and marketing, personnel and training, etc. in an office building. This can have a dramatic effect on the number of lifts selected to serve a building.

- It is important to size each facility properly. Thus, the handling capacities of portals and corridors that lead to stairs, which in turn lead to a lift, should be adequately sized for their anticipated demand. The term 'handling capacity' is used here for passive (non-mechanical) building elements in the same way as it is applied to the mechanical elements. The term 'demand' is used to indicate the level of usage.

- The design and location of portals, corridors and stairs must be coordinated with the mechanical people-handling equipment to ensure the free flow of people, goods and vehicles. For example, if turnstiles are placed in the route to a group of lifts, it is the throughput of the turnstiles that dictates the demand on the lifts.

- The importance of the journey undertaken also needs consideration. An example is a theatre lift in a hospital.

These and other factors lead to an imprecise knowledge of how circulation occurs in a building and permit only empirical methods of design. Much of what this chapter contains should be taken as general guidance only; much of what follows cannot be proved theoretically and many of the conclusions have been empirically derived. Reasons are given for the conclusions so that if new evidence comes to hand (or opinions change) the results can be modified. Regulations may also affect the circulation design, such as fire and safety codes, and these must be taken into account.

2.2   Human factors

The evolution of humankind to its present form has taken a long time. Along the way, various instincts have been implanted within human beings, in order for humankind to evolve and survive. Instinct is still a powerful characteristic of human beings, and fear is its biggest driving force. The fear of harm conditions all of us. The association of one human to another is predicated on trust, or doubt, and this controls how humans behave towards one other. The spatial separation of humans is thus important.

Human beings value personal space. This is measured by a personal buffer zone around each individual. The actual size of the buffer zone varies according to an individual's culture, age, status, gender, disabilities, etc., and even their geographical origin.

The physical dimensions of the human body vary widely. It is a fact that females are generally smaller than males, and people from the Asia/Pacific region of the world are generally smaller than Europeans.

The space an individual occupies can depend on what they are wearing and what they might be carrying, and if they are children, what they are doing.

Fruin (1971) indicates four classes of personal space, when persons are queuing or are occupying a waiting area:

(a) circulation zone
(b) personal comfort zone
(c) no touch zone
(d) touch zone.

Class (a) is where individuals can easily pass between other individuals without disturbing other individuals. The area occupied by each person is some 2.2 m², i.e.: 0.8 persons/m². This is important in the design of lift lobbies and the landings of escalators and moving walks.

Class (b) is where individuals can still pass between other individuals, but may disturb them. The area occupied by each person is some 0.9 m², ie: 2.1 persons/m². This requires the tolerance of one individual to another. Fear can come into play here as individuals would ideally require a bigger comfort zone. An example might be a lift lobby, which is a public place.

Class (c) is where individuals cannot pass between other individuals without disturbing them. The area occupied by each person is some 0.7 m², i.e., 2.4 persons/m², for example, passengers in a lightly loaded lift car.

Class (d) is where individuals cannot pass other individuals. The area occupied by each person is some 0.33 m², i.e., 3.0 persons/m², for example, a crowded lift car.

Fruin does not mention a crowded situation with individuals occupying 0.25 m², i.e., 4.0 persons/m² or the situation in a very crowded lift car or crowded train or bus where individuals occupy some 0.2 m², i.e., 5.0 persons/m².

To determine these zones, Fruin uses an occupancy template. This template was developed by the New York City subway to determine practical standing capacity and also by the US Army. It might, in 1971, have represented a template for 95% of US males. Since then, the US and UK populations have become larger (i.e., overweight/obese) and the template might now be considered to be a ninety percentile (90%). A value for a ninety percentile is statistically a more suitable value, as it is less influenced by extreme maximum values, present when the distribution has a long tail.

This template can be reasonably used for female individuals and children. This is because females usually try to acquire more space in crowded circumstances by folding their arms or enlarging themselves with the objects they are carrying (to avoid touching). Children, on the other hand, are rarely still and therefore demand more space than their weight would imply. It has been observed that, in uncrowded conditions, individual female subjects are comfortable with a personal buffer zone of 0.5 m² (0.8 m diameter circle) and individual male subjects with a personal buffer zone of 0.8 m² (2.0 m diameter circle). To visualise these sizes, a woman's umbrella occupies an area of approximately 0.5 m² and a man's umbrella occupies approximately 0.9 m².

To allow for all these circumstances and other factors, such as body sway, it is recommended that the typical occupancy template be considered to be an ellipse of dimensions 600 mm wide by 450 mm deep, as shown in Figure 2.2. The area of occupancy is thus 0.21 m². This template shape will be assumed to accommodate 90% of all subjects. Note that the actual body template of the individual is inside the ellipse. If this typical occupancy template is used to represent the ninety percentile individual, then larger individuals will be compensated for by smaller individuals. These factors must be borne in mind when designing pedestrian waiting areas, for example lift lobbies, and escalator and moving walk entry and exit landings.

For people who are waiting for an event, the densities shown in Table 2.1 are recommended. These densities are illustrated in Figure 2.2.

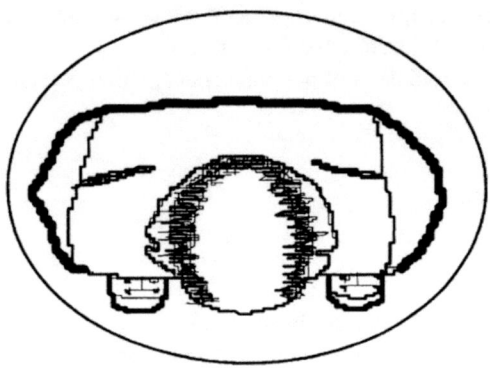

Figure 2.1 Typical occupancy ellipse (showing male subject)

[The ellipse is: 600 mm wide by 450 mm high]

Table 2.1 Density of occupation in waiting areas, e.g., lobbies

Level of density	Comment
Desirable 0.4 persons/m²	Allows individuals to walk freely or to stand still in one place, without any interference from other individuals.
Comfortable 1.0 persons/m²	Allows individuals to walk less freely, with some deviations necessary and for individuals to stand still, without any interference from other individuals.
Dense 2.0 persons/m²	Individuals, who are walking, must now take care not to collide with other persons; and persons waiting are aware that other individuals are present.
'Crowding' 3.0 persons/m²	It is only possible to walk at a shuffle and with care at the average rate of the 'crowd'. There is no or little chance of a contra flow. Individuals waiting are very aware of other individuals.
Crowded 4.0 persons/m²	Walking is almost impossible. Individuals waiting are unhappy to be so close to other individuals (touch zone violated). This density is only possible where persons are placed in a confined space, such as a lift car, or in a rapid transit train.

When considering linear queues, i.e., people 'waiting in line' for a service, it is reasonable to assume two persons per metre length of space. A control barrier can be used to restrain the queue width. The barrier should be at least 600 mm wide. For unrestrained queues, it is necessary to assume they occupy a width of at least 2.5 m.

The sizes shown on the template of Figure 2.2 are suitable for the USA, the UK and some European countries, but would need adjustment if used in other parts of the world.

2.3   Circulation factors

There are a number of factors which affect people's movement as pedestrians and as passengers. They include:

- pedestrian dimensions
- pedestrian velocities
- unidirectional/bidirectional flow
- cross flows

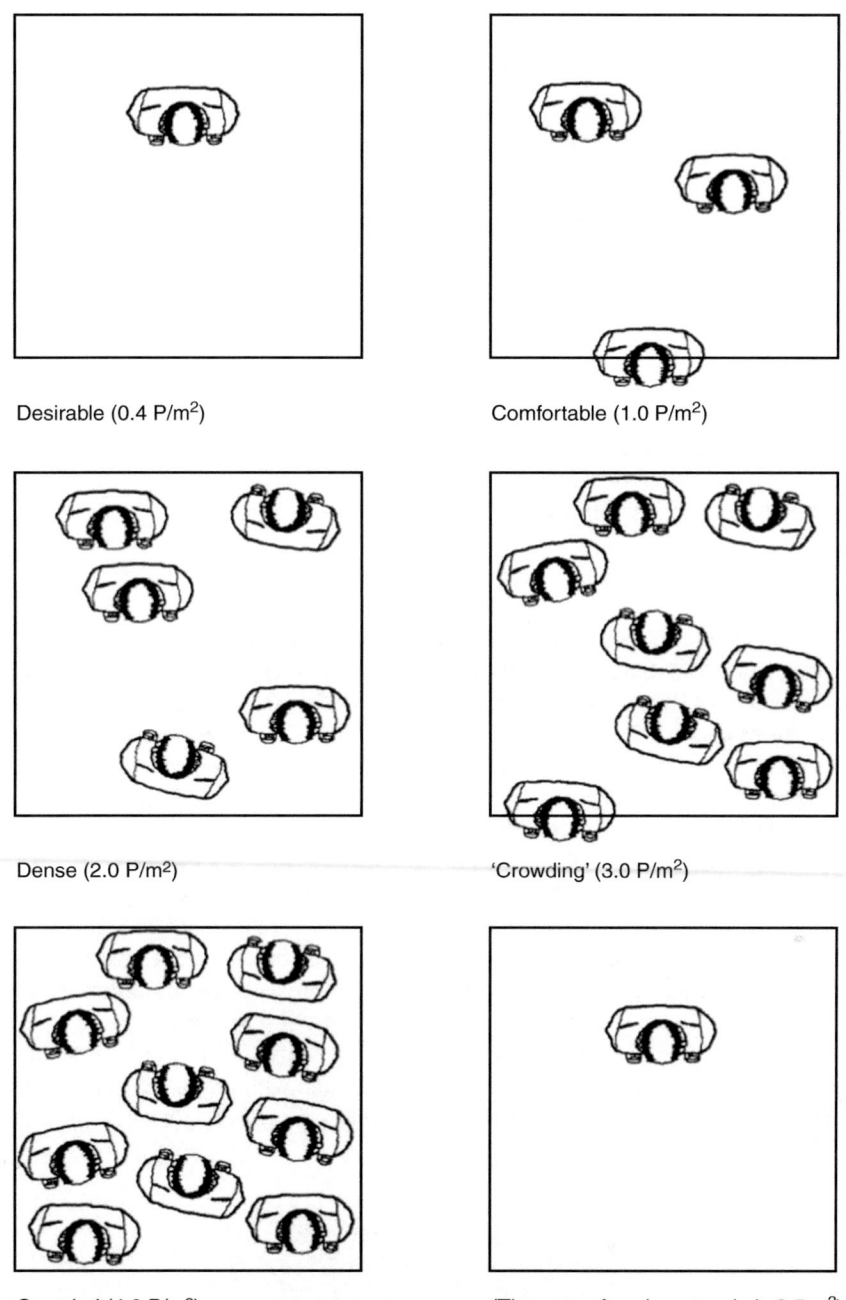

Figure 2.2 Illustration of density of occupation in waiting areas

- patterns of waiting
- site and environmental conditions (air temperature, humidity, floor finish, gradient)
- type of pedestrian (age, gender, group, purpose)
- ability (fitness, disability)
- statutory requirements.

These factors will be considered when considering the various passive and mechanical circulation elements in the following sections.

2.4   Corridor (handling) capacity

Corridors are circulation routes that are constrained by some form of close physical barrier, such as a wall, or a set of cubicles in an open plan office. In general, this causes the density of occupation to be higher than in an unbounded space. The capacity of a straight corridor can be given as:

$$C_c = 3600 \times v \times D \times W \text{ persons/hour} \qquad (2.1)$$

where:
C_c is the corridor handling capacity (persons/hour),
v is average pedestrian speed (m/s),
D is the average pedestrian density (persons/m^2), and
W is the effective corridor width (m).

Equation (2.1) is an empirical relationship with a number of qualifications:

- Pedestrian speed and density are not independent of each other.
- For densities below 0.3 P/m^2, pedestrians can walk freely. Typical walking speeds range from 1.0 m/s to 1.3 m/s. This is called free flow design (see Figure 2.3).
- For densities above 0.5 P/m^2 there is an approximately linear decrease of average walking speed up to a density of about 3.0 P/min, when walking is reduced to a shuffle.
- Throughput peaks at densities of about 1.4 P/m^2. Typical walking speeds range from 0.6 m/s to 0.8 m/s. This is called full flow design (see Figure 2.4).

The flow rates assume a corridor width of 1.0 m. The width of corridor must be at least 900 mm and is assumed to be 1.0 m. Equation (2.1) allows for the flow rate to increase/decrease as the corridor width increases/decreases. This factor must be used with care as small changes in corridor width will have little or no effect. Table 2.2 presents the minimum straight widths of corridors that have been found to be suitable for different purposes.

Table 2.2 is gender specific owing to the average dimensional differences between the genders. Other minimum widths with other combinations, e.g.: man with pram, etc. can be estimated.

Traffic can only flow freely along unrestricted routes. Corridors are rarely free of obstructions. Table 2.3 provides a number of examples.

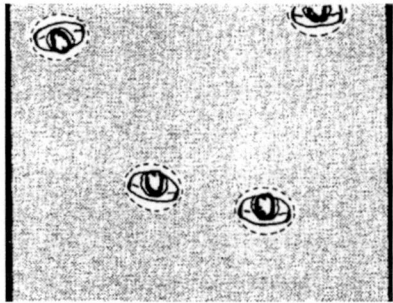

Figure 2.3 Free flow design – 0.3 persons/m^2

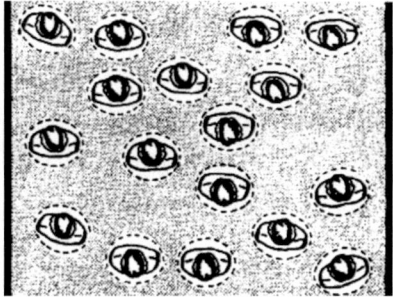

Figure 2.4 Full flow design – 1.4 persons/m²

Table 2.2 Minimum corridor widths

Usage	Minimum width (m)
One-way traffic flow	1.0
Two-way traffic flow	2.0
Two men abreast	1.2
Man with bag	1.0
Porter with trolley	1.0
Woman with pram	0.8
with child alongside	1.2
Man on crutches	0.9
Wheelchair	0.8*

*Very long wheeled vehicles, such as hospital trolleys, require extra width in order to turn at junctions.

Table 2.3 Reductions in corridor widths

Obstruction	Reduction (m)
Ordered queue	0.6
Un-ordered single queue	1.2–1.5
Row of seated persons	1.0
Coin operated machine:	
One person	0.6
Queue	1.0
Person waiting with bag	0.6
Window shoppers	0.5–0.8
Small fire appliance	0.2–0.4
Wall-mounted radiator	0.2
Rough/dirty surface	0.2

Example 2.1

In a hospital corridor, it is necessary for two trolleys to pass each other. Each trolley is pushed by one porter and another person with a bag of equipment walks alongside. What width should the corridor be? If a row of seated persons is encountered, what effect would this have? Determine the probable flow rates at free flow design levels.

From Table 2.2, a trolley and porter occupy 1.0 m width, a man with a bag occupies 1.0 m width. If the traffic is two-way, the minimum clear corridor width will need to be at least 4.0 m.

If a large obstruction, such as a row of seated persons, is encountered, the corridor width would need to be increased by as much as 1.0 m to, for example, 5.0 m. Persons waiting in these circumstances should not be located in a corridor. Unless small obstructions, such as fire appliances, radiators, etc., (see Table 2.3) can be recessed, the width the corridor would need to be slightly increased.

The circulation mix would comprise most people moving slowly and a few others on urgent tasks moving very fast (comparatively). The former would probably have a low speed, perhaps 0.6 m/s and the latter would probably have a higher speed, perhaps 1.5 m/s. A reasonable average would be 2.1 m/s. Using Equation (2.1), the full flow design flow rate (0.3 persons/m²) would be:

$$3600 \times 1.1 \times 0.3 \times 5 = 594 \text{ persons/hour}$$

2.5   Large circulation spaces (malls, concourses, squares, etc.)

If the circulation space is unbounded by the close proximity of walls, roped-off areas, fences etc. then pedestrians can move more easily. The variety of purposes is also wider. The density levels are lower than those for bounded corridors.

An 'uncrowded' density might be 0.2 persons/m², with a typical walking speed of 1.3 m/s (see Figure 2.5).

A 'crowded' density might be 0.45 persons/m², with a typical walking speed of 1.0 m/s (see Figure 2.6). Equation (2.1) can be used to predict a flow rate through a sized space.

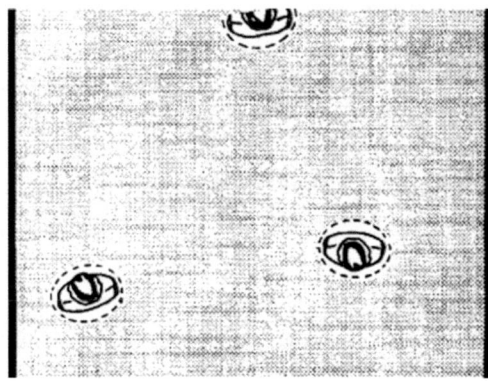

Figure 2.5 Uncrowded flow design at 0.2 persons/m²

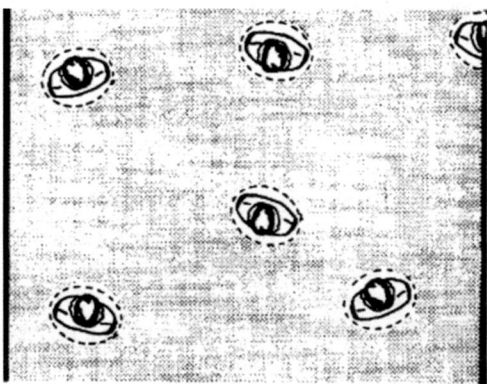

Figure 2.6 Crowded flow design at 0.45 persons/m²

2.6 Portal (handling) capacities

Portals, which are called by various names, i.e., gate, door, entrance, turnstile, etc., form a division between two areas for reasons of privacy, security, access control, etc. They represent a special restriction in corridor width. Their main effect is to reduce pedestrian flow rates. Table 2.4 indicates the probable range of pedestrian flow rates in persons per minute and persons per hour through an opening of 2.0 m.

Note that Table 2.4 indicates flows through a portal of 1.0 m. Most domestic doors are less than this width (approximately 750 mm) and the flow rates would be likely to be the lower values in the range. Doors in non-domestic buildings may be slightly wider than 1.0 m and would permit the higher values in the range to be possible.

2.7 Stairway (handling) capacity

Stairways impose a more stylised and disciplined form of movement on pedestrians. More accurate cones of vision are required for step placement, and assistance is often required by the use of handrails. The movement is more regular as it is disciplined by the steps, and permits higher densities than are possible on the level. Whereas, for free movement during walking on the level a pedestrian requires an area of some 2.3 m^2 (to account for body sway, etc.), a stair-walker only needs to perceive two vacant treads ahead (and room for body sway), and occupies an area of some 0.7 m^2. Thus, free flow design is possible at a density of 0.6 P/m^2 (Figure 2.7) and full flow design is possible at a density of 2.0 P/m^2 (Figure 2.8).

The speed along the slope is about half that on the level, but increased densities are possible. Speed, however, is very much dependent on the slowest stair-walker, owing to the difficulty in overtaking under crowded conditions. Higher speeds in the down direction are very often reduced by the need for greater care resulting in similar speeds in both directions.

The dimensions of a stair limit many aspects of locomotion. For instance, pace length is restricted by tread depth (going). Speed is also affected by the angle of inclination and step riser height. To enable comfortable walking on a stair, a rule of thumb has been established that the sum[3] of the going (*g*) plus twice the rise (r) should lie in the range 550 mm to 700 mm, i.e.:

$$550 < g + 2r < 700 \tag{2.2}$$

This approximately matches the average adult stride on a stairway, which results in a range of riser heights of 100 mm to 180 mm and treads of 360 mm to 280 mm, and a range of possible inclinations from 15° to 33°. A private stair often has a rise of 180 mm and a going

Table 2.4 Portal (handling) capacities

Portal type	Flow (persons/hour)
Gateway	3600–6600
Clear opening	3600–6600
Swing door	2400–3600
Swing door (fastened back)	3600–5400
Revolving door	1500–2100
Waist high turnstile:	
Free admission	2400–3600
With cashier	720–1080
Single coin operation	1200–1800

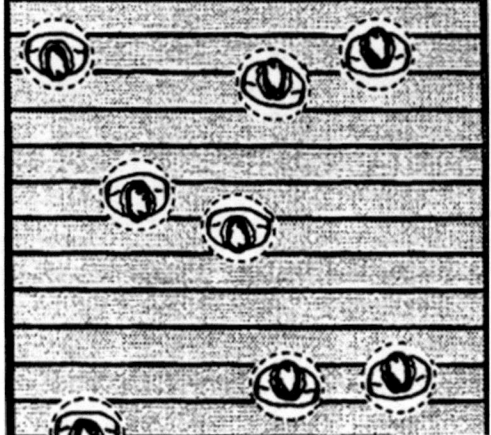

Figure 2.7 Free flow design at 0.6 persons/m²

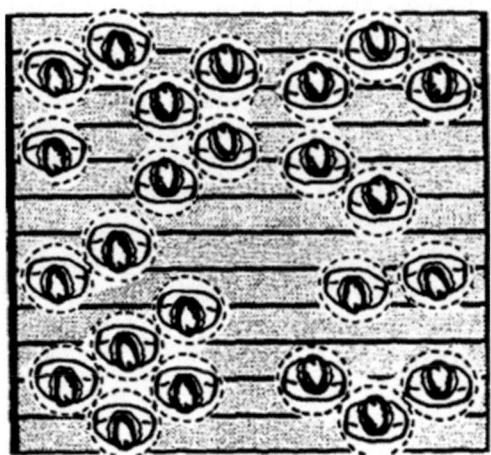

Figure 2.8 Full flow design at 2.0 persons/m²

of 240 mm. In Britain, these dimensions are historical, as two bricks of 3 inches in height, together with two mortar joints of half an inch in height, used to form a stair, producing a step rise of 7 inches, which is almost 180 mm. An efficient inclination has been found to be 27°.

An empirical formula is given in Equation (2.3) for stairway handling capacity (C_s). Table 2.5 indicates typical values for pedestrian stairway speeds along the slope in metres per second, and pedestrian flow rates in persons per minute and persons per hour (bracketed) for each 2.0 m width of stairway.

$$C_s = 0.83 \, (60 \times v \times D \times W) \text{ persons/minute} \tag{2.3}$$

Other symbols are as for Equation (2.1). Note that a stair has 83% of the handling capacity of a corridor.

There are lower densities on stairs serving large unbounded spaces.

An uncrowded stair might have a density of 0.4 persons/m² with a typical walking speed of 0.6 m/s to 0.8 m/s (see Figure 2.9).

A crowded stair might have a density of 0.8 persons/m² with a walking speed of 0.6m/s to 0.8 m/s (see Figure 2.10).

Table 2.5 Stairway (handling) capacity

Traffic type	Free design flow (0.6 P/m²)		Full design flow (2.0 P/m²)	
	Speed	Flow	Speed	Flow
	m/s	persons/hour	m/s	persons/hour
Young/middle-aged men	0.9	1620	0.6	3600
Young/middle-aged women	0.7	1260	0.6	3600
Elderly people, family groups	0.5	900	0.4	2400

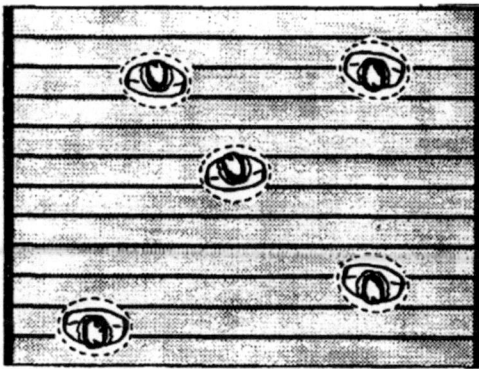

Figure 2.9 Uncrowded stair density at 0.4 persons/m²

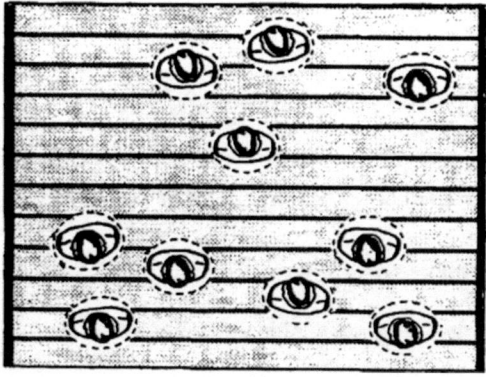

Figure 2.10 Crowded stair density at 0.8 persons/m²

2.8   Escalator (handling) capacity

Escalators provide a mechanical means of continuously moving pedestrians from one level to another. Four factors affect their handling capacity.

- *Speed.* This is measured in the direction of the movement of the steps. Commonly available speeds are 0.5 m/s and 0.65 m/s. Most escalators run at one speed only, although some heavy duty escalators can switch over to the higher speed during heavy traffic. Many escalators have auto-start and/or speed-reduction systems for energy saving measures. A speed of 0.75 m/s is used on the London Underground, and speeds of 0.9–2.0 m/s are used on deep systems in Russia and the Ukraine.
- *Step widths.* Widths of 600 mm, 800 mm and 1,000 mm are available, the latter allowing two columns of passengers to be carried. The hip widths, which are measured between the skirting panels are typically 200 mm wider than the step. Hence, the actual width a person can occupy on a 1,000 mm step width escalator is some 1,200 mm (enough for two people to pass each other).
- *Inclination.* This is usually 30°, but can range from 27° to 35° in some cases. The latter is only available at a maximum speed of 0.5 m/s and a maximum rise of 6 m. The comfortable walking rule for inclination is broken for escalators as the step tread (going) is generally 400 mm, producing a step rise up to 240 mm, in order to achieve the necessary inclination. However, where an escalator can be used for an emergency exit, the rise may not exceed 210 mm. Typically, an escalator has a rise of 210 mm and a going of 400 mm, which, when applied to the stair rule, gives a value of 820 mm. This is outside the range quoted in Section 2.7, and explains why it is much harder to walk on an escalator.
- *Boarding and alighting areas.* These areas must encourage pedestrian confidence and assist the efficient and safe boarding of escalators. It is recommended that at least one and one third flat steps (light duty) to two and one third flat steps (heavy duty) be provided for passengers when boarding/alighting an escalator. The average pedestrian boarding/alighting stride can be assumed to be 750 mm.

The theoretical handling capacity of an escalator (C_e) is given by:

$$C_e = 3600 \times V \times k \times s \text{ persons/hour} \tag{2.4}$$

where:
V is speed along the incline (m/s)
k is average density of people (people/escalator step)
s is number of escalator steps/m.

For the case where the step depth is 400 mm, s becomes 2.5 and Equation (2.4) is:

$$C_e = 9000 \times V \times k \text{ persons/hour} \tag{2.5}$$

The factor k allows for the likely passenger occupation densities.

The construction standard for escalators and moving walks (BS EN 115: 2008 +A1: 2010) suggests (without any proof) the following theoretical maximum handling capacities, which are shown in Table 2.6.

Table 2.6 Escalator theoretical maximum handling capacities (persons/hour) according to EN115: 2008 + A1: 2010

Speed	Step width	Step width	Step width
m/s	1000 mm	800 mm	600 mm
0.50	6000, $k = 1.33$	4800, $k = 1.07$	3600, $k = 0.80$
0.65	7300, $k = 1.24$	5900, $k = 1.00$	4400, $k = 0.75$
0.75	8200, $k = 1.22$	6600, $k = 0.98$	4900, $k = 0.73$

Table 2.7 Escalator practical maximum handling capacity (persons/hour)

Speed	Step width	Step width	Step width
m/s	1000 mm, $k = 1.0$	800 mm, $k = 0.75$	600 mm, $k = 0.5$
0.50	4500	3375	2250
0.65	5850	4388	2925
0.75	6750	5063	3375

It can be seen that the passenger density (k) appears to depend on escalator speed. This might be likely for short-travel escalators, but not for those of some length.

Observations on the London Underground (Mayo, 1966); Al-Sharif, 1996) have shown the occupation densities that occur in practice and these are smaller than EN 115 suggests. In general, only half the available space on an escalator is occupied. At this density, on a 1,000 mm escalator, this would give a standing person a space of 400 mm by 1,000 mm in which to stand, i.e., an area of 0.4 m². This represents the dense level of occupancy of 2.5 persons/m² as given in Section 2.2.2.

A simple view (based on observation) is to assume constant densities of:

$k = 0.5$ person per step for escalators of step width 600 mm
$k = 0.75$ persons per step for escalators of step width 800 mm
$k = 1.0$ person per step for escalators of step width 1,000 mm.

Table 2.7 thus gives the practical maximum escalator handling capacity values in persons per hour for these values for k.

The practice in Britain, Japan and elsewhere of one stationary column and one walking column will not increase an escalator's handling capacity, but will increase the passenger flow rate off the escalator and will decrease an individual's travelling time.

Example 2.2

On the London Underground, it was observed during peak periods that passengers stood stationary on the right hand side of the 1,000 mm escalator at a density of one passenger on every other step. The left hand side was occupied by a walking column of passengers at a density of one person every third step. Assuming the escalator was running at 0.75 m/s and the speed of the walking passengers was 0.65 m/s, what is the passenger flow rate off the escalator?

The flow rate of two stationary columns is given in Table 2.7 as 6,750 persons/hour. As there is only one stationary column, the flow rate for this column will be 3,375 persons/hour.

The occupancy of the walking passengers is one person for every three steps. Therefore, k is 0.33 (one third). But the effective (relative) speed of the passengers is 1.4 m/s (0.75 + 0.65). Then, the flow rate using Equation (2.5) will be:

$$9,000 \times 1.4 \times 0.33 = 4,200 \text{ persons/hour}$$

The total passenger flow rate is thus 3,375 + 4,200 = 7,575 persons per hour.

At any one time, there will be five passengers on six steps, giving a value for k of 0.83, and the actual handling capacity of the escalator will be:

$$9,000 \times 0.75 \times 0.83 = 5,603 \text{ persons/hour}$$

2.9   Moving walks (horizontal and inclined)

Horizontal moving walks have an inclination of 0°, and inclined moving walks have inclinations in the range of 3° to 12°. The running speed is determined by the angle of inclination. The speed is again measured in the direction of movement of the pallets, i.e., along the horizontal. Commonly available widths are 1,000 and 1,400 mm. The latter easily allows two stationary files of passengers, or the possibility of a stationary file and a walking file of passengers on the moving walk. The 1,400 mm wide moving walk considerably assists the movement of shopping trolleys and baggage carts.

The theoretical density of passengers assumed for an escalator with two passengers per 1,000 mm step (with $k = 2$) is 5.0 persons/m^2. In practice, as Section 2.8 indicates, this is never achieved on an escalator, and half this density is more likely, i.e., 2.5 persons/m^2. A moving walk should theoretically permit denser congregations of passengers than an escalator as the space is not rigidly defined by steps. In practice, the probable density will be the 'dense' value at about 2.5 P/m^2.

Table 2.8 indicates practical handling capacities in persons per hour, assuming a density of 2.5 persons/m^2, using Equation (2.1).

EN115 suggests that shopping trolleys and baggage carts can reduce the handling capacity to 80%. It is also suggested that the wider moving walk of 1,400 mm does not increase handling capacity as passengers need to hold one handrail.

2.10   Handling capacity of lifts

The method of sizing a lift installation to serve a traffic demand is given in detail in Chapters 4–8. However, for completeness, some discussion is necessary here. Lifts cannot handle the traffic volumes handled by other facilities, and have a considerable throttling effect on pedestrian movement. For example, the most efficient 8-lift group, comprising 1,600 kg rated

Table 2.8 Handling capacities of moving walks (persons/hour)

Speed	Pallet width
m/s	1000 mm
0.50	4500
0.65	5850
0.75	6750

load lifts serving 14 floors can only provide a handling capacity of 50 persons/minute. This is less than a flight of stairs can provide. And a 3-lift group comprising 800 kg rated load lifts serving 8 floors can only manage 16 persons/minute. Thus, the use of escalators for short-travel systems in buildings is recommended. Fortunately, the very high volumes of passenger demand found in bulk transit systems do not occur when populating or emptying a building. As will be seen later, care must be taken in sizing a lift system for worst-case scenarios.

The density of occupancy when sizing a lift car can be larger as passengers are constrained (by the car walls) and a greater allowance can be made for averaging. The standard BS EN81–20 (2014) provides Table 6 (5.4.2), which indicates 0.1 m² plus 0.2 m² per person up to six persons, then 0.15 m² per person up to 20 persons, and then 0.12 m² per person thereafter. These values would require cars to be very crowded, and it has been observed that lift cars do not fill to their rated (person) loads. This is because the rated capacity is given by a simple formula of dividing the rated capacity by 75 on the assumption that the average passenger weight is 75 kg. Section 5.10.2 deals with this aspect in detail.

It is recommended that a uniform density figure of 0.21 m² per passenger be assumed, when determining the number of passengers a lift car can accommodate when carrying out traffic design. This figure is almost five persons/m², i.e., very dense (see Table 2.1).

Figure 2.11 illustrates the density pattern for a lift with a rated capacity of 17 persons (rated load 1,275 kg) loaded with 16 persons present. It can be seen that the lift is not able to accommodate this number of passengers as the platform area is 2.9 m², which would allow some 14 persons to be accommodated. Even with 14 persons present, the passengers would be in the intimate touch zone discussed in Section 2.2.2. Table 5.8 indicates that the actual car capacity figures are smaller than the rated car capacity suggests.

2.11 Location of facilities

2.11.1 General

Having discussed the various circulation elements and their characteristics, it is now necessary to consider their location. The main principles to bear in mind are to minimise the movements of people and goods, to prevent clashes between people and goods, and to prevent bottlenecks. Thus, the location and arrangement of the passive circulation elements (corridors, portals, etc.) and the active circulation elements (moving walks, escalators and lifts) should take account of:

- the location of entrances and stairs
- the location of lifts and escalators
- the distribution of the occupants in the building.

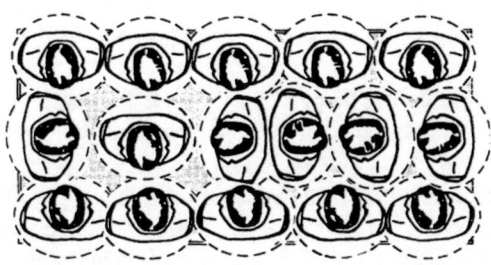

Figure 2.11 1,275 kg lift car occupied by 16 persons

Case Study CS1 gives an example of conflict between various circulation elements.

Ideally, all circulation activities should be centralised in a main core of a building. This is not always possible; sometimes the main lobby is close to the main entrance, and sometimes the building design places the main lobby some distance into the building. This latter case involves occupants and visitors in a long walk to reach the transportation facilities. However, it may be better for occupants to walk to the centre of a building to access stairs and lifts, since their more frequent usage during the day may outweigh the comparative inconvenience during arrival and departure. Generally, the maximum distance to a lift or stair from an occupant's workplace should not exceed 60 m, with a distance of less than 45 m being preferred. Emergency escape routes are usually closer, but do not necessarily form part of the normally used circulatory routes.

2.11.2   *Stairs and escalators*

Where possible, stairs and escalators should not lead directly off corridors, but should be accessed from landing and lobby areas, where people may wait without obstructing a circulation route. Thus, the vertical and horizontal modes of circulation can be allowed to merge smoothly.

If it is the intention to encourage the use of stairs for short journeys to/from adjacent floors (interfloor movement), then the stairs should be clearly visible, adequately signed and reached before entering the lift lobby.

The location of escalators should observe the same recommendations as those for the location of stairways. However, it should be noted that escalators occupy a larger footprint than stairs in order to accommodate their inclination, structure and equipment spaces. It is particularly important that the boarding and alighting areas adjacent to an escalator are not part of another circulation route. This will provide a safe area for passengers to board and alight. This topic is discussed further in Chapter 3.

Escalators are typically used for short-range movement between adjacent floors (except in deep underground railway systems). They are found in offices between principal levels, in shops between trading floors, in shopping centres between malls, and elsewhere, such as railway stations, hospitals, museums, etc. They are usually sited in an obvious circulation path, making it easy for pedestrians to board them.

There are several standard escalator arrangements, as shown in Figure 2.12. Type (c) is typical of a shop as it allows the shop to deliberately lengthen the circulation route to pass goods for sale. This configuration also takes up less space.

2.11.3   *Lifts*

Lifts should always be placed together whenever possible, rather than distributed around a building. This arrangement will help to provide a better service (shorter intervals), mitigate the failure of one car (availability of an adjacent car or cars) and lead to improved traffic control (group systems).

Lift lobbies should preferably not be part of a through circulation route, either to other lifts, or other areas in the building. Lobbies should be provided that are dedicated to passengers waiting for the lifts.

Eight is the maximum number of lifts that it is considered possible to present to waiting passengers, especially if the lifts are large (<1,275 kg). This constraint allows passengers to ascertain the arrival of a lift easily (from the landing lantern and gong signals), walk across the lobby and enter the lift before the doors start to close.

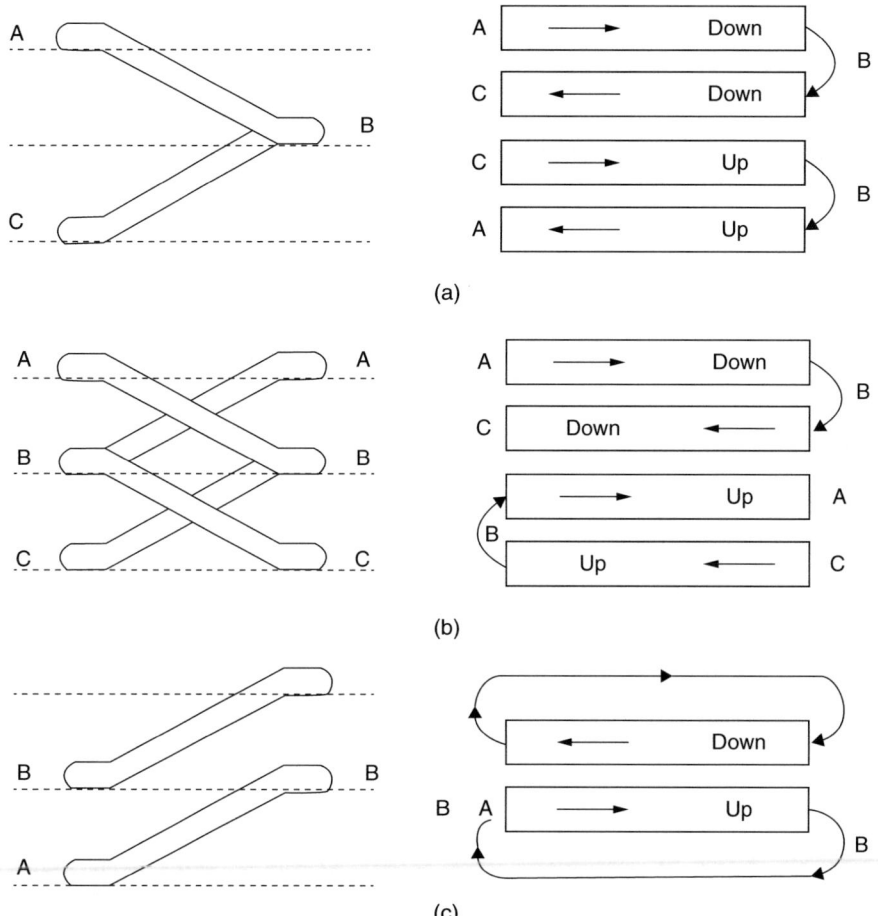

Figure 2.12 Escalator configurations (a) parallel (b) cross-over (c) walk-round

The distance across a lobby is important. If the lobby is too large, passengers have too far to walk and the closure of the car doors has to be delayed (increasing the lobby door dwell time) to accommodate the increased walking time; BS ISO 4190–1: 2010 gives some guidance. For residential buildings, the landing depth (measured in the same direction as the depth of the car) should be at least equal to the depth (d_l) of the deepest car, but not less than 1.5 m. For office buildings, where lifts are located side by side, the landing should be at least equal to 1.5 × d_l and not less than 2.4 m. In office buildings, where lifts are arranged face to face, the distance between facing walls should be at least equal to the sum of the depths of two facing cars, but not more than 4.5 m.

The ideal lobby size would be one that could accommodate one full car load of passengers waiting, and permit the simultaneous disembarkation of one full car load of arriving passengers. This area can be calculated from the information given in Section 2.2.2.

The preferred arrangements from BS5655: Part 6: 2002 of two to four lifts arranged side by side are given in Figure 2.13 and for two to eight lifts arranged opposite each other are shown in Figure 2.14. Note all the lobbies, indicated (L), have separate waiting areas with no through circulation.

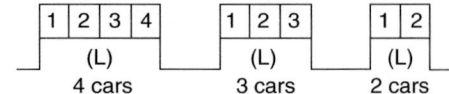

Figure 2.13 Preferred arrangement for lift cars: side by side (or in line)

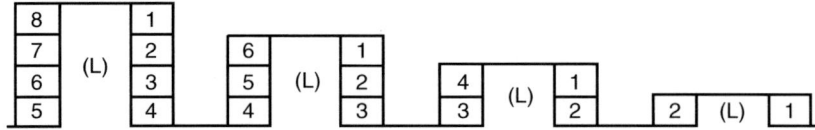

Figure 2.14 Preferred arrangement for lift cars: facing

Example 2.3

Suppose there are four 1,600 kg lifts arranged in a facing (2 × 2) configuration. What distance across the lobby would be appropriate given the car width is 2,100 mm and the car depth is 1,600 mm? The cars occupy a 6.0 m length of lobby (along the front of the cars). Assume the persons waiting do so at the dense level of occupancy. Assume that a 1,600 kg car can accommodate a maximum of 17 persons.

When one of the cars in the group of four arrives, it will disembark 17 persons. Thus, for these passengers to be accommodated at the dense level of occupancy of 2.0 persons/m² (see Table 2.1 and Figure 2.2) in the lobby, an area of 8.5 m² would be required. If it is assumed that there are 17 persons waiting to board, then they will require a further area of 8.5 m². Hence, a total of 17 m² is required. As the lobby length is 6.0 m, then a width of 2.83 m is indicated, which is slightly smaller than the ISO recommendation of the sum of the depths of two facing cars, i.e., 3.2 m. The lobby width should be 3.2 m.

As a 1,600 kg car will generally only be filled to an average occupancy of 80% of its rated load, i.e., 13 persons, and only 13 persons might be waiting, then a 3.2 m width lobby would be more than sufficient.

2.11.4   Stairs, escalators and lifts

In general, lifts are used for 'long distance' travel over a large number of floors and stairs, or escalators for travel over a small number of floors. A judgement is made by the passenger with respect to the waiting time for a lift versus the walking time (and walking effort) with respect to stairs and escalators. Low rise structures, such as shopping centres, sports complexes, conference and exhibition centres, railway stations, airports, hospitals, etc. are good examples of buildings where the provision of stairs and escalators considerably aids circulation. Peters *et al.* (1996a) provide guidance on stair usage, which can be summarised (in round numbers) as shown in Table 2.9.

In office environments, the usefulness of stairs and escalators is lessened, although they can be advantageous where (say) a heavily populated trading floor(s) must be accessed. They are also very useful for access to car parks, relieving the lifts of the duty to travel below the main terminal. Escalators may be used for access to the two lobby levels where double deck lifts are installed.

Table 2.9 Stair usage

Floors travelled	Usage up	Usage down
1	80%	90%
2	50%	80%
3	20%	50%
4	10%	20%
5	5%	5%
6	0%	0%

Table 2.10 Lifts and escalators: division of traffic

Floors travelled	Escalator	Lift
1	90%	10%
2	75%	25%
3	50%	50%
4	25%	75%
5	10%	90%

Table 2.10 offers some guidance to the choice made by passengers between lifts and escalators in offices. The use of escalators is mainly inhibited by the length of time travelling.

The provision of well signed and positioned stairs and escalators can considerably lessen the demands made on lifts. Designers must take these factors into account.

2.12   Facilities for persons with disabilities

"Access Statement: Achieving an inclusive environment by ensuring continuity throughout the planning, design and management of buildings and spaces".

Equality & Human Rights Commission

The discussion so far has assumed that all persons circulating in a building are fully able bodied. However, a large proportion of the population is disadvantaged in some way.

The Equality Act (2010) in the UK and the Americans with Disabilities Act legislation in the USA lay down various provisions and regulations. Approved Document M to the UK Building Regulations (2004) and BS 8300 (2009) provide actual guidance in the UK.

The Equality Act 2010, which replaced the Disability Discrimination Act 1995, as amended in 2005, requires providers to make reasonable changes to improve service for disabled people. There is a legal requirement to make reasonable changes wherever disabled persons would otherwise be at a substantial disadvantage compared to non-disabled people. Reasonable is always debatable. Further guidance can be found in Chapter 11 of CIBSE Guide D: 2015.

The BS EN 81–70 (2003a) standard '*Accessibility to lifts for persons including persons with disability*' states at Annex A: '*Accessibility enables people, including persons with disability,*

to participate in the social and economic activities for which the built environment is intended.'
The standard categorises disabilities into physical, sensory and intellectual, as summarised in
Table 2.11.

All of the disabilities in Table 2.11 can be temporary or permanent. Impaired mobility could
result from a broken leg (temporary), or a missing limb (permanent). Impaired vision could
be as the result of illness (temporary), or due to age (permanent). Intellectual ability may be
impaired as the result of being intoxicated (temporary), or by being dyslexic (permanent).

In addition, some people are self-hindered. For example, a person with a bulky package
or long clothing will be hindered when using an escalator. Another example is a person who
refuses to wear spectacles and cannot clearly see the call buttons for a lift. Sometimes a solution for one disability reacts with that for another. Thus a tall, older person finds it difficult
to read the legends on car operating panels set at a suitable height for wheelchair users. The
BS8300 (2009) standard *'Design of buildings and their approaches to meet the needs of disabled people'* gives some guidance for facilities within buildings, and examples are given in
Table 2.12.

In addition to access arrangements for persons with, in particular, mobility problems, there
is a need for emergency egress in the case of fire or other emergency; DD CEN/TS 81–76
(2011) provides some recommendations. The scope of CEN/TS 81–76 is for the assisted evacuation of persons with disabilities (it does not include for the general evacuation of buildings
and self-evacuation).

Arrangements made to allow persons with disabilities to make use of circulation elements
also assist the able bodied, and should be implemented, wherever possible.

2.13   Concluding remarks

Circulation in buildings is a complex topic as it depends on human behaviour. However, the
basic principles given in this chapter should enable the lift, escalators and moving walks to be

Table 2.11 Categories of disability

Disability category	Sub-category
Physical	Impaired mobility
	Impaired endurance
	Impaired dexterity
Sensory	Impaired vision
	Impaired hearing
	Impaired speech
Intellectual	Learning difficulty

Table 2.12 Examples to assist persons with disabilities

Element	Example
Steps and stairs	Colour contrast at interface
Ramps	Maximum gradient and minimum width
Handrails	Suitable dimensions
Lifts	Manoeuvring space

accessible. There is little point in providing an 8-car group with one turnstile providing access to it.

The circulation case studies give examples of how many problems are designed out.

More advanced pedestrian movement studies might be carried out using pedestrian modelling software such as that provided by Oasys (Arup), Steps (Mott Macdonald) and Exodus (Fire Safety Engineering Group). These studies are out of the range of this book.

Notes (shown in text as[1])

1 Please note that throughout this chapter people are called pedestrians when on foot, and passengers when they are being mechanically transported.
2 Reprinted by *Elevator World* in 1987.
3 'Rise' is the height of the step and 'going' is the depth of the tread.

3 Escalator and moving walk circulation

3.1 Introduction

Escalators are part of the pedestrian flow system and their landings present an interesting problem as there is an interface between the natural mode of movement and the mechanical mode of movement. There are two interfaces, one at the boarding station and one at the alighting station. At boarding, the natural mode of movement, a walking pedestrian on solid ground, changes to the mechanical mode of movement of a stationary passenger on a moving escalator step. At alighting, the standing (passenger) mode alters to the moving (pedestrian) walking mode. Some people walk onto, along and off the escalator and their effect on circulation will be considered later.

It is well documented that most persons boarding an escalator will hesitate as they board. Although they may be walking at a comfortable speed of 1.0 m/s (see Section 2.4) at the interface, they may well take about one second, the equivalent of one metre of walking distance to board (Fruin, 1971[1]). The hesitation time can vary according to how many flat escalator steps are presented before the escalator steps are formed. Human factors such as age, gender, agility, size, disability, purpose, clothing, bags carried, etc. also effect boarding efficiency.

Having boarded, the passenger may then stand and be transported at the escalator rated speed of (say) 0.5 m/s. At the alighting end of the escalator the standing passenger must alight by walking off the escalator. Again, some hesitation occurs with most persons, which again varies according to how many flat escalator steps are presented and the human factors.

The initial discussion will consider a commonly installed escalator with a 1,000 mm nominal step width running at a nominal speed of 0.5 m/s. Such an escalator has a theoretical handling capacity of 100 persons/minute (see Table 2.7). This assumes that one and one third persons are standing on each step. This density is better than that suggested in the previous version of EN 115, where it was suggested that the escalator could accommodate two persons per step.

The more realistic and observed occupancy is one person per step or two persons every other step (see Table 2.8). This level of occupancy could, to some extent, have been naturally imposed by the problems of boarding. The practical handling capacity is thus about half of the theoretical, i.e., 75 persons/minute. It should be noted that the hesitations at boarding often result in an escalator not delivering its potential practical handling capacity, as boarding can only proceed at the speed of the slowest person. This may result in queues developing at the boarding station and the need to provide space to accommodate these waiting persons.

At the level of occupancy of one passenger per step, and assuming a step is 400 mm deep, two passengers occupy an area of 800 mm deep by 1,000 mm wide, i.e., the density of occupancy of the passengers is 2.5 persons/m². Table 3.1, summarises Table 2.1 and indicates that this density, for queuing persons, would be at a density level between 'dense' and 'crowding'.

Table 3.1 Density of occupation by persons waiting

Level	Density
Desirable	0.4 P/m^2
Comfortable	1.0 P/m^2
Dense	2.0 P/m^2
'Crowding'	3.0 P/m^2
Crowded	4.0 P/m^2

The escalator passengers are queuing on the escalator in what is generally accepted as crowded conditions.

When the passengers reach the end of the escalator they are mechanically fed off the escalator. There is no option – they must leave regardless of the space available to accommodate them. At the boarding station, a passenger can choose not to board, but this option is not available at the destination landing. The potential for conflicts with other passengers (caused by collisions between moving and stationary persons, obstruction caused by persons hesitating or standing on the landing, etc.) becomes significant.

Thus, at the alighting interface, the passenger must start to walk off the moving escalator. They become pedestrians and begin to move naturally, and after some hesitation owing to the change of mode from mechanical movement to natural movement, they reach their comfortable walking speed. This is likely to be 1.0 m/s; see 2.4.

There is a similar interface between the natural and mechanical modes at the boarding and alighting points of moving walks (moving walkways or ramps). The problem of boarding onto and alighting from moving walkways is less severe. The alighting/boarding stations are level or slightly ramped onto a set of flat horizontal pallets, which are often wider than escalator steps. Ramps are similar, except the pallets move at an incline of 5°, 10° or 12°. There is less passenger hesitation and uncertainty, and moving walks generally deliver their practical potential handling capacity (Table 2.9).

It is common in Britain for one file of passengers to stand and one file to walk on an escalator. In this case, the inter-passenger distance increases to one every three steps. The escalator handling capacity does not increase, but the transit time for the walking passengers is decreased. The same number of passengers will alight. One advantage with these passengers is that they are less likely to hesitate on boarding and alighting as they do not stop walking, but they may come into conflict with the stationary passengers. Moving walks exhibit a similar scenario.

3.2   Sizing of escalator landings

It is essential to present passengers alighting from an escalator with an opportunity to move freely. They should, therefore, enter a larger space at the landing. But what size should this space be in order to achieve free movement?

It is worth recalling at this point the space required for moving about as indicated in Chapter 2. Pedestrian speed and density are not independent of each other. For densities below 0.3 P/min, pedestrians can walk freely and this is called free flow design. When densities increase above 0.5 P/min, there is an approximately linear decrease of average walking speed up to a density of about 3.0 P/min, when walking is reduced to a shuffle. The throughput peaks at densities of about 1.4 P/min and this is called full flow design.

The European standard for escalators, BS EN 115–1:2008+A1:2010 Safety of escalators and moving walks – Part 1: Construction and installation, Annex A, offers the following guidance:

A.2.5 At the exit(s) of each individual escalator or moving walk a sufficient unrestricted area shall be available to accommodate persons. The width of the unrestricted area shall at least correspond to the distance between the outer edges of the handrails plus 80 mm on each side. The depth shall be at least 2.50 m measured from the end of the balustrade. It shall be permissible to reduce it to 2.00 m if the width of the unrestricted area is increased to at least double the distance between the outer edges of the handrails plus 80 mm on each side.

For succeeding escalators and moving walks the depth of an unrestricted area shall be determined in each individual case depending on e.g. type of use (persons only or persons with transport devices, number of intermediate exits, relative orientation and theoretical capacity).

Where the exit of the escalator or moving walk is blocked by structural measures (e.g. shutters, fire protection gates) an additional stop switch for emergency situations at hand-rail level (taking into account A.2.2) shall be provided with a distance between 2.0 m and 3.0 m before the step/pallet/belt reaches the comb intersection line. This stop switch shall be reachable from inside the escalator/moving walk.

A.2.6 In the case of successive escalators and moving walks without intermediate exits, they shall have the same capacity (see also h) in Table 6).

The wise intention of this clause is to ensure the safety of passengers alighting from a moving escalator (or moving walk). This is achieved by providing each escalator with an unrestricted area or reserved space. Should the reserved space be too small and it becomes crowded with the alighted passengers, or other persons attempting to join a successive escalator, or other persons crossing in front of the escalator to an adjoining escalator, then passengers currently on the moving escalator will be unable to leave the escalator, with significant safety hazards resulting. If circulation is restricted, conflicts occur. These conflicts can be aggravated by passengers who have walked the escalator rather than standing. If these conflicts are likely to cause people to be brought into contact with moving machinery, as is the case here, with the risk of entanglement, then there are considerable safety implications. If efficient and effective circulation resolves such conflicts, then it follows that safety is also improved.

How will Clause A.2.5 help in the understanding of the circulation of persons? Consider Option 1 of EN115 for a single escalator, shown diagrammatically in Figure 3.1. The distance between the handrail centrelines is usually 1.5 m and, as the depth is to be 2.5 m, gives an available area of 3.75 m². If a pedestrian were to walk straight across the reserved space at 1.0 m/s, then the transit time would take 2.5 s. In this time, 3.125 passengers could alight

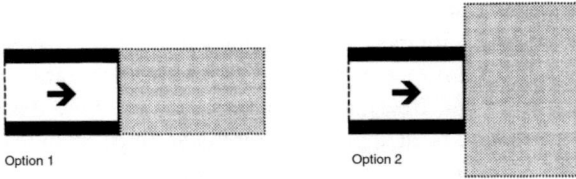

Figure 3.1 EN115 illustration of unrestricted space

(escalator handling capacity of 75 persons/minute). This gives a density of 0.83 persons/m². Free flow movement in a circulation space is possible without the potential for conflict at about 0.3 persons/m², as indicated above.

The density of 0.83 persons/m² is higher, but it would be satisfactory for four main reasons:

1 The movement of persons off the escalator is more disciplined than free movement in an unbounded circulation space.
2 There is likely to be more space available outside the handrail centrelines and hence the reserved space, such as that used for the escalator enclosure and clearance to the building fabric, which increases the width for this option.
3 Persons may leave the reserved space to the side rather than straight ahead, and enter other circulation space more quickly.
4 The escalator passenger perceives a threefold increase in circulation space, reducing the density of occupation from 2.5 persons/m² to 0.8 persons/m².

If Option 2 of EN115 for a single escalator is taken, the transit time to walk 2.0 m would be 2.0 s. During this time, 2.5 passengers would alight. The reserved space they would enter has a width equal to twice the distance between the handrail centrelines, i.e., 3.0 m and a depth of 2.0 m, giving an area of 6.0 m², which is 60% larger than Option 1. This results in a lower density of occupation of 0.42 persons/m².

Option 2 offers more reserved space to the sides than Option 1. This extra space assists the pedestrians to turn to the side (their turning circle), if the forward direction should be bounded by, for example, a wall.

In summary, a 1,000 mm escalator with a rated speed of 0.5 m/s, transporting 75 persons/minute will offer occupation densities of between 0.4 and 0.8 persons/m², dependent on the option selected.

3.3 Mathematical analysis

What happens if escalators of different nominal speeds and nominal step widths are considered? The procedure in Section 3.2 can be expressed mathematically.

The transit time (t_t) for a pedestrian (previously a passenger) to walk at a speed of v_p, in metres per second, across the depth d_l, in millimetres, of the escalator landing is:

$$t_t = \frac{d_l}{v_p} \text{ s} \tag{3.1}$$

The practical handling capacity (C_e) of an escalator (see 2.8) is about 50% of the theoretical:

$$C_e = 0.5 \times 2.5 \ vk \text{ persons/s} \tag{3.2}$$

where v is the nominal speed of the escalator
and k is number of passengers per step.

Assume the flow rate of passengers leaving the escalator is C_e, then the number of pedestrians (Q) on the escalator landing is:

$$Q = C_e \times t_t = 0.5 \times 2.5 vk \times \frac{d_l}{v_p} \text{ persons} \tag{3.3}$$

For Option 1, the area of the landing is given by the distance between the handrail centres X by the depth d_l (taken as 2.5 m):

$$A_l = d_l \times X \text{ m}^2 \tag{3.4}$$

Then, the density of occupation (D_l) on the landing for Option 1 is:

$$D_l = \frac{Q}{A_l} = 0.5 \times 2.5 vk \times \frac{d_l}{v_p} \times \frac{1}{Xd_l} = 1.25 \times \frac{vk}{v_p X} \text{ persons} / \text{m}^2 \tag{3.5}$$

For Option 2, the area of the landing is given by twice the distance between the handrail centres and the depth d_l (taken as 2.0 m):

$$A_l = d_l \times 2X \text{ m}^2 \tag{3.6}$$

Then, the density of occupation (D_l) on the landing for Option 2 is:

$$D_l = \frac{Q}{A_l} = 0.5 \times 2.5 vk \times \frac{d_l}{v_p} \times \frac{1}{2Xd_l} = 0.625 \times \frac{vk}{v_p X} \text{ persons} / \text{m}^2 \tag{3.7}$$

Notice that Option 2 always has half the density of Option 1. Thus, Option 2 will always provide a more efficient and effective, and safer, circulation than Option 1.

Using this mathematics, Table 3.2 can be formed. This shows in its final column the density on the escalator landings. It can be seen that they range from 0.42 persons/m² for Option 2 with a 1,000 mm escalator at a rated speed of 0.5 m/s, to 1.71 persons/m² for Option 1 with a 600 mm escalator at a rated speed of 0.75 m/s.

3.4   Considerations for a design density

What would be a recommended value for the occupation density on an escalator landing, where passengers alight, that allows efficient, effective and safe circulation?

At boarding, a hesitating person or sudden surges of intending passengers can hold up efficient boarding. The intending passengers are inconvenienced by not being able to board, and the escalator may fail to deliver even its practical handling capacity, but there is no danger as they are on 'solid ground'.

The problem is similar at the alighting landing, where a hesitating person can hold up efficient circulation. Here, however, hesitations or sudden surges of persons into the reserved space will bring movement down to a shuffle, with the dire consequence that a jam will occur and passengers on the escalator cannot leave.

It has previously been indicated in Chapter 2 that free flow circulation occurs at a density of 0.3 persons/m², a peak flow occurs at a density of 1.4 persons/m², and a shuffle occurs at 3.0 persons/m². The nearer the occupation density is to 0.3 persons/m², the better the circulation and the safer the movements. These flows are particular to open circulation spaces.

There are four mitigating factors at escalator landings indicated in Section 3.2. The most important is that the passengers are 'more disciplined' as they are on a stair. It was indicated in Chapter 1 that stairs allowed a free flow density of 0.6 P/m² and a full design flow density of 2.0 P/m². As no rule of thumb is available, it is suggested that half the stairway full design density of 1.0 P/m² be adopted.

Table 3.2 Density of landing occupation values for escalators of different rated speeds and step widths

Nominal escalator step width (mm)	Passenger space on escalator (m²)	Passenger density on escalator (P/m²)	Option from EN115	Available area on landing (m²)	Pedestrian transit time on landing (s)	Escalator rated speed (m/s)	Passenger flow off escalator (P/s)	Number of pedestrians on landing	Density of landing occupation (P/m²)
1000	0.40 One person per step	2.5	1	3.75 (1.5 × 2.5)	2.5	0.50	1.25	3.13	0.83
						0.65	1.63	4.08	1.09
						0.75	1.88	4.70	1.25
			2	6.00 (2.0 × 3.0)	2.0	0.50	1.25	2.50	0.42
						0.65	1.63	3.06	0.55
						0.75	1.88	3.76	0.63
800	0.43 Three persons per four steps	2.3	1	3.25 (1.3 × 2.5)	2.5	0.50	1.25	3.13	0.96
						0.65	1.63	4.08	1.25
						0.75	1.88	4.70	1.45
			2	5.20 (2.0 × 2.6)	2.0	0.50	1.25	2.50	0.48
						0.65	1.63	3.06	0.63
						0.75	1.88	3.76	0.72
600	0.48 One person every two steps	2.1	1	2.65 (1.1 × 2.5)	2.5	0.50	1.25	3.13	1.14
						0.65	1.63	4.08	1.48
						0.75	1.88	4.70	1.71
			2	4.40 (2.0 × 2.2)	2.0	0.50	1.25	2.50	0.57
						0.65	1.63	3.06	0.74
						0.75	1.88	3.76	0.86

All numbers are rounded. P is persons

Option 1 presents the worst situation in all cases and Option 2, with its bigger reserved area, is much to be preferred. This might involve increasing Option 1 areas to deal with the whole traffic function.

3.5   Arrangement of escalators: specific cases

A number of arrangements will be considered for a 1,000 mm step escalator running at a rated speed of 0.5 m/s. Option 1 will be considered first, followed by Option 2. Remember, for this situation Table 3.1 shows that the passengers are standing on the escalator at a density of 2.5 persons/m^2.

3.5.1   Arrangement 1: single escalator

Figure 3.1 illustrates this case and the discussion in the section above indicates that no significant problems would be expected.

3.5.2   Arrangement 2: pair of escalators side by side

Option 1. Consider Figure 3.2, where one escalator serves in the up direction and the other in the down direction. In this case, it is possible for persons to cross over from one side of the escalator landing to the other. They may do this through the reserved space belonging to the escalator they are not using. This would increase the density of occupation in that space and the likelihood of conflicts. Extra circulation space should be provided on the landing around the periphery of the reserved space to avoid this situation. This space would ideally extend the depth of the reserved space by 1.0 m, or the overall width by 2.0 m. Another option is to place a stub wall (as illustrated by single lines) between the escalators to protect their reserved space from intrusion.

Option 2. If the pair of escalators are arranged so that the reserved areas did not overlap, then they would comply with Option 2 of EN115 and would be satisfactory. If the two escalators were arranged side by side (abutting) the depth or the overall width should be increased in order to maintain a reserved area of 12.0 m^2.

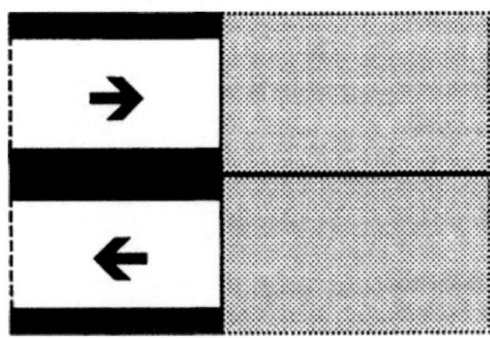

Figure 3.2 Arrangement 2: pair of escalators side by side

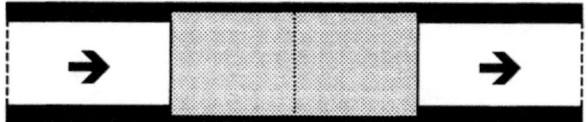

Figure 3.3 Arrangement 3: successive escalators without an intermediate exit

3.5.3 Arrangement 3: successive escalators without an intermediate exit

Option 1. Consider Figure 3.3. Here, the first escalator feeds passengers into the second escalator successively. Both escalators must be the same capacity (EN115, Clause A.2.6). This implies the same rated speed and size. In this case, there are no exits from the midpoint escalator landing, i.e., it acts in a similar way to an intermediate landing on a flight of stairs. The EN115 rule would suggest that at least 4.0 m or 5.0 m of landing would be required. However, the purpose of the midpoint landing is simply to allow passengers to change escalators and the passengers simply alight from one and board the other. There is no possibility of persons leaving or entering.

As the 'corridor' between the successive escalators is the same dimension as that discussed in Section 3.3.2, the pedestrian density will be the same. The distance recommended by EN115 is 5.0 m. This would seem excessive, so what should the inter-escalator spacing be?

Remembering the problems of the mechanical discharge and the natural boarding at alighting, it is important to provide sufficient space between the escalators to accommodate this. It might be achieved by allowing each passenger four walking steps[2] (say) between each escalator. Modern people (Tutt and Adler, 1990[3]) might have an average step of 0.75 m, giving a midpoint landing distance of 3.0 m. This should be considered the minimum spacing. To a great extent, the length of the inter-escalator spacing will be assisted by the number of flat steps provided on each escalator. This minimum spacing should therefore be satisfactory, and there will be few circulation problems in this case.

Option 2. This option is unlikely as the width of the 'corridor' would be too large, but if this option were to be offered, then the 3.0 m spacing would still apply as it is principally dependent on walking patterns.

3.5.4 Arrangement 4: successive escalators with one intermediate exit

Consider Figure 3.4. Here, the escalators have an exit to one side onto a floor landing, allowing passengers to leave the first escalator and others to join the second escalator. The floor landing can be considered as part of the exit area provided it is also unrestricted. If it provides an area of 12.0 m², then a 4.0 m inter-escalator spacing would be suitable, and basically, Option 2 would have been adopted.

For Option 2, the same considerations apply as for Option 1.

3.5.5 Arrangement 5: successive escalators with two intermediate exits

Consider Figure 3.5. Here, the escalators have an exit on each side onto a floor landing, allowing passengers to leave the first escalator and others to join the second escalator. The floor landing can be considered part of the exit area, provided it is unrestricted. The discussion for Arrangement 4 is still valid, and a 4.0 m landing depth might be suitable. However, if there

Figure 3.4 Arrangement 4: successive escalators with one intermediate exit

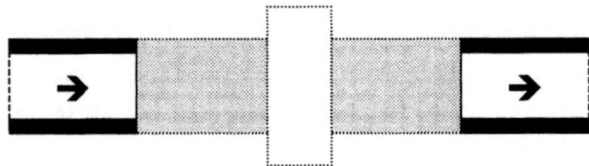

Figure 3.5 Arrangement 5: successive escalators with two intermediate exits

is any possibility of persons (and it is most likely) using the escalator landing to cross from one side of the escalator to the other, extra circulation space of 1.0 m wide (shown unshaded) should be provided. In this case, the escalator spacing should be 5.0 m.

For Option 2, the same considerations apply as for Option 1.

3.5.6   *Arrangement 6: pair of successive escalators side by side with one exit*

Consider Figure 3.6. Here, the pair of escalators serve in both directions, and provided that all passengers leaving the first escalator join the second escalator, the situation would be similar to Arrangement 3. However, this is unlikely, and persons will want to alight from escalators arriving at the escalator landing and persons from the floor will want to join the escalators leaving the floor. Thus, the escalator pair nearest to the exit side of the landing will continually have their reserved space invaded by pedestrians crossing to the other (far pair) escalators. The possibility for conflicts is very great. The EN115 separation of 5.0 m is essential just to ensure satisfactory circulation to the escalators. Moreover, further unrestricted space should be provided next to the escalators.

For Option 2, the same considerations apply as for Option 1.

3.5.7   *Arrangement 7: pair of successive escalators side by side with two exits*

Consider Figure 3.7, which is Arrangement 6 with two exits. As with Arrangement 6, some passengers will leave the first escalator and join the second, and others will leave at the floor landing. New passengers will arrive to leave the floor. Pedestrians from either pair can thus cross into the other pairs' reserved space, thereby increasing the possibility of conflict. The 5.0 m inter-escalator spacing recommended by EN115 is essential.

If there is any possibility (and it is very likely) of persons using the escalator landing to cross from one side of the escalators to the other, more circulation space would need to be provided. This additional circulation space is shown unshaded. In this case, the escalator spacing would be some 6.0 m.

For Option 2, the same considerations apply as for Option 1.

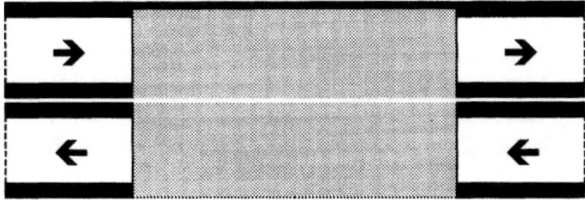

Figure 3.6 Arrangement 6: successive escalators side by side with one intermediate exit

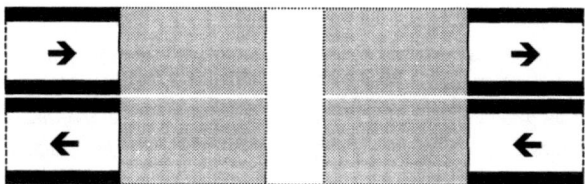

Figure 3.7 Arrangement 7: pair of successive escalators side by side with two exits

3.6 Arrangement of escalators: all types

Section 3.5 considered a number of arrangements for a 1,000 mm step escalator running at a rated speed of 0.5 m/s and Options 1 and 2. The method of analysis can easily be applied to other step widths and speeds. The escalator arrangements specified in Section 3.5 produced results at the most 'comfortable' end of the range of possibilities as far as occupation densities are concerned. Examination of Table 3.1 shows that the other escalator specifications can produce higher landing densities. Care must therefore be exercised when designing an escalator arrangement.

3.7 Another interpretation of EN115, Clause A.2.5

The interpretation of Clause 5.2.1 has another school of thought. In the discussion above, each escalator is considered separately and possesses its own unrestricted space. The reasoning is that if a wall is placed at the distant end of the unrestricted space, then the first escalator complies with the clause. Another escalator can then be provided on the other side of the wall with its own unrestricted space. This would not be sensible for a smooth flow of passengers, so the wall is removed and the inter-escalator spacing is either 5.0 m or 4.0 m, dependent on the option selected.

The other school of thought considers that the unrestricted space can be shared, i.e., an inter-escalator distance of 2.5 m or 2.0 m, dependent on the option selected. This might be satisfactory for successive escalators with no intermediate exits, but not a safe solution for exits with high traffic flows and cross traffic.

The density of pedestrians on the escalator landing given by Equation (3.5) can be simplified to the following formula:

$$\frac{\text{passenger flow}}{\text{unrestricted area}} \text{ persons/m}^2 \tag{3.8}$$

i.e., it is not dependent on the distance between escalators.

Clearly, a zero-depth escalator landing would mean a continuous escalator. There has to be somewhere for the alighting passengers to stand. What is important is that there is enough space to accommodate surges of people, and for the pedestrians to move as freely as possible. The idea of separating escalators of any size or rated speed by 2.0 m or 2.5 m cannot be the intention of EN 115. The absolute minimum should be 4.0 m, and then only where there is a suitable arrangement. The discussions in Section 3.3 should be considered for guidance.

3.8   Moving walks

Moving walks do not present as many problems as escalators as they are either horizontal (moving walkways) or ramped (moving ramps). They are also often wider, i.e., 1,400 mm. Usually, the spacing will be considerable and more than that suggested for escalators as they are often used to move people along rather than upwards. Passengers are not so intimidated at boarding or alighting as it is common for the area in which they are located to be more spacious than escalator 'stairways', thus removing any possible claustrophobic feelings.

Occupation densities on moving walks can be much lower than those on escalators, and the density on the landings could probably be 50% larger than that recommended for escalators.

3.9   Circulation of people with disabilities

The disabled person, as discussed in Section 2.11, may be able to use an escalator or moving walk dependent on their particular disability. Persons with permanent or temporary mobility problems, the partially sighted and the elderly and infirm will be particularly disadvantaged with respect to escalators, but less so for moving walks. The problem for them will be to alight and board safely, and this will increase the time and space they need, which may not be available during peak periods.

Some persons may be temporarily hindered, when carrying bulky items or when wearing clothing that reaches to the floor. The latter circumstance is particularly dangerous. Persons with mobility problems (and those without) will find great difficulty in walking on a stationary escalator as the step height is 230 mm compared to the usual 180 mm. Persons with significant disabilities, mobility problems, or with bulky luggage, etc. should seek other means of circulation, wherever possible.

Mobility-impaired persons in wheelchairs should not attempt to board an escalator and rely on holding onto the handrails for stability. The handrails are not designed for this purpose. Furthermore, although the mobility-impaired person may have developed strong upper-body muscles, should they weaken or suffer cramp, then not only is the person in a wheelchair likely to harm themselves, they will also put other escalator users at risk. Usually, lifts are provided to offer vertical transportation.

Notes (shown in text as[1])

1 Page 102.
2 Two steps equal one pace.
3 Pages 23–28.

Circulation case studies

Contents

The case studies draw on the data tables and discussion of Chapters 2–3 to illustrate the practical aspects of interior circulation.

Case study one

Points of conflict

CS1.1 Building data

Type of building: speculative office – design and build.
Occupant: merchant bank headquarters.
Population: 2,000 persons (above ground), 140 (ground) nil (lower ground, basement).
Number of floors: basement, lower ground, ground, floors 1–6.
Lifts: 6no; 2,000 kg; 1.6 m/s.
Escalators: 2no; 1,000 mm; 0.5 m/s.
Parking: None.

CS1.2 Description of circulation area

As shown in Figure CS1.1, the main entrance (A) has two 2.0 m wide swing doors and two 1.8 m wide revolving doors. The side entrance (B), used for 'drop offs', has one 1.5 m wide swing door and one 1.8 m wide revolving door. A further entrance (H) has a 1.5 m wide swing door, which leads along a corridor (I) beside the escalators to the circulation space (E). This entrance is used as a service route for the ground floor.

The lobby area (C) has some 200 m² clear space.

A set of eight turnstiles (D) leads into a small circulation space (E) of some 15 m² clear space. Leading from this space is a lift lobby (F) of some 30 m², serving a group of six lifts, with doors leading to the ground floor office area at the other end.

From the circulation space (E), a pair of escalators (G) connect to the first floor (one escalator up and one escalator down).

CS1.3 Analysis

Table 2.4 indicates that a swing door will allow a flow of 40–60 persons per minute for each 1 m wide door. Taking the mid value of 50 persons per minute indicates that the two Entrance A swing doors will permit the movement of 200 persons per minute. Table 2.4 also indicates that a revolving door will permit a flow of 25–35 persons per minute. Again taking the mid value, the two Entrance A revolving doors will permit the movement of 60 persons per minute. This gives a total Entrance A entry/exit potential of 260 persons per minute. Entrance B will provide entry/exit opportunities for the occupants. The swing door is 1.5 m wide and will not permit the flow of 100 persons per minute possible for a 2 m wide door. A de-rating should be applied to accommodate this, i.e., to 60 persons per minute. The Entrance B revolving door is the same size as those at Entrance A, and it will permit a movement of 30 persons per minute. This gives a total Entrance B potential of 90 persons per minute.

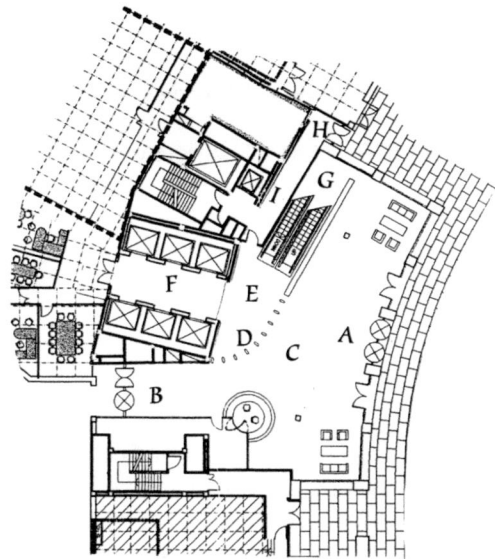

Figure CS1.1 Points of conflict in a merchant bank

Entrance H has been ignored.

Calculations would show the lifts have a 5-minute handling capacity of 240 persons at an interval of 20 seconds.

A single escalator has a theoretical handling capacity of 100 persons per minute; see Table 2.6. However, the practical handling capacity is likely to be 75 persons per minute.

Coin-operated turnstiles allow a throughput of 25–50 persons per minute (Table 2.4). The installed turnstiles use card readers and will operate similarly to coin-operated turnstiles. However, it would be wise to assume a low end throughput of 30 persons per minute. Assuming six turnstiles set in one direction and two in the counter direction, the potential incoming throughput is 180 persons per minute and the outgoing throughput is 60 persons per minute.

There is no parking provided for the occupants, so they must all arrive through the various entrances. The potential population of the building is 2,000 persons and, assuming that a maximum of 80% arrive on any day, with a peak arrival rate of 15%, then the probable 5-minute arrival rate would be 240 persons. This is 48 persons per minute. Surges of twice this size might be expected.

CS1.4 First conclusions

Consider the entry requirement of 240 persons per 5 minutes, i.e., 48 persons per minute. The exit requirement will be assumed to be 50% larger, at 72 persons per minute.

The main entrance provides 260 persons per minute capability. This is more than adequate and would easily allow for surges. The side entrance will also contribute.

The six incoming turnstiles can handle 180 persons per minute. Again, more than adequate for normal and surge conditions.

The escalators serve to the first level only. They can handle all the incoming traffic, which is likely, as only one sixth of the building population occupies the first floor.

The lifts arrive every 20 seconds and fill with 16 people. The lifts can thus handle 48 persons per minute. This is the likely average arrival rate. They would not handle a surge of arrivals on their own. However, some persons would use the escalators and maybe the stairs.

The entrance lobby (C) has 200 m² of clear space and will accommodate 60 persons with free movement (0.3P/ m²) and 280 persons, if crowded.

The circulation space (E) of 15 m² will accommodate 5 to 21 persons, dependent on the occupant density.

The lift lobby (F) has an area of 30 m² and will accommodate some 30 people queuing comfortably at a density of one person per square metre (see Table 2.1).

There is more than adequate provision of entrances, turnstiles, lifts and escalators to serve the maximum arrival rate of 240 persons and to reasonably deal with surges. There is also adequate provision for the exit requirements.

There may not be sufficient circulation space.

CS1.5 Second conclusions

The main problem is that the circulation space (E) after the turnstiles is very small, and the space (F) is the lift lobby, which is a waiting area and not a circulation space.

There are two escalators (G) connecting into circulation space (E). People from the first floor may be passing down the DOWN escalator to reach the ground floor offices, or to exit from the building. People from the ground will be travelling up the UP escalator to reach the first floor.

Note that the DOWN escalator is on the left hand side and people leaving it will immediately conflict with people moving in and out of corridor (I). People have to leave an escalator, but can choose not to board. The left hand escalator should be the UP escalator, to reduce this conflict.

The escalators are so close to the lift lobby (F), that people wishing to alight from the escalators will inevitably come into conflict with people entering or leaving the lift lobby. In addition, people wishing to board the UP escalator will be obstructing the way into the lift lobby, whichever side it is. The escalators should have larger unobstructed boarding and alighting space.

People entering or leaving the circulation space (E) via the corridor (I) will also conflict with both the escalator and lift passengers. This corridor should not access the circulation space at this point.

For the lift lobby, consider a worst-case scenario where two car loads of passengers (32 persons) are waiting in or entering the lift lobby. As indicated above, the lift lobby has an area of 30 m² and so will accommodate some 30 people queuing comfortably. Some people will be moving across the circulation space (E), which could accommodate a further 21 people. So, the lift lobby will be crowded and the vicinity will be busy; ideally, the lift lobby needs to be a little larger.

It will not be workable, however, because the lift lobby is used as a corridor to the ground floor offices (occupancy 140 persons with a potential 5-minute arrival rate of about 21 persons). So, circulation space must be provided to let ground floor staff pass through.

The lift lobby should not be a thoroughfare to avoid conflicts with through traffic.

CS1.6 Points of conflict – lessons to be learnt

- The turnstiles (C) are too close to the lift lobby (F) and the escalators (G).
- The escalators (G) should have larger unrestricted boarding and alighting spaces in circulation space (E).
- The left hand escalator should be the UP escalator.
- The service corridor (I) should be removed and not access the circulation space (E).
- The escalators are placed too close to the other circulation elements.
- The lift lobby (F) needs to be a little larger.
- The lift lobby (F) should not be a thoroughfare and not have a corridor route through it.

Case study two

Commuter railway station concourse

CS2.1 Building data

Type of building: commuter railway station concourse.
Occupant: the public.
Population: variable numbers, but the flow in one direction was estimated at 160 persons/minute, and in the other direction at 80 persons/minute.

CS2.2 Description of circulation area

The area of interest for this case study is the concourse of a commuter railway station. Figure CS2.1 indicates the main elements to be considered. On the left hand side, there is a ticket office. On the right hand side, there is a vending machine, a row of seats, a heating radiator and a fire appliance. Persons will queue for tickets at the ticket office. Although there may be a single ordered queue at the ticket windows, the queue will probably straggle out into the centre of the concourse into an unordered queue. Standing persons after the ticket office can also present an obstruction. The obstruction on the right hand side caused by the legs of people sitting on the seats will have to be avoided. Obstructions will also be caused by persons using the vending machine.

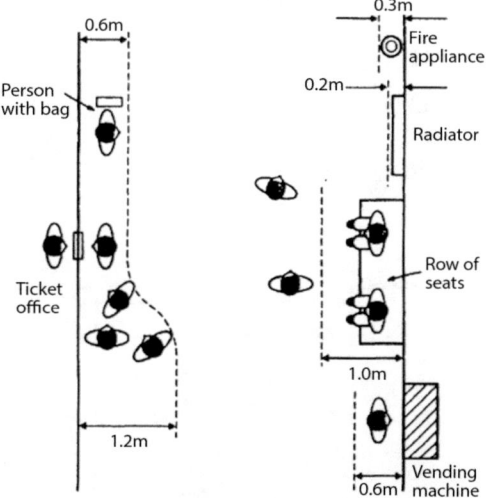

Figure CS2.1 Elements to be considered

Not shown in Figure CS2.1, are the swing doors at the entrance to the concourse from the street and the stairs leading from the concourse into the platform area.

CS2.3   Analysis

Figure CS2.1 indicates some dimensional values for some of the obstructions taken from Table 2.3. On the left hand side, the unordered queue straggles back and becomes 1.2 m, with the front of the queue reducing to 0.6 m at the ticket window. The single standing person also represents a reduction in concourse width of 0.6 m. On the right hand side, the single person at the vending machine causes an obstruction of 0.6 m, and the row of seated persons extends for 1.0 m into the main concourse. The radiator and the fire appliance reduce the effective width of the concourse by 0.2 m and 0.3 m, respectively.

CS2.4   First conclusions

There must be sufficient width of concourse and stairs, and a sufficient number of swing doors to allow the free movement of a total of 240 persons (two-way flow) through the concourse. Table 2.5 indicates that a flow of 60 persons per minute is possible for most people for every one metre width of stair. Thus, at least 4.0 m of stair width will be needed.

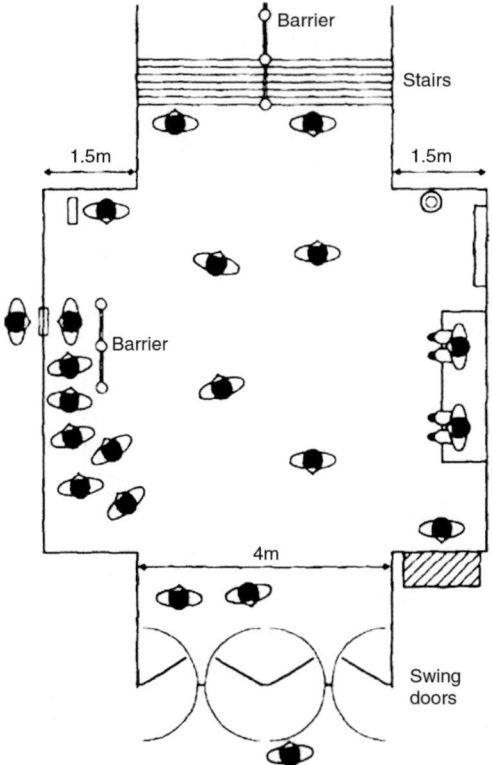

Figure CS2.2 A possible solution

Table 2.4 indicates that 1.0 m of swing door can allow a flow 40–60 persons per minute. Assuming that the commuters would achieve the higher value, then at least four, 1.0 m width doors are needed.

Provided there is a clear 4.0 m width of concourse from the entrance doors through to the stairs, then there should be few conflicts between commuters walking across the concourse.

CS2.5 Second conclusions

Figure CS2.2 shows a possible arrangement. The entrance has been provided with four, 1.0 m wide swing doors, which will allow a flow of up to 240 persons per minute. It will be seen that all the obstructive elements have been moved into 1.5 m deep recesses on each side of a 4.0 m wide unobstructed concourse. A barrier has been positioned by the ticket office window to make the queue more orderly. Standing persons are accommodated in the recessed area. The vending machine has been repositioned on the right hand side and the seats moved back. The radiator and fire appliance now present no obstruction. A central handrail has been placed on the 4.0 m wide stair to impose some discipline in the movement of persons on the stair. It also assists safe passage.

CS2.6 Points of conflict

- Queues at the ticket office.
- Standing persons with bags.
- Persons using vending machines.
- Seated people.
- Radiators and fire appliances.

All these elements prevent the free movement of people.

Case study three

Sizing doorways

CS3.1 Building data

Type of building: concert hall.
Occupant: the visiting public.
Population: 3,500 + staff, etc.
Number of floors: basement, ground, floors 1–5.
Lifts: 5no passenger, 2no goods.

CS3.2 Description of circulation area

The concert hall comprises:

(a) a 2,500-seat theatre
(b) a rehearsal hall accommodating 450 persons
(c) an assembly hall with a capacity for 550 persons
(d) various entertainment suites at Floor 1
(e) a number of workshop and teaching rooms in the basement.

The three halls are connected by a concourse with entrances at the east and west ends. Each entrance has one 2.0 m diameter revolving door and a pair of outward opening swing doors providing a 2.0 m clear opening. The swing doors cannot be fastened back to provide a continuous clear opening. Most people (80%) will arrive and depart via the east entrance, where the main coach and drop off points are situated. If all halls are occupied simultaneously, then some 3,500 people may be present, plus staff, actors, etc. Figure CS3.1 illustrates the circulation area.

The concert hall design team wishes to know if the doorways are suitable to deal with the volumes of people expected to attend the events.

CS3.3 Analysis

Consider two circulation scenarios: before performances (arrivals) and after performances (departure). It is likely that all three halls will be occupied simultaneously on occasions. Thus, some 3,500 persons could arrive and depart at the same time.

Theatre consultants advise that, in a major city, it is likely that the main arrival period will be of some 30 minutes duration, with about 50% (i.e., 1,750 persons) of the audience arriving in the last 10 minutes before a performance. These figures can be considered a worst-case scenario. In a more relaxed environment, the audience arrivals are likely to be over a period of 45 minutes with 50% arriving over the last 15 minutes. The arrival period is affected mainly by the entry arrangements and circulation elements leading to the auditoria.

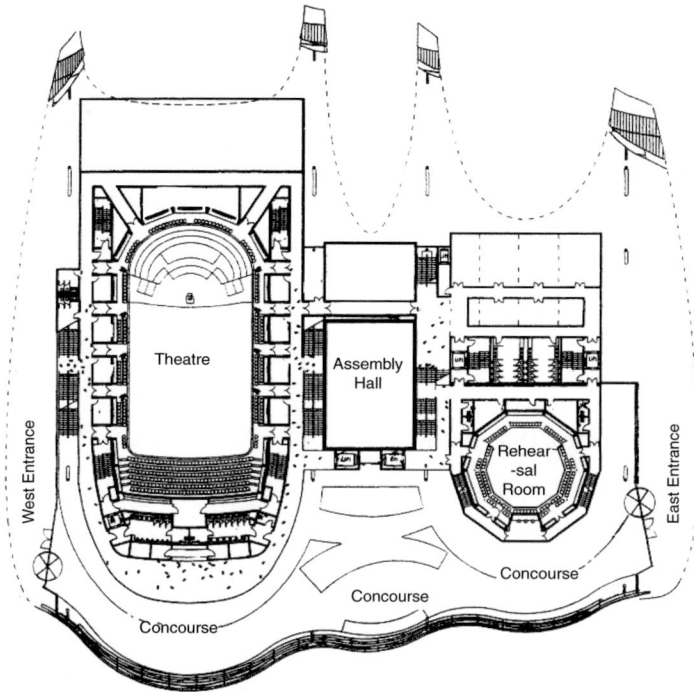

Figure CS3.1 Circulation area in a concert hall complex

Everyone will expect to depart simultaneously. The period over which the departure occurs is affected mainly by the number and size of the exit portals. The flows to these circulation elements will be restricted by the internal circulation elements, e.g., stairs, lifts, corridors and internal doorways. However, it would be reasonable to expect the dispersal rate for 50% of the audience to be achieved over 10 or 15 minutes, i.e., to be no worse than the arrival rate.

CS3.4 First conclusions

The worst-case scenario for arrivals is 1,750 persons over 10 minutes, i.e., 875 persons over 5 minutes. The east entrance has to deal with 80% of these arrivals, i.e., 700 persons over 5 minutes. The equivalent number for the more relaxed scenario is 466 persons over 5 minutes.

The revolving doors will provide a throughput of approximately 150 persons per 5 minutes (Table 2.4).

The swing doors will provide a throughput of approximately 500 persons per 5 minutes (Table 2.4).

The total east entrance door provision is 650 persons per 5 minutes and will satisfy the relaxed scenario, but not the major city scenario.

Removal of the revolving door and its replacement by a swing door would solve the problem, increasing the throughput to 1,000 persons per 5 minutes. However, a revolving door is preferred where there are low traffic levels, such as during the day, as it forms a barrier to the weather. An alternative is to use 'lobbying' to provide a weather shield, i.e., the provision of two sets of swing doors with a lobby area between them. This alternative will, however, reduce the traffic flows for each storm lobby, by as much as one third, to 333 persons per 5 minutes, i.e., a total for two sets of swing doors is 666 persons per 5 minutes. Not quite enough for the worst-case scenario.

CS3.5   Second conclusions

As weather protection was desired, and revolving doors were considered to retard easy circulation, it was decided to install three sets of doors with storm lobbies at the east entrance. This gives a total throughput of 1,000 persons per 5 minutes, which will cope with both the occupancy scenarios.

The main departure demand will be by 2,800 persons, i.e., 80% of 3,500, leaving via the east entrance. They will be able to leave in a period of some 14 minutes. This dispersal rate meets the departure criterion.

CS3.6   Recommendations

Install three sets of swing doors with a storm lobby.

C3.7   Points of conflict – lessons to be learnt

- Consider carefully the throughput rates of the different types of door configurations.
- Take into account weather protection.
- Make reasonable assumptions for arrival and departure processes.
- The model split between different facilities may not be equal.

Case study four

Restaurant access

CS4.1 Building data

Type of building: multi-floor office building.
Occupant: commercial offices.
Population: 6,500.
Number of floors: 33 above ground.
Restaurant: 500 covers.
Lunch time period: 1½ hours.
Lifts: 12no passenger, 2no goods.
Escalators: 4no.

CS4.2 Description of circulation area

Please refer to Figure CS4.1. A restaurant facility is situated on the upper ground (UG) level of a 33 storey building. The entrance to the restaurant is across a 5.0 m wide bridge linking to two lift lobbies containing four car groups placed off a communicating corridor. Two pairs of escalators serve the restaurant end of the bridge between ground and UG, and UG and first floor. The restaurant is reached through a restaurant lobby leading off the bridge.

The entrance to the restaurant lobby is via a pair of single swing doors giving a 2.0 m clear opening and which open towards the bridge. The restaurant lobby has an area of 50 m² and contains a menu board and two card validators. It provides for acoustic isolation, kitchen odour control and, importantly, fire protection.

Access from the restaurant lobby into the restaurant itself is via one pair of single swing doors, giving a 2.0 m clear opening to the restaurant seated area. These doors swing into the restaurant lobby. A pair of single swing doors, giving a 2.0 m clear opening, open into a take-away section. These doors swing into the restaurant lobby.

A further pair of single swing doors, giving a 2.0 m clear opening, provide an exit route from the restaurant seating area. These doors swing into the restaurant lobby.

Some 15% of staff are absent each day and only some 70% of staff use the restaurant.

CS4.3 Analysis

The 15% absentee factor is often used for lift traffic calculations and the 70% participation is likely. There are 500 covers in the seated part of the restaurant, allowing 1,500 to use this area during a 1½ hour lunch period. The remaining staff, some 2,368, use the take-away area during the 1½ hour lunch period. The likely demand (persons) is:

Design population 6500
Absentees (15%) 975

Actual population 5,525
Restaurant 1,500
Take-away 2,368
Participation (70%) 3,868.

The likely scenario is a fairly steady flow into the take-away of 2,368 persons and a peaky flow into the restaurant of 1,500 persons. The circulation elements to be considered are:

- low rise lifts to lift lobby
- high rise lifts to lift lobby
- one pair escalators to/from bridge and G
- one pair escalators to/from bridge to first floor
- corridor from high rise lift lobby to bridge
- swing doors from bridge to restaurant lobby
- one pair entrance doors from restaurant lobby into restaurant
- one pair exit doors from restaurant to restaurant lobby
- one pair entrance/exit doors at take-away service to/from restaurant lobby.

The area at the lobby end of the bridge is very busy and will become very congested as a result of the pedestrian flow over the bridge to the restaurant:

- bidirectional flow across bridge
- bidirectional flow off and onto escalators
- bidirectional flow through bridge to/from restaurant lobby swing doors

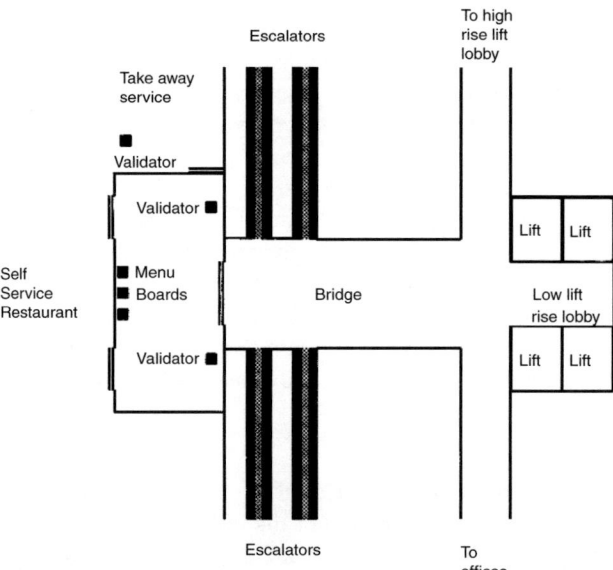

Figure CS4.1 Circulation area for restaurant access

- standing persons in front of card validators
- standing persons in front of menu board
- bidirectional flow through take-away swing doors
- unidirectional flow into restaurant
- unidirectional flow out of restaurant.

CS4.4 First conclusions

Assume a fairly steady flow into the take-away and a peaky flow into the restaurant. If there were to be a steady flow of 3,868 staff over half an hour, this gives an average flow of 43 persons per minute. It is conventional to take twice this rate in order to represent peaks, i.e., 86 persons per minute. At restaurant sitting changeover times, there will be a bidirectional flow, i.e., a possibility of 86 persons in both directions. All this flow must pass over the bridge and through the swing doors of the restaurant lobby.

In general terms (Table 2.4), openings of 1.0 m width will allow the passage of:

60 persons per minute	clear opening
40 persons per minute	a swing door
60 persons per minute	a swing door fastened back.

Thus, the pair of 1.0 m wide single swing doors (2.0 m total) will only permit 80 persons per minute. This is a classical 'pinch point' and the access is not adequate.

The presence of the outwards opening swing doors at the restaurant lobby entrance, which are only some 1.0 m away from the boarding/alighting area for the escalators, is dangerous. Pedestrian confusion and congestion will result. Persons arriving by escalator have no option but to alight at the end of the flight. They cannot stand and wait. On this count, the restaurant lobby doors should be removed and a more open access allowed into the restaurant lobby.

CS4.5 Second conclusions

Please refer to Figure CS4.2. A clear opening should be provided into the restaurant lobby from the bridge of at least 4.0 m clear opening width. This will meet the likely peak bidirectional demand of 172 persons per minute and allow for some congestion. Note also the increase in escalator alighting/boarding area by the use of deflection barriers.

The restaurant lobby is about 50 m² in area, and at a full design pedestrian density of 1.4 persons/m² would accommodate 70 moving persons at any one time. This is too small for the anticipated peak flow. The area should be increased in size to 100 m². This might be achieved by adding the area behind the present position of the menu board back to the structural elements.

The position of the menu board itself will also result in clashes between standing (queuing) pedestrians and moving (walking) pedestrians and it should be relocated in the restaurant lobby. It can be positioned on the new back wall and would then not cause any congestion.

The card validators should be removed from the restaurant lobby and relocated in the restaurant again to reduce any collisions between standing (queuing) pedestrians and moving (walking) pedestrians.

The three sets of lobby doors can then continue to open into the restaurant lobby (the direction is selected for fire escape purposes) without causing problems.

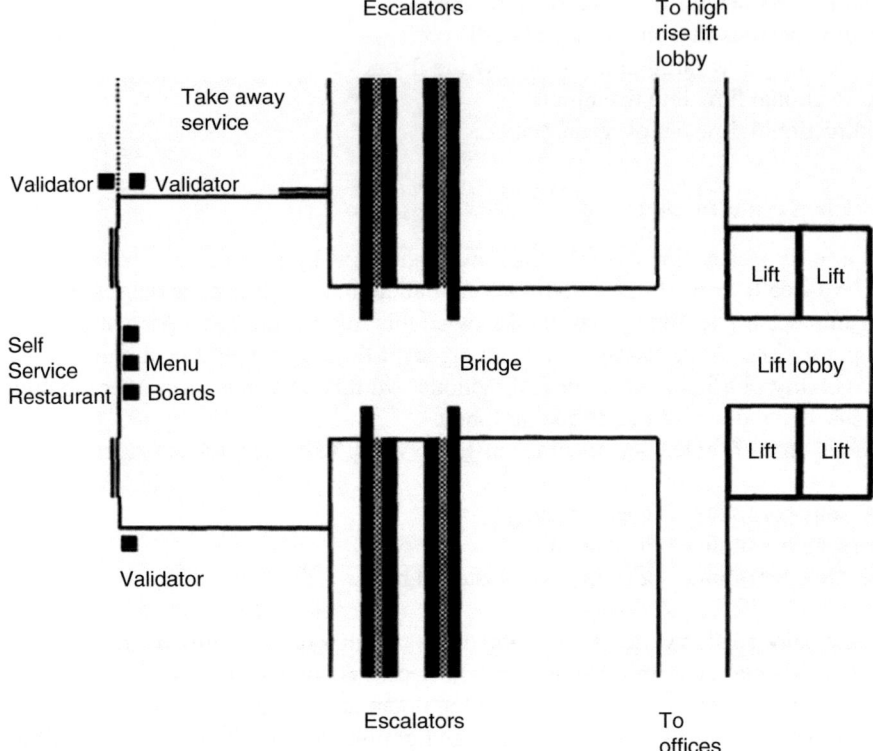

Figure CS4.2 Improved circulation area for restaurant access

CS4.6   Points of conflict – lessons to be learnt

*   The pair of 2.0 m wide single swing doors into the restaurant lobby.
*   The small area of the restaurant lobby.
*   The siting of the menu board in the restaurant lobby.
*   The siting of the card validators in the restaurant lobby.
*   Separate the take-away section from the restaurant section.
*   Safety at the top of the escalators.

Case study five
Escalator landing in an office building

CS5.1 Building data

Type of building: international headquarters.
Occupant: banking and trading.
Population: 8,500 persons.
Number of floors: 4 basements, ground, floors 1–42.
Lifts: 24no passenger, 4no goods.
Escalators: 6no.

CS5.2 Description of circulation area

The area of interest for this case study is the escalator service to ground, second, third and fourth floors. The second, third and fourth floors accommodate financial traders and each have a likely daily population of 434 persons, with a potential daily population of 536 persons.

A pair of 10 m rise escalators (A-up and B-down) located at the east end of the ground floor serve between ground and second floors, reaching a second floor landing at the west end of the second floor.

A pair of 4 m rise escalators (C-up and D-down) located at the east end of the second floor serve between second and third floors and reach a third floor escalator landing at the midpoint of the third floor.

Immediately leading from this third floor escalator landing, a third pair of 4 m rise escalators (E-up and F-down) serve between the third and fourth floors reaching a landing at the west end of the fourth floor.

The two pairs of escalators (C and D, E and F) must be considered as successive sequential escalators as shown in Figure CS5.1. One escalator of the first pair of escalators feeds passengers into an escalator from the second pair in sequence, i.e., in the up direction, C feeds E, and in the down direction, F feeds D.

There is unrestricted access for boarding and alighting the escalators at the ground floor, second floor and fourth floor escalator landings. The third floor escalator landing leads into a 15 m² lobby on the southern side of the escalators from which two sets of glazed, 2 m wide, double leaf, double swing doors lead onto the third floor. There is a wall across the northern end of the third floor escalator landing, so that boarding and alighting must be through the lobby. The doors from the third floor escalator lobby are sometimes locked for security reasons. The distance between the balustrade ends of the two successive escalators (C and D, E and F) at the third floor escalator landing is 3.0 m.

Figure CS5.1 Successive escalators side by side with one intermediate exit

CS5.3   Escalator handling capacity

The rated speed of all escalators is 0.6 m/s[1] and their nominal width is 1.0 m. The theoretical handling capacity from Equation (1.4) is 180 persons per minute and assumes that two persons are standing on each step. This is not practical, as discussed in Section 3.4, and a practical handling capacity will be 90 persons per minute.

As indicated above, there is a likely daily potential for 1,302 persons to occupy the second, third and fourth floors. Suppose that an 'average' 5-minute peak of 15% of the building population is anticipated (similar to that experienced by lifts), i.e., 195 persons per 5 minutes or 39 persons per minute would leave the ground floor. It is unlikely that a smooth arrival process will occur and surges above the average peak might be expected. Common practice is to assume a surge to an average peak ratio of two to one. Thus, Escalator (A) could be required to carry 78 persons per minute. This is below the handling capacity of the escalators, so there should be little queuing. If, however, the daily occupancy of the three floors was the higher value of 1,608 persons, a surge of 97 persons would be possible and there could be some queuing on occasions.

One third of the passengers could stay on the second floor and the remainder (52) could travel up to the third floor escalator landing. Half of these passengers could leave at the third floor and the remainder (26) continue on up to the fourth floor escalator landing. This is a balanced case, and it is quite possible for no one to leave at the second and third floor landings, but for all passengers to travel to the fourth floor.

It would appear that the escalators can handle the average demand and also cope with surges at twice the average demand.

CS5.4   Circulation at the third floor lobby

The third floor lobby has an area of 15 m² and at a full design flow of 1.4 persons/m² could accommodate 21 persons. This should be sufficient to accommodate passengers wishing to board and those who are alighting at this landing. However, if at any time the doors were locked for security reasons, the escalators must stop running as the area could fill in a very short time, which would present an unsafe situation.

Each set of doors leading from the lobby is capable of handling 100 persons per minute. Thus, the two sets can handle more people than the escalators can deliver to the landing.

CS5.5   Circulation at escalator landings

As indicated in the description, the ground floor, second floor and fourth floor escalator landings have unrestricted access, and do not present a problem. The third floor escalator landing is shared by four escalators.

In the worst-case scenario, it is possible for the up escalator (C) from the second floor to third floor to deliver 90 persons per minute (its maximum handling capacity) onto the third floor escalator landing, and the down escalator (F) from the fourth floor to the third floor to deliver the same number of passengers. Thus, a total of 180 passengers per minute travelling to the third floor escalator landing is possible. In addition, some persons might also be leaving the third floor to travel up to the fourth floor, or to travel down to the second floor. This means that over 180 persons could need to traverse the third floor escalator landing every minute.

As Option 1 has been adopted, then using Equation (3.5), the density of occupation on the third floor landing will be 1.0 persons/m². This is the value suggested as the maximum design density in Section 3.4. The passengers therefore move from the escalator at a density of 2.5 persons/m² to a landing density of 1.0 persons/m². If Option 2 had been possible, then the density of occupation would be half the Option 1 value, at 0.5 persons/m².

There is an exit to one side of the pedestrian pathways. Suppose that half the passengers leave the first escalator at the third floor escalator landing and a number of passengers arrive to board either of the second escalators. The passengers walking from the first to second escalators could come into conflict with the boarding passengers.

It is noted that the third floor escalator landing has a 3.0 m distance between the ends of the balustrades of the successive escalators. If this distance is shared between the successive escalators, then each has 1.5 m of unrestricted space instead of 2.5 m. The problem is the opportunity for circulation conflicts, as discussed above, occurring in too small a space. The main concern must be passenger safety. The distance between the successive escalators should be at least 5.0 m as Option 1 has been adopted. Is this sufficient?

The traffic function is complicated as there is cross traffic from one escalator pair to the other; EN155 states: 'consider . . . the whole traffic function. . .'. Thus, it would be sensible to increase the unrestricted landing depth by (say) another 1.0 m. This would be a more acceptable situation.

CS5.6 Conclusions

- The escalators provide sufficient handling capacity.
- The third floor escalator landing lobby and its entry/exit doors are sufficient to meet the circulation demands made on them.
- Where successive escalators are installed, it is important to provide unrestricted areas at landings, at least to meet that suggested by EN115.
- The traffic function must also be considered.

CS5.7 Points of conflict – lessons to be learnt

- Insufficient spacing of successive escalators.
- Ensure that whenever the entry/exit doors to the third floor escalator lobby are locked, that the four escalators are switched off.
- This case study indicates the need to consider safety as well as circulation.

Note (shown in text as[1])

1 This is not a standard nominal speed, see Table 2.7.

Case study six

Escalator landing in court building

CS6.1 Building data

Type of building: court house.
Occupant: court officials, public and prisoners.
Population: not quantifiable.
Number of floors: 4 plus basement.
Lifts: n/a.
Escalators: 4no.

CS6.2 Description of circulation area

The area of interest is the escalator provision from the entrance level (ground) to the court level (level 2) as shown in Figure CS6.1. These two levels are linked by two pairs of successive escalators, with a landing at level 1. The nominal step width of the escalators is 600 mm and they run at a rated speed of 0.5 m/s over an 8 m total rise. The maximum proposed depth of the intermediate landing was 3.0 m, in order to meet footprint requirements.

The client sought advice as to the efficiency and safety of the arrangement.

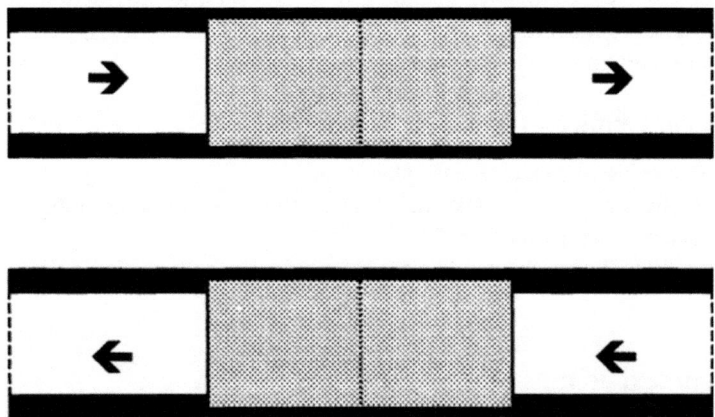

Figure CS6.1 Successive escalators without an intermediate exit

CS6.3 Analysis

The escalators will provide a practical handling capacity of 37 persons per minute. This is likely to be sufficient as lifts and stairs are provided nearby. The depth of 3.0 m on the landing does not comply with the recommendations in EN115, but does follow the suggestions in Section 3.4. This should be satisfactory.

CS6.4 Design change

At a late point in the design it was suggested that the enclosed escalator landing at level 1 be opened up to allow access to the catering facilities located there from both the court and the ground levels. This suggestion was questioned. The client indicated the poor circulation provided to level 1.

The suggestion did not meet the safety requirements outlined in Chapter 3. In addition, the likelihood of persons crossing from one side of the escalators to the other was very strong with consequential conflicts.

A compromise was to restore the central dividing walls, but to allow passengers to leave and join in one direction only. This compromise would be reasonable in the circumstances as the step width of the escalator was 600 mm; hence, all passengers could only alight and board in single file, thus easing the competition for landing circulation space.

CS6.5 Conclusion

The design change could be accommodated.

CS6.6 Point of conflict

- Insufficient spacing of successive escalators for proposed circulation.

Case study seven

Terminus railway station

CS7.1 Building data

Type of building: terminus railway station.
Occupant: passengers.
Population: 500 passengers per train.
Number of floors: 2, platform and concourse (below).
Length of train: 130 m.
Lifts: 1no passenger (disabled access).
Escalators: 8no total (see Figure CS7.1):
 2no up, 2no down at the one quarter point of the platform.
 2no up, 2no down at the three-quarter point of the platform.

CS7.2 Description of circulation area

This case study considers a terminus railway station with the platforms above a concourse area. The trains are 130 m in length and arrive with a 180 s (three minute) headway. Each train accommodates 500 passengers.

A set of two up escalators and two down escalators are situated at both the one quarter and the three-quarter points along the platform length. The escalators run at a rated speed of 0.5 m/s and are 1,000 mm wide.

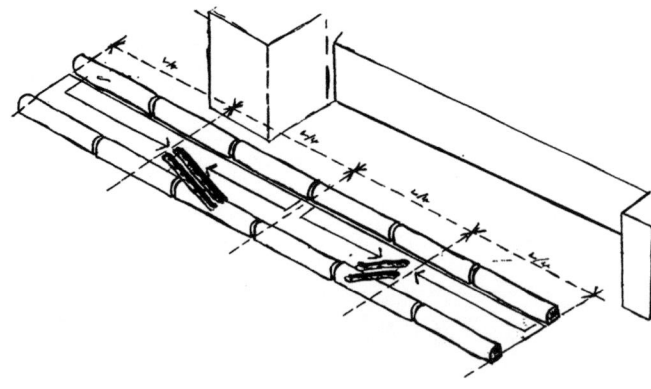

Figure CS7.1 Sketch drawing of original proposal

Can the number of escalators be reduced? The most important time of the day to consider is the morning arrival peak period, with little or no boarding passengers.

CS7.3 Analysis

The most important consideration is to clear the platform before the next train arrives and the next important consideration is to avoid passenger queuing. Assume that the passengers walk at an average speed of 1.3 m/s (3 mph). This is an average between commuters at 1.5 m/s and tourists at 1.0 m/s and is a likely average. Assume 1,000 mm, 0.5 m/s escalators can provide a theoretical handling capacity (Table 2.7) of 75 persons per minute (1.25 persons/second).

Assume half the passengers walk to each escalator set.

Maximum distance a passenger will walk is: 130/4 = 32.5 m
Time for last passenger to reach the escalators is: 32.5/1.3 = 25 s
Passenger arrival rate at the escalators: 250/25 = 10 persons/second
With two up escalators at each position.
Handling capacity of two escalators: 2.5 persons/second
The time for complete clearance will be: 250/2.5 = 100 s
The last passenger will wait: 75 s
The number of passengers queuing will be: (10–2.5) × 25 = 188

Thus, the platform clearance time is 100 s and the platform will be clear by the time the next train arrives.

CS7.4 First conclusions

With bulk transit systems, large numbers of passengers arrive simultaneously. To handle such a situation, it is necessary to spread the load by introducing something for the passengers to do before they reach the mechanical handling equipment. A good technique is to make them walk a reasonable distance. If possible, some sales points should be introduced along the platform to further delay the arrival of passengers to the escalators, but not so much as to fully occupy them during the train headway period of three minutes.

Consider a set of two up escalators at one end of the platform (see Figure CS7.2).

Maximum distance a passenger will walk is: 130 m
Time for last passenger to reach the escalators is: 130/1.3 = 100 s

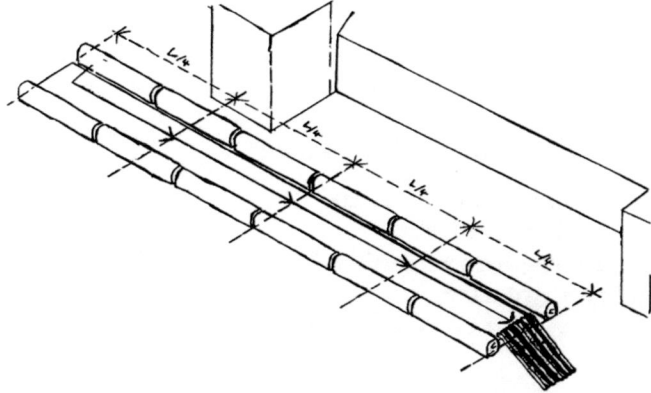

Figure CS7.2 Sketch drawing of final proposal

Passenger arrival rate at the escalators: $500/100 = 5$ persons/second
Handling capacity of the two up escalators: 2.5 persons/second
Thus the clearance time will be: $500/2.5 = 200$ s
The last passenger will wait: 100 s
The number of passengers queuing will be: $(5–2.5) \times 100 = 250$

Thus, the platform clearance time is 200 s and the platform will not be clear by the time the next train arrives. This is not a solution.

Consider a set of three up escalators at the end of the platform.

Handling capacity of the three up escalators: 3.75 persons/second
Thus, the clearance time will be: $500/3.75 = 133$ s
The last passenger will wait: 33 s
The number of passengers queuing will be: $(5–3.75) \times 100 = 125$

Thus, the platform clearance time is 133 s and the platform will be clear by the time the next train arrives.

Consider a set of four up escalators at the end of the platform.

Handling capacity of the four up escalators: 5.0 persons/second
Thus, the clearance time will be: $500/5 = 100$ s
There will not be a queue, or passenger waiting time

The platform clearance time is 100 s and the platform will be clear of passengers by the time the next train arrives.

CS7.5   Second conclusions

Effective load spreading is achieved where the escalators are placed at the end of the platform.

It is not reasonable to design for a peak load. Queuing would only occur at peak times, when some queuing can be tolerated, provided the platform is cleared before the next train arrives. With four escalators, there is no queuing. With three escalators, there would be 125 passengers queuing for up to 33 s. With two escalators, the platform does not clear in time.

The three-up-escalator scenario would be acceptable and reduces the number of escalators that would be required if the distributed proposal had been adopted.

Probably, a set of four escalators would be ideal. The outer escalators could be permanently set for up and down directions with the inner pair operating on a tidal flow basis or switched off outside the peak periods. Remember, escalators are taken out of service for maintenance and inspection (and sometimes fail) and this arrangement would allow this.

CS7.6   Points of conflict – lessons to be learnt

- Platform clearance during a train headway.
- Queues of containable proportions.
- Passenger waiting times that are tolerable.
- Escalator tidal flow possibilities.

Case study eight
Lifts versus escalators

CS8.1 Building data

Type of building: stock exchange.
Occupant: financial traders.
Arrival rate: 375 persons/5 minutes.
Number of floors: ground, 1, 2, 3, 4.
Interfloor distance: 6.0 m (rise).
Lifts Rated speed: 1.6 m/s.
Rated load: 2,500 kg.
Cycle time: 12 s.
Escalator step width: 1,000 mm.
Rated speed: 0.5 m/s.
Handling capacity: 375 persons/5-minutes (practical).

CS8.2 Description of circulation area

During uppeak, do lifts or escalators provide the fastest travel time, when serving a high density, low rise building?

CS8.3 Analysis

The conventional wisdom is that passengers will be influenced to travel either by lifts or escalators mainly by the distance to be travelled, *viz*:

CS8.4 Assumptions

For an escalator with a rated speed of 0.5 m/s, the travel time between boarding and alighting stations is 24 s. Assume that the passenger boarding and alighting time is 2 s. If the escalators

Table CS8.1 Division of traffic between escalators and lifts

Floors travelled	Division of traffic (%)	
	Escalator	Lift
1	90	10
2	75	25
3	50	50
4	25	75
5	10	90

are sequentially arranged, the walk time from one to the next will be assumed to be 2 s. The practical handling capacity of the escalators is 375 persons per 5 minutes.

The lift system is to be sized to provide the same capacity as the escalators. Calculations can show that three lifts are required to serve a single floor above ground, four lifts to serve two floors above ground, five lifts to serve three floors above ground, and six lifts to serve four floors above ground. Further assumptions are: the practical occupancy is 19 persons, the passenger loading time for 19 passengers is 12 s, the interval will be 15 s and an average waiting time is 12 s, regardless of the number of lifts.

CS8.5   Calculations

It is also assumed that the escalator passengers stand and do not walk on the escalator and that the escalators are arranged in either a parallel or a cross-over configuration (see Figure 1.10). Using the assumptions above, calculations can show that the travel time by lift or escalator is as shown in Table CS8.2.

Notes to Table CS8.2. The average waiting time is calculated as 85% of interval. Passengers are assumed to take 19 s to board a lift and 2 s to board/alight from an escalator. The travel time for a lift is the cycle time, and for an escalator is the time on the escalator. The lift delay time is the time spent at intermediate landings for passengers to exit, assuming the lifts stop at all floors. The average passenger alighting time is the time for the remaining passengers to leave a lift at the final destination.

CS8.6   First conclusions

It can be seen that the escalators are quickest when there is only one floor above, equal for two floors above, but slower thereafter. If one line of passengers were to walk on the escalator, their travel time would be reduced by 12 s per rise, i.e., reducing the average travel times by 6 s per rise. However, if a walk-around escalator configuration were to be provided, the escalator travel times could increase by another 25 s per rise.

CS8.7   Second conclusions

Other considerations are space and cost. Probably, the lifts will occupy more space for the low rise scenarios and less for the high rise. Probably, the escalator option is cheaper for the low rise scenario. It is important to note that the lift and escalator installations considered here were both able to provide equal handling capacities. This may not always be the case.

Table CS8.2 Travel times of lifts versus escalators

Travel to:	First floor		Second floor		Third floor		Fourth floor	
Equipment (L = lifts, E = Escalator)	L = 3	E = 1	L = 4	E = 2	L = 5	E = 2	L = 6	E = 3
Average waiting time (s)	13	0	13	0	13	0	13	0
Passenger boarding time (s)	12	2	12	4	12	6	12	8
Average passenger alighting time (s)	6	2	3	4	2	6	1.5	8
Travel time (s)	12	24	24	48	36	72	48	96
Lift delay at stops/walking time (s)	0	0	6	2	8	4	9	6
Total	43	28	58	58	71	88	84	118

Part B
Traffic design

4 Principles of classical lift traffic design

4.1 The need for lifts

Lifts are installed into buildings to satisfy the vertical transportation needs of their occupants and visitors. They are necessary to provide a safe and comfortable means of transportation to the different levels in a building. Some of these requirements are written into statutory regulations, for example, to meet the needs of persons with impaired mobility (BS EN81–70: 2003) or for firefighting (BS EN 81–72: 2003b).

The transportation capacity of the lift group in a building is a major factor in the success or failure of a building as a place to work, in which to live or to receive a service. Like toilets, lifts should be available and easy to use without a second thought. Unfortunately, this is not always the case and speculative buildings often contain an imperfect lift system.

In offices and other commercial buildings, lifts are installed to aid the efficient movement of the occupants around the building when performing their work tasks. This has the benefit of saving time, and hence money.

These financial considerations do not apply for residential property; quite the opposite, money is saved by not providing a lift, and statutory regulations have been framed to ensure suitable lifts are installed. In Britain, for example, it is recommended that a lift be installed in all residences where there are four or more storeys, and that two lifts be installed where a building contains more than six storeys.

The increase in the numbers of high and medium rise buildings since the Second World War has been a challenge to the lift industry. The four decades between 1945 and 1985 saw the acceptance of automatic cars, the introduction of improved drive systems, better traffic control systems and the inclusion of the digital computer in most equipment. Improvements have also occurred in the engineering design and engineering installation of lift systems. The acceptance of traffic design methods has been slower and has only become universally accepted since the early 1970s.

This chapter considers the constraints on a lift traffic design, and introduces the concepts of traffic patterns. This leads to the need to describe the movements of a lift in mathematical terms and the derivation of the round trip time equation.

4.2 Fundamental design constraints

The planning and selection of transportation equipment is a very involved subject. Although the basic calculations are relatively simple, the theory on which they are based is complex. The results obtained need to be tempered with a great deal of working experience of existing buildings in order to ensure satisfactory design results. Without such experience, the reader of this book will not be an expert.

When sizing a lift system for a new building, the major building dimensions should be known. Unfortunately, it is often the case that the architect responsible for the building conception will not have taken professional advice from a lift specialist and may well have fixed the building's core dimensions, thus limiting the space available for the lift system or, even worse, may have defined the number of shafts, their dimensions and travel. This removes one very important degree of freedom from the lift traffic designer. Building circulation, both horizontal and vertical, is the lifeblood of any building; hence, if a successful building is to be designed, it is essential that the architect takes expert advice at conception. This does not imply that the lift designer will take over the core design, but simply that by means of a team approach, various aesthetic and conceptual ideas can be considered early on and optimal solutions offered.

Often the result of a team and professional approach will be a better-sized lift system design, possibly with fewer shafts travelling the whole height of the building, or a rearrangement of service floors and main terminals. The net effect should be a building properly configured for good access with sufficient handling capacity to serve the proposed population and its circulation needs.

Of course, at the low end of the market, there may be only one lift in a building, or its dimensions may be fixed to conform to statutory regulations, or to accommodate the carriage of furniture, etc. But as the lift system moves 'up-market', initial design decisions become more important.

When redesigning for the modernisation of an existing lift installation, the fundamental constraints mentioned above cannot be altered (or not very much) as the building actually exists. However, there is often the advantage that the building population to be served is already known.

4.3   Human constraints

A lift system has to be acceptable to the travelling public. The most important requirement that the public demands is safety. These aspects are covered by the safety standards promulgated at national, regional and worldwide levels, for example by the British Standards Institution (BSI), European Standards Committee (CEN) and the International Standards Organization (ISO). This requirement is most important so that passengers may feel confident about the way they are handled. However, passengers are human and are subject to constraints, which fall into two categories: physiological and psychological, the body and the mind.

4.3.1   *Physiological constraints*

The physiological constraints (the effects of movement on the body) limit the manner in which a passenger may be moved in the vertical plane. The human body is uncomfortable if its internal organs are caused to move within the body frame. This occurs when the body is subjected to acceleration or deceleration, the well-known '*g*' effect. The magnitude of the effect on an individual depends on an individual's age, physical and mental health, and whether the individual is prepared for the experience of sudden movement. It is not clearly established what the level of acceleration is at which permanent harm may be caused to the human body, but it is known, by experience, the levels of acceleration or deceleration that have been found to be generally acceptable when riding in a lift.

1   There is no limit to the velocity at which a passenger may travel in an enclosed lift car, as speed is not noticeable to the passenger. However, very high speed lifts travelling long

distances do cause some passengers ear discomfort due to the pressure changes. Shuttle lifts with speeds of up to 15.0 m/s have been installed. But the values of acceleration/deceleration (rate of change of velocity) should be limited. Until recently, acceleration was permitted up to about one eighth of g_n[1] or 1.5 m/s^2, with values of jerk (rate of change of acceleration) up to 2.0 m/s^3. The effect of an acceleration of one eighth of g_n on a body weighing 80 kg travelling in an upward direction is that it then weighs 90 kg. Likewise, the same body subjected to a deceleration, while travelling in an upward direction, would weigh 70 kg.

At the turn of the century, in order to meet passenger demands of high ride quality, these figures have been reduced to around 1.2 m/s^2 and 1.5 m/s^3. As will be seen, this increases the time to travel between floors and affects traffic handling capacity.

Figure 4.1 shows the relationship between distance, velocity, acceleration and jerk for a lift car movement between adjacent floors.

It is the jerk values (not a very scientific sounding name, sometimes called 'shock'), which causes the most discomfort. If the value of jerk is allowed to exceed 2 m/s^3 for any length of time (tenths of seconds), discomfort will be experienced. Whereas velocity and acceleration/deceleration profiles can be specified and controlled in drive systems, jerk cannot.

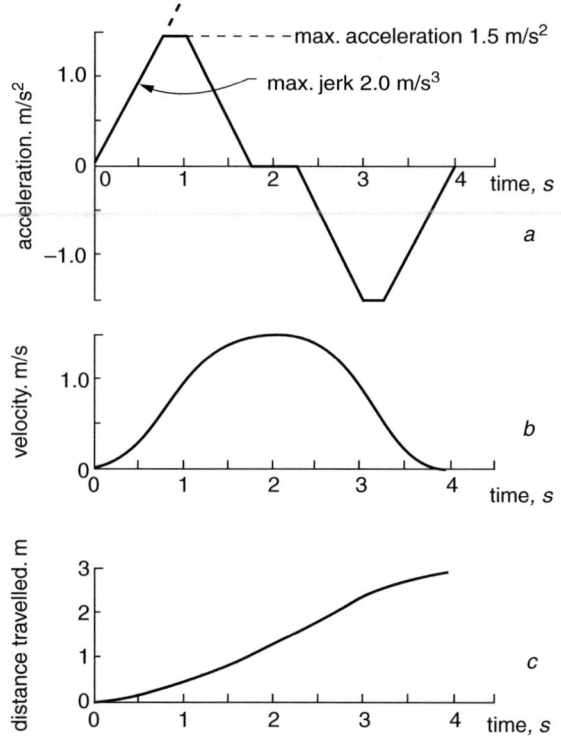

Figure 4.1 Ideal jerk, acceleration, velocity and distance travelled curves for a single floor jump

(a) Acceleration profile: note maximum jerk 2.0 m/s and maximum acceleration 1.5 m/s
(b) Velocity profile: note maximum speed is 1.5 m/s
(c) Distance profile: note the distance travelled is 3.0 m

Constant values of jerk require that the acceleration/deceleration profile increase/decrease at a constant rate, and this is not always practically possible due to drive inadequacies. It is perhaps fortunate that these human constraints do exist as they ease the design of lift drive systems considerably.

4.3.2   *Psychological constraints*

As would be expected, psychological constraints are more subtle. A passenger expects a good service from a lift system. But an individual passenger expects a different grade of service at different times of the day and at different locations. For example, an office worker will not be too annoyed if delayed when travelling up a building to work, but will become very annoyed if delays occur when leaving at night. In contrast, the same office worker would not expect the same grade of service from a lift in a residential block. This constraint can be categorised as the passenger's waiting time constraint. In general, the average waiting time in an office building should not exceed 30 s and in a residential building it should not exceed 90 s. Waiting time is the prime psychological constraint.

A secondary psychological constraint is the transit time, or travel time, in the car after the passenger boards. Here, the passenger is dependent on fellow passengers in the car and other passengers on the landings making calls. A passenger travelling high up a building becomes intolerant of stops after about 90 s of travel. Again, the tolerance level depends on whether the passenger is travelling in the company of friends or colleagues and also on the other passengers' behaviour. For instance, one passenger boarding or alighting is obviously more 'selfish' than two or three transferring at a time. This psychological constraint has been summed up by Strakosch (1967), who stated 'a person will not be required to ride a car longer than a reasonable time'.

There are other psychological effects, such as aesthetic appearance and 'gentle' doors, which add to a passenger's confidence in a lift system and overcome the fears of some persons who are afraid of such machines.

4.4   Traffic patterns

The users of lift systems, the passengers, impose on the lift system the need for it to respond to different traffic patterns. Consider Figure 4.2, which was originally presented by Strakosch (1967) so it dates from the early 1960s. This classic traffic pattern shows, in simple terms, the passenger demand in an office building as represented by the number of individual calls, aggregated for up and down call directions. This office building is subject to a strict time regime of fixed starting, break and leaving times. It illustrates clearly the different traffic patterns of morning uppeak, evening down peak, midday traffic and random (balanced) interfloor traffic.

At the start of the day, there is a larger than average number of 'up hall calls'. This traffic pattern is called the morning uppeak, and is due to the building's occupants arriving to start work.

Late in the day, there is a larger than average number of 'down hall calls'. This traffic pattern is called the evening down peak, and is due to the building's population leaving the building at the end of the working day.

In the middle of the day, there are two separate sets of up peaks and two down peaks. This represents a situation where the occupants of the building take two distinct lunch periods (i.e., 12.00 to 13.00 and 13.00 to 14.00). This pattern is sometimes called two-way traffic.

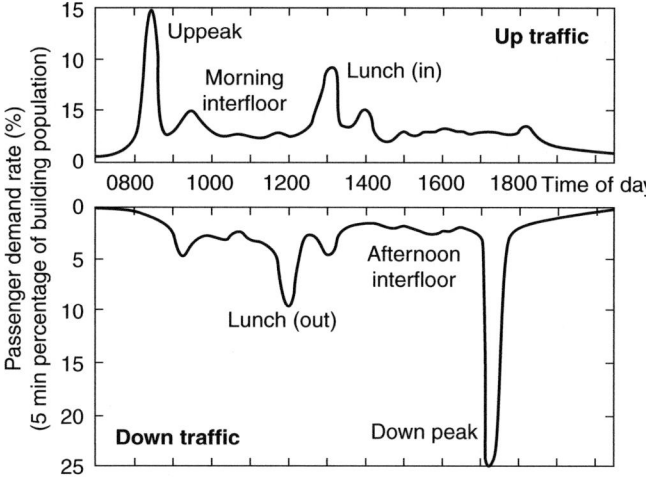

Figure 4.2 Passenger demand for an office building

During the rest of the day, the numbers of up hall and down hall calls are similar in size, and over a period of time are equal in numbers. This traffic pattern is called interfloor traffic, sometimes qualified as balanced interfloor traffic.

In practice, the pattern shown in Figure 4.2 does not exist today, 50 years later. It does, however, serve as a model for discussion. The classical design method assumes this imaginary scenario to size a lift system, and has been found to work.

4.4.1 Uppeak traffic

This traffic condition is shown diagrammatically in Figure 4.3.

> **Definition 4.1:** An uppeak traffic condition exists when the dominant, or only, traffic flow is in an upward direction, with all, or the majority of, passengers entering the lift system at the main terminal[2] of the building.

Uppeak occurs in strength in the morning when prospective lift passengers enter a building intent on travelling to destinations on the upper floors of the building. A smaller uppeak occurs at the end of the midday break(s). It is considered that, if a lift system can cope efficiently with the morning uppeak, then it will cope with other patterns of traffic, such as down peak and random interfloor traffic.

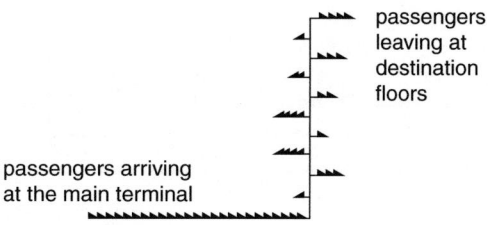

Figure 4.3 Uppeak traffic

The uppeak condition results from employers requiring their employees to arrive at work by a specific starting time. Human nature then exacerbates the condition as the majority of employees feel that in conscience all they must do is to be in a building before the defined starting time and that the employer then has the responsibility to transport them to their work station.

Flexitime[3] (a form of flexible working) goes some way to alleviate the uppeak situation, but unfortunately it is not possible to apply it to all classes of employment (e.g., a call centre). It can also mean that other traffic conditions may become relatively more severe, and if a building designed for flexitime becomes one with a fixed time regime, then the lift system could be seriously undersized.

The arrival rate profile for the classical morning uppeak takes a form as shown in Figure 4.4. Here, the envelope of the curve describes the arrival profile in terms of the instantaneous passenger arrival rate in calls per hour for a period of one hour. Figure 4.4 reveals that the uppeak traffic profile presents a gradual build up prior to the official starting time, and then a more rapid decay afterwards. The lift installation must be able to handle the peak if a satisfactory service is to be provided.

The profile of Figure 4.4 is often idealised by designers in terms of a 5-minute peak value taken as a percentage of the building population (the wide hatched area of Figure 4.4). The industry practice is to size a lift installation to handle the number of passengers requesting service during the heaviest 5 minutes of the uppeak traffic condition (HC5). This is a sound recommendation. To size the lift system to handle the actual peak would require too large a system, which would be very expensive and much of the equipment would be under-utilised during many periods of the working day.

The uppeak traffic condition is often detected by the traffic supervisor so that specific control actions may be taken. Common detection systems determine when a predefined number of cars leave the main terminal, loaded to a predefined level. The duration of the uppeak period detected in this way does not necessarily exist for precisely 5 minutes.

Figure 4.5 is taken from a computer simulation[4] of a peak morning hour. It shows the spatial movements of a number of lifts. Note the increased number of stops during the peak 5-minute period.

This chapter (and most design calculations) use the uppeak traffic condition to size a lift installation as it is simple and easily modelled.

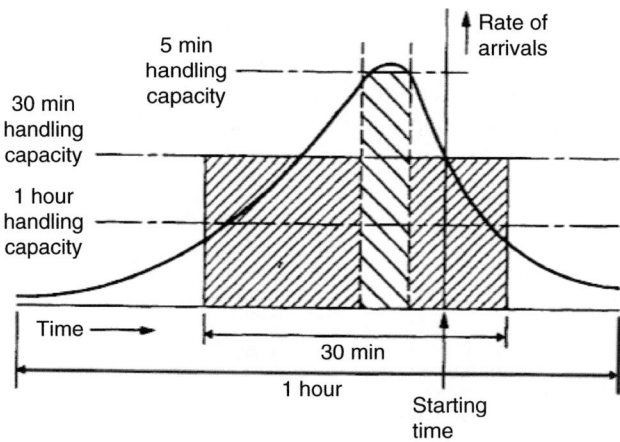

Figure 4.4 Detail of uppeak traffic profile

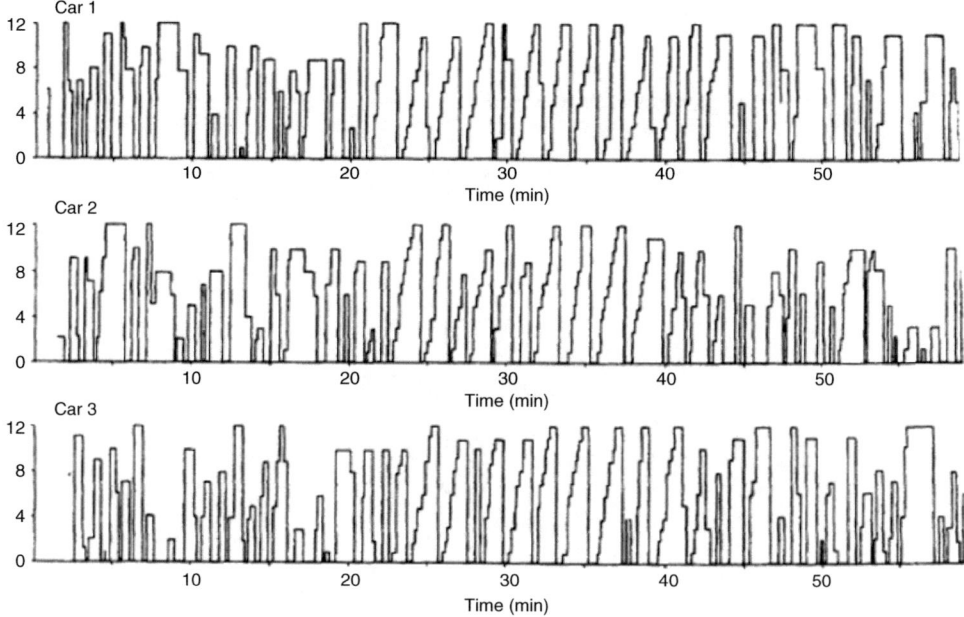

Figure 4.5 Screen shot of the spatial movements of three lift cars during uppeak traffic

4.4.2 Down peak traffic

The traffic condition is shown diagrammatically in Figure 4.6.

Definition 4.2: A down peak traffic condition exists when the dominant or only traffic flow is in a downward direction with all, or the majority of, passengers leaving the lift system at the main terminal of the building.

To some extent, down peak is the reverse of the morning uppeak, and occurs at the end of the working day, and to a lesser extent at the start of the midday break. The evening down peak is more intense than the morning uppeak, with higher demands and with longer durations. Figure 4.7 illustrates these effects.

Figure 4.7 details the down peak traffic profile, showing the larger size and longer duration of the traffic demand. Fortunately, a lift system can be shown to possess 50% more handling capacity during down peak than during uppeak (see Chapter 13). This is because during down

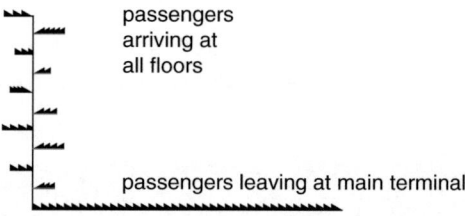

Figure 4.6 Down peak traffic

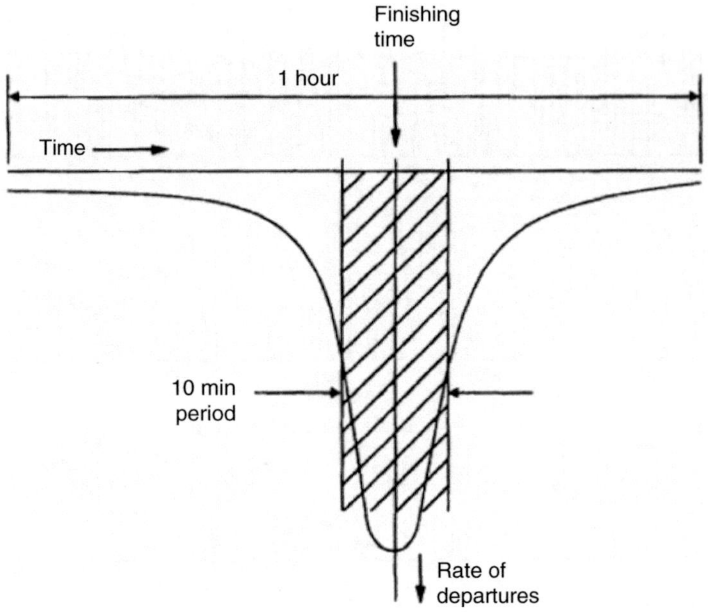

Figure 4.7 Detail of down peak profile

peak, a lift car fills at three, four or five floors and then makes an express run to the main terminal. This reduction in the number of stops results in a shorter round trip time and hence a greater handling capacity during down peak.

The detection of the onset and duration of the down peak traffic condition is usually achieved by similar methods to those used to detect uppeak. Figure 4.8 is taken from a computer simulation of a peak evening hour, again showing the spatial movement of the lifts. Note the smaller number of stops and the express runs to the main terminal floor during the peak 10-minute period.

4.4.3   Two-way and midday (lunch-time) traffic

The two-way traffic condition may not be easily detectable in most buildings. It can arise from the presence of a refreshment floor, which at certain times of the day attracts a significant number of stops and calls. Two-way traffic could thus occur during the mid-morning and mid-afternoon refreshment breaks.

Definition 4.3: A two-way traffic condition exists when the dominant traffic flow is to and from one specific floor, which may be the main terminal.

The midday (lunch-time) period often presents the heaviest demand on a lift system, owing to the simultaneous passenger arrivals and departures of up, down and interfloor traffic at several floors.

Definition 4.4: A midday (lunch-time) traffic condition occurs in the middle of the day and exhibits a dominant traffic flow to and from one or more specific floors, one of which may be the main terminal.

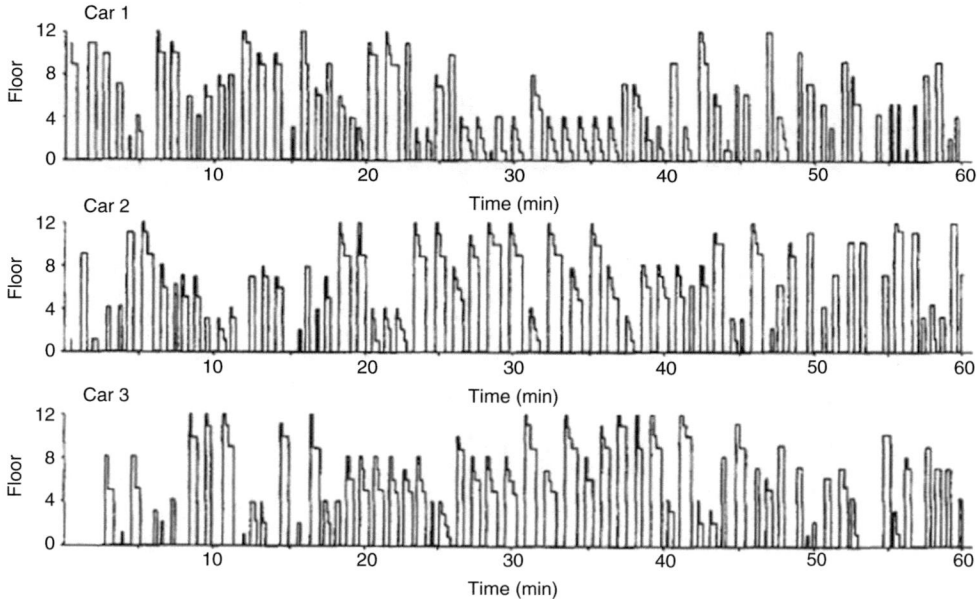

Figure 4.8 Spatial movements of lift cars during down peak traffic

4.4.4 Random interfloor traffic

This traffic condition is the most common traffic situation and exists for the majority of the working day in office buildings.

> **Definition 4.5:** Random interfloor traffic can be said to exist when no discernible pattern of calls can be detected.

Uppeak probably exists for 5 minutes and down peak for 10 minutes, and two- and four-way traffic, if they occur at all, can be considered to be severe cases of unbalanced interfloor traffic. Interfloor traffic is caused by the normal circulation of people around a building during the course of their business. Sometimes, this traffic is called balanced two-way traffic as it involves both up and down trips, and it is balanced because passengers usually return to their original floor after moving about the building.

Figure 4.9 is taken from a computer simulation of an hour of office activity. Note that the figure can be reversed or inverted and still no discernible pattern can be seen in the spatial activity of the lifts.

4.4.5 Other traffic situations

It is common to find office buildings where no dominant traffic flows occur, especially where flexitime is used. Sometimes the uppeak situation occurs twice, as in Figure 4.10, but at a lower intensity. Obviously, traffic patterns are different in institutional and residential buildings, but often dominant patterns similar to those defined above do emerge and hence ease

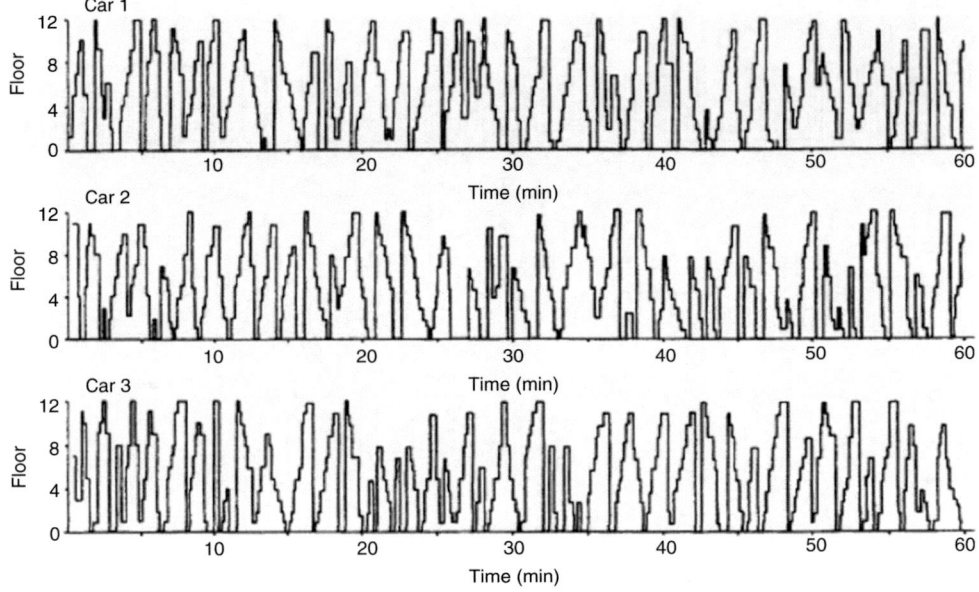

Figure 4.9 Spatial movements of lift cars during balanced interfloor traffic

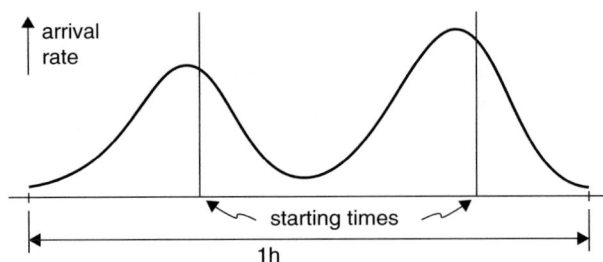

Figure 4.10 Another arrival profile for morning uppeak with two starting times

design procedures. The effect on a lift system of applying a non-smoking regime in a building, where smoking is not permitted inside the building and smokers have to go outside, can have a significant effect. Even today, 24% of people smoke and might crave one smoke per hour.

4.4.6 Summary of traffic conditions

The distinctive 'fingerprints' of uppeak, down peak and balanced interfloor traffic patterns, as represented by the spatial movements of lifts, are:

Uppeak: The lifts arrive at the main terminal, load with passengers, and move up the building, stopping often until the last stop when they express return to the main terminal. During the peak 5 minutes there is a 'staircase' pattern.

Down peak: The lifts stop at a few floors in the building, loading with passengers and then express to the main terminal. After unloading the passengers, the lifts make

Interfloor:

an express run back up the building. There is a small 'staircase' pattern in the reverse direction to the uppeak case.

Interfloor: There is no discernible pattern for a balanced interfloor traffic condition.

4.5 Traffic design

Why is there a need for traffic design? This could be answered as follows:

- To size a lift installation to serve a traffic requirement or meet a capital/recurrent financial requirement.
- To compare competitive tenders.

It is extremely difficult to compare competitive tenders where no standardised methods of specification or common design procedures are used. Each manufacturer and lift consultant often use different methods, and are not keen to explain their approach. Many methods that are published are often sketchy and some are inaccurate. In addition, the use today of modern control systems radically alters some of the design assumptions. An easy to use, acceptable and standard design method will be presented here. This should enable a prospective designer to gain a better understanding of the design procedure and be better able to use it.

Little theoretical or analytical work was carried out into traffic design and control until the 1970s. Simulation techniques were used by earlier workers (Browne and Kelly, 1968), who considered their use essential to investigate better design methods and to develop new traffic supervisory control techniques. Simulation is a tool that is used either where a sufficient mathematical method does not exist, or to support a mathematical method.

By the 1970s, a recognised method of calculation had evolved for uppeak traffic sizing, based on the mathematical determination of average highest reversal floor (H) by Schroeder (1955), average number of stops (S) by Basset Jones (1923) and average number of passengers (P) transported.

The formulae by Barney and Dos Santos (1977) for the calculation of the passenger handling performance of lift systems are now universally accepted. Lift makers often use tables specific to their product range to estimate round trip times, interval, handling capacity, etc.

The sizing of a lift system requires the matching of the demands for transportation from the building's occupants with the handling capacity of the installed lift system. This procedure should also result in an economic solution. The classical procedure used in the traffic design of lift systems is to determine the handling capacity for the uppeak traffic situation. This approach is sensible as the uppeak traffic condition does yield to analytical techniques, although some of the assumptions making this possible are difficult to justify in a real-life situation.

4.6 Uppeak design

What is uppeak or incoming traffic? This has been defined in Section 4.4.1 as Definition 4.1. The idealised profile of Figure 4.4 extends Definition 4.1 to allow a 5-minute uppeak passenger arrival rate to be defined.

Definition 4.6: The uppeak arrival rate (λ) is the number of passengers who arrive at the main terminal of a building for transportation to the upper floors over the worst 5-minute period.

A lift system is thus expected to respond to the peak demand in such a way as to quickly and efficiently transport passengers to their destinations without excessive passenger waiting times occurring or unwieldy queues developing. This implies that the handling capacity of the lift system should be sufficient to carry all those passengers demanding service. So what is handling capacity?

> **Definition 4.7:** The handling capacity (*UPPHC*) of a lift system is the total number of passengers that it can transport in a period of 5 minutes during the uppeak traffic condition with a specified average car loading.

Examination of Figure 4.4 shows that the required instantaneous handling capacity varies from relatively low levels each side of the defined starting time to a high level prior to the starting time. Over a period of one hour, the average arrival rate is low and can be handled by a small number of lifts. However, as soon as the arrival rate exceeds the one hour handling capacity, large queues build up and waiting times become excessive. Only when the arrival rate again falls below the one hour handling capacity can the queues begin to reduce, and even then it will be some time before the queues disappear and the handling capacity again exceeds demand (see Figure 4.11). Thus, it is not satisfactory to size a lift system to handle a one hour average rate of arrival. Conversely, high instantaneous demands obviously cannot be met, except by a large and expensive system. Thus, a compromise is necessary where intending passengers are required to wait a reasonable time for service during the peak demand periods.

A realistic approach then is to define handling capacity for a period of time of less than one hour, but longer than a reasonable waiting time. A period of 5 minutes for the handling capacity definition has achieved general acceptance. From an analytical point of view, it lies between one hour and a reasonable average waiting time, typically 30 s, allowing the smoothing out of short-term transients, but defining a period over which conditions remain reasonably fixed.

Thus, if it is possible to equate the passenger demand as expressed by the 5-minute percentage peak arrival rate with the handling capacity of a lift system, then a suitable configuration will have been designed.

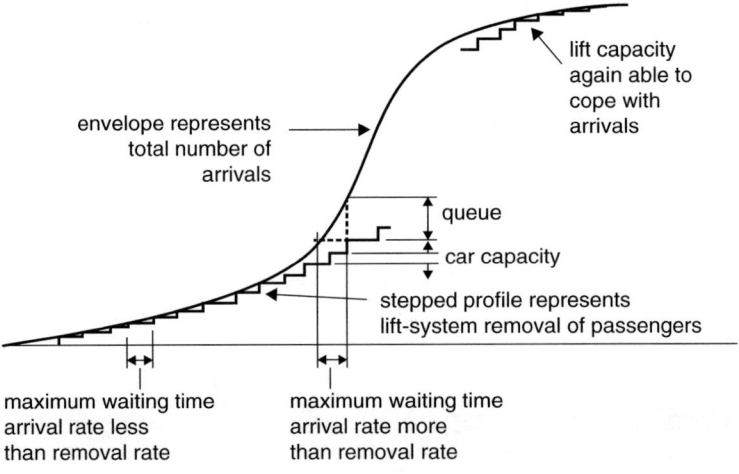

Figure 4.11 Passenger arrival rate and lift system handling capacity

The question now arises of how to calculate handling capacity. This can be answered by considering how a single lift car services the incoming passengers during uppeak. The events are: the lift car comes to the ground floor, picks up passengers and then transports them to their destinations in the upper parts of the building; when all passengers have alighted and the car is empty, it 'express' returns to the main arrival floor. The concepts of a round trip and round trip time thus emerge.

Definition 4.8: The round trip time (*RTT*) is the time in seconds for a single car trip around a building from the time the car doors open at the main terminal until the doors reopen when the car has returned to the main terminal floor after its trip around the building.

A round trip time should not usually exceed two to three minutes (except in very tall buildings) as the majority of this time can represent the transit time for some passengers with destinations on the top floors of a building, which is undesirable.

Now, if it is known how many round trips a single lift car can complete during the peak 5-minute period, then the uppeak handling capacity can be defined.

$$\text{Number of round trips for a single car} = \frac{5 - \text{minutes}}{RTT} = \frac{300}{RTT} \tag{4.1}$$

Therefore, the 5-minute handling capacity (*UPPHC*) for a single car is:

$$UPPHC = \frac{300}{RTT} \times \text{average number of passengers per trip} \tag{4.2}$$

Equation (4.2) is one of the most important equations used in lift traffic design. Before it can be evaluated, it is necessary to determine the average number of passengers carried per trip. If the number of passengers carried is defined as *P*, then Equation (4.2) becomes:

$$UPPHC = \frac{300}{RTT} \times P \tag{4.3}$$

In installations with more than one car, Equation (4.3) becomes:

$$UPPHC = \frac{300}{RTT} \times P \times L \tag{4.4}$$

where *L* is the number of lift cars.

The handling capacity of a lift installation indicates the *quantity of service* a lift system can provide. Passengers are concerned also with *quality of service*, which is how long they must wait. To some extent, the frequency of lift arrivals at the main terminal floor gives an indication of quality. With a single car, the interval between successive arrivals is the round trip time. However, where a lift system contains more than one car, the interval becomes:

$$UPINT = \frac{RTT}{L} \tag{4.5}$$

Figure 4.12 illustrates the relationships between round trip and interval.

Definition 4.9: Interval (*INT*) is the average time between successive lift car arrivals (or departures) at the main terminal floor with cars loaded to any level.

Definition 4.10: Uppeak interval (*UPPINT*) is the average time between successive lift car arrivals at the main terminal during uppeak traffic conditions.

It should be noted that sometimes the term 'waiting interval' is used instead of interval. This is an attempt to create the idea that the inter-arrival period between cars can define the waiting time of a passenger. This is confusing.

Another useful operating performance parameter is to determine the percentage of a building's population handled in the peak 5 minutes. This is called percentage population served.

Definition 4.11: The population served in the uppeak 5 minutes (*%POP*) is defined as the ratio of the uppeak handling capacity and the building population given as a percentage.

$$\%POP = \frac{UPPHC}{\text{Building population}} \times 100 \tag{4.6}$$

As measures of a lift system's operational performance, the three parameters of uppeak handling capacity, uppeak interval, and uppeak percentage population served are the parameters most often quoted by a lift supplier.

Example 4.1

A building is served by three lifts with a round trip time of 150 s. The building population is 400 persons, and each car leaves the main terminal with an average car load of eight passengers. Calculate the uppeak interval, uppeak handling capacity and percentage population served.

From Equation (4.5):

$$UPPINT = \frac{150}{3} = 50\,s$$

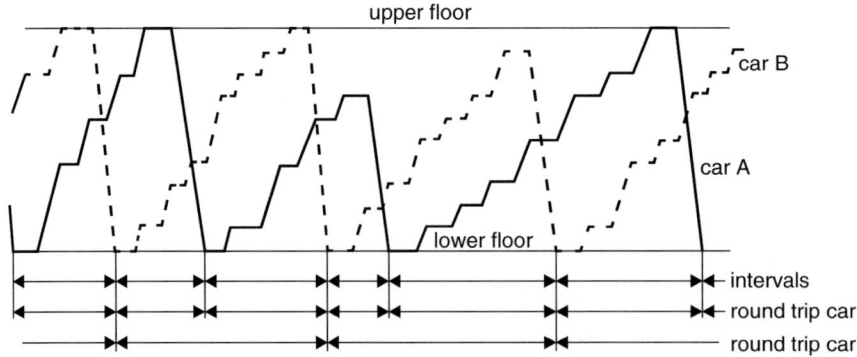

Figure 4.12 Relationship between round trip time and interval

From Equation (4.3):

$$UPPHC = \frac{300}{150} \times 8 \times 3 = 48 \ \text{persons/5 minutes}$$

From Equation (4.6):

$$\%POP = \frac{48}{400} \times 100 = 12\%$$

4.7 Derivation of the round trip time of a single car

Consider the way in which a single lift car circulates around a building during the uppeak traffic condition. The car opens its doors at the main terminal floor and passengers board the car; the doors close. The car then runs to the first stopping floor, going through periods of acceleration, travelling at rated speed, deceleration and levelling. (Travel at rated speed may not occur if the interfloor distance is too small.) At the first stopping floor, the doors open and one or more passengers alight; the doors close. This sequence continues until the highest stopping floor is reached. After the doors have closed, the car is considered to make an express run to the main terminal, thus completing the round trip. This description indicates, and Figure 4.13 illustrates, that a round trip consists of a number of elements.

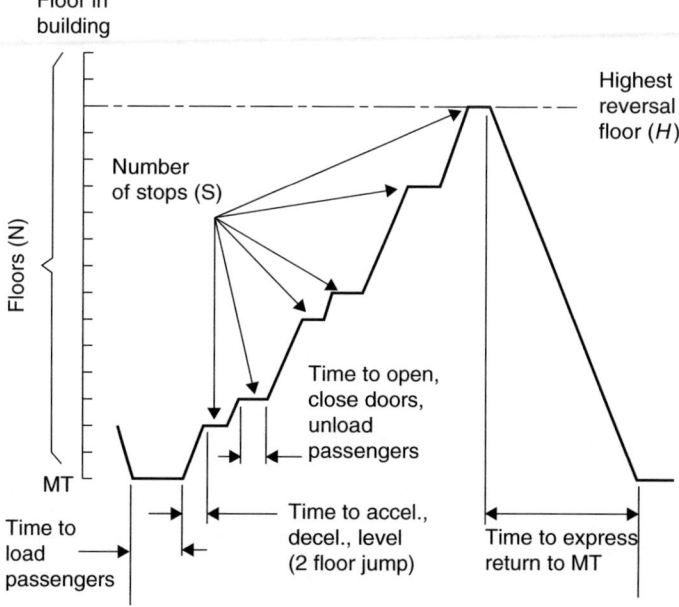

Figure 4.13 The elements of a round trip time

The elements are:

(a) passenger loading time, t_l
(b) passenger unloading time, t_u
(c) door closing and opening times, t_c and t_o
(d) interfloor jump time (for a single floor assuming rated speed reached), $t_f(1)$
(e) time to travel in the upward direction at rated speed for jumps that are greater than a single floor
(f) time to travel from the highest floor to the main terminal.

Items (a), (b), and (c) may be considered to be standing times and (d), (e) and (f) as running times.

The travel sequence of the lift is much affected by the average number of stops made (S), the average highest floor reached (H) and the average number of passengers (P) carried. It is now possible to deduce an expression for round trip time as:

$RTT =$

$Pt_l + Pt_u$
passenger
transfer time

$+ (S+1)(t_c+t_o)$
door operating time

$+ (S+1)t_f(1)$
time to accelerate,
decelerate, level,
etc.

$+ (H - S)t_v$
time to travel
remaining floors

$+ (H - 1)t_v$
time to express to
main terminal floor

(4.7)

Refer to Table 4.1 for the definitions of the parameters. The 'plus one' in the term ($S+1$) occurs to account for the stop at the main terminal floor.

Combining and simplifying:

$$RTT = 2Ht_v + (S+1)t_s + 2Pt_p \tag{4.8}$$

The stopping t_s is an artificial time developed as a mathematical simplification and cannot be measured directly. It is:

$t_s =$ door operating times (t_d) + single floor flight time ($t_f(1)$) – single floor transit time (t_v)

$$t_s = t_d + t_f(1) - t_v \tag{4.9}$$

The performance time (T), can be measured and it is also very useful in determining the dynamic performance of a lift.

$$T = \text{door operating times } (t_d) + \text{single floor flight time } (t_f(1)) = t_d + t_f(1) \tag{4.10}$$

which gives:

$$t_s = T - t_v \tag{4.11}$$

Table 4.1 Further definitions of terms

No.	Time period	Symbol	Description
4.12	Passenger loading time	t_l	The average time for a single passenger to enter a car (boarding time, entry time)
4.13	Passenger unloading time	t_u	The average time for a single passenger to leave a car (alighting time, exit time)
4.14	Passenger transfer time	t_p	The average time for a single passenger to enter or leave a car, i.e.: $t_p = (t_l + t_u)/2$
4.15	Door closing time	t_c	A period of time measured from the instant that the car doors start to close until the doors are locked
4.16	Door opening time	t_o	A period of time measured from the instant that the car doors start to open until they are open 800 mm
4.17	Door operating time	t_d	The sum of the door opening and closing times, i.e.: $t_d = t_c + t_o$
4.18	Car call dwell time	t_{cd}	The period of time that the car doors remain open at a stop in response to a car call, provided no passengers cross the threshold
4.19	Landing call dwell time	t_{ld}	The period of time that the car doors remain open at a stop in response to a landing call, provided no passengers cross the threshold
4.20	Single floor flight time	$t_f(1)$	The period of time measured from the instant that the car doors are locked until the lift is level at the next adjacent floor
	Multi-floor flight time for a jump of n floors	$t_f(n)$	The period of time measured from the instant that the car doors are locked until the lift is level at the nth adjacent floor
4.21	Single floor transit time	t_v	The period of time for a lift to travel past two adjacent floors at rated speed: i.e., $t_v = d_f/v$ where d_f is the interfloor distance and v is the rated speed
4.22	Stopping time	t_s	A composite time associated with each stop, i.e.: $t_s = t_f(1) + t_c + t_o - t_v$
4.23	Performance time	T	The period of time between the instant the car doors start to close and the instant that the car doors are open 800 mm at the next adjacent floor
4.24	Cycle time	t_{cyc}	The period of time between the instant the car doors begin to close until the instant that the car doors begin to close again at the next adjacent floor provided no passengers have crossed the threshold

Equation (4.8) can be modified to include the time parameter T, i.e.:

$$RTT = 2Ht_v + (S + 1)(T - t_v) + 2Pt_p \tag{4.8bis}$$

The three components of Equation (4.8) indicate different aspects of the round trip time. The travel distance relates to the parameter H, the number of stops relates to the parameter S and the number of passengers carried relates to the parameter P. Each of these parameters are dependent on the passenger demand and H and S are dependent on P. Each of the parameters is associated with a time parameter. Equation (4.8) is thus neat, symmetrical, simple and easy to remember.

Equation (4.8) can be expanded to show more detail at the loss of symmetry as:

$$RTT = 2H.d_f/v + (S + 1)(t_f(1) + t_c + t_{o\,-}\,d_f/v) + 2Pt_p \tag{4.12}$$

The expressions for the round trip time presented here may differ from expressions derived by other authors owing to the manner in which the mathematical simplifications have been made, and because of the way in which the various parameters have been defined. Equation (4.8) has received international acceptance.

To evaluate Equation 4.8 requires several numerical values. Chapter 5 provides numerical values for the six parameters (H, S, P, t_v, t_s, t_p) and Chapter 6 shows how to determine passenger demand.

Notes (shown in text as[1])

1 g_n is the acceleration of a body due to gravity, numerically equal to 9.81 m/s^2.
2 The term main terminal is used throughout this book to avoid confusion and is synonymous with ground floor (Britain), first floor (USA), lobby, foyer, main arrival floor and building entrance floor.
3 Flexitime working allows workers to arrive, leave and take refreshment breaks within wide time bands. They are expected to be present during specified periods (core time) and to work a specified number of hours.
4 Details of lift traffic design using digital computer simulation methods are discussed fully in Chapter 17.

5 Evaluating the round trip time equation

5.1 Data sets

Equation (4.8) for round trip time must be evaluated in order to calculate the performance of a lift system. This equation depends on three data sets concerning the building, the lift system and the passengers.

5.1.1 Building data set

The building data set comprises the following:

(a) number of floors
(b) interfloor distance
(c) express jump.

Values for items (a), (b) and (c) are generally decided early in a building's design or are already fixed in an existing building. Sometimes, the lift designer will be able to suggest the number of floors in each building zone in order to design a more effective lift installation. Item (c), the express jump, may be nothing more than a few extra metres between the main terminal floor and the first served floor, or could be some distance if the first served floor above the lobby is an express zone terminal floor high in the building. More often than not, items (a), (b) and (c) will literally be 'set in stone'.

5.1.2 Lift system data set

The lift system data set comprises the following:

(a) number of cars
(b) car capacity
(c) rated speed
(d) flight times between floors
(e) door opening times
(f) door closing times
(g) miscellaneous times (such as car and landing door dwell times, etc.)
(h) traffic control system.

Items (a) and (b) are design variables to meet the required passenger traffic demands and the quality of service. Items (c) to (g) can be established from the lift manufacturer. The effect of (h) cannot always be analysed in precise mathematical terms, but may be assessed empirically.

5.1.3   Passenger data set

The passenger data set comprises the following:

(a) number of passengers boarding from specific floors
(b) number of passengers alighting at specific floors
(c) traffic mode, i.e., unidirectional or multidirectional, etc.
(d) transfer times for passengers entering and leaving cars
(e) passenger actions.

   Items (a) and (b) are dependent on floor populations. During the uppeak traffic condition, item (a) determines the 5-minute handling capacity when all passengers board at one terminal floor and similarly, item (b) defines the activity during down peak traffic conditions. It is these two items that determine the level of duty for a lift system, i.e., the number of starts an hour it will be required to make. Items (c) and (d) are dependent on human behaviour and are not easily predictable. Item (e) is included to cover passenger misbehaviour (door holding, excessive operation of pushbuttons, etc.).
   This data set is the least well defined and is subject to considerable error in its estimation.

5.1.4   Numerical values

The round trip time (*RTT*) equation, Equation (4.8), is:

$$RTT = 2Ht_v + (S+1)t_s + 2Pt_p \tag{4.8}$$

   It was noted in Section 4.7 that the round trip equation comprised six parameters. Three parameters were time independent variables, i.e., *H, S* and *P*, and three parameters were time dependant variables, i.e., t_v, t_s, and t_p. It is these six parameters that must be evaluated in order to determine the round trip time equation. Each variable will now be taken in turn in the following sections, and typical values indicated. Installed systems may have different values.

5.2   Single floor transit time

The parameter t_v requires the values for the interfloor distance (d_f) and the rated speed (v) to be known. This is the time that a car takes to travel past two adjacent floors at rated speed and is defined by Definition 4.21 as the average interfloor distance divided by the rated speed.

5.2.1   Interfloor distance

The average interfloor distance (d_f) is normally determined as the total travel to the highest served floor divided by the number of possible stopping floors above the main terminal. Domestic dwellings average about 3.0 m per floor, and commercial buildings range from 3.0 m to 3.3 m for older buildings to 3.6 m to 4.2 m (or greater) for modern buildings. In the latter case, the increased floor to floor distance is required to accommodate other services (e.g., air conditioning and/or electrical supplies) and various modern technological services (e.g., computer networks and/or telecommunications).

Commercial buildings often introduce a mixed floor pitch for a number of reasons:

- Some floors have increased heights, such as lobby/main terminal floors, service floors, special floors (e.g., those containing a restaurant, lecture room, conference room, VIP suite, etc.).
- Some floors are sometimes unavailable for alighting during periods of the day, such as the first floor (and sometimes the second floor) above the main terminal, service floors, security floors, etc.

It is recommended that an average floor height be assumed, and the irregularities be dealt with separately, as discussed in Chapter 7.

Where a lift is serving a set of floors or zone in a building, which are not adjacent to the main terminal, extra time to make the jump to or from the express zone must be added to Equation (4.8), i.e., $2t_e$, where t_e is the time the lift takes to travel (without stopping) from the main terminal to the express zone terminal:

$$RTT = 2Ht_v + (S+1)t_s + 2Pt_p + 2(t_e - t_f(1)) \tag{5.1}$$

The long 'flight time' t_e can be found using the equations in the Appendix.

5.2.2   Rated speed

The value of the rated speed (v) is usually supplied by the lift maker, who will select it to meet various engineering requirements (i.e., gearing, drive controllers, product line considerations, etc.) and traffic purposes. For instance, goods lifts are generally slower than passenger lifts. Speed, however, is not a dominant factor in Equation (4.8), as illustrated by Example 6.1. It does, however, become significant if the served floors are in an upper zone, where a higher speed will permit the unserved zone to be more rapidly traversed. If a value for v is not provided, it must be chosen by the traffic designer.

The appropriate value for rated speed could be that recommended in the British, European and international standard codes of practice, taken together with experienced judgement. In general, the higher the building rise, the faster the rated speed selected. Often in a zoned building, the rise from an express zone terminal may be small e.g., 10 floors, but the express jump from the lower terminal to the express zone terminal may be large. It is this express jump, which largely determines the rated speed that allows passenger transit times to be kept at reasonable values.

Fire standards (e.g., BS EN 81–72 2003, BS 9999: 2008b) can require a minimum value, i.e., that it shall be possible to travel to the highest floor in the building from the fire control entrance level in less than 60 s. Clearly, this is not possible in very tall buildings and special arrangements must be made in these circumstances.

BS 5655: Part 6 (2011) recommends rated speeds in relation to total travel according to building usage. This can be translated into the time to travel at rated speed (without allowance for acceleration, deceleration or levelling) between the highest and lowest floors (the terminal floors), as shown in Table 5.1. This time is sometimes called the nominal travel time.

For residential lifts, ISO 4190–6: (1984)[1] recommends a maximum (theoretical) time of transit of between 20 s and 40 s to travel, at the rated speed, for a distance equal to the total travel of the lift. The time is graded according to the quality of service required, being 20 s for the best service.

Thus, a simple formula for v is: $D/30 \leq v \leq D/20$, where D is the terminal floor to terminal floor distance.

There is no theoretical upper limit to lift rated speed and it generally does not affect passenger comfort, although some passengers may suffer ear problems resulting from the rapid pressure changes. However, it is limited by practical factors, such as the maximum sheave diameter, rope safety factors (rope bending and fatigue), safety limits (e.g., over travel), etc.

Table 5.2 provides guidance on the selection of the speed of a lift based on the premise that the total time to travel the distance between terminal floors at rated speed should take less than 20 s. In the table, the single floor flight times are shown for three average interfloor distances and are the theoretically derived values. In practice, these theoretical values would need to be increased to allow for the doors to be locked and proved, the brake to lift and other start up delays. The range of values given for acceleration is typical of those found on installed installations. Some installations nowadays limit the acceleration to about 1.2 m/s² in order to provide a good ride quality. This will increase the single floor flight time and reduce the eventual handling capacity.

The figures given in Table 5.2 apply principally to commercial buildings; speeds in residential and institutional buildings may be subject to other design regulations, and similar height and similar function buildings can be installed with a wide range of equipment, e.g., care homes compared to prestige flats.

Table 5.1 Suggested transit times for travel between the terminal floors in different types of building

Building type	Transit time (s)
Large offices, hotels	20
Small offices, hotels	30
Hospitals	24
Nursing and residential homes	24
Residential buildings	20–30
Shops, factories and warehouses	24–40

Table 5.2 Typical lift dynamics

Lift travel (m)	Rated speed (m/s)	Acceleration (m/s²)	Jerk (m/s³)	Single floor flight time (s)		
				3.5 m	4.5 m	5.5 m
<20	1.00	0.4	0.60	6.7	7.7	8.7
20	1.00	0.4–0.7	0.75	6.1	7.1	8.1
32	1.60	0.7–0.8	0.90	5.2	5.8	6.4
50	2.50	0.8–0.9	1.00	5.0	5.5	6.0
60	3.00	1.0	1.25	4.6	5.1	5.6
100	5.00	1.2	1.50	4.3	4.8	5.2
>100	6.00	1.2	1.80	4.1	4.6	5.0

5.3   Time consumed when stopping

The parameter t_s involves evaluation of flight and door times. The time consumed when stopping is given by Definition 4.22 as:

$$t_s = t_f(1) + t_c + t_o - t_v \tag{5.2}$$

which can be expressed in terms of Equation (4.11) as:

$$t_s = T - t_v \tag{4.11}$$

The lift performance time (T) has the most significant effect on the round trip time. It is defined as the period of time between the instant when a stationary lift starts to close its doors and the instant when its doors are open by 800 mm at the next adjacent floor (Definition 4.23). The time to transit two adjacent floors at rated speed (t_v) was dealt with in the previous section.

5.3.1   Cycle time and other times

There can be some confusion regarding the meaning of flight time (Definition 4.20), performance time (Definition 4.23), and cycle time (Definition 4.24), and also when the door opening time is considered to be finished.

Figure 5.1(a) is the usual operating cycle without advanced door opening. This begins with passengers transferring into the car, the doors closing, the interlocks making up and the car moving (brake to brake time). When the car stops moving, the doors open, passengers transfer out of the car, and some passengers may transfer into the car.

Figure 5.1(b) illustrates what happens when the lift is supplied with the advance door opening feature. Here, some of the time of the door opening is shared with the levelling operation. That is, the doors open *before* the lift has finished moving. This feature can reduce the cycle and performance times by between 0.5 s to 1.7 s.

The cycle time is measured from some consistent point in the cycle, i.e., when the doors are closed, or when the doors are just opening. The measurement of cycle time is made with no passengers entering or leaving the car. The cycle time thus includes any car/landing door dwell times, see Section 5.5.1. Figure 5.1(c) shows the operating cycle for a lift responding to a car call, and Figure 5.1(d) shows the operating cycle for a lift responding to a landing call. Cycle time therefore includes time that might be wasted if a car call is registered in error, so that when the lift stops, no one leaves, or when a landing call is registered, so that when the lift stops, no one enters.

Performance time is the most important variable, as this can be controlled and predicted. The components of the performance time must be carefully selected to achieve the correct handling capacity for the lift installation. The lift maker should be contracted at the tender stage to provide this time at the specified values, and the maintenance contractor should be required to keep them at the rated values throughout the life of the installation. Failure to meet the specified times will reduce the handling capacity and may invalidate any traffic design. The equation for the performance time is given by:

$$T = t_f(1) + t_c + t_o \tag{5.3}$$

5.3.2   *The single floor flight time*

The single floor flight time $t_f(1)$ is the time taken from the instant the car doors close to the instant the car is level with the next adjacent floor (Definition 4.20). It is dependent on the distance travelled, the rated speed, the acceleration value and the jerk value. These relationships between distance travelled, velocity, acceleration and jerk are complex and are given in detail in Section 7.4.1 and the Appendix. Flight times need to be obtained for the distance travelled or the number of floors travelled.

Fortunately for designers of lift drives, there are limits on the maximum values of both acceleration and jerk. These constraints are imposed by human physiology, as described in Chapter 4. Passengers are uncomfortable when subjected to acceleration values greater than about one sixth of the acceleration due to gravity (i.e., about 1.5 m/s²). Similarly, the maximum value of jerk commonly used in calculations is about 2.2 m/s³, although there is no drive control on this variable.

As a result of the limits to the drive dynamics, there is little or no difference in single floor flight times for rated speeds in excess of about 5.0 m/s, as the maximum possible values of acceleration and jerk are not reached in the short distance between adjacent floors. Below 5.0 m/s, the flight times are dependent on the type of drive and drive controller, mainly owing to variations in levelling times and non-ideal acceleration deceleration profiles.

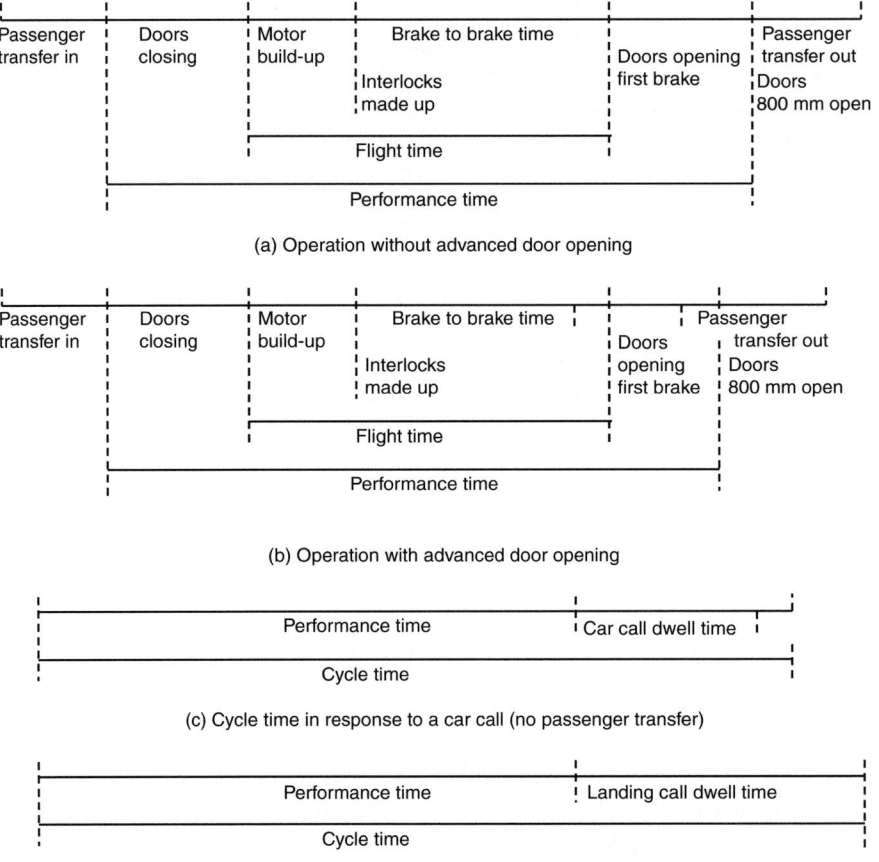

Figure 5.1 Illustration of lift system timings

Table 5.2 indicates the likely range of acceleration values and single floor flight times. The single floor flight times are slightly larger than a theoretical calculation would give to allow for start up delays. Different manufacturers and drive systems will result in variations to the values given. Naturally, the flight times also depend on the interfloor distance. Where the higher speeds have the most effect is where a lift serves a high zone in a building and has to pass non-stop through a number of lower floors.

5.3.3   Door operating times

5.3.3.1   General

Door operating times comprise opening and closing times, and door dwell times. The dynamic operating (opening and closing) times are dependent on a number of factors: door panel velocity, panel arrangement, width and control.

There is a limit to door panel velocity commensurate with passenger safety. The European standards require that the energy present in a moving door should not exceed 10 J. This restriction applies to the closing doors. The restriction has the effect of limiting the maximum door velocities during closing to about 300 mm/s. Doors can operate faster if they are not allowed to touch a passenger. However, this involves the use of complex (and sometimes expensive) passenger detection and door control systems.

There are two horizontal slide door types: side opening and centre opening. Side opening doors have to open the whole width of the doorway, which takes more time. Here, the width is taken as the clear opening width, ignoring any returns or architraves. Centre opening doors open and close more quickly, and the symmetrical reaction against the car frame will reduce car sway.

There are several standard widths. Narrow doors of 800 mm width are usually fitted to cars with a rated load of up to 1,000 kg. Wider doors of 1,100 mm width are fitted to lifts with a rated load of over 1,000 kg persons. Doors of 1,300 mm width or larger are fitted to goods lifts and hospital lifts. Obviously, the narrower the door, the faster the operation.

The control of the door operator can significantly affect door timings in respect of start up, slow down and safety. There are a number of types, usually rated according to their operating speed and method of control of the door operator (open or closed loop). The speeds are not well defined and in broad-brush terms are defined as: low (about 300 mm/s), medium (about 450 mm/s) or high speed (about 750 mm/s). Generally, a low speed operator will be found on a low speed lift and obviously cost less than a high speed operator. Low speed operators are often characterised by delivering the same opening and closing times.

5.3.3.2   Door closing time (t_c)

Door closing always takes place while the lift is stationary, and typical door closing times are given in Table 5.3. The door closing time (t_c) is the time taken from the instant the car doors start to close to the time they are locked up. Remember the energy constraints of the previous section.

5.3.3.3   Door opening time (t_o)

Door opening time is not subject to the energy constraints found in door closing and can be much faster. The doors can operate at any speed, provided the trapping hazard for fingers, clothing, etc. against the door architrave or door lining is negligible. However, as the same door operator will be used for both directions of movement, its capability to reduce the opening times may be small. There are two cases of door opening to be considered: with and without advanced door opening.

Where the advanced door opening time feature is not installed, the door opening time is considered to have ceased, when the passenger transfer may begin. This usually can occur when the doors are open by approximately 800 mm. Thus, in this case, the door opening time is taken to be the time from the instant the doors start to open until the instant the doors are open to a width of 800 mm.

Time can be saved by advanced door opening. Once the car has entered the door zone, some 200 mm from a landing, the doors can be unlocked and opened. Then, an improvement in opening times can be achieved by overlapping the levelling operation with the first part of the opening of the doors, which is called advanced door opening, or pre-opening. This is possible within the door zone, provided that the tripping hazard can be avoided by ensuring that the lift is level at the landing when the doors are 800 mm open.

The door opening value to be used in this case in the evaluation of the parameter ts is the time from the moment the lift is level at the landing, until the doors are open by 800 mm. This time will be less than the measured door opening time for lifts without advanced opening, viz: $t_o - t_{ad}$, where t_{ad} is the time saved by opening the doors during levelling.

Typical door opening times are given in Table 5.3 for normal and advanced opening. The table gives representative values for two door types, two door sizes, and with and without advanced door opening. These values may be used where specific values are not available. A wider range of timings is given in CIBSE Guide D: 2010, Table 7.1.

5.3.3.4   Door weight

The weight[2] of the door is determined by many factors, such as fire resistance, height, width, configuration, impact resistance, etc. A moving door gathers considerable kinetic energy. To protect passengers from injury, the standards require the maximum energy to be limited to 10 J, provided the safety edge is operative. If the safety edge is inoperative, then the energy value must not exceed 4 J. The maximum values of energy acquisition limit the maximum door speed when closing. Typically, a 150 kg door has a maximum speed of 0.23 m/s and a 500 kg door has a maximum speed of 0.13 m/s. For a particular door weight (M), the maximum speed (m/s) at which the doors may move to meet the energy value requirements are given in Table 5.4.

Where the weight of a door is not known, the weight can be approximately estimated by assuming:

A painted hoist way door weighs 35 kg/m^2
Painted car doors weigh 24 kg/m^2
Hangers per door weigh 10 kg
Other hardware (vanes, operating arms, safe edges, etc.) per system weigh 5 kg.

Table 5.3 Typical door closing and opening times (s) for stated door width

Door operation	Opening (advanced)		Opening (normal)		Closing	
Door type	800 mm	1100 mm	800 mm	1100 mm	800 mm	1100 mm
Side	1.0	1.5	2.5	3.0	3.0	4.0
Centre	0.5	0.8	2.0	2.5	2.0	3.0

Table 5.4 Maximum door movement speeds for different weight of doors

Total door weight (kg)	Maximum speed for 10.0 J (m/s)	Maximum speed for 4.0 J (m/s)
150	0.36	0.23
200	0.31	0.20
250	0.28	0.18
300	0.25	0.16
350	0.23	0.15
400	0.22	0.14
450	0.21	0.13
500	0.20	0.13

Example 5.1

A 1,100 mm single panel side opening door has an area of 2.5 m². Calculate the door closing time.

Hoist way door weighs	87.5 kg
Car door weighs	60.0 kg
Hangers weigh	20.0 kg
Other hardware weighs	5.0 kg
Total	172.5 kg

From Table 5.4 (by interpolation), the maximum door speed is 0.34 m/s.

It is the practice to measure door closing times from a point 50 mm from the open and closing jambs. This allows for the acceleration and deceleration time periods at the extremes of travel. Thus, the door will move 1.0 m in 2.94 s.

The actual door closing time will be longer than this to allow for the acceleration and deceleration of the door panels. Add 1.0 s to account for this, giving a door closing time of some 4.0 s.

5.4 Example 5.2: two tenders examined

Two tenders have been received for the provision of a lift system of 800 kg rated load in a 15 storey office block with an interfloor height of 3.3 m. Assume a passenger transfer time of 1.2 s. Compare the two tenders. Table 5.5 gives the received tender information, and Table 5.6 gives data deduced from the given data.

Calculation of the round trip time for Tender A is as follows:

$$RTT = (2 \times 13.7 \times 2.06) + 7.1(6.0 + 1.0 + 3.0 - 2.06) + (2 \times 7.6 \times 1.2)$$

$$= 56.4 + 56.4 + 18.2$$

$$= 131.0 \text{ s}$$

Table 5.5 Tender information

Parameter	Tender A	Tender B
Rated speed (m/s)	1.6	2.5
Door opening time (s)	1.0	3.0
Door closing time (s)	3.0	3.5
Flight time (s)	6.0	5.5

Table 5.6 Deduced data for Example 5.2

Parameter	Tender A	Tender B	Deduced from
Average number passengers (P)	7.6	7.6	$PC = 9.5$ (Table 5.8)
Average highest floor (H)	13.7	13.7	Table 5.11a
Average number of stops (S)	6.1	6.1	Table 5.11a
Single floor transit time (t_v) (s)	2.06	1.32	Definition 4.21
Stopping time (t_s) (s)	7.94	10.68	Definition 4.22
Passenger transfer time (t_p) (s)	1.2	1.2	Design brief
Performance time (T) (s)	10.0	12.0	Definition 4.23

Calculation of the round trip time for Tender B is as follows:

$$RTT = (2 \times 13.7 \times 1.32) + 7.1(5.5 + 3.0 + 3.5 - 1.32) + (2 \times 7.6 \times 1.2)$$

$$= 36.2 + 75.8 + 18.2$$

$$= 130.2 \text{ s}$$

Although the rated speed was different between Tender A and Tender B, other changes altered the values obtained from each of the component parts of the round trip time equation. The result was that both tenders would provide very similar traffic handling systems.

5.5   Passenger transfer times

5.5.1   *Simple transfer times*

The passenger transfer time is the time a single passenger takes to enter or leave a car (Definitions 4.12 and 4.13). This parameter is the least well known of all the components of Equation (4.12), principally because it is dependent on human behaviour. The passenger loading times and the passenger unloading times are not necessarily equal. The transfer time is affected by the direction of the transfer (in or out), the width of the doors, whether a lobby or car is crowded, the shape of the car, the size and type of car entrance, the building type (e.g., commercial, institutional, residential, etc.) and the characteristics of the passengers (e.g., age, agility, purpose, culture, etc.).

Phillips (1973)[3] provides graphs for loading and unloading. These were mathematically described by Jones (1971) as two equations (see Section 8.3.6). Strakosch (1967) offers tables of transfer times related to standard car sizes, which he updates in the 1998 edition to include disability needs. He suggests that the round trip time might include an inefficiency factor to account for passenger transfers.

General rules can be offered. If the car door width is 1,000 mm or less, it may be assumed that passengers enter or exit in single file. For door widths of 1,000 mm and above, it may be assumed that the first six passengers enter or exit in single file and the remainder in double file.

The average passenger transfer time (entry or exit) may be taken as 1.2 s (each way). This time could be reduced for single cars, but increased for groups. The time might be increased for small door openings and reduced for large door openings. For situations where passengers are elderly, or have no reason to rush, the transfer times should be increased to about 2.0 s. ISO 4190–6 considers a passenger transfer time of 1.75 s suitable for residential buildings.

The value to be assigned to t_p is the average for all passengers who are transferring. For example, a 2,000 kg lift may load 16 passengers. Each passenger may have a transfer time of 1.2 s. But the first six passengers take 6×1.2 s, i.e., 7.2 s, then the next ten passengers exit in pairs, i.e., in 6.0 s. The total transfer time for all passengers is thus 13.2 s, and the average transfer time is $13.2/16 = 0.83$ s. This is the figure to be entered in Equation 4.8.

See Table 5.7 for suggested average passenger transfer times (t_p) for offices, taking into account door width and car capacity. Note that it has been observed that passengers often exit lift cars quicker than they enter. This effect can be accounted for by averaging.

5.5.2   Other times affecting passenger transfers

Many lifts are fitted with a number of features that can affect the evaluation of the passenger transfer times. These are:

(a)   car call dwell time
(b)   landing call dwell time
(c)   differential door time
(d)   lobby dwell time.

When a lift arrives at a landing, it opens its doors. The doors are held open for a period of time, called the dwell time, to allow passengers to enter or leave the car before the doors start to close.

In an office building, it is usually sufficient to keep the door open for 2.0 s for a stop in response to a car call as the passengers in a car will be prepared to exit. For a stop in response to a landing call, the door hold open time should be longer, at 3.0 s, as passengers will need to walk to the car entrance.

Where a lift is designated for use by people with disabilities, the dwell times will be considerably extended from the 2 s to 3 s acceptable by people without disabilities. Regulations may require dwell times of from 5 s to 7 s to accommodate disabled persons.

In a residential building or a shopping centre, these times might be set to 7.0 s to allow prams and bicycles to be transported.

Table 5.7 Average passenger transfer times (t_p) for offices

Rated load (kg)	t_p *(s)*
320–800	1.2
1000	1.0
1275	0.9
1600–2500	0.8

The car and landing call dwell times can be reduced if passenger collision detectors are fitted. Once a passenger crosses the threshold, the dwell time can be reduced to 0.5 s, and the passenger detection system can be used to keep the doors open while passengers are present in the threshold. This feature is called differential door timing. Some lift controllers also cause the door dwell times to be reduced to about 0.5 s whenever a new car call is registered, or when an existing car call is registered again, on the car operating panel.

The effect of these dwell times can be significant. For example, if the dwell times are set for too long, and differential timing is not available, the passenger transfer times may well be shorter than the dwell times. Thus, the third term in the round trip equation becomes redundant, and extra time would need to be added to the performance time to compensate, thus increasing the size of the middle term of the round trip equation. If, however, differential door timers are in operation, then the third term of the round trip time equation is required.

Lobby dwell time operates during uppeak periods of traffic and causes a car to remain at the main terminal floor for a period of time after the first car call has been registered. This period of time is often set to prevent a car leaving the main terminal with a small number of passengers. The lobby dwell time should not affect the calculation during uppeak. But if it is not turned off during other traffic conditions, it will delay the movement of the lift.

5.6   Example 5.3: illustration of dwell times

To illustrate the effects of these times, suppose a lift car carrying 12 passengers stops nine times in a 16-floor building, that the lobby dwell time was set to 8.0 s, the car call dwell time was set to 2.0 s, the landing call dwell time was set to 5.0 s and advanced door opening reduced the door opening time by 0.8 s. What effect would this have? Consider each in turn.

5.6.1   Lobby dwell time

This is set at 8.0 s. With 1,100 mm doors, it can be assumed that the first six passengers enter single file and the remainder (6.0) enter double file. The total passenger transfer time would therefore be:

$$6 \times 1.2 + 3 \times 1.2 = 10.8 \text{ s}$$

This time is larger than the lobby dwell time of 8.0 s and therefore has no effect.

5.6.2   Car call dwell time

This is set at 2.0 s. At each floor, an average number of passengers will leave. There are 9 stops and 12 passengers. This gives an average transfer of 12/9 i.e., 1.33 passengers. They will take $1.33 \times 1.2 = 1.6$ s to leave. Thus, the car call dwell time is longer (on average) than the average time for passengers to leave the car. It might be prudent to add 0.4 s to t_s.

5.6.3   Landing call dwell time

This is set to 5.0 s. As the calculation is for an uppeak traffic condition, this time will not operate. However, it is rather long for other traffic conditions, and it should be reduced to about 3.0 s.

5.7   Determination of the number of probable stops (*S*)

During a round trip, a lift car stops at a number of floors for passengers to alight. The round trip time is affected by the number of stops made. Each trip will be different, but what is the average number of stops that can be expected? The term 'probable number of stops' has been avoided as statisticians understand[4] that a parameter labelled 'probable' will yield a number between zero and unity. Thus, the more acceptable term that will be used is 'expected number of stops'. Basset Jones (1923) published a method of calculating the expected number of stops for floors with equal populations. Consider a building with *N* floors above the main terminal. Assume that each floor is equally likely as a destination for passengers.

The probability that one passenger will leave the lift at any particular floor is $1/N$.

The probability that one passenger will *not* leave the lift at any particular floor is the complement of this probability, *viz*:

$$1 - \frac{1}{N} = \frac{N-1}{N} \tag{5.4}$$

Since each passenger is assumed to be independent of all others, the product law of probability gives the probability that *no* passengers from a lift containing *P* passengers will leave the lift at any particular floor as:

$$\left[\frac{N-1}{N}\right] \times \left[\frac{N-1}{N}\right] \cdots \left[\frac{N-1}{N}\right] = \left[\frac{N-1}{N}\right]^P \tag{5.5}$$

Note there are *P* terms. Hence, the probability that a stop will be made at any particular floor is:

$$1 - \left[\frac{N-1}{N}\right]^P \tag{5.6}$$

The expected or average number of stops (*S*) for *N* floors will then be:

$$S = N\left[1 - \left[\frac{N-1}{N}\right]^P\right] \tag{5.7}$$

Values for *S* are dependent on values for *N* and *P* and can be calculated each time using the formulae. It is possible to tabulate these values as shown in Table 5.11.

5.8   Determination of the highest reversal floor (*H*)

Some design procedures assume *H* to be *N*, or in tall buildings, *N*-1. Arbitrary rules are offered (Strakosch, 1998)[5]. It is possible, however, to deduce an expression for *H* with respect to *N* and *P* using probability theory. It is not clear who first derived an expression for *H*, or when, although Schroeder (1955), writing in German, could have been the first. Using the same assumptions and definitions as in Section 5.3, assume a passenger is equally likely to travel to any floor.

The probability that one passenger will leave the lift at any given floor is $1/N$, and so the probability that one passenger will *not* leave the car at a given floor is:

$$1 - \frac{1}{N} \tag{5.8}$$

The probability that none of the P passengers will leave the car at a given floor is:

$$\left[1-\frac{1}{N}\right]^{P} \tag{5.9}$$

The probability of the car travelling no higher than the ith floor is equal to the probability that no one leaves the lift at the Nth, $(N-1)$th, $(N-2)$th, ... and $(i+1)$th, floors is:

$$\left[1-\frac{1}{N}\right]^{P}\left[1-\frac{1}{N-1}\right]^{P}\left[1-\frac{1}{N-2}\right]^{P}\left[1-\frac{1}{N-3}\right]^{P}\cdots\left[1-\frac{1}{i+1}\right]^{P} \tag{5.10}$$

Expanding and simplifying produces:

$$\left[\frac{i}{N}\right]^{P} \tag{5.11}$$

It is now possible to propose that the {probability that i is the highest floor attained} is equal to the {probability that a lift travels no higher than the ith floor} minus the {probability that the lift travels no higher than the $(i-1)$th floor}, *viz*:

$$\left[\frac{i}{N}\right]^{P}-\left[\frac{i-1}{N}\right]^{P} \tag{5.12}$$

Then, the average (or mean) highest floor H is:

$$H=\sum_{1=1}^{N-1}i\left[\frac{i}{N}\right]^{P}-\left[\frac{i-1}{N}\right]^{P} \tag{5.13}$$

Expanding and simplifying, the expected or average highest reversal floor (H) for N floors will then be:

$$H=N-\sum_{1=1}^{N-1}\left[\frac{i}{N}\right]^{P} \tag{5.14}$$

Values for H are dependent on values for N and P and can be calculated each time using the formulae. It is possible to tabulate these values as shown in Table 5.11. This table shows that H is approximately equal to N, where a large capacity lift is installed, but grossly in error, where a small capacity lift is installed.

5.9   Example 5.4: using formulae for *S* and *H*

Calculate the round trip time using the data below and Equation (4.8).
   Given data:

   $N=10$, $d_f=3.0$ m, $v=1.5$ m/s, $t_f(1)=5.0$ s,   door closing $t_c=2.5$, opening times $t_o=2.0$ s

Assumed data:

   P (average car load) $=6.4$ persons, $t_p=1.2$ s.

The data above gives:

   $t_v=3.0/1.5=2.0$ s, $t_s=5.0+4.5-2.0=7.5$ s.

To determine S, use Equation (5.7):

$$S = 10\left[1 - \left[\frac{10-1}{10}\right]^{6.4}\right] = 10(1 - 0.51) = 4.9$$

To determine H, use Equation (5.14):

$$H = 10 - \left[\frac{1}{10}\right]^{6.4} + \left[\frac{2}{10}\right]^{6.4} + \left[\frac{3}{10}\right]^{6.4} + \left[\frac{4}{10}\right]^{6.4} + \left[\frac{5}{10}\right]^{6.4} + \left[\frac{6}{10}\right]^{6.4}$$
$$+ \left[\frac{7}{10}\right]^{6.4} + \left[\frac{8}{10}\right]^{6.4} + \left[\frac{9}{10}\right]^{6.4}$$

$$= 10 - (0.0000004 + 0.00003) + 0.0006 + 0.0028 + 0.012 + 0.049$$

$$+ 0.095 + 0.24 + 0.51)$$

$$= 10 - 0.91 = 9.09$$

Using Equation (4.8):

$$RTT = 2 \times 9.09 \times 2.0 + (4.9+1) \times 7.5 + 2 \times 6.4 \times 1.2$$

$$= 36.4 + 44.2 + 15.4 = 96 \text{ s.}$$

Compare the calculated values of S and H to those given in Table 5.11.

5.10 Determination of number of passengers (P)

5.10.1 Preamble

Industry experts and consultants (Strakosch, 1983, 1998[6]) state that in practice lifts are not observed to fill with passengers to the numbers permitted by the rating plate, but to a lower value between 60% and 70% of rated car capacity, particularly in larger lifts. These observations of lower occupancy values are supported by work done by Day (2001a, 2001b), who found that the comfort of passengers was very important, and that many reasons exist as to why passengers will not fully fill a lift car.

The loading for lift cars is believed to have originated in the early 20th century US-A17 Elevator Codes. Simply put, some 80 years ago (Barney, 1996, 2000a) a male person was assumed to weigh 150 pounds, i.e., 68 kg, and to stand on two feet occupying a platform area of 2.0 square feet, i.e., 0.19 m². The European EN 81 standards of the 1980s decided that all persons would be assumed to weigh 75 kg. This allowed for the fact that people had become larger over the intervening 60 years. Fruin (1971) proposed a body template, based on an ellipse of 600 mm by 450 mm, which has an area of 0.21 m². This value is that suggested in Section 2.2 and Figure 2.1. The Fruin template assumes persons can touch.

Strakosch (1983)[7] said, '*Future development should include capacity loading based on volume rather than weight.*'

ISO/TR 11071–2 1996 says and ISO/TR 11071–2: 2006[8] repeats:

> *While the entire subject of capacity and loading has historically been treated in safety codes as one and the same, it might be more meaningful in the future writing of safety codes to cover loading as a separate issue from capacity. One refers more appropriately to the traffic handling capacity, whereas the other refers to the maximum carrying capacity which has a direct bearing on safety.*

There is a necessary safety requirement limiting the load (mass) that a lift can carry, but there is no problem with this concept when carrying out a traffic design. It is impossible to fill the larger lifts with the number of passengers that this mass represents. When traffic sizing a lift system to handle the passenger demand, safety is unlikely to be compromised.

5.10.2   Evaluating P for application in Equation (4.8)

To evaluate Equation (4.8), a value for the average number of passengers (P) carried during each trip needs to be determined. If passengers arrived efficiently, then as each lift arrived, there would be a number of passengers waiting (P) equal to the probable car capacity of the lift ready to board. Unfortunately, reality is not like that and passengers arrive in a random fashion.

To avoid passengers being left behind (in a queue) to wait for the next lift, it is necessary to assume a lower than 100% utilisation factor for car occupancy. This assumption arises as statistical theory implies that, as the utilisation of a facility increases towards its maximum, the probability of immediate use of that facility reduces or the probability of being left behind increases. Thus, to achieve maximum utilisation of a facility, it is necessary to have a queue of applicants waiting (like at an airport). This is not considered satisfactory for a lift system. Therefore, the design utilisation has to be lower to allow for statistical variations to be accommodated. How much lower than 100% should P be set below actual car capacity?

The probability of the immediate use of a facility is shown diagrammatically in Figure 5.2 with respect to system utilisation. As system utilisation increases, then the probability of a passenger being left behind increases, until at 100% utilisation, there is a high probability of being left behind to queue. The shape of the curve has been shown to apply to such diverse facilities, access to a telephone line, availability of a lavatory, a free bank teller, etc.

Looking at the curve of Figure 5.2, it can be seen that above 70% utilisation, the change in slope increases significantly, and at 90% and above, it increases very rapidly. The ratio of the slope at 90% compared to the slope at 50% is some 25:1. Usually, the 80% point is considered

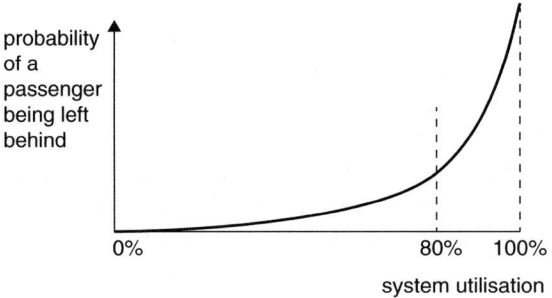

Figure 5.2 System utilisation

to be the 'knee' of the curve for most system utilisation judgements, and this value can be selected for lifts also. Therefore, for lift systems, it is reasonable to consider that the probable number of passengers is 80% of probable capacity.

Traditionally, and as a result of experience, the number of passengers assumed to be carried on each trip is taken as 80% of car capacity. Early explanations justified the 80% value by saying that passengers never loaded a car above 80% car capacity even when queues existed, thus showing remarkable restraint. Other theories to explain the 80% figure proposed are either: circulation difficulties (passengers at the back of the car always want to get out at the first stop); or operational problems (passengers obstructing door closing), both of which have the effect of increasing the round trip time. The 80% de-rating factor appears to have been arrived at by intuition and experience, rather than by theory. In some cases, for example for scenic lifts, the de-rating may be as low as 60%.

This use of an 80% car loading does not mean that cars fill only to 80% of probable rated car capacity on each trip, but that the average load is 80% of rated car capacity. Thus, the design car capacity is the average number of passengers that the lift is assumed to carry on average, when calculating uppeak traffic performance. Setting this parameter at 80% allows for the statistics of passenger arrivals and the effects of lift bunching arising from the group control system. Table 5.8, column 6 gives a list of design capacity values to be used in evaluating Equation (4.8). Strakosch (1983, 1998) offers a similar table (Table 4.4[9]). It is also interesting to note that Table 7 in EN81–20 for goods/passenger lifts puts the rated capacity close to the values given in Table 5.7.

5.10.3   *How many passengers can a lift car accommodate?*

How many passengers can be accommodated in a lift car? Originally, the assumed passenger weight was 150 pounds, i.e., 68 kg, in the UK and the USA, with the passenger standing on 2.0 square feet (0.19 m²).

EN 81–20 (2014) suggests in clause 5.4.2 and Table 8 that taking the rated load (in kg), dividing it by 75 and rounding down to the nearest whole number gives the number passengers that should be declared on the rating plate. The rated load is the maximum safe load a lift can carry and this loading value must not be exceeded.

EN 81–20 (2014) indicates in its Table 6 (5.4.2), that the rated loads a passenger lift may carry should also be related to maximum available car platform size. On this basis, Table 5.8 below shows the maximum available car area (column 2) for some commonly provided rated loads (column 1). Column 3 shows the result of the calculation for number of passengers.

The number of passengers a lift car can carry according to EN 81–20 can be seen to be related to the available car area in a non-linear way (Table 5.8, column 4). Although a lift with a rated load of 450 kg can easily accommodate four passengers, the only way in which a lift with a rated load of 2,500 kg could accommodate 33 passengers would be if the passengers were very small or crushed together, and each only occupied 0.15 m² of a car's platform area. Clearly, passengers do not change size to match the rated load of a lift as seen in column 4 of Table 5.8.

Allowing 0.21 m² per passenger, as recommended in Section 2.2.2, the EN 81–20 lift car platform sizes lead to the probable capacities given in Table 5.8 (column 5). These figures confirm the industry observations by Day (2001a, 2001b) of lower occupancies, particularly in larger lifts.

Definition 5.1: the probable capacity (in persons) is the lift car platform area, in m², divided by the car floor area occupied by a single passenger.

Table 5.8 Number of passengers to be used in traffic design

1	2	3	4	5	6	7
Rated load (kg)	Maximum available car area (m²)	EN 81 number of passengers (persons)	Available area per passenger (m²)	Probable capacity PC (persons)	80% design capacity (persons)	Capacity factor (%)
320	0.95	4	0.24	4.5	3.6	90
450	1.30	6	0.22	6.2	5.0	82
630	1.66	8	0.21	7.9	6.3	79
800	2.00	10	0.20	9.5	7.6	76
1000	2.40	13	0.19	11.4	9.1	70
1275	2.90	16	0.18	13.8	11.0	69
1600	3.56	21	0.17	16.9	13.5	64
1800	3.92	24	0.16	18.6	14.9	62
2000	4.20	26	0.16	20.0	16.0	62
2500	5.00	33	0.15	23.8	19.0	58

Notes to Table 5.8

1 Rated load values, in kg, taken from ISO 4190–1.
2 Maximum area values, in m², taken from EN81–20, Table 6.
3 EN 81 number of passengers is calculated by dividing the value for rated load by 75 as in EN 81–20.
4 Available area/passenger is calculated by dividing maximum available area by EN 81 number of passengers.
5 Probable capacity (PC) calculated by dividing the value for maximum available car platform area by 0.21.
6 Design capacity calculated as 80% of probable car capacity.
7 Capacity factor in per cent, calculated by dividing probable capacity by EN 81 capacity, is used by some designers in calculation and simulation programs (Peters, 1990).

In Europe, this is taken to be 0.21 m² per passenger.

The number of passengers suggested by the Tables 6 and 8 of EN 81–20 should never be used in traffic calculations; the lifts would physically be too small to accommodate the passengers.

Humans are not the same size. They vary over the world. And cultural differences affect personal space. So far, their sizes have been based on US and UK values. It can be said that Asia/Pacific people are smaller, that Scandinavians are the tallest people in Europe, and Europeans from the Latin countries are smaller than North Europeans, etc.

It is interesting to note that the US/UK ratios 68/75 and 0.19/0.21 are almost the same value. It implies that the loading of a human being is 360 kg/m² (75/0.21). If a linear ratio is assumed between area occupied and weight,[10] and their loading is assumed to be constant, then smaller or larger people, by weight, will take up smaller or larger areas of the car platform. Table 5.8 can be drawn.

A European standard PD CEN/TR 81–12: 2014a *Safety rules for the construction and installation of lifts – Basics and interpretations Part 12: Use of EN 81–20 and EN 81–50 in specific markets* suggests the following average passenger weights for different geographical areas of the world in Table 1, Clause 5.9:

These values are used to create Table 5.9, except for the values for 72.5 kg, where values for 80 kg are substituted.

The traffic designer would use the values in Table 5.9 to suit the local operating conditions. The 65 kg column might be used in the Asia/Pacific region, the 68 kg column for Latin Europeans, the 75 kg column for northern Europeans, and the 80 kg column for parts of Russia.

CEN-Table 1 Passenger capacity – examples used to calculate load capacity

Country	Weight of person (kg)
Australia	75
China	75
India	68
Japan	65
Korea	65
South Africa	75
USA	72.5

Table 5.9 Actual car capacity of a common range of ISO 4190–1 passenger lifts

Rated load (kg)	Available car area (m²)	Maximum possible number of passengers of weight/occupancy			
		65 kg/0.18 m²	68 kg/0.19 m²	75 kg/0.21 m²	80 kg/0.22 m²
320	0.95	5.3	5.0	4.5	4.3
450	1.30	7.2	6.8	6.2	5.9
630	1.66	9.2	8.7	7.9	7.5
800	2.00	11.1	10.5	9.5	9.1
1000	2.40	13.3	12.6	11.4	10.9
1275	2.90	16.1	15.3	13.8	13.2
1600	3.56	19.8	18.7	17.0	16.2
1800	3.92	21.8	20.6	18.7	17.8
2000	4.20	23.3	22.1	20.0	19.1
2500	5.00	27.8	26.3	23.8	22.7

(EN 81–12, 2014 gives suggested passenger weights for a number of world zones.) The load plate in the lift would then only indicate the 'recommended number of passengers' or remain silent.

5.11   Examples

Example 5.5

What are the rated, probable and design car capacities for a lift of 2,500 kg rated load?

    EN 81 rated capacity is: 33 persons
    Probable capacity is: 23.8 persons
    Design capacity is: 19.0 persons
    Capacity factor is: 58%

Example 5.6

A lift traffic design requires that the lift accommodates 10.5 passengers on average. What size of lift is indicated?

   Under the design car capacity column of Table 5.8, it can be seen that a lift with a rated load of 1,275 kg load will accommodate up to 11.0 passengers. Thus, a lift of 1,275 kg rated load should be selected.

Example 5.7

A speculative office block of 16 floors above the main terminal is to be built. The interfloor distance is 3.5 m with no express jump. A single 1,275 kg lift with 1,100 mm two panel centre opening (2PCO) doors is to be installed. What will be its characteristics? Assume the passenger transfer time is 1.2 s.

Basic data: $N = 16$, $RL = 1,275$ kg, $d_f = 3.5$ m, $t_p = 1.2$ s

From Table 5.7: $P = 11.0$ and from Table 5.6a $H = 15.1$, $S = 8.1$

The travel will be $16 \times 3.3 = 52.8$ m; so, from Table 5.2, the rated speed could be 3.0 m/s. However, remember that it is a speculative building, so use 2.5 m/s. Hence:

$t_v = 3.5/2.5 = 1.4$ s

From Table 5.2, the single floor flight time could be 5.5 s (0.5 s extra added for start up). From Table 5.3 (advanced opening not considered), the door times could be:

$t_o = 2.5$ s, $t_c = 3.0$ s

Thus:

$t_s = 5.5 + 2.5 + 3.0 - 1.4 = 9.6$ s

Then, solving the RTT Equation (4.8):

$$RTT = 2Ht_v + (S+1)t_s + 2Pt_p \qquad (4.11)$$

$RTT = 2 \times 15.1 \times 1.4 + (8.1 + 1)9.6 + 2 \times 11.0 \times 1.2$

$= 42.3 + 87.4 + 26.4 = 156.1$ s

Therefore:

$INT = 156.1$ s

and hence:

$UPPHC = 300/156.1 \times 11.0 = 21.0$ persons/5-minutes

5.12   Factors affecting the value of the round trip equation

It was noted in Section 5.1.4 that the round trip equation comprised six parameters. Three parameters were time independent variables, i.e., H, S and P, and three parameters were time dependant variables, i.e., t_v, t_s, and t_p. The round trip time is dependent on the values of these six parameters. The smaller they are, the smaller the resulting value of the round trip time, and the higher the handling capacity of the lift system. How can these six parameters be manipulated to achieve this?

5.12.1   The time independent variables

A lift is provided to transport passengers as represented by the variable P, so little can be done to reduce this parameter. Sections 8.5 and 11.6 indicate how values of the variable H and S can be reduced by either limiting the number of floors a lift serves, or by advanced traffic control systems.

5.12.2   The time dependent variables

Examination of the answer to Example 5.5 shows that the first and last (third) terms of the round trip equation are significantly less than the middle (second) term. This is generally so and means that each second saved or added to the middle term reduces or increases the value of the round trip time. This can be developed into a general rule of thumb that for a 1-second change in the performance time, there is a consequential change in the handling capacity of about 5%.

5.12.3   Five lift systems

Consider the following five lift systems, which could be considered to be the four corners of the lift world and its centre of gravity:

$N = 10$ $P = 8.0$ $v = 1.6$ m/s

$N = 10$ $P = 19.2$ $v = 1.6$ m/s

$N = 16$ $P = 12.8$ $v = 2.5$ m/s

$N = 20$ $P = 8.0$ $v = 3.15$ m/s

$N = 20$ $P = 19.2$ $v = 3.15$ m/s

All the systems have:

- a performance time of 10.0 s
- an assumed passenger transfer time of 1.0 s
- an interfloor distance of 3.3 m.

For the five lift systems, determine the effect of:

- increasing the performance time by +1.0 s
- decreasing the rated speed by 10%
- increasing the passenger transfer time by +0.2 s.

Table 5.10 tabulates the effects.

Table 5.10 Effect of changing the time dependent variables

N	P	v (m/s)	RTT(s)	T+ 1.0 s	v – 10%	t_p + 0.2 s
10	8.0	1.6	108	114 (6%)	111 (3%)	111 (3%)
10	19.2	1.6	156	166 (6%)	158 (1%)	164 (5%)
16	12.8	2.5	153	163 (6%)	158 (3%)	158 (3%)
20	8.0	3.15	123	131 (7%)	127 (3%)	127 (3%)
20	19.2	3.15	200	214 (7%)	203 (2%)	208 (4%)

All figures in parenthesis are percentage increases. All figures are rounded.

There are a number of conclusions that can be drawn from Table 5.8. Changing the speed by 10% or the passenger transfer time by 0.2 s has a smaller effect than changing the performance time by 1.0 s. There is little that a designer can do to hasten passengers into or out of a lift, except to provide good signalling and good circulation areas, so the emphasis must be on specifying the performance time as low as possible and keeping it at the specified value by good maintenance.

5.13　Effect of passenger arrival process

5.13.1　*General*

The derivation of the formulae for *H* and *S* assumed that passengers arrived at a constant inter-arrival interval (according to the particular level of arrivals existing) and that the lift arrived at a constant interval to take the intending passengers to their destinations. This effect is illustrated in Figure 5.3(a) where there is a lift system with a rated car capacity of 6 persons and an interval of 20 s, and 6 passengers arrive every 20 s.

No account has been taken of the way in which passengers arrive in a building or the randomness of their destinations. In practice, passengers do not conveniently arrive in batches equal to 80% of the rated car capacity, nor do they register the same number of destinations during each trip. The effect of this randomness is to cause the lifts to take different times to carry out a round trip and they become unevenly spaced. This effect is called bunching. (Buses in the street seem to do this. Although they may be on a 20-minute timetabled frequency, they only appear every hour in threes). This effect is illustrated in Figure 5.3(b) for the same conditions as Figure 5.3(a). Note the overall passenger average waiting times have increased and queues develop.

Note that these 'snapshots' of lift behaviour are a very simple representation of an extremely complex process, which does not bear close scrutiny. This is why it is best to set up statistical models to represent the process, and then to draw general and averaged conclusions from them.

So, if passengers do not obey the constant (sometimes called rectangular) arrival process used to derive Equations (5.7) and (5.14), what process do they obey?

It is generally accepted that passengers arrive into a lift system according to the Poisson probability process. This probability distribution function has been used to describe other phenomena such as the generation of radioactive particles and telephone calls, failures of electronic equipment, and the demands on digital computer central servers. Although this arrival process is not proven with respect to lift systems, work by Alexandris (1977) did go some way

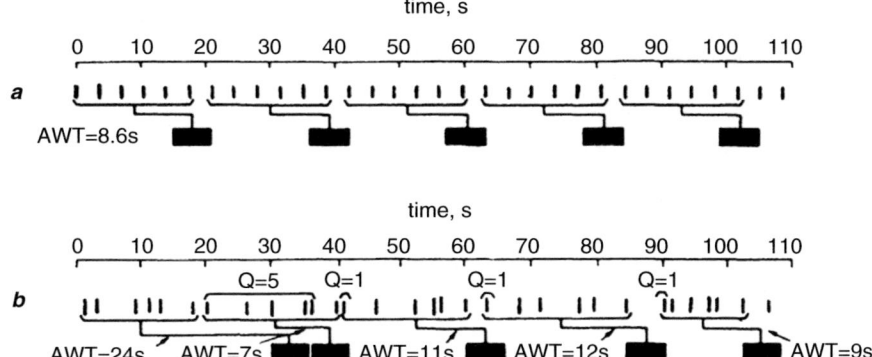

Figure 5.3 A simple representation of passenger arrival and lift car departures

(a) Constant passenger arrivals: constant lift departures. Overall average passenger waiting time 8.6 s
(b) Random passenger arrivals: irregular lift departures. Overall average passenger waiting time 12.6 s

to confirm it. Using observers, Alexandris surveyed three buildings with widely differing lift and other physical characteristics. He came to certain conclusions:

(a) Comparison of the observed and theoretical values calculated for the mean and variance showed a Poisson fit to be reasonable.
(b) The chi-squared goodness-of-fit tests gave evidence that a Poisson arrival rate assumption at least cannot be rejected.
(c) Although there may be other theoretical distributions, which might better accommodate the data, the Poisson distribution must be considered as a good approximation to the actual empirical distribution.

5.13.2 Formulae for S and H using the Poisson probability distribution function

Accepting the Poisson probability distribution function (pdf) as the best representation of the passenger arrival process, what effect does this have on the evaluation of the round trip time equation? Assume that the probability of n calls being registered in the time interval T for an average rate of arrivals λ (in calls per second) is:

$$p_r\left(n\right) = \frac{\left(\lambda T\right)^n}{n!} e^{-\lambda T} \tag{5.15}$$

Tregenza (1972) used this relationship to derive formulae for H and S (shown below with subscript, p). For a building with equal floor populations, he defined pr_0 as the probability of no calls being registered from the main terminal to any floor above during the period of one interval (T):

$$p_{r0} = e^{-\frac{\lambda}{N}T} \tag{5.16}$$

Then, by the same arguments used in developing Equations (5.7) and (5.14):

$$S_p = N\left(1 - p_{r0}\right) = N\left(1 - e^{-\frac{\lambda}{N}T}\right) \tag{5.17}$$

$$H_p = N - \sum_{i=1}^{N} p_{r0}i = N - \sum_{i=1}^{N} \left(e^{-\frac{\lambda}{N}T} \right)^i \tag{5.18}$$

5.13.3   *Comparison between the rectangular and Poisson pdfs*

The expressions for *Hp* and *Sp* obtained for the Poisson pdf can be shown to be almost identical to the expressions obtained for the rectangular probability distribution function. Tregenza suggests that as the arrivals and departures are (or should be) equal they can be defined as:

$$P_p = \lambda T \tag{5.19}$$

In the style of Chapter 4, this can be written:

$$P = \lambda.INT \tag{5.19bis}$$

Thus, to obtain S_p, substitute Equation (5.19) into Equation (5.17) to give:

$$S_p = N\left(1 - e^{-\frac{P}{N}} \right) = N\left(1 - \left(e^{-\frac{1}{N}} \right)^P \right) \tag{5.20}$$

Expanding the exponential gives:

$$e^{-\frac{1}{N}} = 1 - \frac{1}{N} + \frac{1}{2!N^2} - \frac{1}{2!N^3} \text{ etc.} \tag{5.21}$$

Taking the first two terms *only* of Equation (5.21) back into Equation (5.20) produces an equivalent equation to Equation (5.7).

If the third and subsequent terms are included, the Poisson probability values using Equation (5.17) for *S* become smaller than the values obtained using the rectangular probability values from Equation (5.7).

An equation for H_p is obtained by substituting Equation (5.19) into Equation (5.18) and transposing the indices, *viz*:

$$H_P = N - \sum_{i=1}^{N} \left(e^{-\frac{P}{N}} \right)^i = N - \sum_{i=1}^{N} \left(e^{-\frac{i}{N}} \right)^P \tag{5.22}$$

Expanding the exponential by series and taking the first two terms only produces:

$$\sum_{i=1}^{N} \left(e^{-\frac{1}{N}} \right)^P = \left(1 - \frac{1}{N} \right)^P + \left(1 - \frac{2}{N} \right)^P + \left(1 - \frac{3}{N} \right)^P .... \left(1 - \frac{N-2}{N} \right)^P + \left(1 - \frac{N-1}{N} \right)^P$$

Equation (5.23) is identical to the expansion of the summation part of Equation (5.14). Thus, as the remainder of Equations (5.14) and (5.22) are identical, then the values obtained for *H* will also be equal. If the third and subsequent terms are included in the expansion, the Poisson probability values obtained using Equation (5.18) for *H* become smaller than the rectangular probability values obtained from Equation (5.14). See Table 5.10.

Table 5.11 Comparison of rectanngular and Poisson pdf values for number of stops and highest reversal floor.

N	P	S	H	S_p	H_P
5	4.8	3.3	4.6	3.1	4.4
24	4.8	4.4	20.3	4.4	19.5
5	16	4.9	5.0	4.8	5.0
24	16	11.9	23.0	11.7	22.9

5.13.4 Example 5.8: calculating S and H

Calculate values for S and H using the formulae for the rectangular and Poisson pdfs for the following lift systems.

$$N = 5, P = 4.8 \quad : N = 24, P = 4.8: N = 5, \quad P = \quad 16: N = 24, P = 16$$

It should be noted that the Poisson values are always smaller or identical, and the differences will not be significant in practice.

5.14 Summary

Statistical theory has been used to derive two of the parameters of the round trip time equation S and H. Two different probability distribution functions have been considered. The first based on a simple statistical process, and used as long ago as 1923 by Basset Jones, assumes a rectangular input distribution with random arrival times. The second uses the Poisson process to describe arrivals. It is interesting to note that the older procedure (the rectangular pdf) will always produce a more pessimistic result than the Poisson pdf. Thus, if a designer wishes to be a little conservative in a traffic design, then the conventional (rectangular pdf) expressions should be used.

The conventional design assumes the arrival of passengers at constant intervals, served by lifts arriving at constant intervals, as shown in Figure 5.3(a). In practice, neither situation occurs (illustrated by Figure 5.3(b)), where both arrivals and lift departures are random. The problem with randomness in a lift system is that both the random passenger processes will add together to produce queues and large waiting times if an attempt is made to realise 100% handling capacity. To counteract this additive process, it is sensible to de-rate the lift system to allow for periodic overloads. Hence, the use of the 80% factor for determining a value for P when calculating handling capacities.

The P parameter in the *RTT* expression is wrongly assumed to be 80% of rated car capacity by designers. It is wiser to consider P to be 80% of actual capacity.

This chapter (and most design procedures) use the uppeak traffic condition to size a lift installation as it is simple and easily modelled mathematically. It is possible to extend the statistical techniques to derive expressions for the more general case of interfloor traffic. The full significance of quality of service (represented by system handling capacity) and car load versus quality of service (represented by average passenger waiting time) is considered in detail in Chapter 6.

Notes (shown in text as[1])

1 Under revision.
2 Purists will say mass.

3 The first edition of this book is 1938.

4 It is not sensible to proceed further without defining some statistical terms. The *probability* of a variable is defined by a numeric value between zero and unity. Probability then is simply a measure of the likelihood of some occurrence. For example, the probability of stopping at a floor may be 0.1. However, the actual number of stops is called the *expected* number of stops. The expectation is the *mean* or *average* value of some random variable, often termed the *expectance*. It is often of interest to know the spread of possible outcomes about the mean. This variation is essentially measured by the statistical characteristic termed *variance*. If the square root of a value of variance is taken, the value obtained is termed *standard deviation*. If large values of variance occur with respect to values obtained for expectance of some variable, then that variable has wide deviations of value from its mean. Statistical analysis and probability theory are highly mathematical fields and are best dealt with in specialised texts.

5 3rd edition, page 82.

6 2nd edition, page 74; 3rd edition, page 84.

7 Page 108.

8 Clause 8.2.1.

9 Page 85.

10 Purists say mass.

6 Determination of passenger demand and an improved design procedure

6.1 Introduction

The difficulty in planning a lift installation is not in calculating its probable performance, but in estimating the likely passenger demand. Quite often, the building has yet to be built and estimates have to be based on the experience gained with previous similar structures. Existing buildings can be surveyed by observation, or by means of an attached data logger, to determine the current activity. However, even this is prone to error as the building's population may have adapted to poor (or good) lift performance, or changed their behaviour when observed.

It is essential, therefore, that all the parties involved in the planning of a lift installation have a clear understanding of the basis for the planning. For example, it is important that the architect or planner establishes the lift system required at a very early stage and not after the rest of the building has been designed, as often happens.

It is important to remember that the distribution and size of the population of any large building changes regularly. Thus, a tightly planned design may prove inadequate once a building has been occupied for a year or more. To understand the effect of these changes on a design, it is essential to document the criteria and decisions taken at all stages of a design.

6.2 Quality and quantity of service

There are two key factors affecting the demand that a building's occupants will make on a lift system, as indicated in Section 4.6. These are:

- quantity of service
- quality of service.

The quantity of service factor, i.e., how many people will use the lift system over a defined period of time, is represented by the handling capacity. The quality of service factor, i.e., how well the lift system must deal with its passengers, is represented by passenger waiting time. Both factors are interrelated. Both factors depend, among other things, on the type of building and its use and the type of occupier. This makes the design task very difficult for buildings of a speculative nature.

The following sections indicate methods to facilitate the design task by looking at building populations, the likely demand on the lift installation and the quality of service factor. The analysis is relevant mainly to commercial office type buildings. Other buildings will be discussed briefly in this chapter.

The traffic period for evaluation when sizing an office building is based on a 5-minute segment of a pure morning uppeak. During this period of time, it is assumed that there is little or no traffic moving interfloor or down in the building. The lifts are loading passengers at the main lobby, distributing those passengers to various upper floors, and then making an express trip back to the main lobby for the next load.

Today, in practice, there is no pure uppeak traffic demand. The incoming passenger arrivals often barely comprise an 8% of building population arrival every 5-minutes. Thus, it could take an hour to fill a building on this basis. However, there are other traffic movements in the building. These include interfloor traffic to and from magnet floors such as restaurants and gyms, to and from work-related floors, e.g., sales and marketing, and down moving traffic of persons leaving the building to work elsewhere. The demand of such traffic can total up to 12–15% of the building population.

To study such complex traffic patterns requires the use of digital simulation techniques. For calculation procedures (which should always be undertaken before simulation is considered) a pure uppeak traffic pattern can be used. This method can be used to compare any designer's results.

It has been found that if the pure uppeak traffic pattern is sized correctly, all other traffic patterns will also be adequately served. This imaginary concept, if applied properly, works. There are exceptions to this comment. For example, in hotels at meal times, in hospitals at visiting times, in buildings with trading floors (insurance and stock markets), which open at specified times, and at lunchtime in all buildings. Another exception is if the lift installation is using an uppeak 'booster' rather than a conventional traffic controller (see Section 8.5).

6.3   Reprise

The sizing of lift systems to serve the demands of a building's population has interested the lift community since the 1920s. The methods used then were somewhat rough and ready. The problem is to match the demands for transportation from a building's occupants with the handling capacity of the installed lift system. This procedure should also result in an economic solution.

As Chapter 5 indicated, the first attempts to size a lift system to meet passenger demands occurred in the 1920s when Basset Jones (1923) derived a formula for the average number of stops that a lift would make under uppeak traffic conditions. The derivation of a formula for the probable average highest floor for the same condition was not made until 1955 by Schroeder (1955), which was published in German in *Fordern und Haben*). He also produced a three-term formula for the round trip time. Barney and Dos Santos (1975) independently derived a slightly different three-term formula (in English).

Thus, by the 1970s, a recognised method of calculation had evolved for uppeak traffic sizing, based on the mathematical determination of the H, S and P parameters (average highest reversal floor, average number of stops and average number of passengers).

The formulae for the calculation of the passenger handling performance of lift systems are now a universally accepted method of analysis. Lift makers are able to use tables applicable to their product range, based on the formulae, to estimate round trip times, intervals, handling capacity, etc.

During the 1970s, digital computer calculation and simulation packages evolved, which allowed other traffic conditions to be examined. These computer techniques utilise the proven mathematical methods.

6.4   Estimation of population

6.4.1   Passenger data sets

In Section 5.1.3, the passenger data set was found to comprise:

(a) the number of passengers boarding from specific floors
(b) the number of passengers alighting at specific floors
(c) the traffic mode, i.e., unidirectional or multidirectional
(d) the transfer times for passengers entering and leaving cars
(e) passenger actions.

During uppeak traffic, these data are simplified.

Point (a): passengers only load at the lobby.
Point (b): passengers never alight at the lobby.
Point (c): the traffic is unidirectional, i.e., travel is up the building.
Point (d): the passenger transfer times will generally be 'brisk'.
Point (e): there is little opportunity for passengers to misbehave.

To determine the number of passengers who will board and what their demand will be, depends on the building population. The following sections indicate the factors, which need to be considered when estimating the building population.

6.4.2   Purpose of a building

The size of the intended population should be obtained from the building owner or proposed occupier, if possible (and in writing). However, it may be that the population size is not available, or the building is a speculative one, when an estimation must be made.

The number of occupants will vary according to:

(a) the purpose of the building (residential, commercial or institutional)
(b) the quality of the accommodation
(c) the type of occupancy (in the case of office buildings, the type of tenancy).

With regard to point (a), the buildings are generally defined as:

•   residential: where people live, e.g., blocks of flats
•   commercial: where people work, e.g., offices
•   institutional: where people receive a service, e.g., a hospital.

Point (b) implies that the more prestigious the building, e.g., a head office, the more space is available to each occupant. Point (c) is more complex. There are three main types of tenancy:

(i)   diversified
(ii) mixed
(iii) single.

Definition 6.1: a diversified tenancy is a building occupancy condition where no single tenant occupies more than a single floor and no more than one quarter of the tenants of the building are engaged in the same type of business activity.

Definition 6.2: a mixed tenancy anticipates the possibility of multi-floor occupancy by a single tenant or multiple tenants with the same business activity.

Definition 6.3: a single tenancy is a building occupancy condition where a single tenant occupies a substantial portion or zone of the building (say 80%).

The single tenancy situation can present a severe traffic design condition. The group handling capacity with such occupancy can be high (about 14%) for calculation purposes. And some single tenant insurance companies, government entities, or utilities with large numbers of clerical workers can have handling capacity requirements of substantially more than 15% of the population in 5 minutes, if they operate a fixed starting time regime. In these cases, it would be important to establish this demand from the prospective building owner before carrying out any calculations.

In Tables 6.1, 6.2 and 6.3, diversified and mixed tenancies are grouped as multiple tenancy.

6.4.3   Main terminal population

The main terminal population is not normally included in the design population for the following reasons:

(a)   The main terminal floor is the bottom terminal for the lift group, and passengers on this level walk to their workplaces.
(b)   The main terminal floor is occupied in total by a bank or is a retail space. Employees of these types of businesses generally start work later than businesses occupying the majority of the building and, therefore, do not affect the major morning office building traffic peak.
(c)   The main terminal in a building with subterranean or off-site/street parking is served by a separate bank of lifts. In this situation, the persons working on the main terminal floor and parking in the subterranean level would ride this separate bank of lifts to that floor. In the case of traffic from off-site parking or ground level drop off traffic (taxis, etc.), they would walk directly into the main terminal floor and on to their workplace.

The main terminal population may be included in the building population in the following situation. There are subterranean parking levels served by the same group of lifts that serve the upper floors of the building. Thus, persons who work on the main terminal floor and who park in the subterranean levels would use this group of lifts.

6.4.4   Usable area

Most population estimates start from a knowledge of the net usable area, i.e., the area that can be usefully occupied and which excludes circulation space (stairs, corridors, waiting areas), structural intrusions (steelwork, space heating, architectural features, etc.), toilet facilities, cleaners' areas, etc.

The American National Standard ANSI Z65.1–1980 'Standard Method for Measuring Floor Area in Office Buildings' (ANSI/BOMA Z65.1, 1996) gives a useful guide to calculating areas in office buildings. It defines two important terms: rentable area and usable area.

Definition 6.4: the rentable area is determined by measuring between the inside finished surfaces and/or dominant parts of permanent outside walls, excluding any major vertical penetrations (stairs, lift shafts, flues, pipe shafts, ducts and their enclosing walls). No deductions are made for any columns and projections necessary to the building.

Rentable area generally remains fixed for the life of the building and is used to calculate rents.

> **Definition 6.5:** the usable area is determined by measuring between the finished surfaces of the office side of corridors and/or other permanent walls and/or the dominant parts of outside permanent walls and/or the centre of partitions within the rentable area. No deductions are made for any columns and projections necessary to the building.

Usable area indicates the actual occupied area and is important in lift traffic design calculations. Usable area can vary during the life of the building as corridors, partitions, etc. are moved.

In most traffic design cases, it is necessary to calculate and predict the building population from imprecise data. It is necessary always to calculate the usable area and then divide that area by the area allocated per person (in m²) to derive the estimated population.

Where architectural drawings are too schematic to make an accurate estimate of areas, one of the following approximate rule of thumb relationships may be used when the gross area is known:

> Rentable area = 90–95% of gross area
> Usable area   = 75–80% of gross area

or the relationship below if the rentable area is known:

> Usable area   = 80–85% of rentable area
> Sometimes, the term net internal area (NIA) is used. This is basically the area from the inside surfaces of the external walls. No concession is given for penetrations.

Whenever traffic calculations are made, it is important and advisable to indicate (in writing) which estimations have been utilised, and that a check review of the initial study is made once the architectural drawings are developed to a point where an accurate usable area calculation can be made.

In some cases, the building population may be dictated by the owner/client. This is particularly true if the building is being designed for a known occupant.

6.4.5   *Example 6.1: using rules of thumb*

Using rules of thumb above, what are the rentable and usable areas of (a) a tall/slender building and (b) a low/squat building, each having a gross area of 5,000 m²?

(a) This will have a large core compared to the footprint, but the occupants will be close to a lift.
Rentable area = 90% of gross area, i.e., 4,500 m²
Usable area   = 75% of gross area, i.e., 3,750 m²
(b) This will have a small core compared to the footprint, but the occupants may be far from a lift.
Rentable area = 95% of gross area, i.e., 4,750 m²
Usable area   = 80% of gross area, i.e., 4,000 m²

Table 6.1 Estimation of population

Building type	Population estimate
Hotel	1.5–1.9 persons/room
Flats	1.5–1.9 persons/bedroom
Hospital	3.0 persons/bed space*
School	0.8–1.2 m² net area/pupil
Office (multiple tenancy):	
Regular	8–10 m² net area/person
Prestige	10–12 m² net area/person
Office (single tenancy):	
Regular	10–12 m² net area/person
Prestige	12–14 m² net area/person

* Patient plus three others (doctors, nurses, porters, etc.).

6.4.6   Practical population estimations

Table 6.1 gives guidance for estimating the population for a variety of buildings based on surveys and experience of the population to be accommodated.

For offices, the British Council of Offices publishes a *Guide to Specification*, which is concerned with offices and contains guidance on population density. A summary of its contents can be found in Scott (2014).

6.5   Estimation of arrival rate

It is necessary now to determine the percentage of a building's population who will require transportation to the higher floors of a building during the morning 5-minute uppeak. This peak will vary due to effects such as:

(a) the type of building occupancy (different business interests or single tenant)
(b) the starting regime (unified or flexitime)
(c) the location of bulk transit facilities such as bus, train, etc. (distant alighting places will result in a spread of arrivals owing to different walking speeds).

The arrival rate is expressed as a percentage of a building's total population. It is unlikely in many buildings that all the total population is present on any day. The Greater London Council (date unknown) assumed an attendance of 84% (GLC 25). The effective population considered during the uppeak period can be reduced to account for:

(a) persons away on holiday
(b) persons away sick
(c) persons away on company business
(d) vacant posts
(e) persons who arrive before or after the peak hour of incoming traffic.

Thus, the total building population could be reduced by 15% to 20% to account for these factors. Table 6.2 gives guidance of probable peak arrival rates of the remaining occupants.

Table 6.2 Percentage arrival rates

Building type	Arrival rate
Hotel	10–15%
Flats	5–7%
Hospital	8–10%
School	15–25%
Office (multiple tenancy):	
Regular	11–13%
Prestige	13%
Office (single tenancy):	
Regular	13%
Prestige	15%

6.6 Quality of service

6.6.1 Passenger average waiting time (AWT)

The first passenger to arrive at a landing registers a call and then waits for a lift to arrive. Other passengers arrive and wait shorter periods of time.

Definition 6.6: the passenger average waiting time (*AWT*) is the period of time from when a passenger either registers a landing call or joins a queue, until the responding lift begins to open its doors at the boarding floor.

The instant a passenger can enter a car (assuming no passengers are exiting) is when the doors have opened 800 mm. This is because a person feels confident they can pass through a 'gap' of 800 mm, which is about 200 mm wider than the body ellipse of Figure 2.1.

Actual passenger waiting time would be the best indicator of the quality of service that an installed lift system could provide, i.e., the shorter the time the better the service. Passengers tend to be upset if they are made to wait too long, i.e., over 30 s.

However, passenger waiting times cannot be easily measured. Some designers, therefore, use the interval of car arrivals at the main terminal as an indication of service quality. Interval, however, is part of the evaluation of handling capacity, which simply determines the quantity of service. In general terms, interval can be used to indicate the probable quality of service when considering office buildings since an interval of:

* 20 s or less would indicate an excellent system
* 25 s would indicate a good system
* 30 s would indicate a satisfactory system
* 40 s would indicate a poor system
* 50 s or greater would indicate an unacceptable system.

Table 6.3 gives guidance for values of suitable intervals for office and other types of buildings. These values are more comprehensive than those given in some standards, e.g., BS5655, Part 6: As the expectations of passengers, particularly in major city centre offices, have increased, these recommendations have been updated in the current BS5655: Part 6: 2011.

Table 6.3 Uppeak intervals

Building type	Interval (s)
Hotel	30–50
Flats	40–90
Hospital	30–50
School	30–50
Office:	
Regular	25–30
Prestige	20–25

6.6.2   Example 6.2: basic specification

A speculative, regular ten-floor (above the main terminal) building is to be built. Each floor has 1,200 m^2 of usable space. What is the basic specification of the lift system?

Table 6.1 indicates that 10–12 m^2/person should be considered. Assume 12 m^2, which gives 100 persons per floor.

The total population will then be $10 \times 100 = 1{,}000$ persons.

Assume 80% daily occupancy, i.e., this gives a design population of 800 persons.

Table 6.2 indicates 11–15% arrival rate; assume 12.5%.

Then, peak arrival will be 12.5% of 800, i.e., 100 persons.

Table 6.3 indicates an interval of 25–30 s. This is a speculative building so, to save capital expenditure, assume 30 s.

The lift system should be sized to be able to handle 100 persons with a 30 s interval.

6.6.3   Uppeak performance

Caution must be exercised when using interval as a quality of service indicator as passenger waiting time depends on car load. A simple rule of thumb has been to assume that the average passenger waiting time is half the interval. This would be so if passengers were to arrive with equal time spacing, i.e., a rectangular probability distribution function. This rule delivers an imprecise result as it is only accurate when the cars load to less than half full. Lightly loaded cars are an unlikely situation for an uppeak traffic condition.

It has been shown (Barney and Dos Santos, 1977) that a theoretical relationship exists between interval and passenger average waiting time, dependent on the actual percentage car load (passengers in the car). This relationship is depicted in Figure 6.1 and tabulated in Table 6.4. The figure shows the average car load as a percentage of rated car capacity versus performance represented by passenger average waiting time (*AWT*) divided by average interval (*INT*) at that car load. The relationship *AWT/INT* normalises the results.

To a first approximation, the relationship can be used to indicate the probable quality of service of a lift installation. At the conventional assumed car loading of 80%, average passenger waiting time is 85% of the calculated interval. But at a 90% car loading, the average passenger waiting time has extended to 130% of the calculated interval. For loadings greater than 90%, the average passenger waiting time increases rapidly and in theory, at 100%, would be infinite.

For car loads between 50% and 80%, it is possible to develop an approximate equation for the *AWT* as:

$$AWT = [0.4 + (1.8\ P/PC - 0.77)^2]INT \qquad (6.1)$$

where *PC* is the probable car capacity.

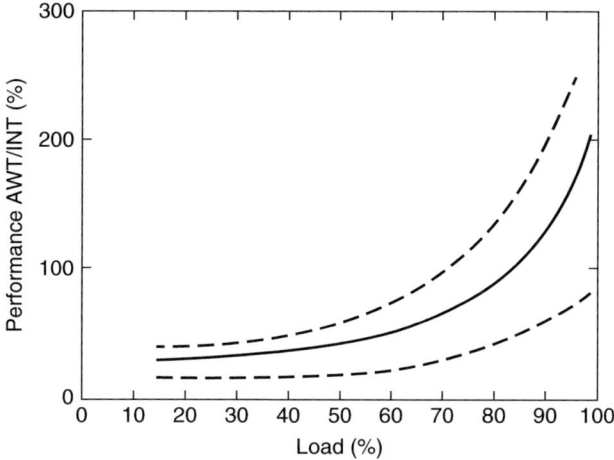

Figure 6.1 Uppeak performance graphical representation

Table 6.4 Uppeak performance numerical values

Car load (%)	AWT/INT (%)	Car load (%)	AWT/INT (%)
30	0.32	75	0.74
40	0.35	80	0.85
50	0.40	85	1.01
60	0.50	90	1.30
70	0.65	95	1.65

6.6.4 Example 6.3: deciding on AWT

Example 6.2 requires a lift system to be sized to be able to handle 100 persons with a 30 s interval. Design such a system and estimate the average passenger waiting time.

If the interval is 30 s, then a lift group must provide ten trips over five minutes.

Each car must load with ten passengers to handle 100 persons in five minutes.

If a lift with a rated load of 1,275 kg is used, the percentage car load is 10/13.8, i.e., 73%.

From Figure 6.1 or Table 6.4, a car load of 73% indicates that *AWT/INT* will be 68%.

The estimated *AWT* will be 0.68 × 30 = 20.4 s. (The formula gives 20.9 s.)

6.6.5 Average (lift) system response time (ASRT)

The average passenger waiting time is calculated by adding all the individual waiting times together and dividing by their number. At present, it is not possible to easily measure these times, unless a squad of observers are employed, or sophisticated (expensive) scanning equipment is installed. It is possible, however, to determine average passenger waiting times from computer simulations and models. Another way of measuring quality of service is to measure the average system response time.

> **Definition 6.7:** the (lift) system response time (*ASRT*) is a period of time that it takes a lift group to respond to the first registered landing call at a floor.

Individual system response times can always be measured, unlike the average passenger waiting time, using external data loggers or the in-built monitoring systems provided in most traffic controllers. Each individual system response time, however, only indicates the maximum waiting time that the first passenger has to wait. Later passengers will wait for less time.

The average system response time (*ASRT*) is the average of a number of individual response times. Care needs to be exercised when determining the average, as some in-built data loggers record system response times over very long periods, e.g., since the lift system was installed. In this case, or even over a 24-hour period, the *ASRT* value will be very low as it would include all periods of time when the lift system is hardly used, e.g., at night, weekends and holidays. The best systems will provide averages over 5-minute periods i.e., over the comparable period used for the calculation of handling capacities.

Definition 6.7 is applicable whether there is a single car or up to eight cars in the group. Strictly, the system response time is measured from the time that the first passenger at a floor registers a landing call until the car doors of the lift servicing that call has opened its doors to a width of 800 mm. In practice, the measurement will be made until the landing call registration is cancelled by the traffic controller. This can be from 5 s to 10 s before the doors are 800 mm open, and this time error must be added to the system response time that is measured to find the actual response times. When presented with data from a logger showing average system response time, it is important to ensure that it is not in error by this value.

It is possible for office buildings to establish criteria for the grade of service provided by an installed lift system. These are best expressed as either the percentage of calls answered in specified time intervals, or the time to answer a specified percentage of calls. That is, to serve the majority of users as well as possible and to some extent ignore the tail of the service distribution. This tail to the distribution has been called [*sic*] the 'forgotten man' problem. Thus, 100% satisfaction should never be quoted.

Table 6.5 indicates the percentage and time values for several grades of service over one hour of peak activity in an office building. An hour of peak activity is taken in order to obtain sensible and realisable results. It should be possible to obtain the grades of service indicated in the table during the worst hour of activity. This might occur during the midday break rather than during the intense, but shorter, uppeak and down peak periods at the beginning and end of the working day.

To allow for this during a peak 15-minute period, say down peak, the next lower grade of service should be possible.

During a peak 5-minute period, say during uppeak, two grades of service lower should be possible.

Table 6.5 Office building average system response time performance

Grade of service	Percentage of calls answered in		Time to answer calls (s)	
	30 s	60 s	50%	90%
Excellent	>75	>98	20	45
Good	>70	>95	22.5	50
Fair	>65	>92	25	55
Poor/unacceptable	<65	<92	>25	>55

To illustrate the use of these criteria, suppose the building being considered has a good system installed. Then over one hour, the system should provide the figures given in the 'good' row (Table 6.5), in down peak the figures given in the 'fair' row and during uppeak the figure given in the 'poor/unacceptable' row. This illustrates that passenger expectations change according to the purpose of their travel.

6.7 Other useful design parameters

6.7.1 Passenger average travel time (ATT)

It is useful to know the average time it would take for a passenger to reach their destination floor (assumed to be half way up the building zone being served) after their allocated lift is ready for boarding, i.e., it has opened its doors. This extra knowledge would help to evaluate the suitability of a planned lift group. It is, however, a secondary quality of service design consideration after average passenger waiting time. This is because passengers travelling to the upper floors of a building zone become annoyed if a lift takes too long to reach their floor. Strakosch (1998) states[1] that for most people, 100 s is a tolerable travel time, which can be further tolerated to some 150 s of travel time if two people exit at each stop. He regards 180 s as the absolute limit. These should be considered maximum values, with average values being about half of these.

> **Definition 6.8:** the passenger average travel time (*ATT*) is a period of time (in seconds) that a passenger spends travelling in a lift, measured from the time the passenger boards the car until the instant that the passenger alights at the destination floor.

A quick rule of thumb, which has been used to evaluate this time, is to use the formula of adding one half of the uppeak interval for a group of lifts to one quarter of the uppeak round trip time for the individual lift in the group, *viz*:

$$ATT = 0.5 \times UPPINT + 0.25 \times UPPRTT \tag{6.2}$$

The figure obtained considerably understates the likely *ATT* as it 'forgets' how quickly a car expresses back to the main terminal floor after the last passenger has alighted. A better rule of thumb, found by comparison to calculations, is to add one half of the uppeak interval to one half of the uppeak round trip time, *viz*:

$$ATT = 0.5 \times UPPINT + 0.5 \times UPPRTT \tag{6.3}$$

A more accurate estimate of how long it takes the average passenger to reach their destination is to modify the round trip time in Equation (4.8) and calculate *ATT* to the midpoint of the local travel for any group of lifts. This means travel for a distance of *H*/2, with the number of stops being *S*/2, and a transfer of *P*/2 passengers boarding the lift and *P*/2 passengers alighting. The resulting formula is given below:

$$ATT = 0.5Ht_v + 0.5St_s + 1.5Pt_p \tag{6.4}$$

If there is any express travel through a number of floors (*Ej*), the formula becomes:

$$ATT = 0.5Ht_v + 0.5St_s + 1.5Pt_p + t_e \tag{6.5}$$

Thus, Equation (6.5) calculates *ATT* to the midpoint of the local and express travel for any group of lifts. This will be to a point half way between the lobby and the high call reversal floor (*H*). The equation also takes account of the passenger transfer times and the express travel. To illustrate this, consider a lift carrying eight passengers, serving a building with 22 floors above the main terminal. What is the position of the average destination floor?

Using Table 5.11a, the column for 8 persons, following this column down to the line corresponding to 22 floors shows that the highest reversal floor is floor 20. The average destination floor is thus floor 10.

6.7.2　Passenger average journey time (AJT)

The primary consideration of passenger average waiting time (*AWT*) can be combined with the secondary consideration of passenger average travel time (*ATT*) to give a passenger average journey time (*AJT*).

> **Definition 6.9:** The passenger average journey time is a average period of time (in seconds) from when a passenger either registers a landing call, or joins a queue, until the responding lift begins to open its door at the destination floor.

NOTE: a passenger is deemed to have alighted, when any passenger detection device is interrupted or the passenger physically crosses the door sills.

Thus, the passenger average journey time is the sum of the average passenger travel time (*ATT*) and the average passenger waiting time (*AWT*). The average passenger travel time is simple to calculate, but the average passenger waiting time depends on car loading, which can only be determined after the car size has been selected. The passenger average travel time, plus one half of the uppeak interval will give a close approximation for evaluation purposes.

$$AJT = 0.5Ht_v + 0.5St_s + 1.0Pt_p + t_e + 0.5INT \qquad (6.6)$$

The passenger average journey time is more accurately obtained by adding the average passenger waiting time to Equation (6.5) and is given by:

$$AJT = 0.5Ht_v + 0.5St_s + 1.0Pt_p + t_e + AWT \qquad (6.7)$$

The average waiting time should be estimated from Table 6.4, according to the car loading.

6.7.3　Summary of AWT, ATT and AJT

The quality of service is particularly important for office buildings. The values given in Table 6.6 indicate the performance times to aim for in a traffic design and the maximum acceptable values.

Table 6.6 Summary of times

Time	Aim for	Poor
AWT	<20 s	>25 s
ATT	<60 s	>70 s
AJT	<80 s	>90 s

Where a passenger uses a shuttle lift (a two-stop lift serving the main entry level and an upper lobby) to first reach the upper terminal floor, and then uses another group of lifts to reach their final destination floor, the values obtained for *AWT, ATT* and *AJT* should be calculated separately for each journey. For example, in a building with a shuttle service to floor 40 and a transfer to a group of lifts serving an upper zone of 20 floors, it is necessary to calculate the shuttle and upper zone time values separately.

6.8   More on uppeak passenger average waiting time, travel time and journey time

6.8.1   Alternative definitions

It is important that traffic designers check the definitions used when comparing traffic designs. Definitions 6.6, 6.8 and 6.9 are used throughout this book. It can be seen that:

$$(AJT) = (AWT) + (ATT)$$

. . . exactly with no periods of time missing.

Now consider the following definitions from NEII-1: 2000,[2] which have been paraphrased a little:

Definition 6.10: the passenger waiting time is the actual time a passenger waits after registering a landing call until the responding elevator door(s) begin to open.

Definition 6.11: the time to destination of a single passenger is measured from landing call registration until the car doors start to open at the destination floor.

Whereas Definitions 6.6, 6.8 and 6.9 are a complete set, Definitions 6.10 and 6.11 are not complete as there is no NEII definition of travel time. The NEII definitions are psychologically sound in the sense that a passenger will consider the waiting time to be over when the lift doors are opening for boarding (Definition 6.10), however, there is a period of time missing between the car doors opening and the passenger entering the lift. Definition 6.10 is close to Definition 6.6, except for the one missing time period.

The passenger will also consider that the travelling time is over when the lift doors are opening for alighting (Definition 6.11), however, there is a period of time missing between the car doors opening and the passenger leaving the lift. Definition 6.11 is close to Definition 6.9, except for the one missing time period.

6.8.2   Another formula for passenger average waiting time

In Section 6.6.3, it was suggested that passenger average waiting time could be obtained easily from Equation 6.1, Figure 6.1 and Table 6.4, but no proof was given. Section 13.3 shows the provenance of Figure 6.1 in Figure 13.2.

Figure 6.1 can be used to complement the conventional design procedures in order to obtain an estimation of average waiting times. The procedure is very simple. For a particular lift configuration and a specified traffic demand, the designer calculates the interval and the percentage load using the procedure in Section 6.11. From the value of the calculated percentage load on the horizontal axis of Figure 6.1, the value for *AWT/INT* can be read from the vertical axis. As the calculated interval is known, the average waiting time can then be evaluated.

Equation (6.1) represents the *AWT/INT* curve in Figure 6.1 for 50%<*P/PC*<80%.

A more accurate, but more complex curve fit covering a wider range of 40%<*P/PC*<95%, was proposed by Kavounas (1993c) as:

$$AWT = 0.2e^x \tag{6.8}$$

where $X = 0.235e^{2.27\frac{P}{PC}}$

6.8.3   *More on passenger average travel time (ATT)*

Equation (6.4) gives a formula to calculate average travel time (*ATT*):

$$ATT = 0.5Ht_v + 0.5St_s + 1.5Pt_p \tag{6.4}$$

So and Suen (2002) have suggested a more accurate formula for the travel time. This is shown as Equation (6.9):

$$ATT_s = 0.5\left[\frac{S+1}{S}\right]Ht_v + 0.5(S+1)t_s + 1.5Pt_P \tag{6.9}$$

When *S* is very large, the So and Suen equation becomes that of Equation (6.4). The difference between the two formulae is not large.

6.8.4   *Example 6.4: comparison of values for ATT*

Consider the following data:

$N = 16$, $P = 12$, $H = 15.2$, $S = 8.6$, $t_v = 1.0$ s, $t_s = 10.0$ s, $t_p = 1.0$ s.

Using Equation (6.4):

$ATT = 68.6$ s

Using Equation (6.9):

$ATT = 74.5$ s

The So and Suen average travel time values are more pessimistic, being a few seconds longer.

6.8.5   *Towards agreed traffic design definitions*

In September 2004, Drs. Barney, Peters, Powell and Siikonen met and agreed a number of traffic design definitions (Barney *et al.*, 2005). These are shown in Table 6.12 and Figure 6.2. It will be noted in this book that passenger average journey time is not defined. It will also be noted that several terms are defined to assist in the use of simulation studies where, for example, waiting time is the sum of walking and standing time, which is important when analysing hall call allocation systems (Chapter 11).

Journey Time	
Time to Destination	
Waiting Time	**Transit Time**
➤ *Call registration* *Doors opening* ◆	*Doors opening* ◆
➤ *Passenger arrives*	*Passenger alights* ◆

Figure 6.2 Illustration of various traffic design timings

6.9   Example 6.5: a speculative building

A speculative building with ten floors above the main terminal floor is to be built. Each floor has 1,500 m² of gross space. The interfloor height is a regular 3.3 m. Assume the passenger transfer time is 1.2 s.

6.9.1   Given data

$N = 10$, $d_f = 3.3$ m, gross floor area 1500 m², $t_p = 1.2$ s.

A speculative building could be occupied by one tenant, i.e., single tenancy.

The usable area could be 80% of gross, i.e., 1,200 m².

Table 6.1 indicates that the density of occupation for a regular building is in the range 10–12 m² per person. As a speculative building, assume worst case, i.e., 10 m²/person. The population will be:

1,200/10 = 120 persons/floor

Total population will be:

120 × 10 = 1200 persons

Assume 80% daily attendance (see Section 6.5) and the design population becomes 960 persons.

Table 6.2 indicates that 11–15% of the population will arrive in the busiest 5-minute period. Assume the worst case is 15%, then the peak arrival rate will be:

960 × 0.15 = 144 persons

Table 6.3 gives a suitable interval as 25–30 s. The building is speculative, so assume 30 s.

Design the lift system to handle the arrival of 144 people with an interval of 30 s. These two values are the user requirements.

6.9.2   Initial sizing

There will be 10 trips in 5 minutes (300 seconds), i.e., a 30 s interval. This means that 14.4 passengers (on average) are transported on each trip. If this represents 80% car occupancy, then the maximum probable number of passengers, *PC*, will be:

14.4/0.8 = 18 persons

The nearest standard car (BS ISO 4190) size is 1,800 kg (maximum probable car capacity 18.6 persons).

6.9.3   Calculation

It is now necessary to evaluate the round trip time equation (Equation 4.8).

$$RTT = 2Ht_v + (S+1)t_s + 2Pt_v$$

The total travel is 10×3.3 i.e., 33 m. From Table 5.2, this suggests a rated speed of 1.6 m/s. Therefore, the single floor transit time is:

$$t_v = 3.3/1.6 = 2.1 \text{ s}$$

Given an acceleration of 0.7 m/s², and a jerk value of 1.1 m/s³, and applying a dynamics program to the Appendix A equations, the likely single floor flight time ($t_f(1)$) could be 5.0 s.

For 1,100 mm centre opening doors, the door opening time is given as 1.8 s with 0.5 s advanced opening, and a door closing time is given as 3.0 s. Thus, assuming a start delay of 0.5 s, a value for T can be obtained as:

$$T = 5.0 + (1.8 - 0.5 + 0.5) + 3.0 = 9.8 \text{ s.}$$

$$t_s = 9.8 - 2.1 = 7.7$$

The number of passengers is 14.4, hence from Table 5.11b, values for H and S can be obtained:

$$P = 14.4; \ H = 9.8; \ S = 7.8$$

Using Equation (4.8):

$$RTT = 2 \times 9.8 \times 2.1 + (7.8 + 1) \ (7.7) + 2 \times 14.4 \times 1.2$$

$$= 41.2 + 67.8 + 34.6 = 143.6 \text{ s}$$

To achieve an uppeak interval of 30 s (or thereabouts) would require five cars.

$$UPPINT = 143.6/5 = 28.7 \text{ s}$$

The uppeak handling capacity will then be:

$$UPPHC = (300 \times 14.4)/28.7 = 150.5 \text{ persons/5 minutes}$$

The design installation would comprise five cars of 1,800 kg rated load, which have a maximum probable car capacity of 18.6 persons and an 80% occupancy value of 14.9 persons. This would deliver an uppeak interval of 28.7 s and an uppeak handling capacity of 150.5 person per 5 minutes. There would be an excess in handling capacity (%POP = 15.7%) and a slightly better interval. Table 6.7 summarises the input data and the results for Example 6.5.

Table 6.7 Data summary for Example 6.5

Input data	
Number of floors	10
Rated load	1800
Actual car capacity	18.6
Number of passengers	14.4
Number of lifts	5
Rated speed	1.6
Building population	960
Interfloor distance	3.3
Express jump	0
Express additional time	0
Single floor flight time	5
Door close time	3
Door open time	1.8
Advance door opening	0.5
Start delay	0.5
Passenger transfer time	1.2

Results	
Number of passengers	14.4
Highest reversal floor	9.8
Number of stops	7.8
Performance time	9.8
Round trip time	143.6
Interval	28.7
Handling capacity	150.5
Percentage population	15.7
Capacity factor (%)	77

6.10 Example 6.6: a high quality building

Suppose the building of Example 6.5 were now to be a high quality building for an international tenant, what system would then be required?

The assumed floor population will now be 12 m²/person, giving a total population of 1,000 persons.

At 80% attendance, the design population becomes 800 persons.

Assume a 15% peak (%POP); the arrival rate will be 120 persons.

The interval required is to be 25 s.

Design the lift system to handle 120 people with an interval of 25 s. These two values are the user requirements.

With 12 trips per 5 minutes (25 s interval), the lifts must accommodate 10 persons on average per trip.

A lift with a rated load of 1,000 kg is too small as the design capacity is 9.1 persons. This means lifts of 1,275 kg rated load should be selected, giving a design capacity of 11 persons.

This gives values for $P = 11.0$; $H = 9.6$; $S = 6.9$.

Keeping the lift dynamic times the same and using Equation (4.8):

$$RTT = 2 \times 9.6 \times 2.1 + (6.9 + 1)(7.7) + 2 \times 11.0 \times 1.2$$
$$= 40.3 + 60.8 + 26.4$$
$$= 127.5 \text{ s}$$

To achieve an uppeak interval of 25 s (or thereabouts) would require five cars.

$$UPPINT = 127.5/5 = 25.5 \text{ s}$$

The uppeak handling capacity will then be:

$$UPPHC = (300 \times 11.0)/25.5 = 129 \text{ persons/5-minutes}$$

Thus, the design installation would comprise five cars of 1,275 kg rated load. This would deliver an uppeak interval of 25.5 s, which is slightly longer than specified, and an uppeak handling capacity of 129 person per 5 minutes, which is much larger than specified at a *%POP* of 16.3%. Table 6.8 summarises the input data and the results for Example 6.6.

Table 6.8 Data summary for Example 6.6.

Input data	Value
Number of floors	10
Rated load	1275
Actual car capacity	13.8
Number of passengers	11.0
Number of lifts	5
Rated speed	1.6
Building population	800
Interfloor distance	3.3
Express jump	0
Express additional time	0
Single floor flight time	5
Door close time	3
Door open time	1.8
Advance door opening	0.5
Start delay	0.5
Passenger transfer time	1.2

Results	Value
Number of passengers	11.0
Highest reversal floor	9.6
Number of stops	6.9
Performance time	9.8
Round trip time	127.5
Interval	25.5
Handling capacity	129
Percentage population	16.3
Capacity factor (%)	80

6.11   An improved design procedure

6.11.1   The iterative balance method (IBM)

In Examples 6.5 and 6.6, the handling capacity was larger than that required, and in Example 6.5, the required interval was not achieved. Clearly, in Example 6.5, there was too much handling capacity and a poorer interval than desired. In effect, there were not enough intending passengers to use the capability of the installed system.

What should the traffic designer do to achieve the two user requirements?

Should the designer modify the component parts of the *RTT* expression to achieve a balance?

The first term can only be altered by changing the rated speed, and the effect would be small. The second term can be altered by changing the single floor flight time or door timings. Thus, a lower specification door gear could produce a matching handling capacity. The third term can be altered by changing the rated car capacity, but this will alter *S* and *H*, and can be counterproductive. Experienced designers will use intuitive procedures incorporating combinations of the above to establish a suitable design to cater for the desired handling capacity.

Using the conventional design method above, a designer would propose initial values for the dynamic parameters, and estimate the probable car capacity of the lifts based on experience. It is then assumed that the average number of passengers (*P*) carried per trip is 80% of the probable car capacity, and values for the expected number of stops (*S*) and the average highest reversal floor (*H*) are evaluated. Hence, the round trip time (*RTT*), interval (*INT*) and handling capacity (*HC*) are calculated.

At this stage, the designer compares the calculated value of *UPPHC* with the number of passengers to be moved in the peak 5 minutes (*%POP*). If the *UPPHC* is greater than or equal to this number of passengers (see Equation (6.10)), then the designer is satisfied that the system will cope with the traffic.

$$UPPHC \geq \%POP \tag{6.10}$$

The configuration will be trimmed if the handling capacity is too large, and should it be smaller than the required value, then the designer must repeat the evaluation for more and (or) bigger and (or) faster cars. However, the values calculated for *RTT, INT* and *HC* are exact only if there is a perfect match between the arrival rate and handling capacity.

The procedure suggested by Tregenza (1972) as Equation (5.19) is significant as he presents the idea of matching the lift handling capacity to the desired handling capacity exactly. This is achieved by not rigidly fixing *P* as a percentage of probable car capacity (*PC*); *P* is allowed to take the most appropriate value. From Tregenza, *P* is equal to λINT, as below:

$$P = \lambda.INT \tag{5.19bis}$$

where λ is the passenger arrival rate in persons per second.

A new design procedure, the iterative balance method (Barney and Dos Santos, 1975), can now be proposed, which can be used with either the conventional or Poisson formulae. For simplicity, it is presented as a series of steps (Table 6.7).

Table 5.11 gives values for integer values of *P* from 5 to 20 persons, and can be used to obtain values for *H* and *S*. It may be necessary to interpolate non-integer values.

The iterative balance method is a classical two point boundary problem, where the start and end results are known, in this case, the balancing interval. To arrive at an answer, a two point boundary problem has to iterate, i.e., change the target value for (INT_T) to converge on the final value (INT_F). To do this, a suitable algorithm has to be chosen. Pick the wrong algorithm and the start and end values diverge rapidly, leading to infinite values, i.e., no solution. The algorithm chosen in step (7), which is simply 'twice the new value minus the old value', does provide convergence.

In step (8), it is important to remember that, where an average car load is much greater than 80%, poor passenger service might result, i.e., long waiting times and queues. The designer must select a suitable car size to meet desired economic and operating conditions. It is important not to make any simplifications, either for the sake of arithmetical ease, or to meet a preconceived idea of a suitable lift installation, at any stage of the calculation until step (8) is reached. Always calculate primary and secondary values as precisely as possible.

The procedure outlined in Table 6.9 allows the *RTT* to be calculated for any values of *P*, *H* and *S*. Only when the initial (estimated) interval in an iteration matches the final (calculated) interval ($UPPINT_f$), are any decisions made.

Note that when calculating the interval at step 5, only integer values of the number of lifts (*L*) can be used. This could be expressed mathematically as:

$$INT = \frac{RTT}{\{round-up\}L} \tag{6.11}$$

6.11.2   *Example 6.7: using the iterative balance method*

Following the steps in Table 6.9, using the iterative balance method to solve Example 6.6:
   Iteration (1)

(1-1) Assume $\lambda = 120$ persons/5 minutes $= 120/300 = 0.4$ persons/s
(2-1) Given $N = 10$; $t_v = 2.1$ s; $t_s = 7.7$ s; $t_p = 1.2$ s. Therefore, $T = 9.8$ s

Table 6.9 New design procedure

Step	Procedure
1	Decide the rate of passenger arrivals (λ) over 5 minutes (user requirement 1).
2	Obtain or decide upon lift system data:
	N number of floors
	t_v the interfloor time
	t_s the operating time
	t_p the passenger transfer time
3	Decide the target interval (INT_T)
4	Obtain:
	P average car load
	H average reversal floor
	S expected number of stops
5	Calculate *RTT*
6	Select *L*, the number of lifts to produce an interval close to that estimated in step (3)
	[round up to integer value of *L*]
7	Compare the target interval in step (3) with the calculated interval in step (6) and if equal or nearly equal go to step 8 or if the calculated interval is significantly different, estimate another value for the interval and then iterate from step (3). A possible new trial could be:
	New $INT = INT$ (step (6)) + [INT (step (6)) $- INT$ (step (3))]
8	Select suitable lifts with a rated load, which allows approximately 80% average probable car capacity.

(3-1) Let $INT_T = 25$ s

(4-1) $P = \lambda INT_T = 0.4 \times 25 = 10$ persons.

From Table 11.5a: H = 9.5; S = 6.5

(5-1) $RTT = 2 \times 9.5 \times 2.1 + 7.5 \times 7.7 + 2 \times 10 \times 1.2 = 121.7$ s

(6-1) Let $L = 5$, then $INT_1 = 121.7/5 = 24.3$ s

(7-1) The value calculated of 24.3 s does not closely match the start value of 25 s. Try a new value *viz*:

Iteration (2)

(3-2) $INT_2 = 24.3 + (24.3-25) = 23.6$ s

(4-2) $P = \lambda INT_2 = 0.4 \times 23.6 = 9.4$ persons

H = 9.5; S = 6.3

(5-2) $RTT = 2 \times 9.5 \times 2.1 + 7.3 \times 7.7 + 2 \times 9.4 \times 1.2 = 118.7$ s

(6-2) Let $L = 5$, then $INT_F = 118.7/5 = 23.7$ s

(7-2) This is a close enough match (error <0.1 s)

(8) The *UPPHC* is $(300 \times 9.4)/23.7 = 119$ persons per 5 minutes.

Taking the average car occupancy as 9.4 persons, the maximum probable car capacities (Table 5.8) of a 1,000 kg car is 11.4 persons, and for a 1,275 kg car is 13.8 persons. The 1,000 kg

Table 6.10 Summary of Example 6.7

Input data	Value
Number of floors	10
Rated load	1275
Actual car capacity	13.8
Number of passengers	9.4
Number of lifts	5
Rated speed	1.6
Building population	800
Interfloor distance	3.3
Express jump	0
Express additional time	0
Single floor flight time	5
Door close time	3
Door open time	1.8
Advance door opening	0.5
Start delay	0.5
Passenger transfer time	1.2

Results	Value
Number of passengers	9.4
Highest reversal floor	9.5
Number of stops	6.3
Performance time	9.8
Round trip time	118.7
Interval	23.7
Handling capacity	119
Percentage population	14.9
Capacity factor (%)	68

lift could be too small, and the 1,275 kg lift much too large. The percentage car occupancy for both cases are:

$$9.4/11.4 = 82\% \text{ and } 9.4/13.8 \times 100 = 68\%$$

Examining the results above, it can be seen that the arrival rate and the lift handling capacity can be made to balance. This was achieved by changing the number of passengers in the car until the resulting system handling capacity was equal (balanced) to the demand represented by the passenger arrival rate. The interval alters to achieve this balance and is less than the specified 25 s, but not a lot smaller.

An acceptable design on economic grounds might be the lifts with a rated load of 1,000 kg as the car capacity is only marginally greater than 80%.

6.12   Iterative balance by spreadsheet

In practice, these calculations would not be carried out manually, but would use a calculation program or a spreadsheet. The advantage of a spreadsheet is that the results are calculated immediately as the input data are changed. Table 6.11 illustrates a spreadsheet method.

Considering Example 6.7. Table 6.11, column 2 indicates the data for Example 6.7 with the performance time as 9.8 s. Column 2 also shows the first iteration and column 3 shows the

Table 6.11 Use of a spreadsheet to solve Example 6.7

Input data	2	3	4	5
Number of floors	10	10	10	10
Rated load (kg)	???	???	1000	1275
Probable car capacity (persons)	???	???	11.4	13.8
Number of passengers (persons)	10.0	9.4	9.4	9.4
Number of lifts	5	5	5	5
Rated speed (m/s)	1.6	1.6	1.6	1.6
Passenger transfer time (s)	1.2	1.2	1.2	1.2

Results	VValue	VValue	VValue	VValue
Number of passengers	10.0	9.4	9.4	9.4
Highest reversal floor	9.5	9.5	9.5	9.5
Number of stops (s)	6.5	6.3	6.3	6.3
Performance time (s)	9.8	9.8	9.8	9.8
Round trip time (s)	121.4	118.0	118.0	118.0
Interval (s)	24.3	23.6	23.6	23.6
Handling capacity (persons/5-min)	124	120	120	120
Percentage population	15.5	14.9	14.9	14.9
Capacity factor (%)	N/A	N/A	82	68
Uppeak average waiting time (s)	N/A	N/A	21	14
				V

Note: the numerical values in the spreadsheet may be slightly different to those obtained manually as the spreadsheet does not round numbers down.

second iteration. Column 4 considers the selection of lifts with a rated load of 1,000 kg and column 5 the selection of lifts with a rated load of 1,275 kg.

6.13 A systematic methodology for lift traffic design: the *HARint* plane

The process carried out in the iterative balance method in Section 6.11 can be seen in the manual calculation carried out in Table 6.7. Once this is programmed into a spreadsheet, this process is hidden and becomes automatic.

Al-Sharif (Al-Sharif *et al.*, 2012; Al-Sharif *et al.*, 2014a) proposed a visualisation of the iterative process called the *HARint* plane. This visualisation is a design methodology that provides a graphical representation of the design. It also allows the designer to understand the optimality or otherwise of the selected design. It is based on defining the two most critical user requirements that represent the quality of service (the target interval denoted as INT_T) and the quantity of service (the percentage arrival rated in 5 minutes denoted as $\%POP$. As explained in Section 6.11, in order for a design to meet the user requirements, the following two conditions must be met:

$$UPPHC \geq \%POP \tag{6.10}$$

$$UPPINT \leq INT_T \tag{6.11}$$

The *HARint* plane uses two axes to represent the two user requirements. The actual interval INT_{act} is represented on the x-axis (abscissa) and the handling capacity ($\%POP$) is represented on the y-axis (ordinate), corresponding to the two user requirements. Each design is thus represented by a pair of values of $\%POP$ and INT_{act} and is also represented as a point on the *HARint* plane. An example of a *HARint* plane representation of a design, where the $\%POP$ is 15% and the required interval is 30 seconds, is shown in Figure 6.3. Shown on it are lines of integer

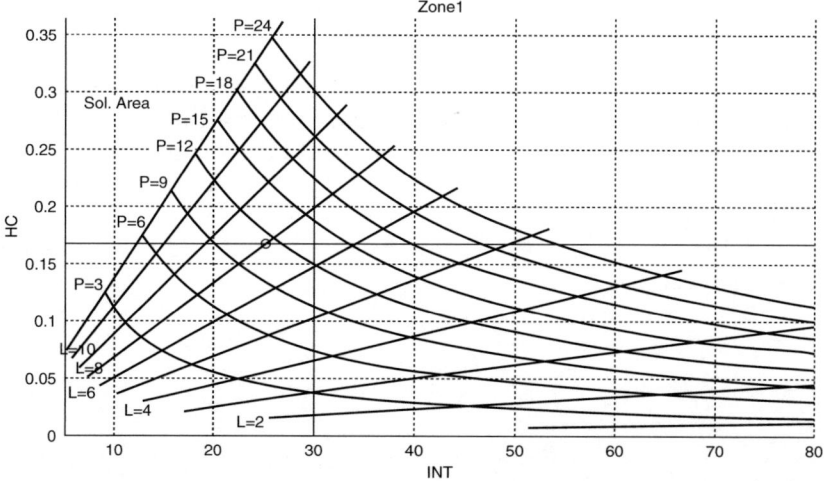

Figure 6.3 *HARint* plane with curves of integer values of *L* and integer values of *P* added and an optimum solution marked with a small circle

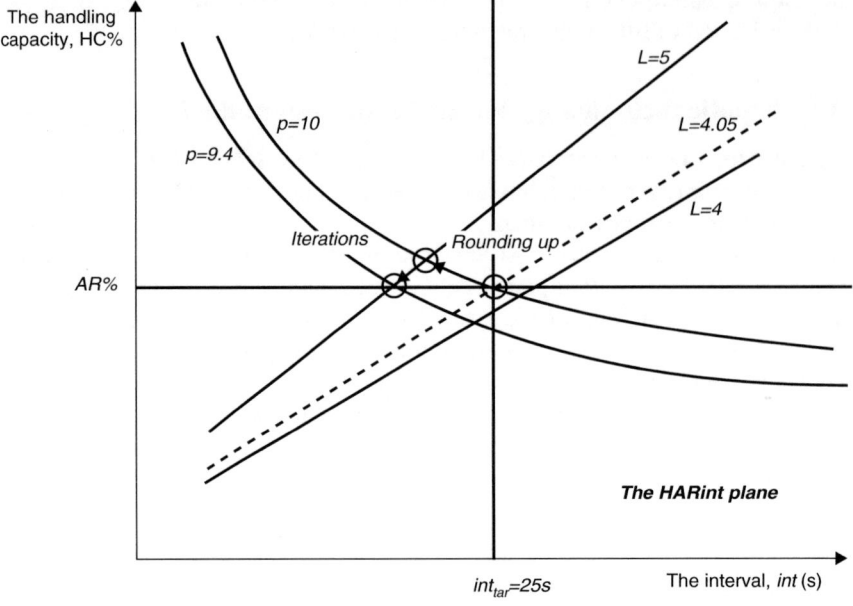

Figure 6.4 The two stages of the calculation are shown by the two arrows

number of lifts (*L*) and curves of integer number of passengers (*P*) boarding the lift in one round trip. The optimum solution is indicated by a large circle.

A specific building, such as that in Example 6.7, has a specified set of parameters. Figure 6.4 shows the iterations performed for Example 6.7. Note that the number of lifts required is five and that the interval is smaller than the user requirement, and thus acceptable.

6.14   Concluding remarks

In this chapter, a number of definitions have been proposed. It is important that each design be prepared against a set of mutually understood definition of terms. Unfortunately, most designers and lift companies have their own set. A set of over 50 terms widely used in the USA are those produced by the NEII (2000). Another set of nearly 2,000 terms and cross references can be found in Barney *et al.* (2009).

An improved design procedure, the iterative balance method, and its visualisation by the *HARint* plane is presented.

Table 6.12 Traffic design definitions (*Note*: all times are in seconds)

Passenger average waiting time *(AWT): period of time from when a passenger either registers a landing call, or joins a queue, until the responding lift begins to open its doors at the boarding floor.*

Note 1: the passenger waiting time continues if a passenger does not enter the responding lift, e.g., it is full (a refusal).
Note 2: the passenger waiting time is zero if the responding lift doors are open when the passenger arrives.

Passenger average walking time*: period of time from when a passenger registers a destination call on a destination input device, until they stand stationary in front of the allocated lift.*

Note 1: the passenger walking time is zero where the destination input device is located in a similar position to the conventional call buttons.
Note 2: the sum of the passenger walking time and the passenger standing time can be considered to be equivalent to the passenger waiting time for systems using conventional call registration devices.

Passenger transit time*: period of time from when a responding lift begins to open its doors at the boarding floor until the doors begin to open again at the destination floor.*

Note: the passenger transit time commences if the responding lift doors are open when the passenger arrives.

Passenger average standing time*: period of time from when a passenger is stationary in front of their allocated lift until the allocated lift begins to open its doors*

Passenger average travel time to destination *(ATTD): period of time from when a passenger either registers a landing call or joins a queue until the responding lift begins to open its doors at the destination floor.*

Passenger average journey time *(AJT): period of time from when a passenger either registers a landing call or joins a queue until the passenger alights at the destination floor.*

Note: the passenger is deemed to have alighted when any passenger detection device is interrupted or the passenger physically crosses the door sills.

Plus:
Definition 6.8: passenger average travel time *(ATT): period of time that a passenger spends travelling in a lift, measured from the time the passenger boards the car until the instant that the passenger alights at the destination floor.*

Source: Barney et al., *Elevator & Escalator Micropedia*, 2009.

Notes (shown in text as[1])

1 3rd edition, page 64.
2 Note: other terms in this publication differ from those used outside the USA.

7 Uppeak traffic calculations

Limitations and assumptions

7.1 Some assumptions in the derivation of the round trip equation

The standard traffic design uses the 'pure' uppeak calculation, i.e., with only an up flow of passengers, to determine the likely performance of a lift system. The calculation of a value for the round trip time using Equation (4.8) relies on a number of assumptions. These are that:

1 the traffic profile is ideal
2 all floors are equally populated or present equal attraction
3 rated speed is achieved for a single floor jump
4 interfloor heights are assumed constant
5 the traffic supervisory system is assumed ideal
6 various lost times, such as passenger disturbance, dispatch intervals, loading intervals, etc. are negligible
7 passengers arrive uniformly in time (rectangular probability function).

It is proposed to deal now with each of these assumptions and their effects on the value obtained for the round trip time (RTT).

7.2 The traffic profile is ideal

Figure 4.4 in Chapter 4 indicated that the 5-minute peak traffic profile would be considered as a rectangle of height equal to the peak arrival rate and of width equal to 5 minutes. That is, it had a square top and nothing happened before or after the peak 5 minutes. During the peak 5 minutes, the lifts are assumed to be evenly distributed about the building, will pick up equal numbers of passengers and make an equal number of stops on each round trip.

In practice, the shape of the arrival profile is as shown by the curved thick line of Figure 4.4. So, there is not at any time a constant arrival rate of passengers. In addition, the lifts will have been carrying out service around the building at the start of the peak 5 minutes and can be in any spatial positions.

Undoubtedly, the past history of the lift positions and the varying arrival rate will make some effect on the calculation.

7.3 Unequal floor populations or demand

In the derivations of the highest reversal floor (H) and the expected number of stops (S), it was assumed all floors held equal attraction for calls originating at the main terminal. In real buildings, this is not always so. Floor populations can vary and result in unequal demand. Alternatively, floor populations may be substantially equal, but work routines may cause unequal rates of population arrival during the peak period. The effect of unequal floor demands is to modify the derivations of H and S.

7.3.1 Number of stops (S) for unequal demand

Consider a building with:

- N floors above the main terminal
- P the average number of passengers present in a lift as it leaves the main terminal
- U the total building population above the main terminal
- U_i the population of floor i.

Using a similar approach to that of Section 5.7, the probability that one passenger will leave the lift at any particular floor i is:

$$\frac{U_i}{U} \tag{7.1}$$

The variable U_i has been represented as the floor population: it could equally represent the passenger demand for floor i.

The probability that one passenger will leave the lift at the first floor is then:

$$\frac{U_1}{U} \tag{7.2}$$

Assuming that the passengers are independent of each other, the probability that one passenger will not leave the lift at the first floor is:

$$1 - \frac{U_1}{U} \tag{7.3}$$

The probability that none of the P passengers in the lift will leave the lift at the first floor is:

$$\left[1 - \frac{U_1}{U}\right]^P \tag{7.4}$$

Thus, the probability that no passengers will leave the lift for the first i floors is: p

$$\left[1 - \frac{U_1}{U}\right]^P + \left[1 - \frac{U_2}{U}\right]^P + \cdots \left[1 - \frac{U_i}{U}\right]^P \tag{7.5}$$

This is synonymous to the lift not stopping at the first i floors.

Then, the probability that stops will be made at the first i floors is:

$$1 - \left[1 - \frac{U_1}{U}\right]^P + \left[1 - \frac{U_2}{U}\right]^P + \cdots \left[1 - \frac{U_i}{U}\right]^P$$

$$= 1 - \sum_{j=1}^{i} \left[1 - \frac{U_i}{U}\right]^P \tag{7.6}$$

Hence, it can be shown (after some algebraic manipulation) that the expected number of stops for N floors is:

$$S = N\left[1 - \frac{1}{N}\sum_{i=1}^{N}\left[1 - \frac{U_i}{U}\right]^{P}\right] \tag{7.7}$$

This is the general case of Equation (5.7) for equal floor demands. Equation (7.6) becomes that of Equation (5.7) when $U_1 = U_2 = U_3 = \ldots = U_i = U/N$.

Hence:

$$S = N\left[N\left[1 - \frac{1}{N}\right]^{P}\right] \tag{7.8}$$

$$= N\left[1 - \left[\frac{N-1}{N}\right]^{P}\right] \tag{5.7}$$

The evaluation by hand of Equation (7.7) for S is obviously tedious, but it is easily calculated by digital computer. Simplifications can be made for cases where there is no population on some floors and equal populations or demand on others, and where groups of floors have the same populations or demand, as for example in 'stepped' buildings.

7.3.2　Examples 7.1 and 7.2

To illustrate some common population distributions/demands, consider the following two examples.

(a) Example 7.1

What is a suitable expression, where there is no population on some floors, or the lift is not permitted to stop at some floors and there is equal population on other floors?

Consider that the lift is not to stop at floor 1 and there is no population on floor 3.

This is best dealt with using Equation (5.7) rather than the general Equation (7.7). Quite simply, the number of served floors is $N - 2$. Hence:

$$S = (N-2)\left[1 - \left[\frac{N-3}{N-2}\right]^{P}\right] \tag{7.9}$$

Thus, it is possible to suggest that instead of thinking of N as the number of floors above the main terminal, it can be considered as the number of floors served above the main terminal, in which case, equations like Equation (7.8) are not appropriate and Equation (5.7) or Equation (7.7) should always be used instead. Thus, to determine the number of served floors N, simply deduct the number of floors not being served from the number of possible floors in a building zone.

(b) Example 7.2

What happens where there are groups of floors with similar population densities or demands?

The lower part of a building comprising B floors has a population of U_B per floor, and the upper part of the building of T floors has a population of U_T per floor. Formulate an expression to cover this case.

The number of floors served *is* N = B + T. It is useful to use Equation (7.8).

$$S = N - \left[B\left[1 - \frac{U_B}{U}\right]^P + T\left[1 - \frac{U_T}{U}\right]^P \right] \tag{7.10}$$

This shows the reduction of N by the deduction of the two probabilities of no stops at floors in the two parts of the building to obtain a value for S.

7.3.3   Highest reversal floor (H) for unequal demand

Using the previous definitions in Section 7.3.1, the procedure is similar to the approach in Section 5.8. The probability that one passenger will leave the lift at a given floor i is:

$$\frac{U_i}{U} \tag{7.11}$$

which becomes the probability of the lift travelling no higher than the ith floor obtained by extension and subsequent algebraic simplification:

$$\left[\sum_{i=1}^{j} \left[\frac{U_i}{U}\right]^P \right] \tag{7.12}$$

Using the same procedure as in Section 5.8, the expectance[1] H is:

$$H = \sum_{j=1}^{N} j \left[\left[\sum_{i=1}^{j} \frac{U_i}{U}\right]^P - \left[\sum_{i=1}^{j} \frac{U_{i-1}}{U}\right]^P \right] \tag{7.13}$$

which becomes:

$$H = N - \sum_{j=1}^{N-1} \left[\sum_{i=1}^{j} \frac{U_i}{U}\right]^P \tag{7.14}$$

Evaluation of Equation (7.14) by manual methods is again somewhat tedious, which may account for some designers guessing H to be equal to N or $N-1$. The error caused by guessing can be large, e.g., for a 6-person lift serving 24 floors H is 20.3 (15% error) giving an error in round trip times of perhaps 5%–10%. Figure 7.1 illustrates the error (vertical axis) for a different number of floors (horizontal axis) and different lift sizes.

Little simplification of Equation (7.14) is possible as the summation is a series of different terms. Only when floors have zero population will terms repeat, giving some simplification in the calculation.

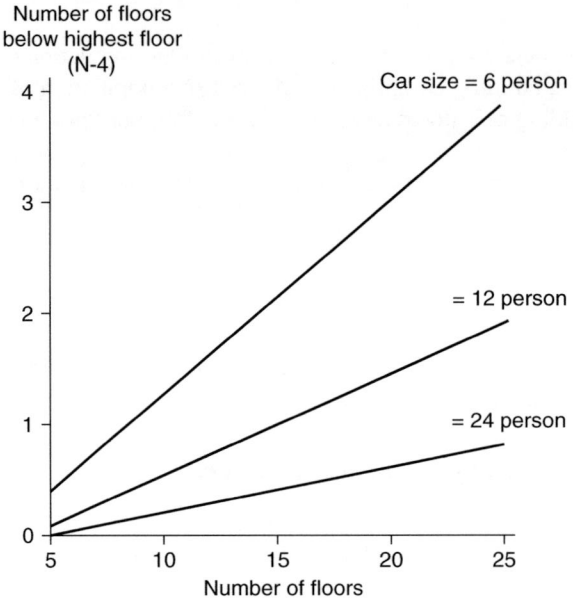

Figure 7.1 Error in the value of *H* when guessed as equal to *N*

7.3.4   Examples 7.3 and 7.4

Two examples are provided to illustrate the calculation of both *H* and *S*.

(a) Example 7.3

Consider a building has ten floors above the main terminal, serviced by lifts carrying an average passenger load of eight persons, and with the following distribution of population:

Floor	1	2	3	4	5	6	7	8	9	10
Population	5	10	25	25	50	50	100	100	100	100

Using a slightly modified Equation (7.7) for *S*:

$$S = N - \sum_{i=1}^{N} \left[1 - \frac{U_i}{U} \right]^{P}$$

$$= 10 - \left[\left[1 - \frac{5}{565} \right]^8 + \left[1 - \frac{10}{565} \right]^8 + 2\left[1 - \frac{25}{565} \right]^8 + 2\left[1 - \frac{50}{565} \right]^8 + 4\left[1 - \frac{100}{565} \right]^8 \right]$$

$$= 10 - \frac{1}{565^8} \left[560^8 + 555^8 + 2 \times 540^8 + 2 \times 515^8 + 4 \times 454^8 \right]$$

$$= 10 - 4.986 = 5.014$$

Using Equation (7.14) for H:

$$H = N - \sum_{j=1}^{N-1}\left[\sum_{i=1}^{j} \frac{U_i}{U}\right]^P$$

$$= 10 - \frac{1}{565^8}\left[5^8 + 15^8 + 40^8 + 65^8 + 115^8 + 265^8 + 365^8 + 465^8\right]$$

$$= 10 - 0.245 = 9.755$$

For comparison, a building with ten floors, serviced by lifts with an average occupancy of eight persons, serving floors with equal demand from Table 5.11a would give:

$$S = 5.7; H = 9.3$$

With the unequal populations in this example: S is smaller at 5.0, but H is larger at 9.8.

(b) Example 7.4

Consider a building has ten floors above the main terminal, with the distribution of the populations reversed to those given in Example (7.3):

Floor	1	2	3	4	5	6	7	8	9	10
Population	100	100	100	100	50	50	25	25	10	5

Calculate values for H and S, using Equation (7.7) for S:

$$= 10 - \left[4\left[1 - \frac{100}{565}\right]^8 + 2\left[1 - \frac{50}{565}\right]^8 + 2\left[1 - \frac{25}{565}\right]^8 + \left[1 - \frac{10}{565}\right]^8 + \left[1 - \frac{5}{565}\right]^8\right]$$

This is the same expression as that obtained in Example 7.3, but order of the terms is reversed. Hence, $S = 5.014$

Using Equation (7.14) for H:

$$H = 10 - \frac{1}{565^8}\left[100^8 + 200^8 + 300^8 + 400^8 + 450^8 + 500^8 + 525^8 + 550^8 + 560^8\right]$$

$$H = 10 - 2.99 = 7.01$$

Comparing again with a building with equal floor populations where $S = 5.7$ and $H = 9.3$, with the unequal populations in this example, both S and H are smaller.

(c) Comments on Examples 7.3 and 7.4

H and S can be seen to be dependent on the relative floor demands in a building. It can also be seen that S is less dependent of the distribution of demand, but that H is more sensitive to it. In addition, the unequal populations will make the *RTT* smaller as the dominant term containing S is always smaller.

7.4   Interfloor flight times and unequal floor heights

7.4.1   *Interfloor flight times*

In fast lift systems, it will take a travel (flight) of several floors for the rated speed to be reached. Hunt (1975) presents an example of speed/time curves for 5.0 m/s rated speed, 1.5 m/s^2 acceleration, where rated speed is only reached during a flight of seven floors or further. The values below illustrate this point:

Floors jumped	1	2	3	4	5	6	7	8
	$t_f(1)$	$t_f(2)$	$t_f(3)$	$t_f(4)$	$t_f(5)$	$t_f(6)$	$t_f(7)$	$t_f(8)$
Flight time (s)	3.6	4.8	5.7	6.5	7.1	7.7	8.3	8.9
Incremental time (s)		1.2	0.9	0.8	0.6	0.6	0.6	0.6

The times above show that only after a flight of between 3 and 4 floors do the incremental times[2] increase by t_v. This implies that, in this case, the acceleration or deceleration process takes some 3.5 floors, invalidating the assumption that jumps of greater than one floor are travelled at rated speed.

The flight times to travel various distances can be measured on an actual system, when calculating installed performance. To obtain estimates of flight times for proposed systems, either a manufacturer's time distance graph should be used, or a calculation can be performed.

The calculation is complex as, not only does speed have to be considered, but values of acceleration/deceleration and values of the rate of change of acceleration/deceleration (jerk) also have to be known. A calculation method for lift dynamics is given in the Appendix.

It is also possible for interfloor flight times to vary according to the load being carried and the direction of travel. This occurs particularly in low cost systems, such as those installed in residential accommodation. Some allowance may need to be made for this variation.

Thus, a problem exists as to the numeric value to attribute to t_s in Equation (4.8), where it was assumed that the rated speed was attained during a single floor jump. This difficulty is exacerbated as it is not known how many single, double, triple, etc. floor jumps are made during each round trip. Phillips (1973) evaluates acceleration and deceleration periods and running time at rated speed as a function of the distance to accelerate from rest to rated speed, and presents typical time/distance curves, but does not indicate how to cater for rated speed not being attained. Jones (1971) assumes rated speed is always reached and Petigny (1972) in his detailed work does not refer to the problem.

Strakosch (1967) takes fixed standard values, chosen according to the rated speed. Gaver and Powell (1971), in order to linearize their expressions for the round trip time, use an approximated solution. They assume that the lift does an express flight to the lowest stopping floor (lowest floor of zone being served by the group of lifts), does all but one of the expected stops, and finally does a direct flight to the highest stopping floor.

Tregenza (1972) modifies Equation (4.8) for *RTT* when the distance required for acceleration and retardation are each greater than half the interfloor distance. He decomposes the term $(S+1)t_s$, considering that from the main terminal, a fraction p_s of all flights produce jumps of a single floor, and the remaining $(1-p_s)$ produce jumps that attain the rated speed. Figure 7.2 illustrates the assumption of rated speed being reached after a jump of one floor:

(a) interfloor flight times, assuming rated speed is reached in a one floor jump
(b) interfloor flight times, no assumptions

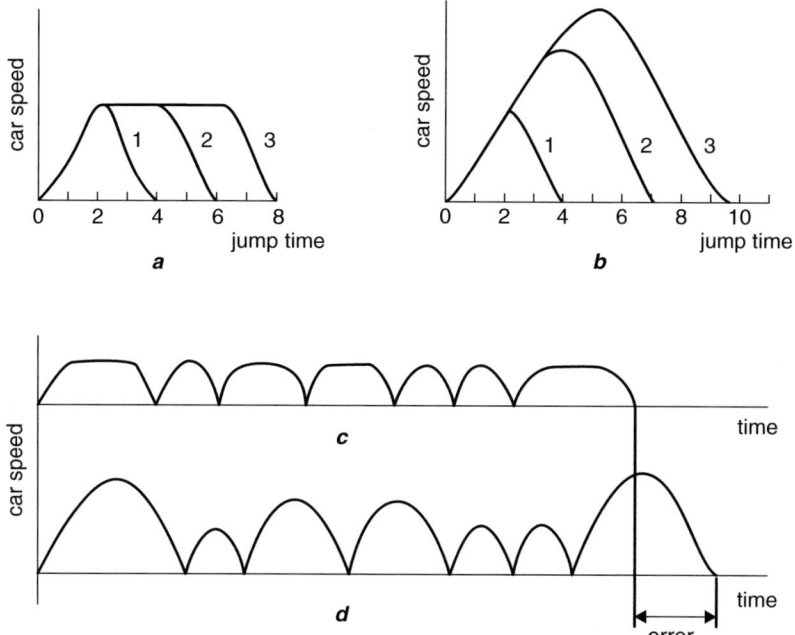

Figure 7.2 Nonlinear flight times for a 13-floor building with 8 stops and jumps of 3×1, 2×2 and 2×3

(c) illustration of flight pattern with eight stops using interfloor flight times as (a)
(d) illustration of flight pattern with eight stops using interfloor flight times as (b).

Note time error between (c) and (d).

It is possible (Roschier and Kaakinen, 1980) by using probability theory, and knowing the population or demand on each floor, to determine the relative numbers of 1, 2, 3, etc. floor jumps. However, the expression is complex.

Another simplistic approach is to determine the most likely number of floors jumped by evaluating the expression *HIS*, and then to approximate to the nearest integer below (desig-nated *LJ*). Some further knowledge of the building, such as any heavily populated or service floors, should enable a designer to estimate whether there will be many jumps in excess of *LJ*. Of course, if the rated speed has been reached for jumps equal to or less than *LJ*, all other floors will be travelled at rated speed. If *LJ* is greater than two, some estimate of the number of these jumps must be made.

7.4.2 Unequal interfloor distances

All the floors in a building may not be of equal height. For example, the entrance floor is often higher to give a spacious aspect. Other floors may be service floors or may accommodate con-ference halls, again causing the interfloor distances to vary. The main effect of unequal floor heights is to increase the travelling time and to further alter interfloor flight times.

It is unlikely that floor heights will vary in a gross fashion, and so probably the easiest method to account for the additional times involved is to adopt the following procedure. Add together the incremental floor distances, extra to the standard interfloor height, multiply by two to account for both directions of travel, and divide by the rated speed to obtain the additional time. The special case of an express jump can also be dealt with in this way.

As an example, consider a building where the interfloor distance for most floors is 3.0 m, but four floors are 5.0 m. Then, the extra distance will be 16.0 m. If the rated speed were to be 4.0 m/s, then the additional time would be 4.0 s.

7.4.3   *Taking account of both interfloor flight time variations and unequal interfloor distances*

A more accurate way to determine the effect of both interfloor flight time variations and uneven interfloor distances is to calculate the average jump for an average floor height. This can be achieved by taking the actual distance to travel to the high call reversal floor H, and dividing it by the probable number of stops S. The procedure could be:

1   Determine H/S to give average interfloor jump.
2   Determine height of building (d_H) to floor H and divide by H to give average interfloor height.
3   Calculate (1) × (2) to give average distance travelled.
4   Look up on a manufacturer's time/distance graph[3] the time to travel the distance found in (3), or calculate it with a dynamics program (see the Appendix).
5   Calculate the assumed time to travel the distance calculated in (3).
6   Calculate the difference between the time obtained in (4) and (5).
7 * Add the time obtained to t_s, when calculating the round trip time.

Peters (1997b) carried out many round trip time calculations (*RTT*) using the flight time to travel the average interfloor distance and compared the results to using the flight time for a standard interfloor height. He found an error of only a few per cent. This was not an unexpected result.

7.4.4   *Example 7.5: estimating additional times*

Use the method in Section 7.4.3 to estimate the additional time to be added to the *RTT* for a lift system, where floors 4, 8, 12, 16 and 20 are 5.0 m high. The other data are:
   $H = 22.2$, $S = 8.0$, $d_f = 3.0$ m, $v = 5.0$ m/s,
   $t_f(1) = 3.6$ s, $t_f(2) = 4.7$ s, $t_f(3) = 5.6$ s, $t_f(4) = 6.4$ s.

1   Average jump: $22.2/8.0 = 2.77$ floors.
2   Average interfloor height: $(22.2 \times 3.0 + 5 \times 2.0)/22.2 = 3.45$ m.
3   Average distance: $3.45 \times 2.77 = 9.6$ m.
4   Travel time: for 3.2 floors (9.6/3.0).

By interpolation from the time values for $t_f(3)$ and $t_f(4)$, the time to jump 9.6 m would be about 5.8 s.

5   Assumed time: for $t_f(1)$ was 3.6 s and for $t_f(3.2)$ will be $3.6 + 2.2 \times 0.6 = 4.9$ s. (0.6 s is from d_f/v, i.e., 3.0/5.0; and 2.2 is from 3.2–1).
6   Difference time: $5.8 – 4.9 = 0.9$ s.
7   Additional time: add 0.9 s to t_s.

7.5   Effect of traffic supervisory system

The conventional calculation procedure based on the *RTT* expression requires that lifts present themselves at the main terminal, evenly spaced by a period of time equal to one interval. Two factors can upset this situation.

First, the uppeak 5 minutes is part of a continuous process, as shown in Figure 4.12. Hence, lifts, in practice, will already be transporting passengers at the onset of the uppeak. As a result, lifts will tend to be randomly dispersed around the building. A severe condition occurs when the distribution of lifts becomes so disturbed that lifts bunch together and move round the building together. The effect of bunching is to reduce the quality of service by making most passengers wait longer for lifts.

Second, the traffic supervisory control system, during most of the day, is arranged to deal with interfloor traffic. The peak periods, uppeak, down peak and lunchtime two-way traffic are generally supervised by special algorithms, which must be switched on when required. The changeover from interfloor to uppeak or down peak control is achieved by monitoring (say) car load, and when this exceeds a predetermined value, the appropriate control algorithm is selected. Thus, the uppeak algorithm must be active just before the peak 5 minutes starts, if it is to be effective.

It is assumed in the round trip calculation that during uppeak, all lifts are express to the main terminal after depositing the last passenger at the highest floor. This is all the uppeak traffic supervisory algorithm can do for uppeak service. However, the time when the uppeak control is switched on is important. If it is too late, only those cars with completed trips will be travelling to the main terminal. Those lifts just starting, or part way through, a round trip can only become available at the main terminal some two or three minutes later, half way through the uppeak period. Some may still be dealing with interfloor or down direction traffic.

It is essential, therefore, to detect the uppeak condition well before it 'takes off' to ensure the full lift system handling capacity is available at the lobby. Some authors (Strakosch, 1967; Tregenza, 1972; Phillips, 1973; Morley, 1962) suggest the *RTT* should be increased to account for running out of schedule by (say) 10%. This so-called 'inefficiency factor' should not form a part of the design calculation, but be considered after a design has been completed. Some uppeak control systems allow down direction and other traffic to be catered for on a timed basis (at least a 3-minute wait). If this feature is incorporated, then the uppeak handling capacity is reduced for the time the delegated lift is not serving uppeak.

Various techniques are employed to improve uppeak performance, such as uppeak sub zoning, sectoring and hall call allocation. These techniques can still be analysed by calculation and will be discussed in Chapter 13.

7.6   Various lost times

The first two terms of the round trip time, Equation (4.8), have been calculated as idealised values. This is not unreasonable as they are basically dependent on electromechanical machinery. The third term relating to passenger transfer times is far more speculative as it is dependent on human behaviour, which is unpredictable at the best of times.

The round trip time can be considerably increased by passengers unthinkingly or maliciously preventing lift cars from continuing on their journey. For example: a passenger may hold a door while finishing a conversation; a tea trolley may be loaded, thereby reducing the car capacity; or, maliciously, a person may enter additional car calls. The effect of all these disruptions is an increase in *RTT*.

It is difficult to quantify these disruptions. Some account might be taken by following Phillips (1973) and Strakosch (1967) and adding 10% to the *RTT*, or it might be better to increase the passenger transfer times as the effect is due to passenger behaviour and is probably proportional to their numbers. It is always worth bearing in mind possible disruption when selecting a lift system. It could be sensible to select one with capacity in hand, in order to counteract this unpredictable effect.

There are four control features that can affect the design calculation. First, some lift control systems cause a lift to remain at the main terminal for a fixed time interval (dispatch interval). Second, others hold a lift at the main terminal for a fixed time from the registration of the first car call (loading interval). These two control features should not affect the running of the lifts provided that:

> dispatch interval < time to load 60% of the probable car capacity
> loading interval < time to load 60% of the probable car capacity

Third, other control systems hold a lift at the main terminal floor with its doors closed until the previous lift is full, or is ready to leave (next car). This feature simply delays the departure of a lift from the main terminal floor.

Finally, some lift systems cause the lift doors to be held open for a fixed time at each stop (door holding interval, or door dwell time) before attempting to reclose the doors. This is provided to allow the movement of passengers into and out of the lift without collision with the doors. This time is often extended to accommodate the movement of pushchairs and bicycles, and for elderly persons and disabled persons to enter safely. Typically, the door dwell time will be longer for a stop in response to a landing call than for a car call, on the assumption that passengers may alight as well as board the lift. In addition, the door dwell time in systems without dispatch or loading interval features may be longer at the main terminal than at other floors. If door dwell times are longer than the time to offload the average number of passengers per stop, than the passenger transfer time t_p should be set to zero.

7.7   Passengers arrive uniformly in time

7.7.1   Passenger arrival process

A number of assumptions in the derivations of the round trip time expression have been discussed in previous sections. These assumptions were concerned with the nonlinear effects of the building and lift system, and also the static population distributions. No account has been taken of the way in which passengers arrive in a building or the randomness of their destinations. In practice, passengers do not conveniently arrive at a building so that 80% of the probable car capacity of a lift is available for them to step into a lift and be transported to their destinations. In addition, passengers do not request the same number of destinations on each trip. The effect of random arrival rates and destinations is to cause cars to become unequally spaced and, in extreme circumstances, to bunch. If large numbers of passengers arrive, queues can also develop. Throughout this chapter, values for *H* and *S* have been described as average values. It is by using these values that the average *RTT* is calculated. Obviously, if the *RTT* has wide deviations from the average, a poor quality service can result. How then, does the *RTT* vary?

7.7.2   Effect of randomness of passenger destinations

Returning to statistical terminology,[4] in Section 5.7, it has been indicated that the probability of a variable is defined by a numeric value between zero and unity. Probability then, is simply a measure of the likelihood of some occurrence. For example, the probability of stopping at a floor may be 0.1. However, the actual number of stops is called the expected number of stops. The expectation is the mean or average value of some random variable, often termed the expectance.

It is often of interest to know the spread of possible outcomes about the mean. This variation is essentially measured by the statistical characteristic termed variance. If the square root of a value of variance is taken, the value then obtained is termed standard deviation. If large values of variance occur with respect to values obtained for expectance of some variable, then that variable has wide deviations of value from its mean.

A mathematical study was made by Gaver and Powell (1971). They considered an uppeak situation, with no interfloor traffic, and assumed that all lifts depart from the main terminal with a constant number of passengers. No analysis was made with car load variations. The expression they deduced allows for lifts to be expressed from the main terminal to a certain floor and then to service that floor and the floors above. They also made some assumptions to try to overcome the problem that the rated speed may not be reached for a flight of one floor. Both cases of equal and unequal floor population are considered. The expression is a linear function of the number of stops and the reversal floor, so the evaluation of the variance of the round trip time was reduced to the calculation of the variance of the two variables (number of stops and reversal floor). The final expression is complex and is best illustrated by examples.

As an example, consider Table 7.1 for a group of lifts serving express zone floors 15 to 26 (12 floor zone) in a building, the lift rated speed being 4.0 m/s.

The variance decreases when the number of passengers increases since there is less randomness in the number of stops. However, two points must be made: the number of floors served is modest and smaller than the number of passengers in the lift; and the group serves an express zone, which means that a significant part of the round trip time, corresponding to the express flights in both directions, is not subjected to variance of round trip time. If a group of lifts with the same characteristics, but serving the 12 floors next to the main terminal, is considered, the variance will be the same but the round trip time will be decreased by the travelling time of two flights of 15 floors. This, for an average interfloor distance and a rated speed of 4.0 m/s, is about 25 s. For such a group of lifts next to the main terminal, the variance becomes more significant.

The values above assume equal floor populations; for unequal floor populations the variance increases. An example given by Gaver and Powell with slightly uneven populations gives an *RTT* expectance of 148 s and a variance of 74 s^2 for 15 passengers in a lift (compare with Table 7.1).

Thus, significant variance of *RTT* does exist for equal and slightly unequal floor populations. Gaver and Powell conclude 'if the distribution varies considerably from uniform we must formulate a new representation of round trip time'. How these variations affect a lift system is not quantified. In the long term, little effect on the carrying capacity will be noticed, but in the short term, queues can build-up, reducing the quality of service. It would appear, therefore,

Table 7.1 Statistical data

Passengers in lift	15	17	19	21
Expectance of *RTT*(s)	149	157	164	170
Variance of *RTT*(s^2)	68	66	62	57
Standard deviation(s)	8.25	8.15	7.89	7.55

that as far as conventional calculations are concerned, randomness of destination need not be considered.

7.7.3   Effect of variations in arrival rate

Passengers do not arrive uniformly in time. It is generally accepted that the arrival of people at a building obeys a Poisson process, i.e., the probability of n calls being registered in the time interval T for an average rate of arrivals λ is:

$$p_r(n) = \frac{(\lambda T)^n}{n!} e^{-\lambda T} \tag{7.15}$$

The assumption of passenger arrival by the Poisson process is not completely proven, and its validation would require extensive data logging on a wide range of lift systems (see Section 5.13).

Petigny (1972) and Tregenza (1972), following different methods of analysis but both considering a Poisson arrival distribution, have related the system behaviour to the arrival of passengers. Petigny's study is purely mathematical and considers a single lift during an uppeak situation. By using probability theory, he deduces Equations (7.7) and (7.14) for S and H for the case of defined floor populations. In addition, by setting up a complex mathematical study involving Markov chain theory, he deduces an expression for the average number of passengers per trip. Although an interesting study, Petigny's work relates to a single lift only.

Tregenza uses a more practical approach and produces an expression similar to Equation (4.8) for RTT, which he writes as:

$$RTT = E_h t_1 + E_s t_s + E_p t_3 \tag{7.16}$$

where $t_1 = 2t_v$ and $t_3 = 2t_p$ and E symbolises expectance.

Rewriting Equation (7.16) in the style of Equation (4.8):

$$RTT = 2H_p t_v + (S_p + 1)t_s + 2P_p t_p \tag{7.17}$$

[The subscript p indicates Poisson pdf.]

Instead of taking P_T as the car capacity or a fraction of it, as in conventional design, he relates the parameters H_T, S_T and P_T as universal functions of an arrival rate parameter p_{ro}. He defines p_{ro} as the probability of no calls being registered from the main terminal floor to any floor above during the period of one interval. For equal floor populations, he produces two equations. These have already been derived as Equations (5.17) and (5.18).

Section 5.13 showed that if the first two terms of the expansion of the equations for H_p and S_p were taken, then H, using a rectangular probability function and Equation (5.14), was identical to H_p, and likewise for S using Equation (5.7) and S_p.

Figures 7.3 and 7.4 show the equations for H and S plotted for both the rectangular and Poisson pdfs. These graphs confirm that if the rectangular pdf is used to derive the H and S equations, their values will always be larger than if the Poisson pdf is used. Thus, the resulting value for the RTT will always be larger, and a design based on it will be slightly conservative. As the equations using the rectangular pdf are simpler to calculate, they are generally used in design calculations.

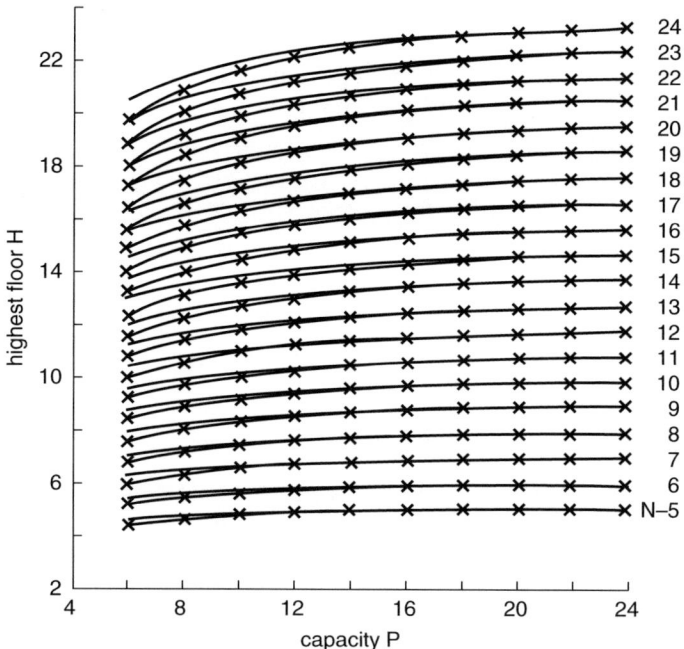

Figure 7.3 Mean highest reversal floor *H* for different numbers of floors and car capacities

solid line: conventional calculations, rectangular probability distribution
line with x: calculation using Poisson probability distribution

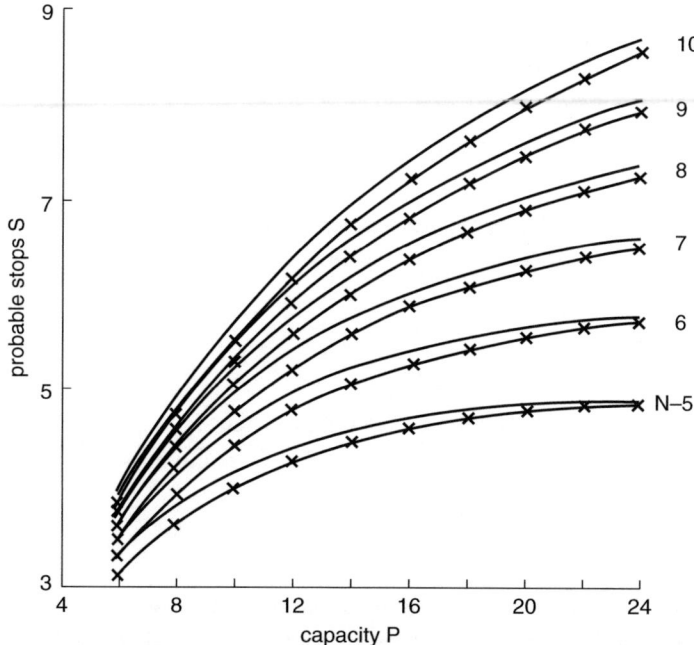

Figure 7.4 Mean number of stops *S* for different numbers of floors and car capacities

solid line: conventional calculations, rectangular probability distribution
line with x: calculation using Poisson probability distribution

The above analysis assumes equal floor populations/demand. Using statistical techniques, expressions can be obtained (Alexandris, 1976) for unequal floor populations. They are:

$$S = N - \sum_{i=1}^{N} e^{-\lambda INT \frac{U_i}{U}}$$

(7.18)

$$H = N - \sum_{i=1}^{N} \prod_{i=N-j+1}^{N} e^{-\lambda INT \frac{U_i}{U}}$$

(7.19)

(The mathematical symbol Π means 'multiply all terms like. . .'.)

7.7.4   Effect of passengers arriving in batches

A special case of passenger arrivals is where the passenger decisions are not independent.

Traditional round trip time calculations assume that passengers arrive in batches of one passenger only and that the decision each passenger makes regarding their destination floor is independent of other passengers. Recent research has shown that passengers, in reality, arrive in batches that have a common destination as shown in an office [1] and on a cruise ship [2]. The average batch size (β) will depend, to a large extent, on the nature of the building, as it is shown that the average batch size in an office is around 1.5 (Kuusinen *et al.*, 2012) and around 1.68 on a cruise ship (Sorsa *et al*, 2012). In a similar way, it would be expected that batch sizes will be higher in shopping centres compared to offices, for similar reasons.

It is important to emphasise that the fact that passengers have arrived in a batch size larger than one is not sufficient, *per se*, to affect the value of the round trip. It is the fact that they arrive in a batch, and that they have a common destination, which is the critical factor that affects the value of the round trip time. This leads to an effective number of passengers smaller than *P* as calculated below:

$$P_\beta = \frac{P}{\beta}$$

(7.20)

Substituting the effective number of passengers in the equation for *S* provides the new value of *S* under batch arrivals, denoted as S_β:

$$S_\beta = N \left[1 - \left(1 - \frac{1}{N} \right)^{\frac{P}{\beta}} \right]$$

(7.21)

The same applies to the equation for calculating the new value of the highest reversal floor denoted as H_β. Substituting the effective value of *P* in the original equation for *H* gives:

$$H_\beta = N - \sum_{i=1}^{N-1} \left(\frac{i}{N} \right)^{\frac{P}{\beta}}$$

(7.22)

However, the new effective value of *P* does not apply to the passenger transfer times as time is still required for the *P* passengers to transfer, regardless of whether they are travelling in a batch or not. So, the amended round trip time (Equation 4.8*bis*) becomes Equation 7.23.

$$RTT = 2 \cdot H_\beta \cdot t_v + \left(S_\beta + 1 \right) \cdot \left(T - t_v \right) + 2Pt_p$$

(7.23)

Another secondary effect noted in [2] is that the batch arrivals have an effect on the car loading. Passengers arriving in batches prefer to stay together. So, they would let a lift car that has too few spaces depart and wait for the next car (thus reducing the effective car loading); they would also try to cram themselves in a nearly full car in order to stay together (thus increasing the effective car loading).

7.8 Comparison of non-linearities and their total effect on the value of the RTT

7.8.1 Categories

In Section 7.1, seven assumptions (1–7) were indicated and now can be categorised as follows:

(a) There are three assumptions, which can be calculated:
   Unequal floor populations (2).
   Nonlinear interfloor flight time (3).
   Unequal interfloor heights (4).
(b) There are three assumptions, which are not determinate:
   The traffic profile is ideal (1).
   The traffic supervisory system is assumed ideal (5).
   Various lost times, such as passenger disturbance, dispatch intervals, loading intervals etc. are negligible (6).
(c) There is one assumption, which does not matter, or has little effect:

   Passengers arrive uniformly in time (7).
   It is now worth considering the magnitude of their effects by looking at an example.
Consider the building used in Example 7.3. It has ten floors above the main terminal, and the floor populations are:

Floor	1	2	3	4	5	6	7	8	9	10
Population	5	10	25	25	50	50	100	100	100	100

The rated speed is 1.5 m/s and the interfloor height 3.3 m. Flight times are:

Jump	1	2	3	4 ...	10
Time(s)	5.1	7.5	9.7	at rated speed	

The height of the main terminal, fourth and tenth floors is 5.0 m. Door closing time is 3.1 s and door opening is 2.0 s. Assume passenger transfer time is 1.2 s per passenger. Calculate the round trip time (*RTT*) and any corrections.

(A) USUAL PROCEDURE

Values for H and S can be obtained from Table 5.1: $S = 5.7$ and $H = 9.3$.

$$t_v = 3.3/1.5 = 2.2 \text{ s}$$

$t_s = (5.1\text{--}2.2) + 3.1 + 2.0 = 8.0 \text{ s}$
$t_p = 1.2 \text{ s}$

Then using Equation (4.8):

$$RTT = 2Htv + (S+1)t_s + 2Pt_p$$
$$= 2 \times 9.3 \times 2.2 + 6.7 \times 8.0 + 2 \times 8 \times 1.2$$
$$= 40.9 + 53.6 + 19.2$$
$$= 113.7 \text{ s}$$

(B)   EFFECT OF UNEQUAL POPULATIONS

From Example 7.3: $S = 5.1$, $H = 9.8$. Then, using Equation (4.8):

$$RTT = 2 \times 9.8 \times 2.2 + 6.1 \times 8.0 + 8 \times 2.4$$
$$= 43.1 + 48.8 + 19.2$$
$$= 111.1 \text{ s}$$

This is 2.6 s smaller, or 2.3% smaller than the value obtained in (a).

(C)   EFFECT OF NONLINEAR FLIGHT TIMES AND UNEQUAL FLOOR HEIGHTS

Using the procedure in Section 7.4.3, and the values for S and H using Equation (4.8) gives:

1   Average jump: $9.8/5.1 = 1.9$ m.
2   Average interfloor distance: $(9.8 \times 3.3 + 2 \times (5 - 3.3))/9.8 = 3.7$ m.
     Floor 10 is not considered as it is above H.
3   Average distance: $3.7 \times 1.9 = 7.0$ m.
4   Travel time: for 2.1 floors ($7.0/3.3$).
     By interpolation between $t_f(2)$ and $t_f(3)$, flight time will be: 7.7 s.
5   Assumed time: for $t_f(1)$ was 5.1 s and for $t_f(2.1)$ will be $5.1 + 1.1 \times 2.2 = 7.5$ s.
     (1.1 s is d_f/v i.e., $3.3/1.5$).
6   Difference time: $7.7\text{--}7.5 = 0.2$ s.
7   Additional time: add 0.2 s to t_s.

The value of *RTT* now becomes:

$$= 111.1 + 6.1 \times 0.2 = 112.3$$

This is 1.4 s smaller or 1.2% smaller than the value obtained in (a).

(D)   DISCUSSION

The corrections do not make much difference in this case. No account has been taken of lost times or the effect of the control system. The above example illustrates the procedure to determine the additional times resulting from non-linearities. The examples do not show very large changes, but each selected system should be considered in this way as it may exhibit significant extra times. In general, however, the effects are likely to be of secondary importance.

It is important to remember that the values of round trip times obtained are average values and they are the sum of three average values for *H*, *S* and *P*. All average values have some variance, or likely range of values. In particular cases, therefore, the *RTT* may deviate greatly from the average. This deviation may be smaller, or greater, than the effects discussed above. However, it is still prudent to determine the magnitude of the non-linearities.

7.9   General Analysis Method

7.9.1   *Mathematical background*

So far, each of the seven assumptions given in Section 7.1 have been analysed individually. Can they be analysed in a single set of equations?

Alexandris *et al* (1979b) considered the problem of analysing traffic patterns other than uppeak. His approach was to model a multi-car system as a bulk service queuing problem. In order to do this he had to make a large number of assumptions. Peters (1990), working from first principles, has been able derive more general formulae. The mathematics is complex and has been reported in Peters (1997a). The General Analysis Method allows round trip time calculations to be performed for any peak traffic flow. This overcomes most of the limitations associated with conventional uppeak calculations. For example, the General Analysis Method will allow assessment of:

1   Office buildings with car parks and basements.
2   Hotel or residential buildings with two-way peak traffic.
3   Shopping centres with heavy interfloor traffic.
4   Offices with restaurants causing heavy peaks at lunchtimes.
5   Double deck lifts.

The General Analysis Method can be applied to most commonly occurring building configurations and passenger traffic demands.

7.9.2   *Basic assumptions*

It is generally accepted that a Poisson process is a reasonable representation of the arrival process of passengers at a lift landing, as Equation 7.15 recalls:

$$p_r\left(n\right) = \frac{\left(\lambda T\right)^n}{n!} e^{-\lambda T} \tag{7.15}$$

Peters (1990) expresses Equation 7.15 as:

$$p\left(n\right)_{i,j} = \frac{\left(\lambda_i INTd_{i,j}\right)^n}{n!} e^{-\left(\lambda_i INTd_{i,j}\right)} \tag{7.24}$$

where:

$p(n)_{i,j}$ is the probability of *n* passengers wanting to travel from floor *i* to floor *j* during the time interval *INT*
INT is the lift system interval.
$d_{i,j}$ is the probability of the destination floor of a call from *i* being the *j*th floor.

In order to determine the probability of an event happening, it is generally easier to calculate the probability of the event *not* happening and subtracting this from unity. This analysis technique can be seen in Chapter 5 (Equations 5.4 and 5.8). So, let:

$$p_{i,j} = p(0)_{i,j} \tag{7.25}$$

which is the probability of no calls from floor i to floor j during the time interval *INT*.

From Equation (7.24):

$$p_{i,j} = e^{-\lambda_i INT d_{i,j}} \tag{7.26}$$

In a pure uppeak, the lifts start and return to the main terminal floor, allowing the probable number of stops (S) and the highest floor (H) to be calculated easily (Equations 5.7 and 5.14). With some traffic patterns, the lifts may never include the main terminal floor in the round trip, i.e., between two reversals of direction. The General Analysis Method allows a probable number of stops (SP) to be calculated between reversals.

By applying Equation (7.26), and after a great deal of mathematics, formulae for the probable number of stops (SP), lowest (LR) and highest (H) reversal floors can be derived.

$$SP = pS_1 + \left(\sum_{j=2}^{N-1} pUS_j + pDS_j \right) + pS_N \tag{7.27}$$

$$H = \sum_{j=1}^{N} jpH_j \tag{7.28}$$

$$LR = (N+1) - \sum_{j=1}^{N} \left[\left((N+1) - j \right) pL_j \right] \tag{7.29}$$

where:

LR	Average lowest reversal floor
pDS_n	Probability of lift stopping at nth floor during down journey
pH_n	Probability of nth floor being the highest reversal floor
pL_n	Probability of nth floor being the lowest reversal floor
pS_l	Probability of lift stopping at lowest floor
pSN	Probability of lift stopping at highest floor
pUS_n	Probability of lift stopping at n^{th} floor during up journey
SP	Average number of stops including main terminal
pi,j	Probability of no calls from i to j during interval INT
λi	Passenger arrival rate at floor i (person/s)

To calculate values from Equations (7.27), (7.28) and (7.29), the floor probabilities need to be known. Often, these can be represented by the populations of each floor. Alternatively, traffic patterns may have been measured. A matrix of demand can then be established such as that shown in Figure 7.5.

Level 14	↑ 0%	↑ 0%	↑ 0%		↑ 17%
Level 13	↑ 17%	↑ 17%	↑ 50%		↑ 5%
Level 12	↑ 0%	↑ 0%	↑ 0%		↑ 5%
Level 11	↑ 17%	↑ 17%	↑ 0%		↑ 3%
Level 10	↑ 0%	↑ 0%	↑ 0%		↑ 8%
Level 9	↑ 17%	↑ 17%	↑ 0%		↑ 7%
Level 8	↑ 0%	↑ 0%	↑ 0%		< **233**
Level 7	↑ 17%	↑ 17%	↑ 0%		↓ 5%
Level 6	↑ 0%	↑ 0%	↑ 0%		↓ 5%
Level 5	↑ 17%	↑ 17%	↑ 0%		↓ 10%
Level 4	↑ 0%	↑ 0%	↑ 0%		↓ 10%
Level 3	↑ 15%	↑ 15%	< **25**		↓ 10%
Level 2	↑ 0%	< **35**	↓ 0%		↓ 0%
Level 1	< **75%**	↓ 0%	↓ 50%		↓ 15%

Figure 7.5 Probability matrix

<*xx* is arrivals (persons/5-minutes) at a floor
↑ is up destination probability (%)
↓ is down destination probability (%)

Figure 7.5 shows a midday peak scenario in an office building where there are restaurant facilities on the top two floors. The peak traffic is a combination of:

1 passengers travelling from their offices to the restaurant for lunch
2 passengers travelling back to their offices after lunch
3 passengers travelling to the ground floor to leave the building
4 passengers returning to the building.

Equations (7.27), (7.28) and (7.29) can then be applied to determine the round trip time (*RTT*) and interval (*INT*).

7.9.3   *The application of the General Analysis Method*

The General Analysis Method can be used to analyse the heavy traffic conditions of uppeak, down peak and midday traffic, but may return anomalous results for low traffic conditions. This is because the underlying statistics are poor, due to the small number of passenger demands. The method provides results comparable to the classical equations given in Chapter 4.

The complex equations for the General Analysis Method can be programmed into a computer. The application of the equations is not presented here, and readers should consult Peters (1990) for full details. Peters *et al.* have further developed the General Analysis Method to deal with double deck installations (Peters *et al*, 1996b)

7.9.4   Application of General Analysis Method to pure uppeak

The definition of *RTT* given in Definition 4.1 refers only to the uppeak traffic condition. The General Analysis Method can be applied to the pure uppeak round trip time calculation, where there are traffic flows to and/or from the main terminal floor.

Example 7.7

Repeat Example 6.7 using the General Analysis Method.

Table 7.2, Column 2 shows the classical results for Example 6.7 and Table 7.2, Column 3 shows the results for the General Analysis Method.

The results in Table 7.2 demonstrate that the interval and capacity factor calculated by the basic *RTT* method and the General Analysis Method are very similar.

The General Analysis Method results for *S* and *H* appear dissimilar. This is because, in an uppeak analysis, the probability of the lift stopping at each of the upper floors is calculated. This provides a value for *S*. In the *RTT* equation, unity is added to *S* to take into account the stop at the main terminal floor. In the General Analysis Method, it is not automatically assumed that the lift stops at the main terminal floor. It calculates the probability of the lift stopping at every floor. Thus, when analysing an uppeak traffic pattern using the General Analysis Method, the value obtained for *S* is greater than from the classical calculation by approximately one; similarly, for the value for *H*. The amended values are shown in parenthesis. When analysing other traffic conditions using the General Analysis Method, the actual number of stops will be presented, not simply those above the main terminal.

With the General Analysis Method, the average lowest reversal floor (*LR*) is also calculated. This is particularly important when analysing buildings with basements or other lightly used floors below the main terminal.

The results shown in Table 7.2 vary by a few percentage points for this particular example. Generally, this is the case for all pure uppeak calculations. Thus, although the General Analysis Method is more mathematically rigorous, the basic calculation method based on Equation (4.8) is confirmed as robust enough to give initial guidance for most design studies. Designers comparing different calculation programs and manual methods should not be unduly concerned when the results are not precisely the same.

Table 7.2 Comparison of results

Parameter	Classical results from Example 6.7	General Analysis Example 6.7
Interval (*INT*)	23.7	23.2 s
Capacity factor (*CF*)	68%	67%
Average number of passengers (*P*) in car	9.4	n/a
Average number of stops (*S*)	6.3	7.0 (6.0)
Average highest reversal floor (*H*)	9.5	10.3 (9.3)
Lowest reversal floor (*LR*)	n/a	1.0
Stopping time ($T^* - d_H/vS$)	7.7	7.8 s
Performance time (*T*)	9.8	n/a

7.10   Other considerations: intended use

The last column of Table 5.8 indicates the likely actual load a lift may be asked to carry based on the possible passenger occupancy. How does this relate to intended usage? ISO 4190–1 (2010) indicates four classes of intended use: residential, health care, general purpose and intensive traffic and ISO 4190–2 (2001) adds goods lifts.

The rated loads (in kg) for each class are:

Residential	320, 450, 630, 1000
Health care	1275, 1600, 2000, 2500
General purpose	630, 800, 1000, 1275
Intensive traffic	1275, 1600, 1800, 2000
Goods	630, 1000, 1600, 2000, 2500, 3500, 5000

The range of the rated loads are intended to match the purpose of their class.

In residential buildings, the range is from the smallest 320 kg, through to 450 kg, which can accommodate a wheelchair, to a 1,275 kg lift, which is both spacious and allows a wheelchair full manoeuvrability. The actual load given in Table 5.8 is close to the rated load and the intended purpose matches well.

The smallest lift (1,275 kg) in the health care range allows full manoeuvrability for a wheelchair and the 2,500 kg lift permits beds and operating trolleys with full attendance. The actual loads carried are likely to be much lower than the rated loads indicated in Table 5.8. This is because the passenger numbers are low, or they have walking aids, or they are in wheelchairs, or on trolleys, or beds. In all these cases, the lift does not have to support loads of the intensity of 360 kg/m². In the health care range, the actual loads will be even lower than the actual loads in Table 5.8, which are already about 75% of the rated loads (column 1 and column 7). Traffic design for health care lifts is not usually required.

The general purpose lifts are intended for small offices and hotels, etc. The range is from the smallest 630 kg, which allows wheelchair access, to 1,275 kg, which gives full manoeuvrability. The actual loads are less than the rated, but are reasonable in case the lifts are used for goods movements. Traffic design will be important for this purpose.

The intensive traffic lifts would be used in office buildings, and range from the smallest (1,265 kg), which allows wheelchairs full manoeuvrability. The range runs from lifts, which are probably too small for most office buildings, and there is a strong likelihood that intending passengers may be left behind, to the largest (2,000 kg), which is probably too big, and where the passenger loading time becomes too long. The actual load is some 75% of the rated load. Traffic design is most important for this purpose.

The goods lifts have a wide range of rated loads to suit all intended usage. Traffic design is not appropriate, but car loading is most important. The loads may have a much higher imposed load than 360 kg/m², which passengers can apply. Therefore, a goods lift should be able to carry the rated load value shown in Table 5.8 for safety reasons.

If passenger loads were to be calculated for passenger lifts by relating them to the platform area on the global basis of 75 kg for every 0.21 m², there would be less confusion for lift traffic designers. However, as ISO/TR 11071–2: 1996 says and ISO/TR 11071–2: 2006 repeats:[5]

> *'Experience shows that it is not realistic to assume passengers are aware of the risk of overloading by the number of passengers in the car'.*

Thus, unless an effective load weighing system with a back-up redundancy device is installed, the rated loads should not be reduced.

7.11   Concluding remarks

Each of the seven major assumptions have been analysed individually and their effect on the calculation of the basic round trip time equation determined. In general, the assumptions have little effect when considering pure uppeak.

For example, statistical theory has been used to derive two of the parameters of the round trip time equation S and H. Two different probability distribution functions have been considered. The first based on a simple statistical process, and used as long ago as 1923 by Basset Jones, assumes a rectangular input distribution with random arrival times. The second uses the Poisson process to describe arrivals. It is interesting to note that the older procedure (the rectangular pdf) will always produce a more pessimistic result than the Poisson pdf. Thus, if a designer wishes to be a little conservative in a traffic design, then the conventional (rectangular pdf) expressions should be used.

However, the consideration of each assumption individually can be tedious. The General Analysis Method combines all the assumptions into an improved mathematical model, allowing not only pure uppeak to be analysed, but also uppeak traffic with other concurrent traffic demands, and other peak traffic conditions, such as midday traffic.

Notes (shown in text as[1])

1 The remarks made in the footnote to Section 5.7 regarding variance, expectance and mean also apply to the above derivations of S and H.
2 Interfloor distance 3.0 m; rated speed 5.0 m/s; then $t_v = 0.6$ s.
3 An example of a time/distance graph can be seen in Phillips (1973), Figure 2.6.
4 Statistical analysis and probability theory are highly mathematical fields and are best dealt with in specialised texts (Papoulis, 1965).
5 Clause 8.1.2.2a.

8 Consideration of lift function, building form, building function, uppeak boosters and disadvantage transportation, and their effect on the round trip time equation

8.1 Introduction

This chapter indicates how the round trip time equation might need to be changed to deal with a number of special situations. For example, a firefighting lift (lift function) installed in a tall building (building form) used as a hospital (building function). Techniques to improve or boost uppeak performance are also discussed. The target equation is Equation (4.8):

$$RTT = 2Ht_v + (S + 1)t_v + 2Pt_p \tag{4.8}$$

8.2 Consideration by lift function

8.2.1 Shuttle lifts (with sky lobbies)

Many tall buildings are divided into several zones: low zone, mid zone, high zone, etc., with service direct from the main terminal floor, situated at ground level. These are called 'local' zones. This becomes impractical with very tall buildings of 70 storeys or more, and shuttle lifts are employed (Schroeder, 1989b) to take passengers from the ground level main lobby to a 'sky lobby' (Browne and Kelly, 1968). Passengers disembark at the sky lobby and then take the local lifts to their final destination. Service is then provided to further low, mid, high zones, etc., using the sky lobby as an upper main terminal floor. The advantage is that the core efficiency is improved (Fortune, 1995, 1996), as the hoist ways can sit on top of each other and extend the whole height of the building (except for the intervening equipment spaces) and occupy the same hoist way 'footprint'. Sometimes, passengers travel down from the sky lobby as well as up (Fortune, 1986, 1990). Most shuttle lifts are single deck, but there are a number of double deck installations. Schroeder (1989a) defines four basic sky lobby configurations:

1 Single deck shuttles, single deck locals, e.g., the first World Trade Center.
2 Double deck shuttles, single deck locals, e.g., Sears Tower.
3 Double deck shuttles, double deck locals, e.g., PETRONAS Towers.
4 Single deck shuttles, single deck, top/down locals, (no example).

Configuration 4 would be difficult to engineer, as offset lobbies would be required. A configuration Schroeder did not consider should be added:

5 Double deck shuttles, single deck, top/down locals, e.g., UOB Plaza.

Examples of these configurations are discussed in Section 8.3.2.

Shuttle lifts are sometimes employed over shorter travel distances, such as from car parks to a main terminal and at underground railway stations, or very long distances to observation platforms. Generally, shuttle lifts serve between two stops only, hence the term 'shuttle', but sometimes serve three stops, i.e., with two sky lobbies.[1]

Shuttle lifts are usually quite large and fast, and provide an excellent service to the sky lobby. Their main disadvantage is that the passengers must change lifts mid-journey, hence increasing their total journey time. When a traffic design involves a change of lift, the two journey times are best quoted separately.

There is no need to modify Equation (4.8). However, note that the value for t_v will be for the travel between the stopping floors. This distance could be 200 m or more.[2]

As the cars are generally large (>2,000 kg), and will fill more fully than is usual for a group of lifts, the passenger transfer times (t_p) when loading and unloading will be more efficient and smaller. The reasons for this are: waiting passengers are 'batched' outside a shuttle entrance expecting its arrival; the lift doors will be 1,100 mm or more wide; and there may be through cars, allowing the separation of the incoming and outgoing passengers. For example, in Section 5.5.1, a lift that was loading 16 passengers, with each passenger requiring 1.2 s to load, gave an average value of t_p of 0.83 s.

The traffic design of a shuttle lift can use Equation (4.8), but as both H and S are known (usually '1'), then Equation (4.1) can be simplified to:

$$RTT = 2T + 2Pt_p \tag{8.1}$$

where: T is the performance time as defined by Definition 4.23.

Example 8.1

Consider a shuttle lift with a rated load of 5,000 kg is transporting passengers from a main terminal to a single sky lobby ($N = 1$) during uppeak with the following data:

$PC = 42.9$ persons (9.0 m²/0.21), $v = 8.0$ m/s,

$t_c = 3.5$ s, $t_o = 3.5$ s, $d_f = 200$ m, $t_f(1) = 30$ s, $t_p = 1.2$ s.

Thus:

$H = 1$, $S = 1$,

$P = 34.3$ (42.9 × 0.83) $Pt_p = 25.0$ s (from discussion above 6 × 1.2 + 14 × 1.2).

$t_v = 200/8 = 25$ s; $T = 3.5 + 3.5 + 30 = 37$ s $t_s = T - t_v = 37–25 = 12$ s

Using Equation (4.8):

$RTT = 2 \times 1 \times 25 + (1 + 1) \times 12 + 2 \times 25.0 = 124.0$ s

Using Equation (8.1):

$RTT = 2 \times 37 + 2 \times 25.0 = 124.0$ s

Such a shuttle lift would have a 5-minute handling capacity of nearly 83 persons. The example illustrates a one-way traffic situation. However, when a shuttle lift is serving a balanced two-way traffic situation, then the lifts would fill fully at each lobby, and the round trip time would increase by a further $2Pt_p$ seconds. If there were two sky lobbies, then the first term of Equation (8.1) would be $3T$, etc.

The number of shuttle lifts that are installed worldwide is not large. Their traffic design is relatively simple, and their application in a building requires expert consideration.

8.2.2   Double deck lifts

Double deck lifts comprise two passenger cabins, one above the other, connected to one suspension/drive system. The upper and lower decks can thus serve two adjacent floors simultaneously. During peak periods, the decks are arranged to serve 'even' and 'odd' floors, respectively, with passengers guided into the appropriate deck for their destination. Special arrangements are made at the lobby for passengers to walk up/down a half flight of stairs/escalators to reach the lower or upper main lobby.

Double deck lifts are often installed in very high rise buildings in the USA and elsewhere, but less so in Europe (see Section 8.3.2). Fortune (1996) indicated that there were 465 double deck lifts in 34 buildings across the world. This might have only doubled over the intervening years.

There are many advantages and disadvantages to double deck operation (Fortune 1996) and special care has to be taken with the lobby arrangements (see Section 8.3.5). One advantage for double deck lifts is that the 'hoist way' handling capacity is improved as there are effectively two lifts in each shaft. A disadvantage for passengers during off-peak periods is when one deck may stop for a call with no coincident landing, or car call, required in the other deck. Fortune (1996) describes special control systems that are available during off-peak periods, such as skip/stop, trailing deck and restricted deck service. This often works well as the off-peak traffic demand is low.

Fortune (1996) expounds the advantages of double deck installations as:

1   fewer lifts
2   smaller car sizes
3   lower rated speeds
4   fewer stops
5   increased zone size
6   quicker passenger transit times
7   30% less core space
8   taller buildings on same footprint
9   smaller lobbies
10   fewer entrances
11   faster installation
12   reduced maintenance costs

and the disadvantages as:

1   few suppliers
2   passenger misuse
3   zone populations must be large
4   balanced demand from even and odd floors
5   interfloor distance must be regular

6   slightly larger hoist ways
7   increased pit and machine room loadings
8   lobby exits need to be larger
9   special facilities for disabled access to 'other' floor.

Kavounas (1989) developed a very succinct analysis of double deck lifts following the direction of the uppeak analysis method described in Chapter 5, starting with Equation (4.8). He makes a number of assumptions:

• The double deck lift serves $2N$ floors above the main terminals.
• Both decks are the same size and carry identical passenger loads (P).[3]
• Both decks experience the same arrival process (pattern).

The expression for the high call reversal floor (H) is not changed, and is given by Equation (5.14) as usual. The expected number of stops (S_d) the double deck lift will make is changed, and can be derived by following the same arguments used in Section 5.5, but because the two decks together carry $2P$ passengers, the expression becomes:

$$S_d = N\left[1 - \left[\frac{N-1}{N}\right]^{2P}\right]$$ (8.2)

The evaluation of this equation could be achieved by using the familiar probable stop table (Table 5.11). However, as $2P$ is likely to be larger (numerically) than 26.4 persons (equivalent to a 33 person rated load), then the evaluation often falls outside the range of the table. Fortunately, Equation (8.2) can be simplified[4] to a simpler expression:

$$S_d = 2S - \frac{S^2}{N} = S\left(2 - \frac{S}{N}\right)$$ (8.3)

For interest, Kavounas also derives expressions for coincident, non-coincident stops and a Figure of Merit.[5]

Definition 8.1: Coincident stops will occur when both decks load or unload at the same time during uppeak traffic.

Definition 8.2: Non-coincident stops will occur when one deck stops to unload without a simultaneous stop for the other deck during uppeak.

If each deck had been independent of the other and not connected together, each would have stopped S times. This leads to the determination of how many stops are coincident (Sc):

$$S_c = 2S - S_d$$ (8.4)

The number of non-coincident stops will then be given by:

$$S_n = S_d - S_c$$ (8.5)

Definition 8.3: The Figure of Merit for double deck lifts is the number of coincident stops to all stops.

The Figure of Merit (*FM*) will be:

$$FM = S_c / S_d \tag{8.6}$$

This Figure of Merit should be as close to unity (100%) as possible. It can be derived from Equations (8.3) and (8.4):

$$FM = \frac{1}{2\dfrac{N}{S} - 1} \tag{8.7}$$

When S is close in value to N, as for example with large lifts serving a small number of floors, then *FM* approaches unity (100%). Siikonen (1998) has used this quality criterion in her analysis of double deck lifts during down peak traffic.

Kavounas also considers the efficiency of passenger transfers. For loading, this is unchanged and is given by Pt_p as usual. For unloading with equal numbers of passengers exiting at each stop from both decks, it will also be Pt_p. But in the worse cases, where there are no coincident stops, it will be twice this value at $2Pt_p$. In practice, the total transfer time will lie between these two values. After some consideration, Kavounas gives the unloading transfer time (t_u) as:

$$t_u = P\left[2 - \frac{S}{N}\right]t_p \tag{8.8}$$

Using Equations (8.3) and (8.8), a modified expression for Equation (4.8) can be formed:

$$RTT = 2Ht_v + \left[S\left[2 - \frac{S}{N}\right] + 1\right]t_s + Pt_p + P\left[2 - \frac{S}{N}\right]t_p \tag{8.9}$$

Equation (8.9) has not been completely simplified in order to indicate the changes clearly. If very large rated car capacity double deck lifts serve a small number of floors, then $S \rightarrow N$, making the modifying term $(2 - S/N)$ equal to unity and Equation (4.8) is regained. Strakosch (1983)[6] used this assumption for his analysis of double deck traffic by considering that the probable number of stops equal the number of floors served. For the fully loaded condition, the simplistic Strakosch procedure gives comparable results to Kavounas.

Peters (1995a, 2000), using his General Analysis Method, has also analysed double deck calculations, and has derived complex formulae for the general traffic case. For the special case of double deck lifts used during uppeak, Peters (1995a) compares his approach with Kavounas, and comments: 'The results show a high degree of consistency for uppeak analysis'. The simplicity of the Kavounas method is to be recommended for pure uppeak calculations.

The traffic design for double deck lifts is relatively simple; however, when Equation (8.7) is evaluated, the values of t_v are for a double interfloor distance, $t_f(1)$ are for a double interfloor distance, and N is half of the $2N$ floors served. The handling capacity will be twice that calculated for one deck. The application of double deck lifts in large buildings requires expert attention.

Example 8.2

Consider a 16-floor building to be served by six double deck lifts with a 1,275/1,275 kg rated load. The interfloor distance is 3.9 m, and the rated speed is 2.5 m/s. The flight time for a 3.9 m jump is 4.9 s, and for a 7.8 m jump is 6.4 s. The door closing time is 2.7 s, and the door opening time is 2.0 s. Assume the passenger transfer time is 1.0 s.

For double deck lifts, each deck will serve eight floors from lower and upper lobbies.

From Table 5.8, the 80% design capacity is 11.0 persons.

From Table 5.11a: then $H = 7.7$, $S = 6.2$.

Also, $t_v = 7.8/2.5 = 3.1$ s; $t_s = 2.7 + 2.0 + 6.4 - 3.1 = 8.0$ s

Using Equation (8.9):

$$RTT = 2 \times 7.7 \times 3.1 + [6.2 (2-0.775) + 1] 8.0 + 11.0 \times 1.0 + 11.0 (2.0-0.775) \times 1.0$$

$$= 47.7 + 68.8 + 24.5 = 141.0 \text{ s}$$

$INT = 23.5$ s, $HC = 140$ (one deck) = 280 (both decks) persons/5 minutes

In comparison, consider single deck lifts serving all 16 floors from one lobby.

From Table 5.6: taking $P = 11.0$, then $H = 15.1$, $S = 8.1$
and $t_v = 3.9/2.5 = 1.6$ s; $t_s = 2.7 + 2.0 + 4.9 - 1.6 = 8.0$ s.

Using Equation (4.8):

$$RTT = 2 \times 15.1 \times 1.6 + [8.1 + 1] 8.0 + 2 \times 11.0 \times 1.0$$

$$= 48.3 + 72.8 + 22.0 = 143.1 \text{ s}$$

$INT = 23.9$ s, $HC = 138$ persons/5 minutes

The double deck group therefore delivers over twice (280/138) the handling capacity of a single deck lift group with a similar interval. Thus, in general, a double deck installation provides over twice the single deck provision. Case Study CS14 illustrates such a traffic calculation.

8.2.3   Firefighting lifts

Firefighting lifts are discussed in Section 6 of the CIBSE Guide D: 2015, which provides an extensive list of further references. A wide range of regulations apply across the world, but the most comprehensive is BS EN 81–72: 2003. All the regulations apply at the discretion of the local fire authorities. Generally, firefighting lifts are required in UK buildings, which have space that can be occupied more than 18 m above and/or 9 m below the fire access level for every 900 m² of building footprint, or part thereof. They must serve all occupied levels, have at least a 630 kg rated load, and be capable of reaching the highest (or lowest) level served in less than 60 s. They must not be used for the transport of goods and must be unobstructed at all times.

Firefighting lifts are often single lifts situated around the floor plate. Their size is often the lowest possible permitted (630 kg), and their speed is often the lowest possible to reach the highest occupied floor in 60 s from the fire service access level. The handling capacity is therefore low, and as usually only a single lift is present at each location, the interval is equal to the round trip time. Firefighting lifts should not generally be considered as part of the vertical transportation provision, but they do provide a useful addition to the vertical transportation services of a building. For instance, in a building with a large floor plate, occupants may be much nearer to a firefighting lift than the main group, and may use it in preference, despite its poorer performance.

Sometimes, a firefighting lift is part of a group and extra precautions are necessary to ensure its fire integrity, *viz*: protected stairways, additional doors, etc. These precautions may affect the traffic circulation to these lifts, and should be taken into account when calculating the handling capacity of such a group. In particular, the firefighting lift may be smaller than the other passenger lifts and may be equipped with additional doors and car operating panels. The speed, door dynamics, etc., may be different and the handling capacity of the group will need to be determined by calculating the round trip time for each lift, adding these together, and dividing the total by the number of lifts to obtain the average interval for the group.

No changes are required to Equation (4.8) in order to calculate the performance of a standalone lift. The design procedure may need to be reviewed to take into account any different equipment specification, where the firefighting lift is part of a group.

8.2.4   Goods lifts

The need for goods lifts has increased substantially in recent years. Despite the computer revolution, the amount of paper into a building and scrap paper leaving a building has increased. It is also quite common to find, in any type of building, one or more floors under refurbishment, with the requirement to bring in materials and equipment, and to remove rubbish and debris.

All buildings should be served by an adequate number of goods lifts of a suitable size. This will ensure that the passenger lifts are used for their designed purpose and are not 'abused' as goods transporters to the detriment of passenger service and probable damage to the car interiors. Where passenger lifts are used as goods lifts, either generally, or in an emergency, the interiors should always be protected with drapes hung on custom hangers.

It is recommended that all office buildings contain at least one dedicated goods lift, particularly if the building is designed for single tenant occupancy. The following points should be noted:

- For office buildings, provision should be made for one dedicated goods lift for floor areas up to 30,000 m², or part thereof.
- For larger buildings, an additional goods lift should be provided for each additional 40,000 m² gross floor area.
- Dedicated goods lifts should have a minimum capacity of 1,600 to 2,000 kg.

Dedicated goods lifts can be provided as single units or in small groups, and should not be considered as part of the passenger vertical transportation provision. It is likely that they will be slower than the regular passenger lifts and, due to their condition, not attractive to most passengers.

Sometimes, a goods lift is part of a passenger group, and extra precautions should be made to protect it from damage. The goods lift may be larger and of a different rated speed than the adjacent passenger lifts, and the handling capacity of the group will need to be determined by calculating the round trip time for each lift, adding these together, and dividing the total by the number of lifts to obtain the average interval for the group.

No changes are required to Equation (4.8) in order to calculate the performance of a standalone goods lift. The design procedure may need to be reviewed to take into account any different equipment specification, where the goods lift is part of a group.

8.2.5   *Observation lifts*

Observation, panoramic or scenic lifts contribute to the vertical transportation system of a building. They are often installed in offices, hotels and shopping complexes to provide a feature or visual impact, and they can draw a large percentage of 'pleasure' riders in the latter two locations. In offices, the whole or part of a group may be comprised of observation lifts. Shuttle lifts may be of this type.

In shopping centres particularly, the lifts may be provided individually or in small groups. Here, they will generally have lower rated speeds, hence, longer flight times, and the door times may also be longer in order to operate the glass doors safely. There may also be long door dwell times to allow children, the elderly and pushchairs to cross the threshold safely. The car interiors are also often shaped for aesthetic and viewing purposes rather than easy circulation in the car, and can have a narrow entrance leading to a wider area at the back of the car. This has the effect of making passenger transfers inefficient.

All these factors can reduce the traffic handling performance from that obtained from a conventional lift. Some designers use 60% of actual capacity instead of 80% when carrying out an analysis.

However, in shopping centres, observation lifts are not the main vertical transportation facility, which is generally provided by escalators. They are invaluable, however, to the elderly, persons with mobility problems, persons carrying goods and parents with pushchairs.

In other locations, where observation lifts are part of a group, they may not have the same specification as the adjacent conventional lifts. The individual round trip times would need to be calculated separately, added together, and divided by the number of lifts to obtain the average interval for the group.

No changes are required to Equation (4.8), but the design procedure may need to be reviewed to take into account different equipment specifications, inefficient passenger transfers and long door dwell times.

8.2.6   *Lifting platforms and lifts for persons with restricted mobility*

Lifting platforms comprise a platform, which travels a few metres in order to accommodate small level changes in a building. This permits persons with disabilities, who are ambulant or in wheelchairs, to move between these levels. These lifting platforms may to be used under supervision, and do not contribute to the general vertical transportation facilities.

Lifts for disabled persons may be regular lifts, which have been set aside for the use of persons with disabilities and which may not be available to non-disabled persons, i.e., they may be card, key or code entry controlled. These lifts contribute little to the general vertical transportation facilities as their usage is low. Where a lift, which is part of a group, is designated for disabled persons' use, it may have slower door times, longer dwell times, etc., when used by a disabled person.

No changes are required to Equation (4.8), but the design procedure may need to be reviewed to take into account usage by disabled persons, for example, BS EN 81–70: 2003.

8.3   Consideration by building form

8.3.1   *Tall buildings*

In modern high rise buildings, each lift is not usually required to service every level as this would imply a large number of stops during each trip, and consequently, long journey times for the passengers in the car.

Examination of Table 5.11 indicates that for a particular car occupancy of a lift, the number of stops (S) increases as the number of floors served (N) increases. As the round trip time in

Equation (4.8) is dominated by the central term, which includes S, the effect is to increase the round trip time, which in turn increases the uppeak interval, the passenger waiting time and the passenger journey time. A similar deterioration of performance occurs for the other traffic conditions.

The solution is thus to limit the number of floors served by the lifts. This introduces the concept of zoning; this is where a building is divided so that a lift or group of lifts serves a designated set of floors. A rule of thumb is to serve a maximum of 15–16 floors (particularly in office buildings) with a lift, or a group of lifts. Nowadays, with the introduction of hall call allocation traffic control systems, the number of served floors can be increased.

Definition 8.4: a zone is a number of floors, usually contiguous, in a building served by a group or groups of lifts.
Note: a zone may be interrupted by a number of unserved floors, such as a retail sector of car park levels.

Definition 8.5: a local zone is a zone adjacent to or close to the main terminal for that zone.

Definition 8.6: a high rise zone is a zone situated above any local zone in the middle or at the top of a building.

The use of the term 'zone' is sometimes used to mean 'sectors' in some traffic control systems. Sectors[7] are used in the implementation of a traffic controller algorithm, and the term is also used to define a designated, but smaller, set of floors in a building. There may be as many sectors in a zone as there are cars to serve that zone.

There are two forms of zoning: stacked and interleaved, as illustrated in Figure 8.1

8.3.1.1 Stacked zones

A stacked zone building is where a tall building is divided into horizontal layers, in effect, stacking several buildings on top of each other, with a common 'footprint' in order to save ground space, see Figure 8.1 (a). It is a common and recommended practice for office and institutional buildings.

Each zone can be treated differently with regard to shared or separate lobby arrangements, grade of service, etc. The floors served are usually adjacent, although some buildings may have split sub zones, where the occupants of each sub zone are associated with each other and can be expected to generate some interfloor movements. The number of floors in a zone, the number of lifts serving a zone and the length of the express jump all affect the round trip time. The round trip time can be calculated by adding a time equal to the time (t_e) taken to jump through the unserved floors in both directions. The changes required to Equation (4.8) are given by the last term in Equation (5.1), *viz*:

$$RTT = 2Ht_v + (S + 1)t_s + 2Pt_p + 2(t_e - t_f(1)) \tag{5.1}$$

This is illustrated in Example 8.3.

Example 8.3

Consider a 36-floor building served by 16-person cars. All zones are to have an equivalent grade of service. Design one system with equal numbers of cars, and another system with equal numbers of floors. Assume identical values for T, t_p, etc.

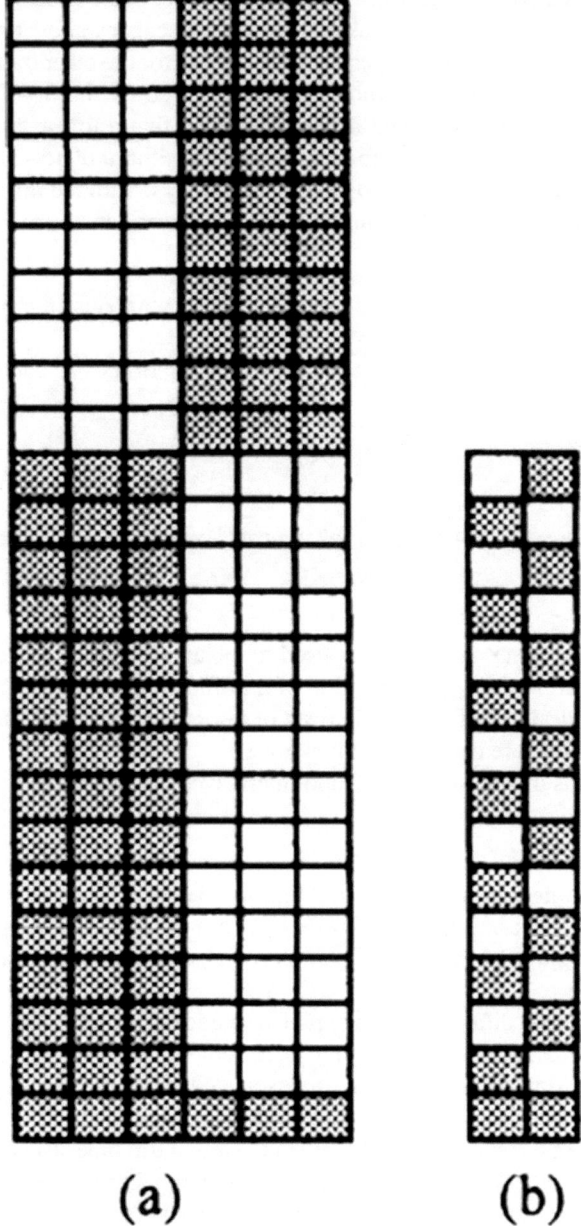

(a)   (b)

Figure 8.1 (a) Stacked and (b) interleaved zones

In order to accommodate the express jump to the first served floor for the zones not immediately above the main terminal, the calculation of the *RTT* will use Equation (5.1). This is a
modification of Equation (4.8). Table 8.1 shows a design for an equal number of cars, with a
decreasing number of floors in each zone. This is an attempt to equalise the round trip times at
the main terminal (MT) to compensate for the express jumps (MT to 15; MT to 27) by reducing

Table 8.1 Equal numbers of cars

Zone	No. of floors	Served floors	No. of cars	Speed (m/s)	UPPINT (s)
Low	14	1–14	6	2.50	24.0
Middle	12	15–26	6	3.15	24.3
High	10	27–36	6	4.00	23.5

the number of floors to be served. Table 8.2 shows a system serving an equal number of floors, with an increasing number of cars to compensate for the time to transit the unserved express jumps. There are a wider range of values for *UPPINT*. It is not possible to fine tune the number of lifts as there are fewer of them. It is usually easier to adjust the number of floors per zone as there are more of them, than the number of lifts per zone.

The introduction of groups of lifts to serve different zones of a building requires more space at the main terminal level. If the building footprint is small, the occupied space is also small, with the lower levels of the building filled with hoist ways. The positioning of the groups is important, and adequate signs should be displayed to quickly and simply direct the passengers to the correct group. The arrangement of lifts within each group is given in Section 2.10.3.

Unless the machine rooms for all zone groups are placed at the top of the building, the machine rooms will be found just above the last served floors in each zone. Thus, if a building has the same size of floor plate at every floor, then the space occupied by the hoist ways and machine rooms reduces at the higher levels, making more rentable space available at the higher levels. As zone boundaries are changed, the zone populations change, not only by the number of floors, but also by the addition or removal of rentable space. This must be considered in the traffic calculation. Remember that hoist ways and machine rooms will occupy at least two levels above the last served floor in a zone.

Traffic Case Study CS11 provides an example of a multi-zoned building.

8.3.1.2 Interleaved zones

An interleaved zone is where the whole building is served by lifts, which are arranged to serve either the even floors or the odd floors, see Figure 8.1(b). This has been common practice in public housing, and has been used in some office buildings. So, for example, in a 16 floor residential building, one lift may serve: G, 1, 3, 5, 7, 9, 11, 13, 15, while another lift serves: G, 2, 4, 6, 8, 10, 12, 14, 16.

The effect is to reduce the number of stops a lift makes because there are fewer floors to be served. This also reduces the capital costs because there are fewer openings and landing doors to

Table 8.2 Equal numbers of floors

Zone	No. of floors	Served floors	No. of cars	Speed (m/s)	UPPINT (s)
Low	12	1–12	5	2.50	26.9
Middle	12	13–24	6	3.15	23.9
High	12	25–36	7	4.00	21.1

install. The service to passengers, however, is poorer than with a duplex serving all floors because there is only one lift to take them to their dwelling. Tenants tend to solve this by calling both cars at the main terminal and if it is the 'wrong' one, walking a flight of stairs to their floor (if they do not have mobility problems). Thus, cars are unnecessarily brought to the main terminal.

Furthermore, it can also be shown (Barney, 1988) that fewer landing doors perform 70% more operations than where the lifts stop at all floors. This is likely to cause an increased level of call backs to the most troublesome component of a lift system. Interleaved zoning is not recommended for the above reasons and is a '. . . proven disaster in the US' (Strakosch, 1988).[8]

When calculating the *RTT* for interleaved zoning, remember that the interfloor distance is twice the standard spacing, leading to an increased 'single floor flight time' and the number of floors served is also halved (*N/2*).

There are no changes required to Equation (4.8), but care must be taken to use the correct values for the various variables.

Example 8.4

Consider a 16 floor residential building with two 800 kg lifts (*PC* = 7.6 persons). Each lift has the following characteristics: t_v = 2.0 s; t_s = 10.0 s; t_p = 1.5 s.

If the lifts operate as a duplex: *P* = 7.6; H = 14.6; *S* = 6.2 (from Table 11a) then:

$$RTT = 2 \times 14.6 \times 2 + 7.2 \times 10 + 2 \times 7.6 \times 1.5 = 141.8 \text{ s}$$

$$INT = 70.9 \text{ s}$$

$$HC = 32.2 \text{ persons/5 minutes}$$

If the lifts operate as a simplex: *P* = 7.6; H = 7.5; *S* = 5.2 and

$$RTT = 2 \times 7.5 \times 2 + 6.2 \times 10 + 2 \times 7.6 \times 1.5 = 114.8 \text{ s}$$

$$INT = 114.8 \text{ s}$$

$$HC = 19.9 \text{ persons/5 minutes (one lift)} = 39.8 \text{ persons/5 minutes (both lifts)}$$

The effect of providing a 'skip stop' installation is to increase the handling capacity by some 24% and increase the interval by some 38%. This might not be considered to be a satisfactory performance.

8.3.1.3   Transfer floors

Most tall and very tall buildings (see Section 8.3.2) provide some means to travel between zones and stacks. This is sometimes achieved by overlapping zones (PETRONAS Towers), introducing extra stops (Sears Tower) or shuttle lifts (World Trade Center). A common served floor (other than the main terminal or sky lobby) is important where there are common facilities to be accessed, e.g., restaurant, travel bureau, sports facilities, post room, reprographics, etc.

The effect on traffic handling can be disruptive. In general, it is important to restrict access to such floors during uppeak and down peak, although the object of such a floor would be defeated at other times, i.e., at the midday break or during interfloor traffic. During up or down peak traffic, the transfer floor should be either included in the zone below or the zone above, whichever produces similar interval times at the main terminal floor. It is difficult to calculate the effect of a transfer floor, and simulation techniques are often employed. Traffic Case Study CS11 offers a unique solution.

8.3.2　Very tall buildings

8.3.2.1　History

The Council on Tall Buildings and Urban Habitat keep a survey of the 100 tallest buildings in the world.

In 1999, they reported 63 in North America, 30 around the Pacific Rim, 4 in Europe (1 in the UK) and 3 others.

In 2014, they reported 20 in North America, 48 around the Pacific Rim (34 in China), 29 in the Middle East, 1 in Europe (1 in the UK) and 2 others.

Significant changes in location has occurred over 15 years away from North America to the Middle East and the Pacific Rim. The top segment have got taller with little change in the bottom segment.

Very tall buildings come with problems (Barney, 2003b) In an emergency due to fire or other event, there is a need to provide safe firefighter lift access and safe passenger egress, particularly for mobility-impaired occupants (Barney, 2003a)). There is also a limit to the rated speed, not from engineering problems, but for the comfort of passengers due to the rapid changes of air pressure (So, 2014; de Grout, 2014). Without pressure-controlled cars, the limit to the rated speed may be as low as 12.5 m/s.

8.3.2.2　Definitions

Very tall buildings might be defined as those buildings over 30 to 40 storeys high. This height can be related to nature, as the tallest tree ever measured was 132.6 m.

Generally, if service can be provided from the ground level main terminal floor to every floor in the building, this is a tall building.

Once shuttle lifts to sky lobbies are required, then the building could be called very tall. Fortune (1997) defines a tall building as a 'skyscraper', i.e., 'A high rise building with more than one zone of elevators' and a very tall building as a 'Mega High Rise building', i.e., 'A building with one or more sky lobbies and in excess of 75 floors'.

8.3.2.3　Traffic design considerations for very tall buildings

Very tall buildings, sometimes described as 'monumental' buildings, are few in number compared to the totality of buildings worldwide, and their traffic design requires expert

Table 8.3 100 tall buildings

Year	Top 50			Bottom 50		
	Low	High	Range	Low	High	Range
1999	260 m	450 m	190 m	230 m	260 m	30 m
2014	321 m	828 m	507 m	287 m	320 m	33 m

consideration. The design involves a mixture of zoning, shuttle lifts and double deck lifts. The formulae, which have already been given, deal with these traffic systems.

As a general rule, a building of 60 floors can be served with four groups of lifts (a practical limit) serving 4 zones from a main terminal lobby. If double deck lifts are used, this permits up to 80 floors to be served from a main terminal lobby. Above an 80 floor building, sky lobbies must be used, with shuttle lifts to serve them. This permits service to 120 to 160 floors with one sky lobby and 180 to 240 floors with two sky lobbies. Remember also, that the maximum practical number of lifts that can be grouped together is eight cars with four facing four, unless a hall call allocation traffic algorithm is used. Other special zoning concepts exist, which are not considered here (Aoki, 2010)

There are no fundamental changes required to Equation (4.8). However, care must be taken to apply the appropriate formulae, e.g., Equation (8.1) for shuttle lifts; Equations (8.2), (8.3) and (8.7) for double deck lifts, and Equation (5.22) to deal with express jumps. Care must also be taken to use the correct values for the various parameters.

8.3.2.4   Techniques to service very tall buildings

Very tall buildings employ many techniques to provide lift service. The techniques are best visualised by considering four of the tallest buildings in the world. These are: PETRONAS Towers, The Sears Tower, the first World Trade Center and the United Overseas Bank.

(A) DOUBLE DECK SHUTTLES TO DOUBLE DECK LOCALS

PETRONAS Towers, Kuala Lumpur, Malaysia (1996) are 452 m high with 88 stories above ground. Uniquely, the two towers are situated side by side and are joined part of the way up by a high level bridge. The buildings are divided into two stacks.

Stack 1 has two groups of six double deck lifts serving the main terminal and floors 8–23 at 4 m/s and the main terminal and floors 23–37 at 5 m/s. Floor 23 acts as a transfer floor.

Stack 2 is served by a group of five double deck shuttle lifts at 6 m/s to sky lobbies at floors 41/42. From the sky lobbies, there are three groups of six double deck lifts serving three zones: floors 44–61 at 3.5 m/s; floors 61–73 at 7 m/s; floors 61/62 and 69–83 at 7 m/s. Floors 61 and 62 act as transfer floors.

(B) DOUBLE DECK SHUTTLES TO SINGLE DECK LOCALS

The Sears Tower, Chicago, USA (1974) is 436 m high with 103 stories above ground. The building is divided into three stacks.

Stack 1 has three groups of six single deck lifts serving the main terminal and floors 5–10 at 2.5 m/s; the main terminal and floors 10–17 at 3.5 m/s; the main terminal and floors 17–23 at 4 m/s, and one group of five lifts serving the main terminal and floors 23–28 at 5 m/s.

Stack 2 has sky lobbies at floors 33/34, served by eight double deck shuttle lifts at 7 m/s. From the sky lobbies, there are three groups of six single deck lifts serving floor 33 and floors 35–42 at 2.5 m/s; floor 33 and floors 42–49 (two lifts also serve floor 27) at 3.5 m/s;

floor 34 and floors 49–57 at 4 m/s; and one group of five single deck lifts serving floor 34 and floors 58–63 at 5 m/s.

Stack 3 has sky lobbies at floors 66/67, served by six double deck shuttle lifts at 8 m/s, which also can stop at floors 33/34 for service to stack 2. From the upper sky lobby, there are three groups of four single deck lifts serving floor 66 and floors 68–74 (two lifts also serve floor 63) at 2.5 m/s; floor 66 and floors 75–81 at 3.5 m/s; floor 67 and floors 82–87 at 4 m/s; and one group of five single deck lifts serving floor 67 and floors 88–102 at 5 m/s. In addition, there are two observation lifts serving the main terminal and floor 103 at 9 m/s.

(C) SINGLE DECK SHUTTLES TO SINGLE DECK LOCALS

The World Trade Center, New York, USA (1972, destroyed by terrorists, 11 September 2001) comprised two towers, and was 416 m high with 110 floors. It was divided into three stacks.

Stack 1 had 4 groups of six single deck lifts, serving the main terminal and floors 9–16 at 4 m/s; the main terminal and floors 17–24 at 5 m/s; the main terminal and floors 25–32 at 6 m/s; and the main terminal and floors 33–40 at 7 m/s.

Stack 2 had a sky lobby at floor 44, served by eight single deck shuttle lifts at 8 m/s. From the sky lobby, there were four groups of six single deck lifts serving floors 46–54 at 2.5 m/s; floors 55–61 at 4 m/s; floors 62–67 at 4 m/s; and floors 68–74 at 5 m/s.

Stack 3 had a sky lobby at floor 78, served by eight single deck shuttle lifts at 8 m/s. From the sky lobby, there were four groups of six single deck lifts serving floor 78 and floors 80–86 at 2.5 m/s; floor 78 and floors 87–93 at 4 m/s; floor 78 and floors 94–99 at 4 m/s; floor 78 and floors 100–107 at 5 m/s.

(D) TOP/DOWN SERVICE

A top/down lift installation is where a sky lobby is used to serve building zones or stacks, both in the conventional up direction, but also in the down direction (Schroeder, 1989a). This does mean that passengers may (psychologically) be concerned that they have travelled up a building only to be required to then travel down to their destination. This is a subtle defeat of Rule 3, and the technique has been applied in a few buildings. An example is the United Overseas Bank, (Fortune, J. W., 1986)), which is 280 m high with 66 storeys, divided into three stacks.

Stack 1 has a group of six single deck lifts serving the main terminal and floors 7–20.

Stack 2 has a sky lobby at floor 37, served by the lower deck of six double deck shuttles. From the sky lobby, a group of six single deck lifts serve (down) floors 20 and 23–36.

Stack 3 has a sky lobby at floor 38, served by the upper deck of six double deck shuttles. From the sky lobby, a group of six single deck lifts serve (up) floors 41–59.

This building can be categorised as double deck shuttles to single deck locals.

8.3.2.5   *Very tall buildings: a postscript*

A monstrously tall building, which was never built, was Frank Lloyd Wright's mile high (1,580 m) building (Fortune 1992). He proposed the building in 1956 to be built in Chicago, USA. This building would have 528 storeys and would have accommodated 130,000 occupants. Wright proposed to install 76 quintuple (5) deck lifts. Fortune estimates that more than twice that number would have been needed to obtain current day performance standards. This building would have been twice as high as the world's current tallest building, the Burj Khalifa.

8.3.3   *Basement service*

The provision of lift services to basements has been considered by convention to be either a costly nuisance, or a mere appendage to the normal lift system serving between the main entry terminal and the floors above. However, basements are provided in buildings for many purposes, at many levels of usage, and cannot be ignored. Pearce (1995) provides a comprehensive analysis of basement service.

Passengers requiring lift service in or to the basements are generally those arriving by car, or as a result of a facility located in the basement area, such as restaurants, leisure facility, health clubs, etc. The number of passengers arriving by car and requiring lift service from the basements is limited by a number of factors. These include: the number of vehicles that can be admitted to the basements during the uppeak period; the number of car spaces in the basements; and the expected number of persons in each car entering the basements. Planning authorities generally place limits on the number of car parking spaces to be provided in a building. Quite often, this represents only 10% of the total building occupancy.

Service to floors above the main terminal is treated as the important part of the lift design since it provides transportation for almost all of the building population. Basements are usually a very small addition to this, but can increase the passenger waiting time, and thus cause annoyance to waiting passengers, either when a full lift bypasses the main terminal or the lift stops partly full, allowing only a small number of the passengers to enter. Strakosch (1967)[9] gave advice, which is equally relevant today, but is still being ignored by building developers and designers, as an economy measure:

> All elevators in a group should serve the same floors. This is a common-sense rule that is often violated for false economy. If, for example, only one car out of a group of three serves the basement, people wishing to go to the basement from an upper floor have only one chance out of three that the next elevator that comes along will take them to the basement. Conversely, people in the basement must wait three times as long for elevator service than upper floor passengers. Ideally, all cars should serve the basement, but if not, a special shuttle elevator must be considered, to run only between the main floor and the basement.

There are three basic service arrangements for basements. The first arrangement is that not all lifts serve the basements, which is not satisfactory as Strakosch (1967) comments:

> The expedient of providing a separate call button at an upper floor to call the single car that serves the basement has often been tried and never proved satisfactory. The average person will operate both the normal call button and the basement call button, take the first

car that comes along, and cause the basement car to make a false stop. Such false stops will add up in lost elevator efficiency over the years to more than pay for the cost of the extra entrances on all elevators.

The second arrangement is where all lifts serve all the basement levels. It is the most expensive arrangement, but the most versatile arrangement with one or two basements, although it would involve a time penalty on the overall system. The availability of lifts to serve the basements can be limited during peak traffic conditions by modern control systems.

Finally, separate basement service is appropriate in a building with multiple basements, and where there are high rise groups, or high speed lifts serving high zones, as it would prove to be less expensive than the alternative of all lifts serving the basements. An illustration of these arrangements is provided in Figure 8.2.

Figure 8.2 shows, from left to right: no basement service; part basement service (1 lift); full basement service; and a basement service to two levels in the main group, and a separate basement shuttle service group.

Peters (1997a) has analysed a system of four lifts, 2 of which serve the basement. He found the two lifts not serving the basement had an interval of 40 s, whereas the two lifts serving the basement had an interval of 45 s. He concludes the interval at the lobby would be 21 s and at a basement 45 s. See Example 8.12.

In the uppeak traffic condition, the presence of a served basement will introduce at least one extra stop (if the main terminal is bypassed on the way down) and probably two extra stops (if the lift stops at the main terminal on the way down and again on the way back up). The second circumstance will arise if passengers press both buttons at the main terminal 'in order to make the lift come quicker' as they invariably do. The outcome is passenger loading delays at the main terminal as passengers try to decide whether to enter a down-going lift or wait for an up-going lift. Another effect of service to the basement area during uppeak, is that cars arrive at the main terminal already partly full, thus causing more confusion.

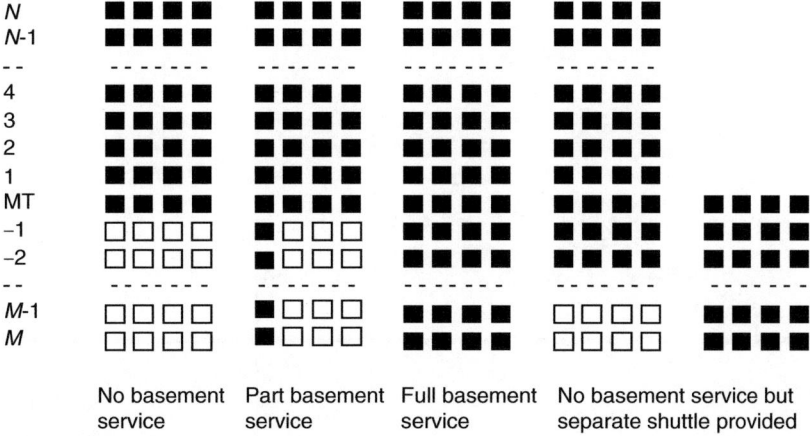

Figure 8.2 Basement service arrangements

For this example, a 4 lift group is shown as an illustration.
M refers to the number of floors served below the main terminal floor.

Barney (1993) in the first edition of CIBSE Guide D wrote:

The time penalty for the extra stops can be between 10 s and 20 s and say between 5 s to 10 s for the increased passenger loading times i.e. some 15 s to 30 s to be added to the RTT. In the case of a 16 person lift serving 16 floors the RTT could be 150 s and the extra times would add 10% to 20% to the RTT.

During down peak, any trip down below the main terminal to, say, leisure facilities will add one extra stop and the extra time to transit the extra interfloor distance; say 10 s to 12 s. This will reduce the handling capacity during down peak.

Designers will need to take account of these factors when sizing an installation with served levels below the main terminal.

An important factor is the ratio of arrivals from the basements to the arrivals at the main terminal, i.e., whether the ratio is 10:90 or 40:60, etc. The ratio will depend on the building size. Three possible ways to account for this in traffic design are:

1   Add an estimate of the time penalty for the extra stops to the *RTT*, as above.
2   Calculate the probable stops and reversal floor in a similar way to the upward service calculation (Nahon, 1990). The resulting additional time can then be added to the normal RTT.
3   Increase the value of *N* in the use of Equations (4.8) or (5.1) by the number of basement floors to obtain a value for the RTT.

Barney and Pearce (2000d, 2001a) found that the method suggested by Nahon gave the most accurate results. Nahon assumed that passengers boarding at the basement floors can be modelled as if they were alighting. This is a reasonable assumption. He did not give details of his method. The Nahon method provides a separate calculation for any number of basement levels based upon the uppeak model, which is then added to the above ground round trip time. This method also allows for the normally different basement interfloor heights.

To understand the method, it is necessary to modify Equation (4.8) to incorporate Nahon's method. Consider the spatial movement of a lift serving above and below the main terminal, as shown in Figure 8.3.

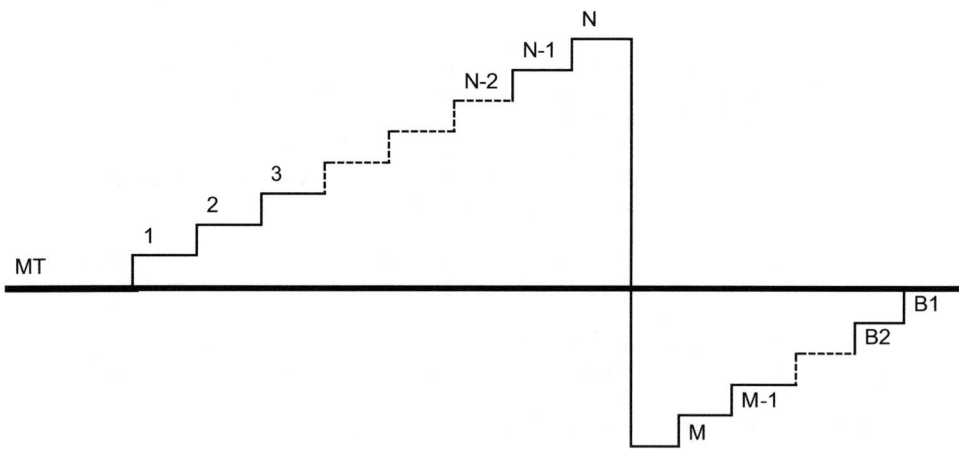

Figure 8.3 Spatial movements of a lift serving above and below ground

The round trip consists of a number of elements resulting from passengers travelling from the basement floors or from passengers joining a lift at the main terminal:

a) passenger loading time t_l
b) passenger unloading time t_u
c) door closing and opening times t_c and t_o
d) interfloor flight time (for a single floor assuming rated speed is reached) $t_f(1)$
e) time to travel in the upward direction at rated speed for jumps greater than a single floor t_a
f) time to travel from the highest floor to the main terminal t_e
g) passenger loading time for xP passengers
h) passenger unloading time for xP passengers
i) door closing and opening times
j) interfloor flight time (for a single basement interfloor distance assuming rated speed is reached)
k) time to travel in the upward direction at rated speed for jumps greater than a single floor
l) time to travel from the main terminal to the lowest floor

where:
x is the percentage of passengers boarding at the basement floors and present in the lift on its arrival at the main terminal.

The terms (a) to (f) have been considered in Section 4.7 and give rise to Equation (4.8). Terms (g) to (l) are additional terms as a result of basement service. The addition of items (a) to (l) translates into the equation:

$$RTT = 2Ht_p + (S + 1)t_s + 2Pt_p + 2H_M t_{vm} + S_M t_{sm} \qquad (8.10)$$

where:
H_M is the average lowest reversal floor
S_M is the average number of stops below the main terminal
t_{vm} is the interfloor transit time below the main terminal (s)
t_{sm} is the stopping time below the main terminal (s)

It will be noted that the first three terms are the same as Equation (4.8). The last two terms represent the time penalty to serve the basements.

The effect of serving floors below the main terminal has an effect on the main traffic patterns. During interfloor traffic, there will be no appreciable deterioration in service. But during uppeak and down peak, the loss of cars below the main terminal will affect service. The Nahon calculation method is very plausible as it does mathematically calculate the effect of basement service. It does, however, depend on the theory that a 'reversed' uppeak is reasonable.

Equation (8.10) can be used to calculate an incoming 'uppeak' traffic pattern, where most of the passengers arriving at the main terminal travel to floors above the main terminal in the conventional way, and others travel to floors below the main terminal. Care must be taken to select the correct values for basement demand (see Traffic Case Study CS10).

N = 10	N = 10	N = 16	N = 20	N = 20
P = 8.0	P = 19.2	P = 12.8	P = 8.0	P = 19.2

Example 8.5

Five lift systems are presented in Section 5.12.3. They represent installations in the 'four corners of the (lift) world' and its 'centre of gravity', i.e.:

Data assumed:

$L = 1$, $T = 10$ s, $t_p = 1.0$ s, $d_f = 3.3$ m (above MT), 2.5 m (below MT).
$v = 1.6$ m/s (for $N = 10$), 2.5 m/s (for $N = 16$), 3.15 m/s (for $N = 20$).
$M = 1, 3, 5$, $x = 10\%, 20\%$.

Table 8.4 indicates the changes to the round trip time serving one, three or five basement floors (M) with two levels of basement traffic demands (x) at 10% and 20%, i.e., cars are 10% or 20% occupied on arrival at the main terminal. The extra time incurred will increase the service interval at the main terminal and reduce the overall handling capacity. For the systems considered in Table 8.4 the additional times range from 7 s to 29 s.

8.3.4   Multiple entry levels and entrance bias

Some buildings have more than one main entrance, and each entrance may be served by its own group of lifts. Other buildings may be designed with the main entry points at more than one level (Pearce, 1995). The effect of more than one main terminal is disruptive, and in the interests of efficient circulation, buildings should not be designed in this way. If there is more than one entrance, means should be provided to bring the two routes together at a single lift lobby. If this is not possible, then the lift system sizing should take into account the extra times incurred for stopping and loading at multiple entry floors. This could be achieved by using Equation (8.8).

Table 8.4 Additional times to serve one, three or five basements

Basement demand	N/P	RTT(M = 0) (s)	RTT(M = 1) (s)	RTT(M = 3) (s)	RTT(M = 5) (s)
Nahon @ 10%	10/8.0	108	+7	+10	+14
Nahon @ 20%	10/8.0	108	+7	+14	+19
Nahon @ 10%	10/19.2	156	+8	+14	+21
Nahon @ 20%	10/19.2	156	+8	+20	+28
Nahon @ 10%	16/12.8	153	+8	+13	+17
Nahon @ 20%	16/12.8	153	+8	+17	+24
Nahon @ 10%	20/8.0	123	+8	+11	+14
Nahon @ 20%	20/8.0	123	+8	+15	+20
Nahon @ 10%	20/19.2	200	+8	+16	+22
Nahon @ 20%	20/19.2	200	+8	+20	+29

Another difficulty is deciding whether the building population will use each entrance (and their associated group of lifts) equally. In the absence of any guidance, the solution is to assume an entrance bias of 60% and size the lift groups to meet this demand at an extra cost.

There are no changes required to the round trip time equation. Any separate lift systems may not be equally loaded, and some extra handling capacity may be needed. Extra time may be needed to take care of multiple passenger loading and unloading times. Traffic Case Study CS13 illustrates some aspects of multiple entrances.

8.3.5   Lobby design

The design of lift lobbies, especially at the main terminal, where there is most activity, is a neglected area. Pearce (1996) suggested that many lobbies are considered 'a trivial element in lift design'. Very often, the design of a lift lobby is considered against the requirements of aesthetic design, security, noise limitation, smoke penetration and fire precautions, rather than the movement of people. If a lift lobby becomes too crowded, especially during uppeak, then intending passengers often reach a lift and cannot board it because it is too full, or cannot reach it and it leaves partially empty. The latter event reduces the handling capacity, and the former may delay the dispatch of the lift as passengers may need to leave as it has become not only crowded, but also overloaded.

Lobby arrangements have been discussed in Section 2.11.3, where guidance regarding lobby size is indicated, and in circulation Case Study 1, where the number of passengers who can be accommodated are calculated. Pearce (1996) suggests that sometimes this guidance is breached for the convenience of the machine room layout or structural efficiency. The shape of lobbies is also important. For example, a group of lifts whose doors are fitted to a convex shaped lobby would be most inefficient, whereas if the doors were fitted to a concave shaped lobby the lobby design would be nearly perfect.

What effect does lobby design have on the round trip equation? Section 2.11.3 suggested that if lifts are served from large lobbies, the lobby door dwell time may need to be increased. The Australian Computer Aided Design Service (Pearce, 1996) suggests:

> *Should a lobby be narrow or the lift cars a shape that restricts passenger movement or the lift doorways are fitted with very thick walls it would be reasonable to allow an increase on loading and unloading times, typically 10%.*

It would appear, therefore, that the effect of lobby design cannot be completely quantified and that the best guess, where a problem is anticipated, is to increase the passenger transfer time (t_p) used in Equation (4.8) by 10% to account for the inefficiency. In severe situations, some designers add 10% to the round trip time.

The lobby design for double deck lifts presents problems as there are two lobbies, one above the other. Fortune (1995) indicates several solutions and the care to be taken in the lobby design. One arrangement is by using split level lift lobbies reached by stairs, pedestrian ramp or escalators from an entrance lobby, positioned midway between the lift lobby levels. Another arrangement is where the entrance lobby is also either the upper, or the lower lift lobby, and the other lift lobby is by stairs, pedestrian ramp or escalators.

8.3.6   *Effect of door width and car shape*

This effect is not specifically due to building form, but it could be, and so it is discussed here. For example, the available core shape may influence the well shape, or an observation lift of an eccentric design may be installed. The main effect will be on the efficiency of passenger transfers. Phillips (1973) has a graph for evaluating passenger transfer times (made into equations by Jones, 1971); Strakosch (1967) has tables giving suggested times, which in later editions, he terms transfer inefficiencies. The estimation of passenger transfer time is often simply common sense (see Section 5.5).

Table 8.5 shows the car platform sizes and door widths that are commonly available and recommended in ISO4190–1. The table shows four classifications: residential, health care, general purpose and intensive traffic, which are discussed below, in turn.

8.3.6.1   *Residential lifts*

These lifts are generally small, i.e., up to rated loads of 1,000 kg and fitted with narrow doors in the range 700 mm to 900 mm. The car width is never larger than 1,100 mm and therefore loading/unloading will generally be single file. This will influence the value used for t_p, which will generally be at the high end of the range at 2 s per passenger.

8.3.6.2   *Health care lifts*

These lifts are generally large and spacious, i.e., rated loads from 1,275 kg to 2,500 kg, in order to accommodate wheelchairs, beds and trolleys. The doors are wider, in the range 1,100 mm to 1,400 mm, to accommodate movement of the equipment. Calculations of round trip times are not particularly relevant in this situation.

8.3.6.3   *General purpose lifts*

These are generally at the low end of the rated load range. Lifts with rated loads of up to 800 kg have doors of 700–800 mm size, thus restricting loading/unloading to single file. The car

Table 8.5 ISO4190 car platform and door sizes

	Residential				*Heath care*			
Rated load (kg)	320	450	630	1000	1275	1600	2000	2500
Platform width (mm)	900	1000	1100	1100	1200	1400	1500	1800
Platform depth (mm)	1000	1250	1400	2100	2300	2400	2700	2700
Door width (mm)	700	800/900	800/900	800/900	1100	1300	1300	1300/1400

	General purpose				*Intensive traffic*			
Rated load (kg)	630	800	1000	1275	1275	1600	1800	2000
Platform width (mm)	1100	1350	1600	2000	2000	2100	2350	2350
Platform depth (mm)	1400	1400	1400	1400	1400	1600	1600	1700
Door width (mm)	800/900	800/900	900/1100	1100	1100	1100	1200	1200

platform shape allows circulation in the car. Lifts with rated loads above 800 kg generally have doors of 1,100 mm or larger, and this allows double file entry after (say) the first six passengers have transferred. The platform shape allows circulation in the car.

8.3.6.4   *Intensive traffic lifts*

These lifts are generally larger than 1,275 kg and are fitted with doors of 1,100–1,200 mm width. This allows double file passenger transfers and the car platform shape allows easy circulation around the car.

8.3.6.5   *Unusual car shapes*

Some lifts may be very deep, but narrow. This will hinder circulation in the car and slow passenger transfers as passengers jostle to move in/out. This type of platform shape occurs frequently with observation/scenic lifts, which tend to provide as much rear surface as possible in the car in order that passengers obtain a good view. Other lifts may be wide but shallow. This means that the doorway will be obstructed by passengers at the sides of the car. All these factors should be considered when carrying out round trip calculations.

8.3.7   *Effect of large floor plates*

Some floor plates are the size of football fields. Here, the discussion indicated in Section 2.10.1 should be considered. No special changes to Equation (4.8) are necessary with respect to the calculation method.

8.3.8   *Effect of building facilities*

There will be facilities in buildings that will distort traffic movements. Examples are restaurants (positioned at the top of the building, in the basement, and even half way up the building); drinks and sandwich machines; leisure club facilities (swimming pools, gymnasia); toilet facilities; post rooms; trading floors, etc. These floors will provide a powerful attraction at different times of the day, and must be considered in the traffic design.

No special changes are necessary to Equation (4.8) with respect to the calculation method. Traffic Case Study CS12 illustrates the effect.

8.4   Consideration by building function

8.4.1   *Airports*

Most airports are arranged on two main levels with the arrival level below the departure level. There may then be other levels above and below providing various services (e.g., baggage handling, catering) and facilities (offices). Another common characteristic is an adjacent, underground or elevated railway station.

Passengers at airports with baggage will use the trolleys provided. Most airports have sufficiently large halls and corridors, and no problems should arise when the trolleys are used on one level. However, when the passenger wants to move from one level to another, difficulties can arise. Escalators should not be used, although moving ramps are ideal. Lifts are the main means of vertical movement.

Generally, each baggage trolley will be attended by two persons plus their baggage. The weight of the baggage is generally restrained by the 20 kg allowance most (economy) passengers are allowed, plus some 5 kg of hand luggage. Thus, a trolley will weigh (including its own weight) some 75 kg, i.e., equivalent in weight to one person. However, it will occupy the space taken by three or four persons. Thus, the total weight of two passengers and their trolley will be some 225 kg and occupy the space of some five people. This occupancy and loading requirement must be taken into account.

Consider a lift of rated load 3,500 kg. According to EN81–20, Table 6, the maximum available car area for an electric traction lift must then be 6.60 m². According to the body template (Figure 2.1), only 31.4 passengers can be accommodated. They will weigh only 2,357 kg, some 67% of the rated load. If each pair of passengers and their baggage trolley occupy 1.05 m² (five human spaces) then the 3,500 kg lift can accommodate 12.6 passengers and 6.3 trolleys. The total load will be 1,400 kg. This is 40% of the rated load. Thus, it can be seen that in these circumstances the lift is very unlikely to be overloaded.

If a hydraulic lift with rated load of 2,000 kg were to be installed, then the platform area from EN81–20, Table 7 would be 6.64 m². It is still possible to accommodate 12.6 passengers and 6.3 trolleys. This is still a load of 1,400 kg, which is some 70% of the rated load. This would indicate that hydraulic lifts are most suitable for this environment as they can be installed with a smaller rated load. Their poorer dynamic performance would not be a significant disadvantage. The number of up starts required by hydraulic lifts may be a limiting factor rather than the traffic calculation.

There are no changes required to Equation (4.8), but care will need to be taken in the assumptions of car occupancy levels.

As with shopping centres (Section 8.4.10), the movement of baggage trolleys from one level to another is a problem. A solution to this problem is to install moving ramps, and this greatly improves circulation.

8.4.2   Car parks

Car parks can be attached to shopping centres, offices, airports, railway stations, etc. They are often multi-storey, although those at out of town shopping centres and railway stations may be a single level. The pedestrian demand is more likely to be restrained by the vehicle entry and exit ramp handling capacities. Another factor is the vehicle occupancy, which is likely to be two persons per car.

For offices, the peak demand is often in the evening, when building occupants are attempting to reach their cars. The office lifts, which may not serve the car parking levels, will bring large numbers of people to the lobby. Those persons with cars will then make a significant demand on any lifts serving the car park levels. The demand on the car park lifts is similar to that experienced by the main lifts during the morning peak period, but this demand is downwards.

Once the occupants have reached their vehicles, they may then spend significant time reaching the vehicle exit. Another factor is the vehicle occupancy, which is likely to be low at about 1.2 persons per car, unless car pools are in operation. However, there will still be a demand for an efficient lift service. The car park lifts must therefore be designed to meet this demand.

The traffic design should use Equation (4.8) if the car park lifts are separate to the main lifts, and Equation (8.10) if the lifts are part of a basement service.

8.4.3 Department stores

This category applies to large departmental and chain stores. These stores will have many entrances, some of which may open to a main street while others open into shopping mall areas. The opportunity therefore exists for 'leakage' into and out of shopping centres. Many stores will own lifts and escalators inside their occupancies. These facilities may be used by shopping centre shoppers to move between mall levels. Thus, store facilities enhance those provided by a shopping centre or mall to the mutual advantage of both. Many of the considerations concerning shopping centres discussed in Section 8.4.10 apply to department stores.

8.4.4 Entertainment centres, cinemas, theatres, sports centres and concert halls

Buildings providing these functions can specialise in one of the activities, or many of them. Many sports centres are low rise and do not require lifts. Town centre buildings such as cinema complexes, concert halls and theatres will be of higher rise. Such complexes generally use escalators as the main vertical transportation element. Lifts provided in these circumstances do not have to meet a large demand and may only have to satisfy the requirements for disabled persons and firefighting.

There are no changes required to Equation (4.8), but care must be exercised in its application.

8.4.5 Hospitals

The building form is important, i.e., whether the building has a small footprint and is tall (US practice) or has a large footprint and is low (UK practice). In the former case, where lifts are used as a primary circulation element, their proper operation is vital, particularly when dealing with operating theatre emergencies. In Britain, most hospitals are designed on a 2–3 storey low rise principle, although many city hospitals have high rise elements.

The principal corridors are sized to accommodate bed and trolley movements, and therefore present no difficulties when handling pedestrian movements. Lifts are provided mainly as a means of moving bed-bound patients from floor to floor.

The traffic designer will need to understand the *modus operandi* of the hospital before finalising a design. Factors to be considered include: numbers of staff and shift patterns; numbers of visitors and visiting hours; location of theatres, X-ray, etc.; distribution and deliveries of food, beverages, supplies, etc.; and waste disposal, patient emergency evacuation, porterage, etc. It is important that patient bed lifts are separate from the visitor and staff lifts to avoid cross infection. The Health Technical Memorandum (HTM 08–02: 2008) gives some general guidance for UK hospitals.

There are no changes required to Equation (4.8).

8.4.6 Hotels

Lifts play an important part in the circulation of guests and service staff in a hotel. Escalators should be employed for short-range movements, e.g., to connect function levels with the lobby.

The traffic patterns in hotels are complex, and are not comparable to the morning and afternoon peaks in an office. The most demanding times are at check out (08:00 to 10:00) and check in (17:00 to 19:00). At these times, heavy two-way traffic occurs, with guests going to and from rooms and restaurants, and in and out of the hotel. Calculations should therefore assume equal numbers of up and down stops at these times.

At most times, lifts are unlikely to load to more than 50%. The lift sizes should be at least 16 persons in order to accommodate luggage and provide guests with uncrowded travel conditions. As a rule of thumb, assume one lift for every 90–100 keys. This rule must be used with care as it would not be suitable for a low rise hotel with 30% of its rooms at the entrance level. Neither would it be suitable for a high rise hotel with a small footprint. There are also differences between the operational needs of a 'transit' hotel near to airports, etc. where guests stay one night, and hotels used by longer-term and holiday guests.

It is recommended that the service traffic (baggage, goods, room service, messengers, etc.) be served by a secondary vertical transportation system, leaving the main lifts for the guests. As a rule of thumb, there should be one passenger/goods lift for every two passenger lifts.

There are no changes required to Equation (4.8).

8.4.7   Railway stations

Railway stations may be served mainly by stairs and pedestrian ramps, although some, particularly the deeper underground stations, will use escalators. Generally, railway stations, whether above or below ground, have poor provision of lifts. This will change as the requirements to assist persons with limited mobility are applied.

When passengers want to move from one level to another with hand baggage, difficulties arise. When baggage trolleys are used, these difficulties increase. As with shopping centres (Section 8.4.10), a solution to this problem is to install moving ramps, and this also greatly improves circulation.

Lifts should be considered to be the main means of vertical movement in this case. There are no changes required to Equation (4.8).

8.4.8   Residential buildings

The customary basis for the traffic design of a residential building (apartments/flats) is to determine the number, car capacity and speed of passenger lifts necessary to adequately handle a 5-minute, two-way traffic period based on the type of occupancy. The estimation of the population in a residential building is usually based on the number of bedrooms and the occupancy per bedroom. Suitable rules of thumb for the number of persons occupying a flat are given in Table 8.6.

The commonly used design period for a residential building is the afternoon 5-minute, two-way traffic condition, which is considered the most demanding traffic period. During this period of time, people are both entering and leaving the building. The lifts are loading passengers at the main lobby, distributing these passengers to various upper floors, reversing at the uppermost hall call, stopping in the down direction for additional passengers, and transporting them to the main lobby. In low income housing, where many children and adults are leaving for school and work at the same time, the morning down peak may also be very heavy. Table 8.7 gives guidance.

Table 8.6 Occupancy factors for residential buildings

Type	Luxury	Normal	Low income
Studio	1.0	1.5	2.0
1 Bedroom	1.5	1.8	2.0
2 Bedroom	2.0	3.0	4.0
3 Bedroom	3.0	4.0	6.0

Table 8.7 Design criteria: residential buildings (5-minute, two-way)

Type	Interval (s)	Handling capacity
Low income	≤50–70	≥5–7%
Normal	≤50–60	≥6–8%
Luxury	≤45–50	≥8%

Often in residential and low income (local authority) flats, one passenger lift is generally arranged to allow furniture movement, coffins and stretchers and to handle other service needs. Luxury flats may include a separate goods lift for furniture, tradespeople and domestic help. These goods lifts are usually 'hospital shaped' with capacities of around 2,000 kg.

There are also requirements that each flat shall have access to an alternative lift during maintenance or out of service conditions. This has often been achieved in the past for low income blocks of flats, by using two simplex lifts operating on an interleaved (skip stop) basis. This solution is not recommended today. Another solution for low income residential blocks is high level walkways to an alternative lift.

There are no changes required to Equation (4.8), but care must be exercised in its application.

8.4.9   *Residential care homes and nursing homes*

Homes generally have a low traffic requirement, which can be catered for by a single lift. Larger homes might acquire a second lift, giving security of service in the event of breakdown or maintenance.

There is little opportunity to use the round trip time equation.

8.4.10 *Shopping centres*

Circulation in shopping centres can be considered as circulation in large circulation spaces (see Section 2.5). Shopping centres are generally built on two or three levels of retail with several levels of car parking. Lifts do not play a major part in the transportation arrangements, which are usually centred on escalators, although inclined moving walks are being installed in increasing numbers, too. Often scenic, or observation lifts are provided not only for transportation, but as an enjoyable experience. These lifts are usually hydraulic with slow flight times and slow door times. The traffic handling in a shopping centre is eased by the enjoyment aspects and the many modes of movement available.

Attention must be given to the transportation of disabled persons and pushchairs. Often, 'pram' lifts are provided. The shape of these lifts is important, as is their positioning on the concourse.

The use of car park lifts is determined mainly by the maximum rate of entry of vehicles and the average occupancy of each vehicle. These figures are usually determined from an associated (road) traffic study.

As a rough guide to lift provision, assume one person of lift capacity is required for each 100 m² of gross lettable retail area. So, a 4,000 m² store would require two 2,000 kg lifts (see Table 5.8). Lifts should always be located in pairs and not singly in order to provide a reasonable interval of 40–60 s and security of service during breakdowns and maintenance. It is unlikely that the lifts will fill to more than 50%, and if trolleys are available this could be an optimistic value.

There are no changes required to Equation (4.8), but care will need to be taken in the assumptions made regarding vehicle occupancy levels.

A major problem in multi-level shopping complexes is the movement of shopping trolleys from one level to another. Escalators can be designed to accept a trolley, but are often too dangerous in use as goods on or in them can dislodge and injure other escalator users. A commonly applied solution to this problem is to install inclined moving walks and this greatly improves circulation and they are generally safer in use.

8.4.11 Universities and other education buildings

University buildings can be classified as institutional buildings where the occupants receive a service. Where universities occupy city sites, many have tall buildings (10–20 stories) and even those on out of town sites follow suit in order to reduce land use and keep a compact campus. Most buildings are mixed function: lecture rooms, laboratory and offices, although some buildings may specialise as lecture blocks. There are hourly cycles of 10 minutes of demand before and after each 50-minute lecture, tutorial or seminar session. In between the (change over) peaks, the activity levels are low.

Often, a university campus will have a mixed collection of office-type buildings, halls of residence, catering services and factory-like units containing teaching and research equipment (laboratory reactors and the like). The office-type buildings can be treated in the same way as detailed in Chapters 4–7. The halls of residence can be treated in a similar way to hotels, although perhaps at lower levels of demand and performance. The catering services can be attached to either the office-type buildings or halls of residence, and should be treated similarly to those provided in office facilities or hotel facilities. The research buildings will probably be low rise and be subject to special movement provisions associated with the equipment installed, i.e., barriers to radioactive areas.

The dominant feature of a university campus is the lecture change over periods. To install lifts in a tall building to suit this demand is not cost effective (and universities do not have large capital budgets). Somehow, the demands made for 10 minutes every hour must be reduced. A solution is to try to rearrange the activities in the building to reduce the load on the lifts. An example in Tutt and Adler (1990)[10] illustrates the relationships in a small firm. A set of relationships can be formed for a university building. Some suggestions are:

- Place lecture facilities in the lower levels, say basement, ground and three to four floors above the entrance level. The general public will generally ascend one to two floors up and descend two to three floors down. Students (who are mainly young and fit) will probably go one more floor. For this activity to occur, it is essential to provide wide, well lit and visible stairs.
- Laboratory, bulk service facilities (computer clusters, libraries) and student administration (registrars, bursar, career advisory, etc.) can be placed from the fourth floor upwards. These are either used for periods longer than one hour (laboratories), shorter than one hour (administration) or randomly (libraries).
- Offices should be placed at the top of the building. Their occupants will generally use the lifts on a more random basis.

There are no changes required to Equation (4.8), but care must be exercised in its application.

8.5   Improving uppeak handling capacity (boosters)

8.5.1   *Rationale*

Sometimes, the traffic designer specifies too few lifts, or the architect refuses to provide sufficient space for the number of lifts required, or the building population increases, and the installed lift system cannot handle the uppeak traffic demand. Several techniques are available to improve the uppeak handling capacity of an installation (Barney, 1992). These techniques, sometimes called 'boosters', have to be used carefully (Barney, 2000c) as they generally cause other traffic conditions to deteriorate (see Section 8.5.4).

8.5.2   *Conventional uppeak traffic control*

Under conventional traffic control, once the uppeak traffic condition has been detected (e.g., by load weighing, number of car calls registered, etc.) all cars are returned to the main terminal floor after the last passenger has disembarked at the high call reversal floor (H). Down landing calls are ignored or serviced on an occasional basis. This traffic control system is available from most manufacturers.

Equation (4.8) is used to calculate the round trip time (RTT). Using Equation (4.5), the interval (INT) can be found and using Equation (4.3) or Equation (4.4), the handling capacity (HC) can be found. The passenger service interval ($PSINT$) will be equal to the interval (INT) as each passenger enters the first car to arrive.

Definition 8.7: The passenger service interval is the period of time, in seconds, between lift arrivals serving the destination floor of the passenger.

Example 8.6

Consider a 16-floor building served by six lifts with the following data:

$L = 6$, $N = 16$, $P = 10$, $t_v = 1.0$ s, $t_s = 8.0$ s, $t_p = 1.0$ s.

For $P = 10$, from Table 6.11a: $H = 15.0$, $S = 7.6$

Using Equation (4.8) the $RTT = 119$ s.
Using Equation (4.5) the $INT = 20$ s.
Using Equation (4.3) the $HC = 151$ persons/5 minutes.
Using Equation (6.4) the $ATT = 48$ s.
Using Equation (6.7) the $AJT = 58$ s.

This example will be used to compare the other four traffic algorithms discussed below. It will be seen that there are more stops compared to the other four traffic algorithms. In addition, all the cars must travel to a reversal floor, high in the building.

8.5.3   *By sub zoning*

In sub zoning systems, the building zone is divided into two sub zones, and the lift group is divided into two subgroups for the duration of the uppeak period. The cars are permanently allocated to a sub zone, and passengers are directed, by illuminated signs, to the subgroup that serves their floor. The sub zones may not contain equal numbers of floors, nor may equal numbers of lifts serve each sub zone. The technique works well with at least six lifts in the group and has been offered by a number of lift manufacturers.

The effect is to reduce the number of floors served and hence the number of probable stops each lift can make. This reduces the value of the middle term in the round trip time equation (4.8), and increases the handling capacity. The disadvantage is that passengers will have to wait longer for a lift serving their sub zone. The traffic designer should always attempt to provide similar interval times at the main terminal for each sub zone.

The calculation would use Equation (4.8) for the lower sub zone. The upper sub zone is calculated in the same way, except allowance has to be made for the express jump through the lower sub zone by using Equation (5.1).

Example 8.7

Consider the building in Example 8.6 with a lower sub zone of nine floors and an upper sub zone of seven floors.

Lower sub zone:
For $P = 10$ from Table 5.11a: $N = 9$, $L = 3$ $H = 8.6$, $S = 6.2$.

Using Equation (4.8) the $RTT = 95$ s.
Using Equation (4.5) the $INT = 32$ s.
Using Equation (4.3) the $HC = 95$ persons/5 minutes.
Using Equation (6.4) the $ATT = 39$ s.
Using Equation (6.7) the $AJT = 55$ s.

Upper sub zone:
For $P = 10$ from Table 6.8a: $N = 7$, $L = 3$, $H = 6.7$, $S = 5.5$.

Using Equation (5.1) the $RTT = 111$ s.
Using Equation (4.5) the $INT = 37$ s.
Using Equation (4.3) the $HC = 81$ persons/5 minutes.
Using Equation (6.4) the $ATT = 49$ s.
Using Equation (6.7) the $AJT = 67$ s.

Table 8.8 compares the original building (Example 8.6) with the sub zoned building (Example 8.7). The table indicates that the handling capacity has been increased from 151 to 176 persons/5 minutes, i.e., by 16%. This is the equivalent of one extra lift. The passenger

Table 8.8 Comparison for uppeak sub zoning from Example 8.7

Zone	N	L	INT	HC	ATT	AJT
Original building	16	6	20	151	48	63
Lower sub zone	9	3	32	95	39	55
Upper sub zone	7	3	37	81	49	67

All are figures rounded

service interval (*PSINT*) will be equal to the interval of the lifts serving the respective sub zone. There is an increased passenger waiting time (20 s to 32 s (lower zone) and 37 s (upper zone)) as indicated by the increased interval times. But the passengers take less time travelling in the lifts in order to reach their destinations (48 s to 33 s (lower zone) and 37 s (upper zone)). The average journey time for passengers travelling to floors in the lower sub zone is smaller and those travelling to floors in the upper sub zone will take a little longer time (63 s to 67 s).

8.5.4   By sectoring

Uppeak sub zoning can be extended by dividing the building into more than two sub zones or sectors. The number of sectors can be made equal to (or slightly less than) the number of cars. Cars are not permanently assigned to a sector. As cars arrive at the main terminal floor, they serve the sectors strictly in turn. Passengers will have to wait longer for service, but the group interval is smaller. Passengers are directed to cars serving their floors by destination signs above the cars, and have to continually scan the destination panels placed outside each lift entrance until they find their desired destination indicated. Where there are more lifts than sectors, it allows some lifts to travel back to the main terminal as the others travel up the building. Each sector generally contains the same number of floors, except the highest may have fewer floors and the lowest may have more floors. The number of floors in each sector is small, e.g., 3/4/5, and consequently, the round trip time is reduced and the handling capacity increased. The passenger service interval (*PSINT*) will be equal to the interval of the group multiplied by the number of sectors. One lift manufacturer (Powell, 1992) has proposed this system.

   To analyse the system, the procedure used for sub zoning could be extended, and each sector calculated individually. The individual round trip times could then be averaged, and an average interval found for all the lifts in the group. This group interval would then need to be adjusted by multiplying it by the number of sectors to obtain the passenger interval. An alternative is to consider a notional sector placed centrally in the served zone and to calculate its round trip time, interval, etc.

Example 8.8

Consider the building in Example 8.6. The building will be divided into four sectors of four floors each. There will therefore be six lifts 'sharing' four sectors. The notional sector for a calculation will be considered to be floors 7–10.

   The sectored building:

For $P = 10$ from Equation (5.14): $H = 3.9$, and Equation (5.7): $S = 3.8$.
Using Equation (5.1) the $RTT = 80$ s.
The group interval $= 13$ s.
The passenger service interval will be four times the group interval, i.e., 56 s.
Using Equation (4.3) the group $HC = 211$ persons/5 minutes.
Using Equation (6.4) the $ATT = 37$ s.
Using Equation (6.7) the $AJT = 65$ s.

   The number of stops is reduced as each lift now only serves four floors. Table 8.9 indicates that the handling capacity has been increased by 46% from 151 to 211 persons/5 minutes. This

Table 8.9 Comparison for uppeak sectoring from Example 8.8

Zone	N	L	INT	HC	ATT	AJT
Original building	16	4	20	151	48	63
Sectored building	16	4	56*	211	37	65

All figures are rounded. *$PSINT = 4 \times 14$ s $= 56$ s

is the equivalent of nearly three extra lifts. The passengers have to wait on average 2.8 times longer, as indicated by the increased interval time (20 s to 56 s) for a lift serving their destination. However, they spend less time in the lift (48 s to 37 s), although their average journey times are slightly longer (63 s to 65 s).

A sectoring system can be improved by applying dynamic sectoring where the number of floors in a sector is not fixed, but changes according to the number of passengers in a lift (car capacity factor). Thus, if lifts travel to a particular sector above a threshold capacity, consideration can be given to changing the sector boundaries. This technique will increase the handling capacity, but could further confuse passengers when they try to locate a lift travelling to their floor. It is also more complex technology.

8.5.5   By call allocation: single zone

Research in the 1970s (Closs, 1970, 1972), reported by Barney and Dos Santos (1977), and implemented by a lift manufacturer (Schroeder, 1990b, 1990c), provided a different approach. This was a new signalling arrangement whereby passengers registered their destinations before they entered a lift. This system is generally termed call allocation, sometimes hall call allocation, destination call allocation, adaptive call allocation, etc.

If a keypad is provided (see Figure 11.5) at the main terminal floor, passengers can register their destination floor, and a more efficient allocation can be made. Cars are not permanently allocated to specific floors. Passengers are notified (on the keypad) which car will take them to their destination immediately they have registered their call.

The call allocation traffic supervisor allocates passengers to lifts in a way that ensures a smaller number of stops. This causes the round trip time to become shorter, the passenger waiting time to generally become longer, and the passenger journey time to generally become shorter.

The basis of the algorithm is to consider not just one lift (conventional procedure) but several lifts at once. The number of lifts considered can vary from two to four. The effect of a larger number is longer waiting times. A compromise is usually made at between two and three lifts. The number of lifts considered is termed 'the look ahead factor', k. The conventional calculation of H and S is thus not possible. Fortunately, Schroeder (1990d) offers a solution.

The conventional formula for S is given by Equation (5.7) and the modified formula by Schroeder is given by Equation (8.11):

$$S = N\left[1 - \left[\frac{N-1}{N}\right]^P\right] \tag{5.7}$$

$$S = \frac{N}{k}\left[1 - \left[\frac{N-1}{N}\right]^{kP}\right] \tag{8.11}$$

It will be noted that Equation (8.11) implies that each lift will only serve N/k floors, but with a large lift carrying kP passengers. Similarly, the conventional formula for H is given by Equation (5.14) and the modified formula by Schroeder is given by Equation (8.12):

$$H = N - \sum_{i=1}^{N-1} \left[\frac{i}{N} \right]^P \tag{5.14}$$

$$H = N - \sum_{i=1}^{N-1} \left[\frac{i}{N} \right]^S \tag{8.12}$$

The value obtained for H in Equation (8.12) is dependent not on the passengers carried (P), as in the conventional case, but on the number of stops (S) made. The passenger service interval ($PSINT$) will be k multiplied by the group interval.

Example 8.9

Consider the building in Example 8.6 with a call allocation traffic controller using a look ahead factor (k) of 2.
For $P = 10$ using Equation (8.11) $S = 5.8$[11] and using Equation (8.12) $H = 14.1$.[12]

Using Equation (4.8) the $RTT = 102.6$ s.
Using Equation (4.5) the $INT = 17$ s.
Using Equation (4.3) the $HC = 175$ persons/5 minutes.
Using Equation (6.4) the $ATT = 40$ s.
Using Equation (6.7) the $AJT = 57$ s.

Table 8.10 indicates that the handling capacity has increased by 16% from 151 to 175 persons/5 minutes, the equivalent to one extra lift. The passengers have to wait 1.7 times longer, as suggested by the increased interval time ($PSINT$), i.e., from 20 s to 34 s, for a lift serving their destination. However, they spend a little less time in the lift (48 s to 40 s) and their journey times are shorter (63 s to 57 s).

8.5.6   By call allocation: sub zoning

A further technique with call allocation is to use dynamic sub zoning. Here, the building is divided into two sub zones. The boundary of the sub zones can change according to the demand to each of the sub zones, determined by the individual car loadings. The intending passengers will be unaware of the changing boundary as they are always told at call registration which

Table 8.10 Comparison for call allocation from Example 8.9

Zone	N	L	INT	HC	ATT	AJT
Original building	16	6	20	151	48	63
Call allocation	16	6	34*	175	40	57

All figures are rounded. *k = 2, so twice the calculated interval lift they are to travel in. The passenger service interval will be k multiplied by twice the group interval.

The calculation will be the same as that employed for a single zone hall call allocation system, but with a reduced number of floors. In the case of the higher sub zone, account must be taken of the time to pass through the lower sub zone.

Example 8.10

Consider the building in Example 8.6 with a call allocation traffic controller using a look ahead factor (k) of two with a lower sub zone of nine floors and an upper sub zone of seven floors.

Assume three lifts per sub zone.

Lower sub zone:

Using Equation (8.11) $S = 4.0$[13] and using Equation (8.12) $H = 7.8$.[14]
Using Equation (4.8) the $RTT = 76$ s.
Using Equation (4.7) the $INT = 25$ s.
Using Equation (4.6) the $HC = 119$ persons/5 minutes.
Using Equation (6.4) the $ATT = 30$ s.
Using Equation (6.7) the $AJT = 55$ s.

Upper sub zone:

Using Equation (8.11) S is 3.4[15] and using Equation (8.12) H is 5.9.[16]
Using Equation (4.8) the $RTT = 85$ s.
Using Equation (4.7) the $INT = 28$ s.
Using Equation (4.6) the $HC = 105$ persons/5 minutes.
Using Equation (6.4) the $ATT = 36$ s.
Using Equation (6.7) the $AJT = 64$ s.

Table 8.11 indicates that the handling capacity has increased by 50% from 151 to 224 persons/5 minutes, the equivalent to three extra lifts. The passengers have to wait 2.7 times longer, as suggested by the increased interval time (*PSINT*), i.e., from 20 s to 50/56 s, for a lift serving their destination. However, they spend less time in the lifts (48 s to 30/36 s) travelling to their destination floor, and their average journey times are shorter or the same (63 s to 55/64 s).

Table 8.11 Comparison for sub zoned call allocation system from Example 8.10

Zone	N	L	INT	HC	ATT	AJT
Original building	16	6	20	151	48	63
Lower sub zone	9	3	50*	119	30	55
Upper sub zone	7	3	56*	105	36	64

All figures are rounded. *$k = 2$, so twice calculated interval

Table 8.12 Comparison of the four 'boosters' and the original (underlying) system

System	INT	HC	ATT	AJT
Original building	20	151	48	63
Subzoning	33	176	44	61
Sectoring	52	211	37	65
Basic call allocation	34	175	40	57
Call allocation with sub zoning	52	224	33	60

8.5.7 Some conclusions

The four techniques described above: sub zoning, sectoring, call allocation and call allocation plus sub zoning all increase the handling capacity of the underlying lift installation. There is always a penalty to pay in terms of increased waiting times and journey times. The travel times, however, are shorter. Table 8.12 compares the four techniques where, for the sub zoned systems, the handling capacities have been added and the intervals, travel times and journey times have been averaged across the sub zones.

The techniques substantially increase (boost) the handling capacity of the underlying system for the uppeak traffic condition. If a building is under lifted, or the population increases, then the techniques can be used to improve the handling of the morning peak traffic.

The sub zoning and basic call allocation techniques offer about 16% extra handling capacity with shorter travel times and similar journey times. The waiting times increase, however, by 65% (interval changes from around 20 s to 33 s). The call allocation technique can hide most of this increased time from the passenger. This is possible as the call registration panel can be some distance from the lift that the passenger will travel in. The passenger will therefore take some time to walk to the lift entrance. This walking time can be used to overlap the actual service time to the passenger. The walking time can be (say) 10 s, but it should not be any longer than the group interval in case an assigned lift is delayed.

The sectoring and call allocation with sub zoning offer about 50% more handling capacity, a reduced travel time, but a 25% longer journey time. The waiting time increases considerably by 250% (interval from 20 s to 50 s). The call allocation technique can hide only some of this time from the passenger.

If a lift system has sufficient handling capacity, but the passenger waiting, travel and journey times are unacceptable, then the techniques can be used to reduce these times. For example, if with sub zoning the required handling capacity was still to be 151 persons/5 minutes (Table 8.12) then the interval (waiting time), travel time and journey times would fall. A further benefit in these circumstances is that the car loads will also be reduced, giving the passengers a more pleasant ride.

Uppeak boosters can improve the overall performance of a system for uppeak, either by increasing the handling capacity or by improving the passenger times. However, such techniques do not improve the performance of the other major traffic conditions. They stay the same.

8.6 Dealing with a group of lifts with different specifications

It is usual for all the lifts in each group to have the same specification, i.e., speeds, capacities, number of floors served, operating times, etc. Occasionally one or more lifts in a group will be different. Examples are:

- Firefighting lift of smaller and slower specification.
- Goods/passenger lift of larger and slower specification.

- Lifts that serve a number of basement floors in addition to the general floors.
- Lifts that serve a number of penthouse floors in addition to the general floors.

To illustrate the calculation, consider two examples.

Example 8.11

In a group of four lifts, one lift is used for goods. The data are:

$N = 12$, $P = 16$, $H = 11.7,5 = 9.0$.

Regular lifts: $t_v = 1.0$ s, $t_s = 10.0$ s, $t_p = 1.2$ s.

Goods lifts: $t_v = 2.0$ s, $t_s = 13.0$ s, $t_p = 1.0$ s.

The round trip time, from Equation (4.8), for the regular (*reg*) lifts is 161.8 s.
The round trip time, from Equation (4.8), for the goods (*gds*) lift is 208.8 s.
The average round trip time will be:

$$\frac{3 \times RTT_{reg} + 1 \times RTT_{gds}}{4} = \frac{3 \times 161.8 + 208.8}{4} = 173.6 \text{ s}$$

and the interval will be 43.4 s.

Example 8.12

Peters (1997a) analysed a system of four lifts, two of which served the main terminal and two served one basement (see Section 8.3.3). He found the round trip time of the two lifts not serving the basement was 80 s, and the two lifts serving the basement (*bas*) was 90 s. Following the same procedure as for Example 8.11 above, the average round trip time will be:

$$\frac{2 \times RTT_{reg} + 2 \times RTT_{bas}}{4} = \frac{2 \times 80 + 2 \times 90}{4} = 85 \text{ s}$$

The interval at the main terminal will be 85/4, i.e., 21 s. However, the interval at the basement will be 90/2, i.e., 45 s.

8.7   Transportation for disabled people

The discussion so far has assumed that all persons circulating in a building are fully able bodied. However, a large proportion of the population are disabled in some way.

Local regulations sometimes require specific actions to be taken. For example, the Equality Act 2010 in the UK and the Americans with Disabilities Act legislation in the USA lay down various provisions and regulations. Generally, the regulations require equipment providers to make reasonable changes to improve service for disabled people. The European standard EN 81–70: 2003 '*Accessibility to lifts for persons including persons with disability*' states at Annex A:

> *Accessibility enables people, including persons with disability, to participate in the social and economic activities for which the built environment is intended.*

The factors likely to affect traffic design are:

1   Cars may need to specific sizes, e.g., the platform size of a lift with a rated load of 1,275 kg provides a motorised wheelchair with full manoeuvrability.
2   Door sizes shall be at least 900 mm clear opening.
3   Door dwell times may need to be considerably extended from the 2 s to 3 s acceptable by the able bodied to from 5 s to 7 s.
4   Door operating times, particularly closing times, may need to be longer.
5   Rated speeds may need to be lower to achieve better levelling.
6   Signage may need to be larger and clearer.
7   With call allocation destination systems, the walking times may need to be longer.
8   With call allocation destination systems, an empty car may be taken out of group control to be provided to persons with impaired mobility.

These factors will affect various terms in the round trip time equation.

8.8   Concluding remarks

This chapter has examined a number of special situations and how the round trip time equation might need to be changed. Various methods to deal with them have been discussed. Some techniques to improve or boost uppeak performance have been analysed. See Table 8.13 for a summary.

Table 8.13 Summary of special situations

Situation		Comments	Equation
8.2.1	Shuttle lifts & sky lobbies	Long travel distances affect t_v. Increased passenger transfer times with two-way traffic.	4.8, 8.1
8.2.2	Double deck lifts	Double 'values' for t_v, $t_f(1)$; 'half values' for N. Handling capacity twice the value for one deck.	8.9
8.2.3	Firefighting lifts	When standalone use Equation (4.11). When part of a group may have different specification.	4.8
8.2.4	Goods lifts	When standalone use Equation (4.11). When part of a group may have different specification.	4.8
8.2.5	Observation lifts	Low performance.	4.8
8.2.6	Platform lifts and lifts for the disabled	When standalone not considered as part of circulation. When part of a group may have different specification.	4.8
8.3.1.1	Stacked zones	Express jumps to be considered.	5.1
8.3.1.2	Interleaved zones	'Double values' for t_v, $t_f(1)$; 'half values' for N.	4.8
8.3.1.3	Transfer floors	Decide which zone a transfer floor belongs to during peak traffic.	4.8, 5.1
8.3.2	Very tall buildings	Complex situation. Apply general, sky lobby and double deck equations.	4.8, 5.1, 8.9
8.3.3	Basement service	Use for service from basement and to basement. Select percentage demand carefully.	8.10
8.3.4	Multiple entry and entrance bias	Apply basement equations. Care in deciding entrance demand split.	4.8, 8.10

(Continued)

Table 8.13 (Continued)

Situation		Comments	Equation
8.3.5	Lobby design	Adjust t_p to account for inefficiencies.	4.8
8.3.6	Door width & car shape	Adjust t_p to account for inefficiencies.	4.8
8.4.1	Airports	Lifts have lower performance. Car occupancy levels affected by baggage trolleys.	4.8
8.4.2	Car parks	Demand affected by vehicle ramp capacity. Critical demand during evening for outgoing passengers.	4.8, 8.10
8.4.3	Department stores	Lifts have lower performance. Car occupancy levels affected by prams and shopping.	4.8
8.4.4	Entertainment buildings	Difficult application of Equation (4.11).	4.8
8.4.5	Hospitals	Lifts critical in tall hospitals.	4.8
8.4.6	Hotels	Highest demands at check out and check in. Low car occupancy.	4.8
8.4.7	Railway stations	Poor provision of lifts.	4.8
8.4.8	Residential buildings	Highest demand late afternoon.	4.8
8.4.9	Care and nursing homes	Low usage.	n/a
8.4.10	Shopping centres	Lifts have lower performance. Car occupancy levels affected by prams and shopping.	4.8
8.4.11	Educational buildings	Highest demand when teaching periods begin/end.	4.8

Experience is required to correctly design these installations.

Notes (shown in text as[1])

1 Sears Tower, Chicago, USA.
2 PETRONAS Towers, Kuala Lumpur, Malaysia.
3 Kavounas counsels that a more accurate expression could be obtained if the variances of the number of passengers carried on each deck were to be considered. The improvements would only be secondary as most double deck cars will be designed to fill to capacity during up peak, thus reducing the variances.
4 To prove this, replace S in Equation (8.3) by the expression for S given by Equation (5.7), expand combine and simplify to obtain Equation (8.2).
5 Fortune (1996) provides a comprehensive glossary of double deck lift terms.
6 Page 338.
7 See Definition 10.5.
8 *Elevator World*, page 102, November 1988.
9 Page 35.
10 Page 121.
11 kP is 20. Use Table 5.11b, look down column for 20 persons and across row 16 and divide value by 2.
12 Index will be 5.8. Use Table 5.11a, interpolate between columns 5 and 6 to read a value opposite row 16.
13 kP is 20. Use Table 5.11b, look down column for 20 persons and across row 9 and divide value by 2.
14 Index will be 4.0. Use Table 5.11a, extrapolate from column 5 to deduce a value opposite row 9.
15 kP is 20. Use Table 5.11b, look down column for 20 persons and across row 7 and divide value by 2.
16 Index will be 3.4. Use Table 5.11a, extrapolate from column 5 to deduce a value opposite row 7.

9 General philosophy of lift traffic design by calculation

9.1 Introduction

This chapter deals with traffic calculations only, computer simulations are dealt with in Chapters 16 and 17.

Generally, many traffic designs can be carried out by calculation, hence the importance of understanding the calculation method.

The traffic design procedure should always start by applying the classical calculation methods discussed in Chapters 4–8. It is often possible to modify these methods to take into account some unusual features in a building, or an unusual passenger demand. This may be all that is needed to produce a satisfactory traffic design.

It is important to remember that traffic design is not an exact science and another designer will make different decisions and assumptions, which will affect the outcome. Each design will have validity.

Remember also, that defining a distance to the nearest millimetre or a time to one hundredth of one second is not necessary. Any traffic design that is implemented will be used by human beings, who tend to modify their behaviour to suit the environment in which they exist. Thus, errors of as much as ten or so per cent are often not significant. Gross errors, however, of say one lift too few (under lifted) will be noticed, as is evident during peak periods when a lift is out of service. In contrast, too many lifts (over lifted) are a waste of capital and recurrent expenditure.

Where the design requirements are more complicated, computer simulations may need to be carried out. Simulation models allow all traffic conditions to be examined more exactly.

It is important to remember, however, that all calculations are based on a mathematical model and deliver an average answer to a design. The designs obtained from a mathematical model and a simulation model may be quite different as a result.

This chapter describes a methodical (philosophical) procedure for the traffic design of lifts. It takes the form of illustrative examples. The reader does not have to study Chapters 4–8 to apply the procedures. Links back to these chapters are indicated for further study and a close in-depth understanding.

Calculation programs embedded in an Excel spreadsheet implementing the iterative balance method (IBM; see Sections 6.11 and 6.12) will be used in this chapter to automate the underlying equations. The designer of any spreadsheet that is developed must be familiar with the principles on which it is based as described in Chapters 4–8.

In association with this chapter, Traffic Case Studies 9–14 should also be consulted.

9.2   Generalised design approach

Chapters 4 to 8 have introduced the concept of the round trip time. Calculating the round trip time is the building block of lift traffic analysis and design. This section will introduce the overall approach to design.

Any compliant design must meet the two user requirements: quality of service as represented by a target interval (int_{tar}), and the quantity of service matching the passenger demand arrival rate (λ), also expressed in relation to the building population as $\%POP$, by the handling capacity of the lift system.

Having introduced the method for calculating the round trip time, it is now possible to introduce the general approach for finding the required number of lifts (L). It is assumed that the L lifts will operate in one group. Under conventional group control, any passenger requesting service can board the first arriving lift of the L lifts in the group.

Once the round trip time has been evaluated, dividing it by the target interval stipulated by the user will provide the number of required lifts in the group (L), i.e.:

$$L = \frac{RTT}{INT_{tar}} \qquad (9.1)$$

The solution to Equation (9.1) rarely results in an integer value, so a 'rounding up' is applied to the nearest integer number of lifts. This recognises the fact that a whole number of lifts must be selected. From this value of L, it is possible to find the actual interval that will be achieved in practice by using Equation (4.5), i.e.:

$$UPPINT = \frac{RTT}{L} \qquad (4.5)$$

And the handling capacity can be calculated from Equation (4.4) as follows:

$$UPPHC = \frac{300}{RTT} \times P \times L \qquad (4.4)$$

There is a no guarantee that the quantity of service requirement will be achieved, i.e., the inequality shown below is not necessarily true:

$$UPPHC \text{ or } \%POP \geq (\lambda \times 300)/U \qquad (9.2)$$

It is important to note the effect of the assumed probable car capacity on the solution. The selection of the probable car capacity (PC) will determine the value of the average number of passengers (P) boarding the lift car in each round trip. The number of passengers will, in turn, affect the value of the round trip time and hence the required number of lifts in the group (L).

As has been discussed in the earlier chapters, the car will not fill up to its maximum probable capacity (see column 5 of Table 5.8) and it is customary to assume that it will only fill up to 80% of its maximum probable capacity (see column 6 of Table 5.8).

Setting the value of P to 80% often results in a non-integer (fractional) number. This is acceptable because the number of passengers P boarding the car is an average of all the various round trips. The parameter P is a random variable and the calculations presented here are using its average value.

9.3   Rule-based lift traffic design

It will be seen in Example 9.5, that the traffic designer was intuitively applying rules. Rule-based design methods have been presented for lift traffic design (Alexandris, 1988; Cho *et al.*, 2000; Alani *et al.*, 1995) and for software configuration (Sanjeev, 1991).

For example, in order to address the problem of excessive handling capacity, the following rule may be used:

if   {the handling capacity is much larger than the expected arrival rate}
and   {the actual interval is much smaller than the target interval}
then   {reduce the number of passengers}

This is what happened between Design Stage 1 and Design Stage 2 in Example 9.5.
To address a large handling capacity and very small interval, the following rule may be used:

if   {the handling capacity is larger than the expected arrival rate}
and   {the actual interval is much smaller than the target interval}
then   {reduce the number of lifts and increase the number of passengers}

The application of this rule can be seen in Example 9.5 between Design Stage 2 and Design Stage 3.
If handling capacity is high and the interval is high, the following rule may be used:

if   {the handling capacity is larger than the expected arrival rate}
and   {the actual interval is larger than the target interval}
then   {reduce the number of passengers}

The application of this rule can be seen in Example 9.5 between Design Stage 3 and Design Stage 4.
The sequence of the application of rules and the type of rules are often personal to the experienced traffic designer. However, the above illustrates one approach.

9.4   Simple design examples

In this section, it is assumed that the traffic designer must start with pre-defined lifts of a specified rated load from an architect's or developer's design.

Simple design examples are presented here that will provide a simple methodology for lift traffic design. The four examples assume the following conditions in the building:

1   Equal floor heights.
2   Rated speed attained in one floor journey.
3   Single entrance.
4   Uppeak traffic only.

Example 9.1 and 9.2 assume equal floor populations and Example 9.3 and 9.4 do not.

9.4.1   *Example 9.1: equal floor populations*

A building has six floors above the main entrance floor. The lift cars to be used have a rated load of 1,000 kg. The building has a total population of 540 persons and equal floor populations. Assume the following parameters:

Floor to floor height: 3.5 m
Rated speed: 1.0 m/s
Rated acceleration: 1.0 m/s^2
Rated jerk: 1.0 m/s^3
Passenger transfer time: 1.2 s
Door opening time: 1.8 s
Door closing time: 2.7 s

User requirements:

Interval = 30 s; $\%POP$ = 15%

It will be assumed that the car will fill up to 80% of its maximum probable capacity (11.4 persons). So, it will be assumed that the number of passengers travelling during the round trip time equation is 9.1 passengers.

As the floor populations are equal, the formulae for the probable number of stops, Equation (5.7), and the highest reversal floor, Equation (5.14), can be used. They do not need to be calculated as these values are tabulated in Table 5.11. However, using a spreadsheet in which the equations are solved, the following results are obtained for an installation of three lifts; see column 1 of Table 9.1.

RTT = 100.4 s
$UPPINT$ = 33.5
$\%POP$ = 15.1

The installation of three 1,000 kg lifts provides the user requirement for the quantity, but not the quality of service. As the actual interval is larger than the target interval, and the handing capacity is almost exactly equal to the arrival rate, one of the design criteria has been met, and thus, the system design is not compliant.

9.4.2   Example 9.2: equal floor populations

The following example also applies to a simple building and attempts to illustrate the same concepts as those presented in Example 9.1, but using a slightly larger building with a larger population and a higher rated speed.

A building has ten floors above the main entrance floor. The lift cars to be used have a rated load of 1,275 kg. The building has a total population of 800 persons and equal floor populations. Assume the following parameters:

Floor to floor height: 4.5 m
Rated speed: 1.6 m/s
Rated acceleration: 1.0 m/s^2
Rated jerk: 1.0 m/s^3
Passenger transfer time: 1.2 s
Door opening time: 2.0 s
Door closing time: 3.0 s

User requirements:

Interval = 30 s; $\%POP$ = 15%

It will be assumed that the car will fill up to 80% of its probable capacity, giving an effective average number of passengers (*P*) in a round trip of 11.0 passengers.

Using the spreadsheet method used for Example 9.2, the results are shown in column 2 of Table 9.1 for an installation of five lifts.

$RTT = 139.9$ s
$UPPINT = 28.0$
$\%POP = 14.7$

The installation of five, 1,275 kg lifts achieves the two user requirements.

9.4.3 Example 9.3: unequal floor populations, greatest population in upper floors

This example is an application of the case where the floor populations are unequal, and illustrates how the calculation method can be adapted. Different formulae for the value of *S*, Equation (7.7), and *H*, Equation (7.14), have to be used.

Consider the population distribution given in Example 7.3 of 565 persons:

Floor	1	2	3	4	5	6	7	8	9	10
Population	5	10	25	25	50	50	100	100	100	100

The lift system is to have the same parameters as Example 9.2 and same user requirements.

Floor to floor height: 4.5 m
Rated speed: 1.6 m/s
Rated acceleration: 1.0 m/s²
Rated jerk: 1.0 m/s³
Passenger transfer time: 1.2 s
Door opening time: 2.0 s
Door closing time: 3.0 s

User requirements:

Interval = 30 s; *%POP* = 15%

For simplicity, assume that the lifts load to an average (80%) of eight passengers. How many lifts are required?

Example 7.3 indicates that: $H = 9.8$ and $S = 5.01$.

Using a spreadsheet to calculate the performance is shown in column 3, Table 9.1:

$RTT = 119.9$ s
$UPPINT = 30.0$
$\%POP = 14.2$

The interval requirement has been achieved but the handling capacity is low. However, note that the capacity factor is 70%.

An increase in the average number of passengers to 8.6 persons raises the capacity factor to 75%, and returns the following data in column 4, Table 9.1):

Table 9.1 Data for Examples 9.1 to 9.4

	1	2	3	4	5
Input data	Value				
Number of floors	6	10	10	10	10
Rated load (kg)	1000	1275	1000	1000	1000
Actual car capacity (persons)	11.4	13.8	11.4	11.4	11.4
Number of passengers (persons)	9.1	11.0	8.0	8.6	8.0
Number of lifts	3	5	4	4	4
Rated speed (m/s)	1	1.6	1.6	1.6	1.6
Building population (persons)	540	800	565	565	565
Interfloor distance (m)	3.5	4.5	4.5	4.5	4.5
Express jump (m)	0	0	0	0	0
Express additional time (s)	0	0	0	0	0
Single floor flight time (s)	5.5	5.4	5.4	5.4	5.4
Door close time (s0)	2.7	3	3	3	3
Door open time	1.8	2	2	2	2
Advance door opening	0.5	0.5	0.5	0.5	0.5
Start delay	0.5	0.5	0.5	0.5	0.5
Passenger transfer time	1.2	1.2	1.2	1.2	1.2
Results	Value	Value	Value	Value	Value
Number of passengers	9.1	11.0	8.0	8.6	8.0
Highest reversal floor	5.8	9.6	9.8	9.8	7.0
Number of stops	4.9	6.9	5.0	5.0	5.0
Performance time	10.0	10.4	10.4	10.4	10.4
Round trip time	100.4	139.9	119.9	121.4	104.2
Interval	33.5	28.0	30.0	30.3	26.1
Handling capacity	82	118	80	85	92
Percentage population	15.1	14.7	14.2	15.0	16.3
Capacity factor (%)	80	80	70	75	70

$RTT = 121.4$ s
$UPPINT = 30.3$
$\%POP = 15.0$

The interval increases slightly to 30.3 seconds, thus achieving both user requirements with an installation of four, 1,275 kg lifts.

9.4.4   Example 9.4: unequal floor populations, greatest population in lower floors

Consider the population distribution given in Example 7.4 of 565 persons:

Floor	1	2	3	4	5	6	7	8	9	10
Population	100	100	100	100	50	50	25	25	10	5

Note this is a reverse distribution of Example 9.3.

The lift system is to have the same parameters as Example 9.4 and the same user requirements.

For simplicity, assume that the lifts load to an average (80%) of eight passengers. How many lifts are required?

Example 7.4 indicates that: $H = 7.01$ and $S = 5.01$.

Using a spreadsheet to calculate the performance is shown in column 5, Table 9.1:

$RTT = 104.2$ s
$UPPINT = 26.1$
$\%POP = 16.3$

Both user requirements have been (over) achieved again with a low capacity factor of 70% with an installation of four, 1,275 kg lifts.

9.4.5   Discussion

In the examples, sometimes the resulting design was compliant and sometimes one user requirement was and the other was not. Example 9.3 illustrated that changing a parameter in this case could achieve both requirements. However, this is not necessarily always the case. It is to be hoped that the architect or developer allows the traffic designer the opportunity to select an appropriate number of lifts of a suitable car capacity (rated load). The next section discusses this.

9.5   Advanced example showing the different options available

In order to illustrate the problem of arriving at a suitable design, consider the following example.

Example 9.5: an illustrative example

This example indicates a possible sequence of decisions.

Given data

Number of floors (N) = 8
Interfloor distance (d_f) = 3.6 m
Total population (U) = 600 persons

User requirements

Target interval (int_{tar}) = 30 seconds
Arrival rate ($\%POP$) = 12% of the building population in five minutes

Assumed data

Equal floor heights and floor populations, single entrance, uppeak (incoming) traffic.
Passenger transfer times = 1.2 s

Designer's decisions

The traffic designer, from experience, has selected five 1,275 kg lifts as they provide full manoeuvrability to wheelchair users. This implies 1,100 mm doors and the following door times from Table 5.3.

Door opening time (t_o) = 2.5 s
Door closing time (t_c) = 3.0 s

As the total travel distance is 28.8 m from Table 5.2, a suitable speed is 1.0 m/s (although 1.6 m/s might be better). There could be an engineering reason for not selecting a rated speed of 1.6 m/s, for example, insufficient head room for the higher speed. This gives kinematics (from Table 5.2) of:

Acceleration $(a) = 0.70$ m/s^2
Jerk $(j) = 0.75$ m/s^3

From a dynamics program, the single floor flight time is 6.0 s
As a starting point, assume a probable car capacity of 13.8 persons (Table 5.8). Using the 80% loading factor, this gives the number of passengers (P) equal to 11.0 passengers.
Table 9.2, column 1 shows:

Interval $= 27.7$ s (too low)
$\%POP = 19.8\%$ (very high)

Reduce the number of passengers to eight persons.
Table 9.2, column 2 shows:

Table 9.2 Data for Example 9.5

	1	2	3	4
Input data	*Value*			
Number of floors	8	8	8	8
Rated load (kg)	1275	1275	1275	1275
Actual car capacity (persons)	13.8	13.8	13.8	13.8
Number of passengers (persons)	11.0	8.0	11.0	7.0
Number of lifts	5	5	4	4
Rated speed (m/s)	1	1	1	1
Building population (persons)	600	600	600	600
Interfloor distance (m)	3.6	3.6	3.6	3.6
Express jump (m)	0	0	0	0
Express additional time (s)	0	0	0	0
Single floor flight time (s)	6	6	6	6
Door close time (s0)	3	3	3	3
Door open time	2.5	2.5	2.5	2.5
Advance door opening	0.5	0.5	0.5	0.5
Start delay	0.5	0.5	0.5	0.5
Passenger transfer time	1.2	1.2	1.2	1.2
Results	*Value*	*Value*	*Value*	*Value*
Number of passengers	11.0	8.0	11.0	7.0
Highest reversal floor	7.7	7.5	7.7	7.4
Number of stops	6.2	5.3	6.2	4.9
Performance time	11.5	11.5	11.5	11.5
Round trip time	138.5	122.8	138.5	116.6
Interval	27.7	24.6	34.6	29.1
Handling capacity	119	98	95	72
Percentage population	19.8	16.3	15.9	12.0
Capacity factor (%)	80	58	80	51

Table 9.3 Review of Example 9.5

Design Stage	P (persons)	L	int_{act} (s)	HC% (persons/5-minutes)	Comments
1	11.0	5	27.7	19.8	Excessive handling capacity
2	8.0	5	24.6	16.3	Too low interval
3	11.0	4	34.6	15.9	Too high interval
4	7.0	4	29.1	12	Acceptable design

Interval = 24.6 s (too low)
%POP = 16.3% (too high)

Reduce the number of lifts to four and increase the number of passengers to 11 persons Table 9.2, column 3 shows:

Interval = 34.6 s (too high)
%POP = 15.9% (too high)

Reduce the number of passengers to seven persons Table 9.2, column 4 shows:

Interval = 29.1 s (OK)
%POP = 12.0% (OK)

From this example, it can be seen that in some cases, it is possible to reduce the number of lifts by one lift, while still achieving a suitable interval.

It should also be noted that the capacity factor is now 51% and smaller lifts could be used, for example, 800 kg ($P = 7.6$). This might not be possible if persons with restricted mobility are to be catered for.

There is little point in increasing the rated speed to the next preferred value of 1.6 m/s, as the user requirements have been met.

Example 9.5 has illustrated the range of options available to the designer, and how the use of personal judgement and trial and error can be used to arrive at a feasible solution.

9.6   Zoned designs

9.6.1   Buildings with many floors

All the designs presented so far have been based on one single zone (i.e., the lifts in the group serve all the floors in the building). This is not always possible as the number of floors increases beyond 20 floors, when it becomes necessary to split the building into zones. This section presents a detailed design example.

9.6.2   The building to be considered

This is a case where an incoming occupant to an existing 31-storey building seeks reassurance that the installed lift installation meets their requirements.

The building is to accommodate: trader floors offices associated with the traders at the lowest level, a prestigious mid zone used as offices, and a less prestigious upper zone used as a call centre.

The building comprises three zones: low (floors 1–9), mid (floors 10–20) and high (floors 21–30).

The zonal boundaries are shown in Figure 9.1

The intended occupancy figures and floor activity provided by the prospective occupier are given in Table 9.4.

A facilities floor is to be situated at floor 7 in the low zone and provides restaurant, snack bar, reprographics, etc. for the whole building. Stops are possible to this floor by mid zone and upper zone lifts. It will be assumed that about 5% of the passengers arriving during the 5-minute uppeak period will stop at the facilities floor.

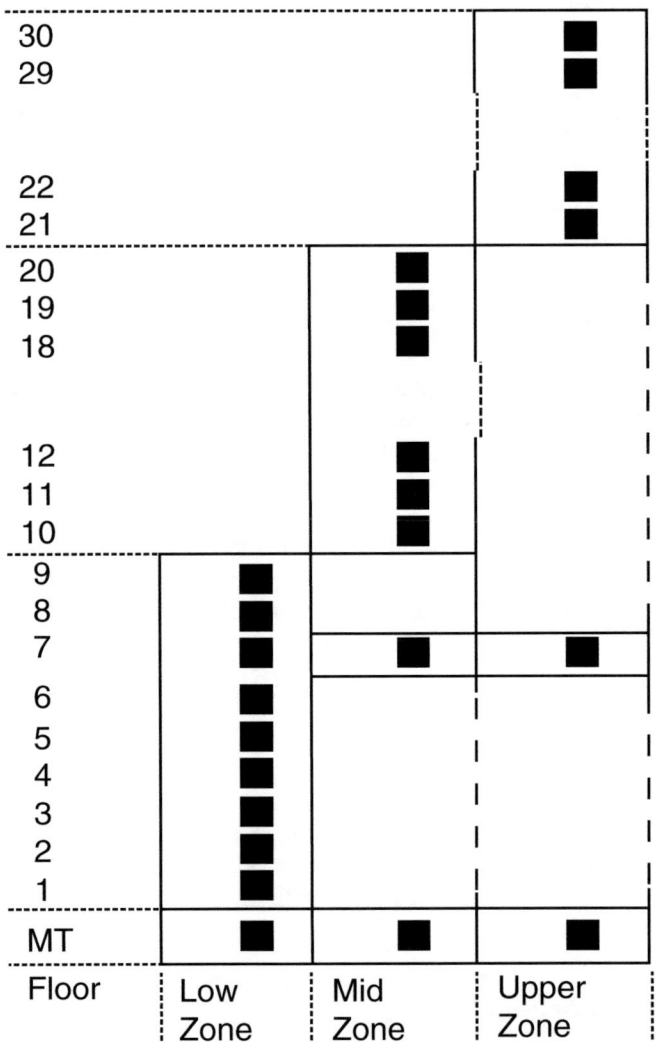

Figure 9.1 Zonal arrangement in the building to be considered

Table 9.4 Building data

Floor	Activity	Population
1–3	Traders	600/floor
4–6	Office	160/floor
7	Facilities	80/floor
8–9	Office	160/floor
10–20	Office	153/floor
21–30	Call centre	160/floor

Elevation: floor 1: 11.7 m, floor 10: 49.5 m, floor 21: 98.8 m

Table 9.5 Measured lift data

	Low zone	Mid zone	High zone
Number of floors	9	11	10
Floor ID	1–9	10–20	21–30
Interfloor distance (m)	4.2	4.2	4.2
Express jump (m)	11.7	49.5	98.8
Rated speed (m/s)	3.15	4.0	6.0
Acceleration (m/s^2)	0.9	1.0	1.0
Jerk (m/s^3)	1.3	1.5	1.5
Flight time (s)	5.1	4.8	4.8
Express time (s)	7.9 (11.7 m)	16.0 (45.3 m) *	22.4 (94.6 m) *
Door opening time (s)	1.8	1.8	1.8
Door closing time (s)	2.6	2.6	2.6
Start delay (s)	0.3	0.3	0.3
Advance door time (s)	0.5	0.5	0.5

* Note: the express time is not to the lowest floor of the zone, see explanatory text

There are no transfer floors between the mid and upper zones, except at floor 7 (or ground). This allows the building to be let as three separate zones, or to one tenant with appropriate security control of the access to each zone.

The lifts serving the zones are:

Low zone,	6 lifts, rated load 2,500 kg
Mid zone,	8 lifts, rated load 2,000 kg
High zone,	8 lifts, rated load 2,500 kg

Site measurements were made of the three lift installations, and the data obtained are shown in Table 9.5.

9.6.3 Classical design of low zone by calculation method

The lower zone comprises nine floors (floors 1–9) above the ground, with an escalator service provided to the lower three floors (floors 1–3). The peak hour for arrivals is from 09:00 to 10:00, and the traffic design calls for user requirements of a 15% uppeak demand with a 25 s interval. This zone contains the facilities floor at floor 7.

As the first three floors are served by escalators, the lifts might deal with 60 persons travelling to floor 1, 150 persons travelling to floor 2 and 300 persons travelling to floor 3, if the guidance of Table 2.11 is followed.

Thus, there are a total of 510 persons travelling to floors 1–3. Floors 4–6 and 8–9 have populations of 160 persons per floor, i.e., 800 persons. Some passengers will have their workplace on the facilities floor 7 (say 80), and other persons in the low zone may call there during the uppeak period (say 80). It will be assumed that 160 persons travel to floor 7. Thus, the total number of passengers to be transported by the low zone lifts is 1,470 persons, which for a 15% peak arrival demand, is 220 persons.

As the number of lifts, their rated speed and load, dynamics, etc. are known, the iterative balance procedure embedded in a spreadsheet (Section 6.11) will be used to calculate the performance. Table 9.6 (column Low) shows the calculation using the IBM program.

The six car group provides an interval of 24.3 s and a car occupancy of 17.8 persons.

A 6 car group will be satisfactory if lifts with a rated load of 2,500 kg are installed. They permit a maximum probable design occupancy of 23.8 persons. It is likely that the number of stops will be slightly lower than the number (7.9) stated in Table 9.5, as the demand for floor 1 will be lower and the demand for floor 3 will be higher than the demand for the other floors.

Table 9.6 Data for Example 9.5

	Low	Mid	High
Input data	Value		
Number of floors	9	12	11
Rated load (kg)	2500	2000	2500
Actual car capacity (persons)	23.8	20	23.8
Number of passengers (persons)	17.8	16.2	19.3
Number of lifts	6	8	8
Rated speed (m/s)	3.15	4	6
Building population (persons)	1470	1760	1680
Interfloor distance (m)	4.2	4.2	4.2
Express jump (m)	11.7	45.3	94.6
Express additional time (s)	7.9	16	22.4
Single floor flight time (s)	5.1	4.8	4.8
Door close time (s0)	2.6	2.6	2.6
Door open time	1.8	1.8	1.8
Advance door opening	0.5	0.5	0.5
Start delay	0.3	0.3	0.3
Passenger transfer time	1	1	1
Results	Value	Value	Value
Number of passengers	17.8	16.2	19.3
Highest reversal floor	8.9	11.7	10.8
Number of stops	7.9	9.1	9.3
Performance time	9.3	9.0	9.0
Round trip time	145.9	169.0	183.6
Interval	24.3	21.1	23.0
Handling capacity	220	230	252
Percentage population	14.9	13.1	15.0
Capacity factor (%)	75	81	81

9.6.4   Classical design of mid zone by calculation method

The mid zone comprises 11 floors (floors 10–20), the peak hour is from 08:30 to 09:30, and the traffic design calls for a 13% arrival rate with a 25 s interval.

The population of each floor is 153 persons, i.e., a total of 1,683 persons. The demand to reach floor 7 might be some 77 persons during the uppeak period, making the total mid zone population 1,760 persons, and the 13% uppeak demand will be 230 persons/5 minutes.

As the number of lifts, their rated speed and load, dynamics, etc. are known, the iterative balance procedure embedded in a spreadsheet (Section 6.11), will be used to examine the performance

To accommodate the facilities floor at floor 7, it is necessary to modify the calculation data. Floor 7 is two floors below the first stopping floor (floor 10) in the mid zone. To accommodate this and enable a calculation to be carried out, it will be assumed that floor 7 is positioned as floor 9. This will introduce an extra stop, increasing the floors served from eleven to twelve floors ($N = 12$). The express jump must also be reduced by 4.2 m (= 45.3 m), as shown in Table 9.6.

Table 9.6 (column Mid) shows that an 8 car group will provide the handling capacity of 230 persons per five minutes required at an interval of 21.1 seconds with an average load of 16.2 passengers in the cars. The occupancy level in the car is 81%.

The traffic design for the mid zone is complicated by the presence of the facilities floor just below the lowest floor in the zone. There will be an error caused by the amendment of the design data to accommodate the facilities floor traffic. However, the error should not be significant. There is a good case for study by simulation.

9.6.5   Classical design of upper zone by calculation method

The upper zone comprises ten floors (floors 21–30) and the occupants are required to be at their desks by 09:00 to provide a call centre service. A 15% peak arrival rate should be assumed. A 30 s interval would be acceptable for this zone.

The floor population is 160 persons per floor, i.e., a total of 1,600 persons. The demand to reach floor 7 might be some 80 persons during the uppeak period, making the total upper zone population 1,680 persons.

As the number of lifts, their rated speed and load, dynamics, etc. are known, the iterative balance procedure, embedded in a spreadsheet (section 6.11), will be used to examine the performance.

Floor 7 is 14 floors below the first stopping floor (floor 21) in the upper zone. To accommodate this, and enable a calculation to be carried out, it will be assumed that floor 7 is positioned as floor 20. This will again introduce an extra stop ($N = 11$), and again the express jump must be reduced by 4.2 m (= 94.6 m), as shown in Table 9.6.

Table 9.6 (column High) shows that an 8 car group provides the handling capacity of 15% and a lower than specified interval of 23.0 s with 19.3 passengers in the cars. This occupancy level is higher at 81%. The designer might assume that, as the floors are populated with call centre staff, there would be absences owing to illness, etc., which would reduce the demand.

The traffic design for the upper zone is complicated by the presence of the facilities floor being so far below the lowest floor in the zone. Again, there will be errors owing to the method of accommodation used to enable a calculation to be carried out. This again is a good case for study by simulation. The traffic design for the upper zone is considered further in the Simulation Case Studies CS18–21. These studies look at uppeak, down peak, interfloor and midday activity.

9.9.6   Discussion

The equipment chosen for the lift systems considered in this chapter will provide better than average performance with reasonable door timings and modest drive dynamics.

The results indicate very good interval times and that the handling capacities required are met. The values obtained for the intervals (see Tables 9.6 and 9.7) are often lower than those specified, but are low in order to achieve the specified handling capacities. No intervals are lower than 15 s when passenger loading inefficiencies might occur, where more than one lift is loading at the same time at the main terminal.

9.9.7   Escalator service

Finally, the design is not complete until the usage of the escalators at the lower three floors is considered. Can the escalators adequately serve these lower three (trader) floors? The escalator usage was assumed in the calculation of the lift system to be as given by Table 2.11, i.e.:

- 90% of passengers travelling to floor 1
- 75% of passengers travelling to floor 2
- 50% of passengers travelling to floor 3.

Then, the lowest escalator must be able to transport all the persons not using the lifts up to floor 1. That is 540 persons to floor 1, 450 persons to floor 2 and 300 persons to floor 3, making a total of 1,290 persons. Assuming a 15% peak arrival, similar to that experienced by the lifts, this is 193.5 persons in 5 minutes. A 1,000 mm wide escalator with a rated speed of 0.5 m/s has a handling capacity of 75 persons per minute, i.e., 375 persons/5 minutes. Thus, the escalators will be able to meet the demand and will be relatively lightly loaded, especially those connecting floors 1/2 and 2/3.

Table 9.7 Comparison of results by calculation

Zone	N	L	Load	%POP	Interval	P
Low	9	6	2500 kg	15.0	24.3 s	17.8
Mid	11	8	2000 kg	13.0	21.2 s	16.3
Upper	10	8	2500 kg	15.0	23.0 s	19.3

Traffic case studies

Contents

A note on calculation

In these case studies, the underlying mathematics described in earlier chapters is not presented, instead a simple spreadsheet is used based on these mathematics. Readers can compose such a spreadsheet for themselves or obtain a copy (to use at their own risk) from the book's web page at www.routledge.com.

In the spreadsheet extract shown in CSA below, the upper section is the *Input data* and the lower section is the *Results*. The spreadsheet can be used in several ways.

Suppose all the data about a lift installation are known and its characteristics are to be evaluated. Enter all the data as shown in the upper part of column '2-Capacity' and read off the round trip time, the interval, the handling capacity and the percentage population served from the lower part. Note also that the car capacity factor is 80%.

Suppose the percentage population required is known (say 12%) but the car size has not been determined. In column '3-%POP', the number of passengers (to 5) in the car is adjusted until %POP is 12%. Then, a suitable lift can be selected; in this case, it would be a lift with a rated load of 450 kg (see Table 5.8).

Table CSA Illustration of a spreadsheet

Input data		
Number of floors	16	16
Rated load (kg)	950	950
Actual car capacity (persons)	10	10
Number of passengers (persons)	8	5
Number of lifts	4	4
Rated speed (m/s)	3.0	3.0
Building population	500	500
Interfloor distance (m)	4	4
Express jump (m)	0	0
Express additional time (s)	0	0
Single floor flight time (s)	5.5	5.5
Door close time (s)	3	3
Door open time (s)	2	2
Advance door opening (s)	0	0
Start delay (s)	0.6	0.6
Passenger transfer time (s)	1	1
Results	*Value*	*Value*
Number of passengers	8.0	5.0
Highest reversal floor	14.7	14.0
Number of stops	6.5	4.4
Performance time (s)	11.1	11.1
Round trip time (s)	128.0	100.2
Interval (s)	32.0	25.1
Handling capacity (P/5-minute)	75.0	59.9
Percentage population	15.0	12.0
Capacity factor (%)	80	50

Case study nine

Tall building with separated lift lobbies

CS9.1   Introduction

This case study considers a slender 32 floor building, built in the 1970s. There were two groups of three lifts situated at each end of the building floor plate, with one group serving a zone of odd floors, and the other group serving a zone of even floors.

The design of the building considered its shape and form rather than the practicality of its building operations. The ideal shape which allows efficient vertical transportation is 'compact'. The use of a design, which places the load bearing structure at the ends of the building, means the centre of the building is too weak to support the lift systems. Thus, the lifts are placed at the ends of the building, contrary to good circulation practice. The main entrance is central to the building. Thus, the building is not conventional.

The lift dynamics were poor and two panel side opening doors were fitted to the cars. The owner wanted to take the 30-year-old building 'upmarket' and thus needed to improve the vertical transportation system.

CS9.2   The original system

The original skip/stop system was designed circa 1970, probably to design criteria for high prestige buildings, i.e.: 15% uppeak handling capacity; a 30 s or less uppeak interval; a density of occupation of one person per 14 m²; and a 100% daily attendance. The rentable size of each floor is 420 m². As the building is tall and slender, i.e., an inefficient design, the usable area is 80% of this, i.e., 336 m² (Section 6.4.4). This implies 24 persons per floor, i.e., 384 persons in each zone.

The data for the original system are:

Odd zone:	serving 16 floors: MT, 1–31
Even zone:	serving 16 floors: MT, 2–32
Number of lifts per zone:	3
Rated load:	950 kg
Platform area:	2.10 m²
Rated speed:	3.0 m/s
Interfloor distance:	6.0 m (between served floors)
Doors:	1,100 mm wide, two panel side opening
Door opening time:	2.0 s
Door closing time:	4.2 s
Single floor flight time:	6.5 s (6.0 m)

Table CS9.1 Spreadsheet results

1-Input data	2-Both	3-Low	4-High	5-Low	6-High
Number of floors	16.0	18.0	14.0	18.0	14.0
Rated load (kg)	950	950	950	950	950
Actual car capacity (persons)	10.0	10.0	10.0	10.0	10.0
Number of passengers (persons)	8.0	8.0	8.0	6.1	6.4
Number of lifts	3.0	3.0	3.0	3.0	3.0
Rated speed (m/s)	3.0	3.2	4.0	3.15	4.0
Building population	384	432	336	432	336
Interfloor distance (m)	6.0	3.0	3.0	3.0	3.0
Express jump (m)	0.0	0.0	59.0	0.0	59.0
Express additional time (s)	0.0	0.0	19.4	0.0	19.4
Single floor flight time (s)	6.5	4.2	4.2	4.2	4.2
Door close time (s)	3.7	2.3	2.3	2.3	2.3
Door open time (s)	2.0	1.8	1.8	1.8	1.8
Advance door opening (s)	0.0	0.4	0.4	0.4	0.4
Start delay (s)	0.5	0.3	0.3	0.3	0.3
Passenger transfer time (s)	1.0	1.0	1.0	1.0	1.0
Results	*Value*	*Value*	*Value*	*Value*	*Value*
Number of passengers	8.0	8.0	8.0	6.1	6.4
Highest reversal floor	14.7	16.5	12.9	16.1	12.6
Number of stops	6.5	6.6	6.3	5.3	5.3
Performance time (s)	12.7	8.2	8.2	8.2	8.2
Round trip time (s)	154.6	102.6	128.3	88.6	117.4
Interval (s)	51.5	34.2	42.8	29.5	39.1
Handling capacity (P/5-minute)	46.6	70.2	56.1	62.0	49.1
Percentage population	12.1	16.2	16.7	14.3	14.6
Capacity factor (%)	80	80	80	61	64

The lift can accommodate a maximum of ten persons (2.10/0.21), therefore, assume P = 8.0 persons. From the data, it is possible to calculate the original performance. Both zones will perform similarly. The output from the spreadsheet is shown in Table CS9.1, column 2-Both.

The lift system cannot serve the original design requirements. The handling capacity is 12.1% and not the 15% required, and the interval is 51.5 s and not the 30 s required. The only way that the lift system could cope would be for: either the building to have a lower population than indicated; or the demand to be less than 15%; or the daily attendance to be less than 100%; or a combination of all possibilities. For example, if the attendance was 85%, and the peak arrival rate was 13%, then the 5-minute peak arrivals would be 42 persons (384 × 0.13 × 0.85), lower than the system capability of 46.6 persons/5 minute. This is maybe what happened, i.e., the handling capacity was sufficient, but the interval was unacceptable.

Eventually, the population of the building increased as it aged, and the lift system was unable to cope. Thus, 30 years later, a new design was required.

CS9.3 A new design

A slender building of 32 floors will always be difficult to 'elevate' as (a) it is not quite tall enough for three zones and (b) up to five lifts could be required in each group in order to achieve the necessary quality of service, i.e., with respect to the interval and the average passenger waiting time. In 1972, the British Code of Practice CP407 indicated: '*For excellent quality of service, 1 lift is required for every three floors . . . interconnected in one group*'. Strakosch stated in 1967 that a skip/stop arrangement was not to be recommended.[1]

The building was provided with two lobbies, so this gives the opportunity to divide the building into a stacked configuration served from the two lobbies. The installation of additional lifts would assist, but this is impossible owing to the building structure. Thus, two groups of only three lifts arranged to serve two zones can be proposed. The lift and door dynamics (rated speed, door configuration, door times, etc.) and passenger handling facilities (passenger detection) should also be improved. This was achieved by increasing the rated speeds, improving the motion profile and by fitting centre opening doors.

A transfer floor needs to be provided at the zone interface. The transfer floor should be placed as high as possible in the building in order to compensate for the express jump to the first served floor in the high zone, without compromising the quality of service to either zone. Calculations show the best position for the transfer floor to be floor 19. This will therefore be the highest floor in the low zone and the lowest floor in the high zone. During uppeak traffic, only the high zone lifts would serve floor 19, thus, the last stopping floor in the low zone would be floor 18. The data for the new system are:

Low zone:	3 lifts serving 18 floors: MT, 1–19 (1–18 uppeak)
High zone:	3 lifts serving 14 floors: MT, 19–32 (19–32 uppeak)
Rated load:	950 kg/12 persons
Rated speed:	3.15 m/s (low zone) 4.0 m/s (high zone)
Interfloor distance:	3.0 m (between served floors)
Express jump:	59 m (high zone)
Doors:	1,100 mm wide, two panel centre opening
Door opening time:	1.8 s
Advance opening time:	0.4 s
Door closing time:	2.3 s
Single floor flight time:	4.2 s (3.0 m interfloor distance)

From these data, it is possible to calculate the new performance using the spreadsheet. Both zones will perform differently. Table CS9.1, column 3 Low and column 4 High shows that the quality of service represented by the interval at the main terminal is better for both zones, but still not as required. The round trip and interval times are still too long for this type of building. But the underlying handling capacities of 16.2% and 16.7% are not really required.

The floor populations are still set at 24 persons per floor, which returns percentage population values of 16.2% and 16.7%, respectively. Suppose now that the density of occupation had changed to one person/10 m², then the floor populations would be 34 persons and the zone populations would then be 612 and 476 persons, respectively. If the daily attendance was only 85%, and the arrival rate were now to be 12% (which is more typical in the twenty-first century), then the handling capacities required would be 62 and 49 persons/5 minutes and not 70.2 and 56.1.

The spreadsheet can be used in an iterative way to show this. Progressively reduce the number of passengers being carried until the required handling capacity is reached, and then examine the interval.

Both zones will perform differently. Table CS9.1, columns 5 Low and 6 High show the results. The low zone now provides a 29.5 s interval, but the interval for the high zone is too high at 39.1 s.

CS9.4   Commentary

The performance of the low zone is now within the desired specification of handling capacity and interval. The high zone interval is improved, but fails to meet the desired interval. The number of passengers travelling in the cars is much lower (6.1 and 6.4) than the available capacity, giving a more enjoyable journey for the passengers. Little more could be done to improve the installation without structural changes, for example to add another lift to each group.

Case study ten

Basement service in a low rise building

CS10.1 Introduction

Basements were considered in Section 8.3.3, and Table 8.4 indicated a range of results. Now consider a practical example of a nine storey building where there are three basement parking areas. The relevant data provided are:

LB   43 car parking spaces
B 59   car parking spaces
LG   74 car parking spaces + 40 cycle spaces
G   1,160 m² [83]
1   1,170 m² [84]
2–5   1,270 m² [91]

The areas given are net internal and are all net usable areas. The density of occupation is taken as one person per 14 m² to give the population figures shown as [nn] above. What lift system should be installed?

CS10.2 Discussion

The total building population is 440 persons. The population on floors 1–4 will be 357 persons. They will all reach their work destinations using the lifts, whether they arrive via the basements, or via floor G. The population on floor G is 83, some of whom will arrive via the basement car parks, thus using the lifts and some will arrive directly to floor G via the main entrance and not use the lifts at all. There are 176 car parking spaces. Assuming 1.2 passengers per car, this would indicate a daily arrival via the car parks of 211 persons. Add to this the 40 cyclists, gives a total daily basement arrival of 251 persons. This implies that 189 persons (440–251) arrive via the G floor entrance.

The effect of some persons arriving directly at the G floor and not using the lifts can be considered to balance other persons travelling to the G floor from the basements, and therefore alighting at the G floor. Therefore, it is reasonable to take the population of floors G–4 of 440 persons when calculating the percentage population figures.

The arrival of cars into a car park is generally limited by the number of entry ramps. The vehicle entry rate is given as one vehicle every 10 s, which is the time to negotiate the entry barrier control. Thus, in the basement, there could be a maximum 5-minute arrival rate of 36 persons. Some of these arrivals may work on the ground floor, but will use the lifts to get there. If a 15% peak is assumed, then some 28 persons (15% × 189) arrive at the G floor over the peak 5-minute period for travel to floors 1–4. The peak arrivals at the ground floor can occur in

Table CS10.1 Spreadsheet

1-Input data	2-First	3-Second
Number of floors	5	5
Number of basement floors	3	3
Rated load	1000	1000
Actual capacity	11.4	11.4
Number of passengers	9.1	7.6
% basement passengers	56.0	56.0
Number of lifts	2	3
Rated speed	1.6	1.6
Building population	440	440
Interfloor distance	3.8	3.8
Basement distance	2.7	2.7
Express jump	0	0
Express additional time	0	0
Single floor flight time	5.1	5.1
Basement flight time	4.4	4.4
Door close time	2.6	2.6
Door open time	2	2
Advance door opening	0.4	0.4
Start delay	0.5	0.5
Passenger transfer time	1	1

Results	Value	
Number of passengers	9.1	7.6
Highest reversal floor	4.9	4.8
Number of stops	4.3	4.1
Performance time	9.8	9.8
Round trip time	110.1	103.5
Interval	55.0	34.5
Handling capacity	50	66
Percentage population	11.3	15.0
Capacity factor (%)	80	67

a worst case scenario at the same time as the peak arrivals from the basements. The arrivals at the three basement floors (36 persons) are larger than the numbers arriving at the ground floor (28 persons). Thus, lifts arriving at the ground floor will be 56% occupied. This is calculated as 36/64, where 64 represents the total peak 5-minute arrivals at the ground floor.

A modified spreadsheet (to include basements) is used, which is based on the mathematics of Section 8.3.3. The table indicates the first design with two lifts. The percentage basement passenger occupancy at the lift arrivals at floor G was taken as 56%. The rated speed was 1.6 m/s, which is appropriate to the height of the building. The planned rated load was 1,000 kg with an actual capacity of 11.4 persons and a design occupancy of 9.1 persons. The relevant other parameters are shown in Table CS10.1.

CS10.3 Commentary

The first design shows a very long interval of 55.0 s and a handling capacity of 11.3%. This indicates that three lifts are needed. Designs with three lifts and a car occupancy of 9.1 persons gave too high a handling capacity. Once this was reduced to 7.5 persons, a suitable handling capacity of 15% was achieved at an interval of 34.5 s. The rated load could be reduced to 800 kg.

In this case, the basement demand is very high and will distort service from the main terminal floor (floor G) to floors 1–4, particularly as the lift cars will be, on average, 56% full on arrival at the ground floor.

Case study eleven
Tall building with total connectivity

CS11.1 Introduction

The building to be considered is a 54 level triangular shaped tower with 51 office floors from Level 4 to Level 54 for a single prestige tenant (Godwin, 1993). The levels are arranged in 'villages' of three floors with a garden at the lowest floor. Three villages share a central atrium of nine floors. As the villages progress up the building, they rotate by 120°, producing a cycle of three villages before the original orientation is regained. Three cores are positioned at the apexes of the triangle, and include the lift hoist ways and toilet pods. Each core was designed to accommodate six lifts.

CS11.2 The brief

1   To minimise or eliminate the necessity for transfer levels.
2   Users to alight at a core furthest from the garden.
3   Male and female toilet pods to be accommodated in two of the three cores.
4   Levels 4 and 47 are common facilities accessible to all occupants.
5   Levels 48–54 to be a secure area at lower levels of occupancy (total 300 persons).
6   Each floor is to accommodate a nominal 60 persons.
7   Lifts to be visible from the outside of the building.

CS11.3 The design

Designs do not appear spontaneously, but in a progressive way, as this did (Godwin, 1993). The actual core that each passenger must use is defined by design brief point (2), which requires a passenger to alight at a core furthest from the garden. This means that every third village is in the correct orientation. This leads to the idea of a non-contiguous 15 floor building zone comprising three floors, skip six floors, progressing up the building. As this is not a conventional zone of contiguous floors, it would be important that occupants of each village group have common relationships (Tutt & Adler, 1990), i.e., sales with marketing, and training with human resources, in order to reduce inter-zonal traffic.

To meet the requirement of design brief point (1), the idea of secondary lift lobbies emerges. There are two secondary lobbies on each floor. They have to accommodate the toilet pods, one male and one female per floor. However, space can be made available to enable access to three of the six lifts in each of the secondary cores, one for a floor orientated 120° 'ahead' and one for a floor orientated 120° 'behind'. This secondary circulation route would accommodate interfloor movements only, and car calls would not be possible. The secondary core lifts would only respond to landing calls, car calls would be inhibited.

The service to the secure floors 48–54 is ignored, as it would be served by two separate lifts via a screening desk in the blue core at level 47. Thus, it was possible to envisage each group of lifts serving some 15 floors in each village group, having substantially the same population. The groups of floors are identified as 'red', 'yellow' and 'blue' village groups. The floors served are shown in Table CS11.1.

The associated interfloor service is not shown in Table CS11.1 but it is illustrated in Table CS11.2. This shows a segment of the building from floor 12 to floor 22. The arrangement repeats up to floor 46.

Passengers need some initial orientation for their interfloor travel. Consider a passenger at floor 14 (red group). If the passenger wished to travel to: another red group floor they will go to a red lobby (with six lifts); to a yellow group floor they will go to a yellow lobby (with three lifts); and to go to a blue group floor they will go to a blue lobby (with three lifts).

CS11.4   Commentary

This unique design permits travel within each zone using six lifts, and travel to the other two zones with three lifts. The interfloor demand is thus balanced. The design requires some extra 45 entrances per lift group, i.e., 135 in total. Some space is taken to create the secondary lobbies. The design brief is substantially fulfilled.

This design was not applied (Jappsen, 2000), as the number of office floors was reduced from 51 to 46. The requirement for a secure service to the upper floors was also removed. The

Table CS11.1 Floor service arrangements

Village group	No. of groups	Floors served floors					Total	
Red	4	5–6	13–15	22–24	31–33	40–42	47	16
Yellow	4	7–9	16–18	25–27	34–36	43–45	47	17
Blue	4	10–12	19–21	28–30	37–39	46	47	15

Note that primary interchange floors are provided at Levels 4 and 47.

Table CS11.2 Segment of design strategy: floors 12–22

Floor	Red group	Yellow group	Blue group
22	■ ■ ■ ■ ■ ■ ■ ■ ■		■ ■ ■
21	■ ■ ■	■ ■ ■ ■ ■ ■ ■ ■ ■	■ ■ ■
20	■ ■ ■	■ ■ ■ ■ ■ ■ ■ ■ ■	■ ■ ■
19	■ ■ ■	■ ■ ■ ■ ■ ■ ■ ■ ■	■ ■ ■
18	■ ■ ■ ■ ■ ■	■ ■ ■ ■ ■ ■	
17	■ ■ ■ ■ ■ ■	■ ■ ■ ■ ■ ■	
16	■ ■ ■ ■ ■ ■	■ ■ ■ ■ ■ ■	
15	■ ■ ■ ■ ■ ■ ■ ■ ■		■ ■ ■
14	■ ■ ■ ■ ■ ■ ■ ■ ■		■ ■ ■
13	■ ■ ■ ■ ■ ■ ■ ■ ■		■ ■ ■
12	■ ■ ■	■ ■ ■ ■ ■ ■ ■ ■ ■	■ ■ ■

original design had an empathy with three: triangular building, 3 sides, 3 apexes, 3 cores, 3 floors to a village, 3 villages to an atrium of nine floors, 6 lifts to a zone of 15 floors, 3 lifts for interfloor movements. The new design was a triangular building: 3 sides, 3 apexes, 3 cores, 4 floors to a village, 3 villages to an atrium of 12 floors, 5 or 6 lifts to a zone of 15/16/17 floors, 3 transfer floors for interfloor movements.

Godwin (1993) remarked '*Never follow blindly conventional solutions to nonconventional buildings*'. This case study illustrates a unique solution to the interfloor movement problem.

Case study twelve
Medium rise trader building

CS12.1 Introduction

Most buildings do not have the luxury of total interconnectivity as described in Case Study 11. This case study considers a, yet to be built, regular (not prestige) trader building where the lower three floors are occupied by trading floors, and the upper nine floors contain office accommodation occupied by multiple tenants. The building is designed with three zones: a lower zone (floors 1–3) served by escalators; a middle zone (floors 4–7) and an upper zone (floors 9–12) served by lifts with a common facilities floor between these two lifted zones.

The client asked (in 2002) two questions: for the original design to be confirmed for two population densities of one person per 10 m² and one person per 14 m² on uppeak traffic; and the effect on interfloor traffic of moving a common facilities floor from floor 8 to floor 6. The trading floors are populated at one person per 7 m². The client asked for the British Council of Offices' best practice criteria[2] to be applied, *viz*: an interval of 30 s and a peak arrival rate of 15%.

CS12.2 The original system

Table CS12.1 The basic building data

Floor	NIA (m²)	Population @ 1/7	Population @ 1/10	Population @ 1/14	Elevation	LZ	MZ	UZ
12	3500		350	250	55.6			
11	3500		350	250	51.4			
10	3500		350	250	47.2			
9	3500		350	250	43.0			
8	3500		350	250	38.8			
7	3500		350	250	34.6			
6	3500		350	250	30.4			
5	3500		350	250	26.2			
4	3500		350	250	22.0			
3	3000	430			16.5			
2	3000	430			11.0			
1	3000	430			5.5			
Ground	2000	n/a	n/a	n/a	0			
Equipment						E	L1	L2

NIA: net internal area. Shaded area indicates floor service
Population: 1/7 = one person/7 m², 1/10 = one person/10 m², 1/14 = one person/14 m²
LZ: lower zone. MZ: middle zone. UZ: upper zone
Equipment: E=escalators 3 pairs, 1,000 mm, 0.5 m/s. L1=6, 1,600 kg, 2.5 m/s. L2=6, 1,600 kg, 3.15 m/s

CS12.2.1 Escalator provision

The demand at the ground floor for the first set of escalators will be to populate floors 1–3 with 1,290 persons. If the intending passengers arrive as a 15% demand over 5 minutes, then this is 194 persons. The practical handling capacity of a 1,000 mm wide escalator with a rated speed of 0.5 m/s is 4,500 persons/hour (Table 2.8), i.e., 375 persons/5 minutes. This is nearly twice the likely demand and thus the escalators can easily accommodate a sudden 2:1 surge in demand.

CS12.2.2 Lift provision

Floor 8 is the common facilities floor where restaurant, travel agency, reprographics, post room, etc. are located. In view of the small number of floors each lift group is serving, and the large rated load of the lifts, it is reasonable to assume the occupants of floor 8 are split equally between the two groups. Then, the population to be served to floors 4–7 and half of floor 8 is 1,575 persons at 1/10 and 1,125 at 1/14. There will be an identical population demand in the upper zone. Detailed information was not provided regarding the lift equipment except the rated load, rated speed and number proposed (see legend to Table CS12.1). The equipment values selected for operating times, etc. can be seen in the spreadsheet Table CS12.2.

Table CS12.2 Performance figures

1-Input data	2-MZ-15	3-HZ- 15	4-MZ-10	5-HZ-10	6-MZ-ef	7-HZ-ef
Number of floors	5	5	5	5	5	5
Rated load (kg)	1600	1600	1600	1600	1600	1600
Actual car capacity (persons)	16.9	16.9	16.9	16.9	16.9	16.9
Number of passengers (persons)	13.5	13.5	13.5	13.5	12.8	12.4
Number of lifts	4	4	4	4	3	4
Rated speed (m/s)	2.5	4	2.5	4	2.5	4
Building population	1125	1125	1575	1575	1071	1071
Interfloor distance (m)	4.2	4.2	4.2	4.2	4.2	4.2
Express jump (m)	0	38.8	0	38.8	0	38.8
Express additional time (s)	0	13.8	0	13.8	0	13.8
Single floor flight time (s)	4.63	4.63	4.63	4.63	4.63	4.63
Door close time (s)	2.6	2.6	2.6	2.6	2.6	2.6
Door open time (s)	1.8	1.8	1.8	1.8	1.8	1.8
Advance door opening (s)	0.5	0.5	0.5	0.5	0.5	0.5
Start delay (s)	0.3	0.3	0.3	0.3	0.3	0.3
Passenger transfer time (s)	1	1	1	1	1	1

Results	Value	Value	Value	Value	Value	Value
Number of passengers	13.5	13.5	13.5	13.5	12.8	12.4
Highest reversal floor	4.9	4.9	4.9	4.9	4.9	4.9
Number of stops	4.8	4.8	4.8	4.8	4.7	4.7
Performance time (s)	8.8	8.8	8.8	8.8	8.8	8.8
Round trip time (s)	84.8	109.8	84.8	109.8	83.0	107.0
Interval (s)	21.2	27.4	21.2	27.4	27.7	26.7
Handling capacity (P/5-minute)	191	148	191	148	139	139
Percentage population	17.0	13.1	12.1	9.4	13.0	13.0
Capacity factor (%)	80	80	80	80	76	73

For the best case occupancy of 1,125 persons, Tables CS12.2, columns 2 and 3 show that the BCO recommendation for a 15% handling capacity is obtained for the mid zone but not the high zone. The intervals of some 21 and 27 seconds are most satisfactory. This means the client's original BCO requirements are nearly met. For the worse case occupancy of 1,575 persons, Table CS12.2, columns 4 and 5 show that the handling capacities are well short of requirements.

Today, this type of building would expect (according to CIBSE Guide D: 2015 and Tables 6.2 and 6.3) to meet a 13% arrival demand (mid value in the range) and a 25 s interval (lower value in the range).

However, there are some other factors that could be considered. The floor areas used in the calculations above were the net internal areas, and these can be reduced by (say) 15% to obtain net usable area (Definition 6.5). In addition, 100% attendance of the building population is unlikely on any day and therefore the effective population (Section 6.5) can be reduced to (say) 80% of possible population. This means the effective population of each zone can be reduced to 68% (0.85 × 0.80), i.e., to 1,071 in the 1/10 persons/m² case and 765 persons in the 1/14 persons/m² case.

Columns 6 and 7 of Table CS12.2 show that the 13% handling capacity requirement has been met with three and four lifts, respectively. However, the intervals are slightly longer than specified.

The passenger occupancy of the cars is lower than the maximum actual possible (13.5 persons), which will provide a more comfortable rise.

CS12.3 Commentary

- The escalator provision is excellent.
- One group of three lifts and another of four lifts would mostly meet the modern recommendations.

CS12.4 Moving the facilities floor

Assume that the characteristics of floors 6 and 8 are completely interchangeable. The previous situation was:

Middle zone: G, 4–7, 8 (50%)
Upper zone: G, 8 (50%), 9–12
The new situation is:
Middle zone: G, 4–5, 6 (50%), 7–8
Upper zone: G, 6 (50%), 9–12

The number of stops in the middle and upper zones would remain unchanged. The highest reversal floor in the middle zone would be slightly higher as floor 8 now has a bigger demand. Thus, the performance of the lifts would be little changed.

Persons from the upper zone wishing to reach floors in the middle zone would have to transfer at floor 6 and move back up the building to reach floors 7–8. Similarly, persons from floors 7–8 wishing to travel to the upper zone would first have to travel down to floor 6 and then back past their originating floors to enter the upper zone. This would increase the travel time of such passengers, and is clearly illogical.

The main effect is to destroy the symmetry of the original arrangement. This is a subjective rather than an objective view.

Case study thirteen
Multiple entrance modernisation

CS13.1   Introduction

A rent review between a long-term tenant and an owner is often contentious, and generally carried out by agents, rather than the principal parties. The main bone of contention is where a tenant has improved or proposes to improve the building to the advantage of both parties and wishes this to be reflected in the ensuing rent. This case study considers such a building prior to such a modernisation. Factors to be considered in this building are: that it has two entrances; two cores serving different numbers of floors; and an escalator service to the lower four floors.

CS13.2   The building and its population

The building, built in the 1980s, is rectangular in shape with a main entrance in the middle of a long side. The two cores (A and B) serve twelve floors and are located to the left and right of the main entrance. The first four floors are served by escalators through an atrium. Four lifts in each core are arranged in line. The vertical transportation facilities in this building thus defy the principles of good circulation and lift location practice. However, as occupants must walk some distance to each group, this has the effect of smoothing any peak flows that occur, and easing the demand on the lift system.

When built, the likely density of occupation of the floors was one person per 14 m². The 1980s performance criteria, if the recommendations of BS CP407 (1972) were followed, were probably a 15% handling capacity at an interval of 30 s. Today, the density is one person per 10 m². The lift system is performing badly, and clearly was not designed to service the larger numbers of occupants. The tenant wishes to bring the building up to modern standards.

To perform a traffic analysis of this building, three uppeak scenarios should be considered:

(1) Worst condition, where all the 830 possible occupants use the lifts to reach their destinations (Scene 1).
(2) Probable condition, where only 670 of the possible occupants use the lifts and some of the attendees use the escalators to reach floors 1–4 (Scene 2).
(3) Likely condition, where only 603 of the possible occupants use the lifts, as there is a 90% attendance and some of the attendees use the escalators to reach floors 1–4 (Scene 3).

A number of other assumptions must be made:

1   The net internal (rentable) areas will be reduced by 90% to calculate the net usable area (see column 3 of Table CS13.1).
2   The lifts in cores A and B will serve floors 1–12 equally, i.e., at 50%.
3   When calculating the performance with escalator usage, it will be assumed the escalators transport:

90% of passengers: ground–floor 1
67% of passengers: ground–floor 2
33% of passengers: ground–floor 3
10% of passengers: ground–floor 4.

Figures shown in square brackets [*nn%*] in Table CS13.1 are passengers using the lifts.
(4) Stair circulation will not be considered.

Table CS13.1 summarises the main data about this building.

CS13.3 The system to be improved

The equipment installed in the mid-1980s was not (even then) up to date, being a Ward-Leonard drive with poor door control and inadequate signalling fixtures. All twelve lifts are identical except for the number of floors served. Measurements made of the unmodified system showed reasonable door timings, very poor motion dynamics and excessive door dwell times. These latter deficiencies lead to a cycle time of 18.6 s, although the performance time is reasonable at 13.0 s.

The spreadsheet Table CS13.2, column 2 shows the equipment data[3] and performance for cores A and B, and indicates the underlying handling capacity as 71 persons/5 minutes at an interval of 54.7 s.

The 71 persons/5 minute handling capacity represents a percentage handling capacity of 8.6% for Scene 1, 10.6% for Scene 2 and 11.8% for Scene 3. Only if the building population were to be 473 persons, would the original criterion of 15% be meet. At this population, the interval would still be 51.4 s, which means the other original criterion would not be met.

CS13.4 The improved system

A modern system employing a variable voltage, variable frequency (VVVF) control can deliver excellent performance times of some 9.5 s over the 4.5 m interfloor distances. This, combined with landing call dwell times set at 3 s, car call dwell times set at 2 s, and with passenger detectors capable of initiating differential door timing, will make the cycle time dependent on

Table CS13.1 Building data

Floor	Rentable area (m^2)	Usable area (m^2)	Population ($1/10\ m^2$)	Scene 1	Scene 2	Scene 3
1	1777	1600	160	80	8 [10%]	7 [9%]
2	1777	1600	160	80	27 [33%]	24 [30%]
3	1777	1600	160	80	54 [67%]	48 [60%]
4	1777	1600	160	80	72 [90%]	65 [81%]
5	1777	1600	160	80	80	72
6	1777	1600	160	80	80	72
7	1333	1200	120	60	60	54
8	1333	1200	120	60	60	54
9	1333	1200	120	60	60	54
10	1333	1200	120	60	60	54
11	1333	1200	120	60	60	54
12	1111	1000	100	50	50	45
Totals			1660	830	670	603

Table CS13.2 Performance data

1-Input data	2-A+B	3-Scene 1	4-Scene 2	5-Scene 3
Number of floors	12	12	12	12
Rated load (kg)	1590	1590	1590	1590
Actual car capacity (persons)	16.3	16.3	16.3	16.3
Number of passengers (persons)	13.0	11.5	7.6	6.2
Number of lifts	4	4	4	4
Rated speed (m/s)	2.75	2.75	2.75	2.75
Building population	830	830	670	603
Interfloor distance (m)	4.5	4.5	4.5	4.5
Express jump (m)	0	0	0	0
Express additional time (s)	0	0	0	0
Single floor flight time (s)	5.82	4.96	4.96	4.96
Door close time (s)	2.7	2.7	2.7	2.7
Door open time (s)	1.9	1.9	1.9	1.9
Advance door opening (s)	-7.28	0.5	0.5	0.5
Start delay (s)	0.9	0.4	0.4	0.4
Passenger transfer time (s)	1	1	1	1
Results	*Value*	*Value*	*Value*	*Value*
Number of passengers	13.0	11.5	7.6	6.2
Highest reversal floor	11.6	11.5	11.1	10.8
Number of stops	8.1	7.6	5.8	5.0
Performance time (s)	18.6	9.5	9.5	9.5
Round trip time (s)	218.7	127.7	104.6	94.7
Interval (s)	54.7	31.9	26.2	23.7
Handling capacity (P/5 minute)	71	108	87	79
Percentage population	8.6	13.0	13.0	13.0
Capacity factor (%)	80	71	47	38

Table CS13.3 Summary of performance for improved system

Scene	Scene 1	Scene 2	Scene 3
P	11.5	7.6	6.2
INT (s)	32	26	24
HC (P/5 min)	108	87	79
%POP (%)	13	13	13

passenger movements rather than equipment timers. Improved signalling will enable passengers to quickly identify the next lift.

The spreadsheet Table CS13.2, columns 3, 4 and 5 give the results, and Table CS13.3 summarises the results. All scenarios can provide handling capacities of 13% and with intervals of less than 32 s. In one of the three cases, the intervals are lower than the modern performance criterion of 25 s. None of the cars fill to the design capacity of 13 persons.

CS13.5 Commentary

The underlying equipment can be modernised to meet the modern performance criteria. Even for the worst case scenario, the performance would be most acceptable. There is room for negotiation on rents.

Case study fourteen
Double deck installation

CS14.1   Introduction

Double deck installations are few in number compared to the millions of single deck installations. This case study indicates how a double deck solution solved a problem.

The building under consideration is a 16 storey headquarters for an international bank. Before the Pacific Rim financial crisis in the late 1990s, it was intended to populate the building in a prestigious manner with one person per 14 m² of usable space. This resulted in an actual population of 1,250 persons. The building was in the course of erection, and the core had been constructed, when the client decided to retrench and place more staff in the building at a density of occupation of one person per 10 m². At the same time, the footprint of the building was increased by 50%. Thus, the potential population rose to 2,625 persons, without the opportunity to provide more hoist ways in the core. What to do?

CS14.2   The original design

The original design was for a group of six, 1,600 kg lifts with a rated speed of 4 m/s. The data are shown in Table CS14.1. This table indicates that a 168.2 persons/5 minute handling capacity, i.e., 15%, is possible with a 24 s interval, when serving an effective population of 1,125 persons. (The 1,125 figure is 90% of the possible population of 1,250 persons.) This is an appropriate performance (<25 s interval, 15% handling capacity).

If the possible population rises to 2,625, i.e., an effective population, at 90% of 2,363 persons, even eight lifts could not cope. Eight lifts would have a handling capacity of 224 persons (8/6 × 168), i.e., 9.5%. An answer is to install double deck lifts.

CS14.3   Double deck design

Using Equation (8.9) embedded in the spreadsheet calculation, the performance of six double deck lifts serving eight floors each is given in column 4 of Table CS14.1. The table shows that each deck can serve 1,181 passengers, with a total handling capacity of 347 persons/5 minutes (14.7%) at an interval of 23 s. Thus, slightly more than twice (347/168.2 = 2.06) the uppeak performance has been obtained, but with a small reduction in the percentage handling capacity. This is acceptable.

CS14.4   Consequences of using double deck lifts

The double deck installation serving eight floors provides 2.06 times the handling capacity of a single deck serving 16 floors. Thus, the lift installation will now be able to handle the uppeak traffic demand. There will be little effect on the down peak traffic, mainly due to the

Table CS14.1 Performance data

1-Input data	2-Original	3–8-lifts	4–6-DD
Number of floors	16	16	8
Rated load (kg)	1600	1600	1600
Actual car capacity (persons)	16.9	16.9	16.9
Number of passengers (persons)	13.5	13.5	13.5
Number of lifts	6	8	6
Rated speed (m/s)	4	4	4
Building population	1125	2625	2363
Interfloor distance (m)	4	4	8
Express jump (m)	0	0	0
Express additional time (s)	0	0	0
Single floor flight time (s)	4.72	4.72	6.36
Door close time (s)	2.8	2.8	2.8
Door open time (s)	2	2	2
Advance door opening (s)	0.5	0.5	0.5
Start delay (s)	0.4	0.4	0.4
Passenger transfer time (s)	1	1	1
Results	*Value*	*Value*	*Value*
Number of passengers	13.5	13.5	13.5
Highest reversal floor	15.3	15.3	7.8
Number of stops	9.3	9.3	6.7
Performance time (s)	9.4	9.4	11.1
Round trip time (s)	144.4	144.4	140.0
Interval (s)	24.1	18.1	23.3
Handling capacity (P/5 minute)	168	224	347
Percentage population	15.0	8.5	14.7
Capacity factor (%)	80	80	80

small number of floors served by each deck and the likelihood that the cars will fill at three or four floors and then travel to the main terminal. There may be difficulties for interfloor traffic, but as the traffic levels are low, one deck may suffice by using the principle of a trailing deck (section 8.2.2).

Notes (shown in text as[1])

1 Page 38.
2 BCO recommendations in 2002 for lifts were very brief. They were intended for the lay person, i.e., architect, owner, developer, etc., not a lift industry specialist, as a means towards initial equipment sizing. The BCO Guide: 2014 is considerably changed to relate to the current practice contained in the CIBSE Guide D: 2015, and readers are referred to it.
3 Notice that in order to set the cycle time to 18.6 s, the advance door opening time is set to a negative value of −7.28 s and the passenger transfer time (valid at the main terminal floor only) is set to 0.5 s.

Part C

Traffic control

10 Legacy traffic control

This chapter introduces the concept of lift traffic control. It presents the concept of single lift control and some of the rules associated with it. It then introduces the concept of lift group control and the need for it. An overview is then given of three classical lift group control algorithms that were used in the era preceding the use of microprocessors in lift traffic control. It is useful to examine the group control algorithms that were implemented in hard-wired relay logic controllers. Many of the ideas are still valid today and have formed the basis of modern traffic control systems. Their study in this chapter is illuminating.

Overall, this chapter will show that appropriate automatic traffic control systems can enable a single lift, or a group of lifts, to operate at very high efficiency, provided this is matched by high-performance, reliable electromechanical equipment. The chapter concludes by presenting additional features of lift group traffic control.

Chapter 11 will build on this chapter by discussing the principles of computer-based traffic control systems. Chapter 12 indicates systems available today.

10.1 Background

The overall control of lift systems raises two different engineering requirements:

*   First, some means of commanding a car to move in both up and down directions, and to stop at a specified landing must be provided.

For a single lift to operate, it requires power, logic and speed control. Logic control (as opposed to speed control) provides a means of commanding a car to move in both up and down directions, and to stop at a specified landing. It governs the activities of the lift in moving between floors, opening and closing its doors, and responding to car calls. These control systems are not discussed in this book.

*   Second, in a group of lifts working together, it is necessary to coordinate the operation of the individual lifts in order to make efficient use of the lift group.

There are very few cases where one lift can provide the handling capacity required in a building, and a number of lifts are needed. These lifts must operate as one coordinated group, as opposed to a number of single lifts placed in close proximity to each other. As a group, as soon as a landing call is registered, the group controller must allocate the landing call to one, and only one, of the lifts in the group. This is traffic (passenger) control.

10.2 History

Individual lift control is a basic necessity and, as such, was present from the very beginnings of lift usage. Early hydraulic and steam driven lifts were operated on 'hand-rope' control (Strakosch, 1967): the operating device was a rope that ran the length of the shaft and actuated a valve in the basement. As shafts were not fully enclosed, a passenger requiring lift service at a particular landing could reach in and operate the rope to summon the lift. Although practical because it needed no attendant, this type of operation was very unsafe.

The advent of electrical lifts brought the electrical car switch operation, where a lift attendant had complete control of the car by moving a handle-operated switch to drive the car in the up or down directions. The attendant (manual) single car control is the simplest form of control that can be used. It was the attendant who decided the floors where stops would be made, initially, by relying on direct observation of the landing halls through the entrance gates to know whether service was required. Later, in order to increase the efficiency of service, and to allow for closed hoist ways, signalling systems, able to *give* the attendant the information required about the traffic demand were introduced in the car. At the same time, other features concerning passenger safety and comfort were introduced and required by law, e.g., landing door interlocks[A] and car gates.

Attendant control has almost disappeared today. In some buildings 'It may be desirable to have the attendance of a commissionaire or lift operator as part of the atmosphere of service associated with the building' (Fletcher, 1954). An attendant is sometimes employed today in special buildings for security reasons.

Aware of the limitations of this type of control, lift engineers developed automatic electrically operated drive control and signalling systems for single lifts. Automatic drive and signalling control of lifts made rapid progress. It allowed better acceleration, slowing and levelling of cars, higher operating speeds and superior traffic handling capacities. The automatic control not only saved considerable labour for the attendant, but also brought better results than control dependent on human decisions.

The advent of automatic doors allowed the development of fully automatic pushbutton (FAPB) lift systems, and completely eliminated the need for an attendant. Automation is particularly indispensable when a group of lifts is to be controlled because, in addition to the individual lift control, it is necessary to interconnect the lifts and provide a group control system capable of operating the lifts efficiently under various traffic patterns.

In broad terms, it is possible to identify five generations in the history of the control of groups of lifts, as shown in Table 10.1.

Table 10.1 Generations of control

Era	Dates	Traffic control type
I	1850–1890	Simple mechanical control
II	1890–1920	Attendant and electrical car switch control
III	1920–1950	Attendant/dispatcher and pushbutton control
IV	1950–1975	Group control:
		IVa scheduled traffic control to 1960
		IVb demand traffic control from 1960
V	1975–1990	Computer-based group control
VI	1990 –	Hall call allocation

Notice that these generations are approximately equivalent to human generations. Also notice the change of technology: mechanical (I), electromechanical (II and III), electrical (IV), electronic (V) and better signalling (VI).

10.3 Scheduled versus demand or 'on call' traffic control

In Table 10.1, under Generation IV, the terms 'scheduled' and 'demand' traffic control are mentioned and require some explanation.

Lift traffic control systems need to respond to the necessity of providing efficient control for a group of lifts in order to service a common set of landing calls. The traffic control system is required to present an improved performance, compared with the lifts working independently. A major problem that the traffic control system has to solve is to keep the lifts in the group equally distributed along the shaft height, especially during heavy demand periods. Under busy traffic conditions, the lifts have a tendency to bunch together and to 'leap-frog' each other, that is, to stop at alternate floors, with one lift frequently overtaking the other.

The first solution to this problem was provided by the scheduling systems where the lifts are dispatched from the terminal floors at convenient time intervals, and made to run like buses between these terminal floors. Scheduling systems do not work 'on call', in the sense that the lifts are dispatched in response to a demand, but rather they are dispatched from the main terminal floors according to a schedule, without the necessity of a call being registered. Improvements were introduced in the scheduling system in order to provide a number of different operating programs for the dispatch of lifts. Such programs are designed to deal with the various patterns of traffic flow and intensity of traffic encountered over the working day. However, scheduling systems suffer from a major disadvantage: the lifts spend a considerable amount of time at the top and bottom terminals waiting for a dispatch interval to expire, hence reducing the potential handling capacity. The stop at the top terminal is often pointless, and the lifts do no useful work while waiting for dispatch. In addition, lifts frequently make fruitless journeys between terminal floors and thus run unnecessarily when traffic is light.

Because of these disadvantages, lift traffic control engineers have developed non-scheduling control systems, which only respond when landing calls are made; the 'on call' system. There are several of these systems available on the market. They may behave quite similarly for certain traffic patterns, for example, under uppeak, but they can also present dissimilar performance under other traffic conditions, such as down peak or local traffic demands. The main features available in the most representative conventional implementations of lift group control systems are analysed and discussed in this chapter.

10.4 The five rules of lift traffic control

There are five important rules (Closs, 1970) that must be adhered to in the operation of a lift. They are listed below:

Rule 1 A car may not stop at a floor where no passenger enters or leaves a car

Rule 2 A car may not pass a floor at which a passenger wishes to alight

Rule 3 A passenger may not enter a car carrying passengers and travelling in the reverse direction to the passenger's required direction of travel

Rule 4 A car may not reverse direction of travel while carrying passengers

Once the passenger is in the car

Rule 5 Car calls always take precedence over landing calls.

Rule 5 ensures that passengers already in the lift reach their destinations, whatever the demand on the landings may be. Rule 4 means that a lift must continue to serve the last car call in the direction of travel before reversing direction. These rules must also be obeyed by all group traffic control systems.

It might be thought that a more optimal solution could be to ignore Rule 2 in order to provide a more equal service to all landing and car calls. This might involve a down travelling lift collecting an up travelling passenger at a floor where a down-travelling passenger alights, travelling down to a lower floor to allow another passenger to alight, and then to travel up to the desired destination of the up travelling passenger. Closs (1970, 1972) showed this to be untrue and that the best solution is to always collect calls ahead and only reverse when the last passenger has exited the lift. This generally results in calls being answered in a different order to their registration. This procedure is not necessarily the mathematically optimum solution.

Rule 4 could be violated by absent-minded passengers and Rule 5 could be violated by the search procedure for the optimum path stopping a car and then reversing direction, hence also violating Rule 2.

10.5   Single lift traffic control

10.5.1   Single call automatic control

The simplest form of automatic lift control is single call automatic control. Single call push-buttons are provided on the landings. This form of control is also termed non-collective or automatic pushbutton (APB) control. The passengers directly operate the lift by pressing landing and car buttons, so no attendant is necessary. Car calls are given absolute preference over landing calls, which are only answered if the lift is available. If a passenger in the car presses a call pushbutton corresponding to the required destination floor, the lift moves direct to this floor bypassing any intermediate floors at which landing call buttons have been pressed. When a landing call pushbutton is pressed and the lift is free, the call is immediately answered. If the lift is in use, a landing signal indicates 'lift busy', thus hopefully reducing the intending passenger's frustration.

This type of control is only suitable for short travel passenger lifts serving up to four floors in, for example, small residential buildings with a light traffic demand. It provides a very low carrying capability, as most of the time the lift carries a single passenger. It can also produce large passenger waiting times, owing to the many trips bypassing other passenger requests. The automatic pushbutton control is, however, suitable for goods lifts, particularly when a single item of goods can fit in the lift at one time.

10.5.2   Collective control

Passengers today do not expect an attendant to be present in a lift. They were removed long ago for economic reasons, except in special circumstances. Passenger-operated collective control provides better performance than attendant operation. Fletcher (1954) refers to a survey made in one installation, where a group of four identical lifts were arranged so that two of them operated as duplex passenger control and the remaining two operated with attendants. The duplex system provided a handling capacity 20% greater than the attendant controlled lifts.

The most common form of automatic control used is collective control. This is a generic designation for those types of control where all landing and car calls made by pressing push-buttons are registered and answered in strict floor sequence. The lift automatically stops at landings for which calls have been registered, following the floor order rather than the order in which the pushbuttons were pressed. Collective control can either be of the single button, or of the two pushbutton types.

10.5.2.1 Non-directional collective control

Single pushbutton collective control provides a single pushbutton at each landing. This pushbutton is pressed by passengers to register a hall call irrespective of the desired direction of travel. Thus, a lift travelling upwards, for example, and detecting a landing call in its path will stop to answer the call, although it may happen that the person waiting at the landing wishes to go down. The person is then left with the options to either step into the car and travel upwards before going down to the required floor; or to let the lift depart and re-registers the landing call. Owing to this inconvenience, this type of control is only acceptable for short travel lifts.

10.5.2.2 Down collective control (up-distributive, down collective)

Single pushbutton call registration systems may, however, be adequate in buildings where there is traffic between the ground floor and the upper floors only, and no interfloor traffic is expected. A suitable control system is the down collective control (sometimes called up-distributive, down collective) where all landing calls are understood to be down calls. A lift moving upwards will only stop in response to car calls. When no further car calls are registered, the lift travels up to the highest landing call registered, reverses its direction and travels downwards, answering both car and landing calls in floor sequence.

10.5.2.3 Full collective control (directional collective)

The two pushbutton full collective control (also designated directional collective control) provides each landing with one up and one down pushbutton, and passengers are requested to press only the pushbutton for the intended direction of travel. The lift stops to answer both car calls and landing calls in the lift direction of travel, in floor sequence. When no more calls are registered in the lift direction ahead of the lift, the lift moves to the furthest landing call in the opposite direction, if any, reverses its direction of travel, and answers the calls in the new direction. This control system is suitable for single lifts or duplexes (two lifts) serving a few floors with some interfloor traffic. Typical examples are small office buildings, small hotels and blocks of flats.

Directional collective control applied to a single lift car is also known as simplex control. Duplex (two lifts), triplex (three lifts) or quadruplex (four lifts) control systems are available for groups of directional collective controlled lifts interconnected to work as a team. This is the simplest form of group control (see Section 10.3).

10.6 Group traffic control

10.6.1 Some definitions

A single lift will not always be able to cope with all the passenger traffic in a building. Thus, a number of lifts may be installed, often side by side, and the problem arises of interconnecting the lifts. Many buildings have lifts installed where two lifts operate independently, although they are placed together. Such lifts do not operate efficiently as people will tend to register landing calls at both lifts, thus motivating the two lifts to answer the same call. This causes extra trips and false stops, and under heavy demands, it produces bunching of lifts.

Where a number of lifts are installed together, the individual lift control mechanisms should be interconnected, and there should also be some form of automatic supervisory control provided. In such a system, the landing call buttons are common to all the lifts that are interconnected, and the traffic supervisory controller decides which landing calls are to be answered by each of the lifts in the group.

> **Definition 10.1:** a group of lifts is a number of lifts placed physically together, using a common signalling system, and under the command of a group traffic control system.

The lifts then serve a common set of landing calls, but are allocated calls according to a set of rules. The purpose of the group traffic control system is to coordinate the operation of the individual lifts. This is essentially intended to maximise the transport capability with the given facilities, to improve passenger service in terms of shorter waiting times, and to make available a number of features to deal with specific traffic situations. With the development of high rise buildings and other large scale structures, group supervisory control of lifts is most important as a facility providing a central function of modern architecture.

> **Definition 10.2:** a group traffic control system is a control mechanism to command a group of interconnected lifts with the aim of improving the lift system performance.

The use of a group controller improves the efficiency of the lifts in the group. The total handling capacity of the group controlled lifts is more than the sum of the handling capacity of the individual lifts acting alone. For this reason, it is strongly recommended that lifts are placed in one group and located in a central location in the building, rather than being split as individual lifts in widely spread parts of the building. The additional advantage of such an arrangement is that if a lift is taken out of the group or goes out of service, passengers can still access the other lifts in the group (without even noticing that one of the lifts is out of service).

Legacy group control systems were expected to provide more than one program or control algorithm in the traffic control system to allocate lifts to landing calls. The appropriate operating program is determined by the pattern and intensity of the traffic flow encountered by the lift system. The selection of the proper control algorithm can be done manually when an attendant operates a key switch on a control panel. However, in the more complex systems, the operating program is automatically selected by a traffic analyser, which assesses the prevailing traffic conditions. Such a detection mechanism is based on either the measurement of car loads by means of weighing devices installed in the car floor, or on the counting of landing and/or car calls or on timing devices. Once a particular traffic condition applies, the control system may introduce only a few changes or adaptations in the control policy to cater for the specific circumstances inherent to the new traffic situation, or it may switch to a substantially different control algorithm.

> **Definition 10.3:** a group traffic control algorithm is a set of rules defining the traffic control policy that is to be obeyed by the lift system when a particular traffic condition applies.

The complexity of the group traffic control system is related to the number of lifts in the group and the features that are required from the lift system. It varies from very simple two lift control systems to very sophisticated schemes, with up to eight lifts. The control systems can vary from very simple, single program systems to complex multi-program systems where a number of control algorithms are available to cover such conditions as uppeak, down peak, heavy sector demand, heavy floor demand, balanced traffic, off peak and night service.

The traffic control system complexity depends, not only on the number of available control programs, but also on the complexity of the algorithms themselves. A lift system with a large variety of control algorithms is not necessarily the best system as some problems may arise in the transfer of control from one algorithm to another, and an effective redistribution of lifts takes some time, making response to transient changes in traffic requirements very difficult to achieve consistently.

10.6.2 Call allocation

The primary function of a group of lifts is to answer the car and landing calls belonging to the group in the most appropriate way. To make this possible, it is obviously necessary that any car or landing calls, once registered, are memorised until they are answered. The control system must know, at any instant, the demand placed on the group of lifts in order to take the most suitable action to deal with the particular traffic pattern. Thus, a car call station is necessary for each lift in the group. The landing call system is common to all lifts in the group, and must include one up and one down pushbutton at each floor. An obvious exception is made for the terminal floors, when one pushbutton is sufficient. For architectural balance or for ease of call registration, the landing call buttons are frequently replicated two or more times at each floor. However, they are all interconnected in parallel.

A traffic control system must distribute the lifts equally around the building zone in order to provide an even service at all floors. It is also important that only one lift be dispatched to deal with each landing call. Thus, an allocation policy is necessary to determine which lift answers each particular landing call.

> **Definition 10.4:** landing call allocation is the procedure by which a lift is assigned to service a particular landing call and prevents other lifts from starting to move, or continuing their travel, in response to that landing call.

The simplest method used to allocate landing calls to specific lifts is by sending the nearest lift. This works particularly well for the duplex, triplex or quadruplex control systems operating under an interconnected directional collective control system, described in Section 10.5.2.3. A common method used to provide call allocation is by grouping the landing calls into sectors within each building zone, and allocating lifts to each sector. If the supervisory control system parks any idle lifts in vacant sectors, then a good distribution of lifts is achieved and better performances may be obtained.

> **Definition 10.5:** A sector is a group of landings, or of landing calls, considered together for lift allocation or parking purposes.

There are several classical ways of grouping landings or landing calls into sectors. The number of sectors in a building zone is generally dependent on the number of lifts in the group of lifts serving the zone. Two main methods can be used to group landing calls into sectors: static and dynamic sectoring.

10.6.2.1 Static sectoring

Where sectors are defined statically, a fixed number of landings are grouped together to constitute a sector. One of the existing schemes considers a number of levels in each sector and includes both up and down landing calls within the common sector limits.

> **Definition 10.6:** a common sector is a fixed sector that is defined for both up and down landing calls originating from a number of (usually) contiguous landings.

A common design with this type of sectoring, where there are three to four lifts, is to define as many sectors as there are lifts. This allows a lift to be assigned to each sector under light to medium traffic conditions, and therefore provide equalised service to all the sectors. A good service to the main terminal floor is also considered to be an important feature of the lift system. Where there are five or more lifts, there may be one more lift than there are sectors, thus providing a 'floating' lift to cover sudden demands.

As an example, consider a group of five lifts serving a local zone of 16 floors above the main terminal floor. Suppose that an equal floor demand is expected. A suitable partition might consist of defining four equal sectors of four floors each, with the main terminal considered as a single floor sector. This ensures preferential service to the main terminal floor and an even service to the other sectors.

An alternative method of static sectoring arranges the (usually) contiguous up landing calls into a number of up-demand sectors, and contiguous down landing calls into an independent number of down-demand sectors, thus defining directional sectors.

> **Definition 10.7:** a directional sector is a fixed sector that includes a number of (usually) contiguous landing calls defined for one direction only.

Consider the same local zone discussed above, again served by a group of five lifts. For this type of scheme, it might be appropriate to define as many down sectors as there are lifts, and one less up sector than the number of lifts. A solution is where the main terminal floor is again given an up sector to itself in order to maintain a good service for the incoming passengers. The boundaries of the down sectors do not need to be the same as the boundaries of the up sectors.

In the above discussion, the main terminal floor is the bottom floor of the building. If one or more basement floors exist, they can be grouped into a low priority sector or, alternatively, the main terminal could be taken as the highest floor of the lowest sector and some restrictions to the basement service would be included.

The limits of static sectors are often dictated by the type of tenancy, the existence of unequal traffic demands within the different floors in the zone, or the need for preferential service at predetermined floors. Thus, the sectors may contain a different number of floors. A sensible policy to adopt when defining the sectors is to try to equalise the amount of traffic originating at each floor, biased by the relative importance of the quality of service provided to the sector. For example, if a single floor sector is defined for an executive floor, the number of calls registered in this sector is certainly very small; however, a very good service is required at such a floor and may justify the procedure.

If equal traffic intensities are expected at each landing, then sectors of the same size should be provided. It often happens that this is not possible because there is not an integer relationship between the number of floors and the number of sectors. A possible solution for common sectors and down directional sectors is either to make the lowest sector (main terminal not included) larger or the highest sector smaller. This is based on the assumption that more lifts pass through the lowest sector than any other, owing to the domination of traffic to and from the building entrance, and hence, any spare car capacity will be taken up by stopping in this sector. In addition, passengers will frequently walk up one floor and down one or two floors, rather than wait for a lift. In the case of up directional sectors, it may be appropriate to make the highest up sector larger as, under normal conditions, less up traffic originates in this sector in comparison with lower up sectors.

10.6.2.2 Dynamic sectoring

In a dynamic sectoring scheme, the number of sectors and the position and limits of each sector depend on the instantaneous status, position and direction of travel of the individual lifts. Thus, dynamic sectors are not defined at the design stage, but are defined during normal lift operation, and are continuously changing. The sector of each lift extends from the lift to the next lift ahead, which is either idle or travelling in the same direction. When a sector reaches a terminal floor, it continues in the opposite direction. A stationary idle lift (parked lift) is the boundary for one up and one down sector. Fully loaded lifts and idle lifts travelling for parking or any other special reason are not considered in the sector definitions. It may happen that two or more lifts are located at the same floor with an identical direction of travel. In this case, one single floor is defined for the lifts, and it is allocated to the lift with the highest reference symbol.

Definition 10.8: a dynamic sector is a sector that includes a variable number of floors defined by the position of moving and idle cars.

10.6.3   Assignment of lifts to sectors

There are several ways of assigning lifts to demand sectors (Chan & So, 1995). The dynamic sector concept assigns a lift in the sector definition itself, alleviating the necessity for a separate allocation algorithm. Thus, incoming landing calls in a sector are automatically allocated to the corresponding lift. Static sector systems, however, require a separate method of allocation.

In a static sector scheme, a simple method of assignment consists in allocating a lift to a sector if the lift is present in that sector and the sector is not allocated to another lift. Special rules can then be developed for de-assignment, e.g., a lift leaving a sector or by being fully loaded, etc. Special rules can be developed to assign a lift to a sector, e.g., to an adjacent unoccupied sector, by a timer, by priority, etc. Some of these possibilities will be discussed in the next sections.

10.6.4   Information for passengers: signalling

For an efficient boarding of a lift, it is important that passengers are given information. This would include:

* Landing and car call registration pushbuttons should illuminate when operated to indicate the call has been accepted.
* The illumination of the landing and car call registration pushbuttons should be extinguished when the lift arrives at the requested landing.

- The arrival of a lift at a landing should be signalled by a visible directional (arrow shaped) lantern on the landing at least 4 s before the lift starts to open its doors. It would be convenient to colour code the up and down directions.
- The arrival of a lift at a landing should be signalled by an audible gong on the landing at least 4 s before the lift starts to open its doors. It would be convenient to code the up and down directions by the number of gong strokes.
- Arrow shaped direction indicators should be positioned in the rear of all lifts visible from the landing.
- Direction and floor position indicators should be provided in the car above the doors.
- To assist persons with disabilities, Braille and raised characters should be provided on tactile fixtures, speech announcements may be provided on the landings and in the car, and colour contrasting legends and numerals should be provided.

BS5655: Part 6: 2011 suggests '*that the purpose of every pushbutton and indicator should be clearly understood by all passengers*'. BS ISO 4190–5: 2006 gives some guidance on fixtures.

10.7   Examples of legacy group traffic control

The previous sections considered the features required for lift group traffic control. A number of examples of possible methods of lift group traffic control are now considered in some detail, ranging in complexity from a simple fixed logic system to classical traffic control schemes. Many of these traffic control systems discussed here have been commercially installed by companies such as Otis (VIP 260), Westinghouse (Mark 4) and Schindler (Aconic), and are still operating today. The descriptions here are deliberately generic, rather than proprietary, and may be at variance to actual implementations.

10.7.1 Simple control: 'nearest car' (NC)

The simplest type of group control is the directional collective control described in Section 10.2.2(c). It is suitable for a group of two or three lifts, each operating on the directional collective principles and serving seven or so floor levels. The assignment of lifts to landing calls is achieved by the 'nearest car' (NC) control policy.

A single landing call system with one up and one down pushbutton at each landing, except for the terminal landings, is required. The NC control is expected to space the lifts effectively around the building in order to provide even service, and also to park one or more lifts at a specified parking floor, usually the entrance lobby floor (main terminal). Other features, which might be included, are the bypassing of landing calls when a lift is fully loaded, and the possibility of taking a lift out of the group for a special trip under independent car control for inspection, maintenance or because the car is faulty. The remaining lifts will continue to provide service to all floors.

Car calls are dealt with according to the directional distributive control principles. Landing calls are dealt with in the normal way by reversal at highest down and lowest up calls. Thus, the lift answers its car and landing calls in floor sequence from its current position and in the direction of travel to which it is committed.

The only group traffic control feature contained in this simple algorithm is the allocation of each landing call to the lift that is considered to be the best placed to answer this particular call. The search for the 'nearest car' is continuously performed until the call is cancelled after being serviced. The distance (*d*) between a particular landing call and a lift is 'measured' in levels:

$$d = |\text{car floor} - \text{landing floor}| \text{ levels}^{[B]} \tag{10.1}$$

and can vary from zero to *N* in a building of *N* + 1 floors.

From the distance between landing call and car, and taking into account the landing call and the car directions, a figure of suitability (*FS*) is evaluated according to four rules:

Rule (a): if the lift is committed to move towards the landing in the same direction as that required by the landing call, then a position bias is given to the lift, such that the lift appears to be one floor nearer to the landing call. It is:

$$FS = (N + 1) - (d - 1) = (N + 2) - d \tag{10.2}$$

Equation (10.2) applies in the special case of a lift and a call being at extreme ends of the shaft, even though the lift is not technically moving in the same direction. Then:

$$FS = (N + 2) - N = 2 \tag{10.3}$$

Rule (b): if the lift is committed to move towards the landing in the opposite direction as that required by the landing call, then:

$$FS = (N + 1) - d \tag{10.4}$$

The figure of suitability is therefore its maximum value equal to *N* + 1 when the lift is at the calling floor.

Rule (c): if the lift is committed to move away from the landing call, it is considered to be 'in service', and the figure of suitability is arbitrarily fixed as one, irrespective of the value of distance.

Rule (d): a lift that is idle, i.e., it has no car or landing call commitments, has a figure of suitability given by Equation (10.4).

The allocation procedure assigns each landing call to the lift presenting the highest figure of suitability for the particular call. If the figures of suitability are the same for two or more lifts, the call is assigned to the nearest of these lifts or, in the case of equality of distance, to the first lift that reached that figure of suitability some time earlier. Lifts that are fully loaded are not considered in the allocation procedure. However, if all the lifts are fully loaded, the landing calls are temporarily allocated to the 'best' placed lift. A lift in these circumstances will not stop at such landing calls unless the lift becomes less than fully loaded as the result of a passenger alighting before reaching such a landing call. When a lift is not assigned to any landing call and no car calls are registered, it is made to stop at its current floor, with its doors closed. If no parking algorithm is provided, the distribution of lifts in the building will be poor, as stationary lifts are ready to move as soon as they become assigned to a landing call.

Consider a triplex control system serving a building of seven floors (main terminal and six floors). At a particular time instant, the lift positions and status, and the registered car and landing calls are as illustrated in Figure 10.1.

To determine which lift is allocated to each landing call, the figures of suitability for each pair of landing call and lift are indicated in Table 10.2. To see how these figures of suitability are derived, consider the evaluation for the down call at floor 4.

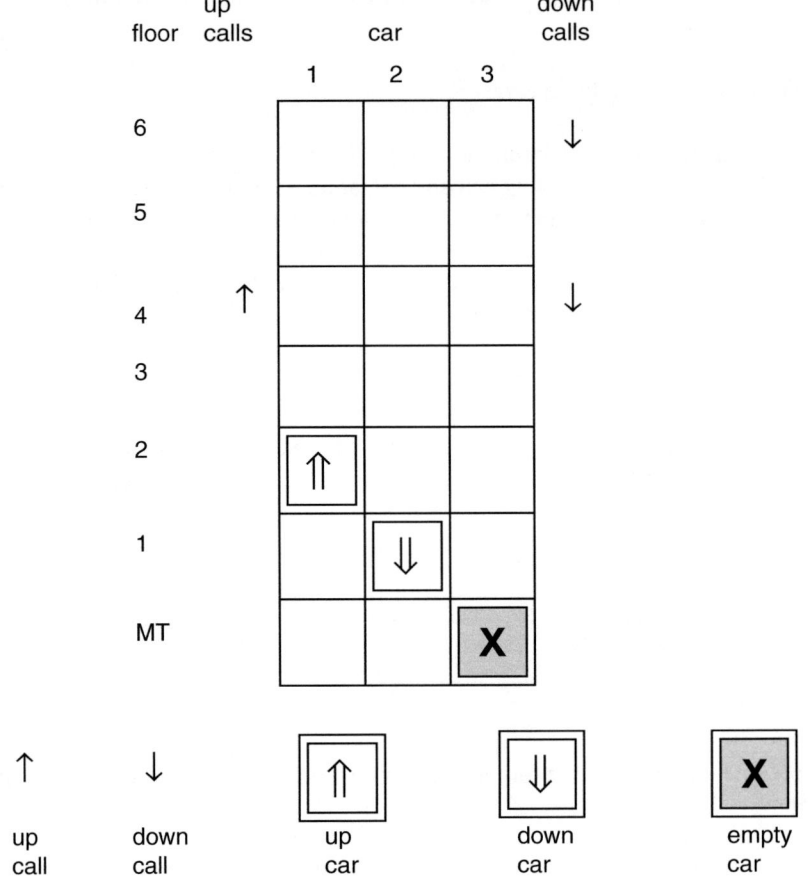

Figure 10.1 Example operation of a triplex traffic control system

Table 10.2 Figures of suitability for the example in Figure 10.1

Call	Car 1 – UP @ floor 2	Car 2 – DOWN @ floor 1	Car 3 – EMPTY @ floor MT
4-UP	6	1	3
6-DN	3	1	1
4-DN	5	1	3

Car 1 – is moving towards the call
  – opposite directions for call and travel (Rule (b))
  – distance = 2 floors; therefore FS = 7 − 2 = 5
Car 2 – moving away from the call, FS = 1 (Rule (c))
Car 3 – not committed to direction (Rule (d))
  – distance = 4 floors; therefore FS = 7 − 4 = 3.

The maximum figures of suitability in Table 10.2 are included for each landing call. They show that all the landing calls in the example are allocated to car 1.

No other special features are included in the NC traffic control system. It provides a single program control algorithm designed specifically to cater for interfloor traffic in low rise buildings. The allocation procedure has some similarities with dynamic sectoring, however it does not take great advantage of the direction of the landing call. For instance, in the example, the allocation of the down call at floor 4 to car 1 is highly inconvenient, as this lift is going to stop at least at two floors (4-UP, 6-DN) and maybe at floor 5 if the 4-UP caller requests it, before answering the down call at floor 4.

A much better policy would be to allow the idle lift stationed at the ground floor (car 3) to move up and help the first lift in this particularly unbalanced traffic situation. Actually (see Table 10.3), the down call at floor 4 will eventually be served by car 3, as it will be allocated to the call as soon as car 1 leaves floor 4, travelling upwards (assuming that no more landing calls have been registered). Meanwhile, the call has been waiting for an unnecessary time.

Owing to these types of difficulties, this simple algorithm is suitable only for buildings with a small number of levels where the distribution of lifts is not very critical. Under such conditions, this algorithm provides reasonable interfloor traffic performances. This algorithm is not appropriate to serve down peak traffic. The allocation procedure will cause the lifts to travel to calls too high in the building, neglecting calls lower in the building, and thus giving better service to the higher floors. Again, in a low rise building, and assuming that the occupants at the lower floors use the stairs rather than the lifts, the system may be acceptable.

It is under uppeak conditions that this algorithm presents its greatest disadvantages. Most control systems express the lifts down to the main terminal as soon as they discharge their last passenger while the uppeak condition applies. Unlike such systems, this simple algorithm uses the normal allocation procedure, assigning a lift to the ground floor only when a call is registered there. Lifts will park at upper floors for some time, thus providing a poor uppeak service.

10.7.2   *A fixed sectoring common sector system (FS)*

A suitable fixed sectoring common sector control system can be devised for dealing with off peak traffic and can be complemented with special features to cater for heavy unbalanced traffic. The FS system divides the building into a number of static demand sectors (Figure 10.2) equal to the number of lifts. A sector includes both the up and down landing calls at the floors within its limits. A lift is allocated to a sector if it is present in that sector, and the sector is not committed to another lift. Fully loaded lifts are not considered for allocation.

An assigned lift operates on the directional collective principle within the limits of its range of activity. The lift responds to the landing calls registered inside its sector, and to landing calls in adjacent vacant sectors above. It also responds to landing calls below the car sector if the lift is assigned to the lowest occupied sector. If landing calls are detected behind a lift that is serving its sector and within the sector, then this sector is considered to be vacant in order to allow the lift located below to answer these calls.

Table 10.3 Revised figures of suitability

Call	Car 1 4 – UP @ floor 4	Car 2 – EMPTY @ floor MT	Car 3 – EMPTY @ floor MT
6-DN	5	1	1
4-DN	1	3	3*

* Reached this value first

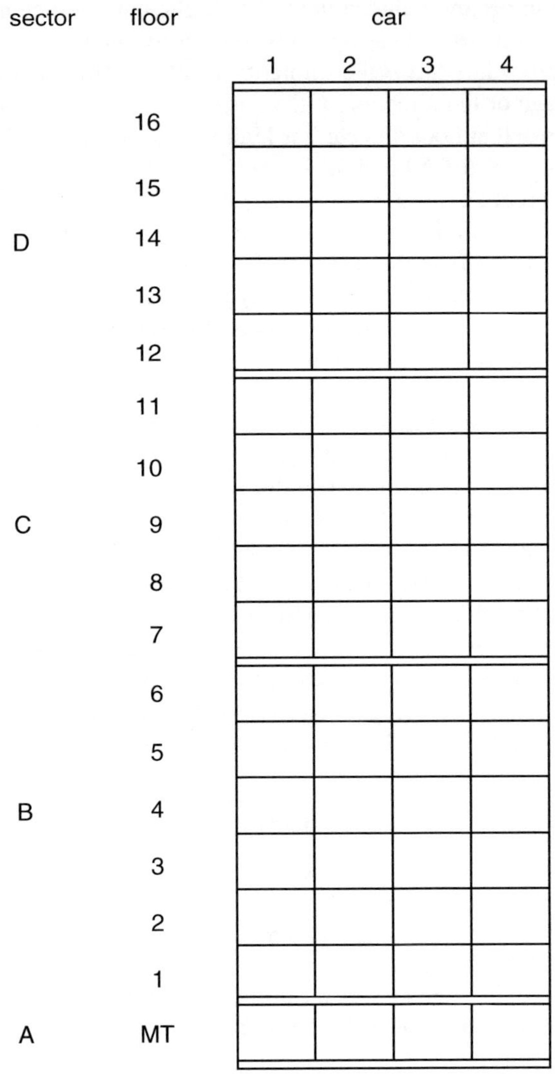

Figure 10.2 Illustration of the assignment of floors to fixed common sectors

The de-assignment of a lift from its sector takes place when the lift leaves the sector. A lift picks up calls ahead when travelling in either direction, even if it is not assigned to the sector. So, the lift is allowed to start or continue its movement in response to:

(a) registered car calls
(b) landing calls registered inside its sector
(c) landing calls in contiguous unoccupied sectors above and adjacent to the sector to which the lift is assigned
(d) landing calls registered behind the lift assigned to the contiguous sector above, if this lift is travelling upwards
(e) all the landing calls registered below the car sector if the lift is assigned to the lowest occupied sector.

The de-assignment of a lift from its sector in this scheme implies that the lift leaves the sector or becomes fully loaded within the sector. A lift is allowed to pick up landing calls on its way even if it is not assigned. Fully loaded lifts do not, however, stop in response to landing calls.

The FS system, by distributing the lifts equally around the building, presents a good performance under balanced interfloor traffic. It also performs well for uppeak and unbalanced interfloor traffic conditions. It lacks a proper procedure to cater for sudden heavy demands at a particular floor. Under heavy down peak traffic conditions, a poor service may be provided to the lower floors of the building, owing to problems in recycling the lifts to unoccupied sectors.

To consider how an assignment of floors to sectors might be made, consider a 16-floor building (including the ground floor) served by four lifts as shown in Figure 10.2.

There will be four sectors:

Sector A	Ground
Sector B	Floors 1–6
Sector C	Floors 7–11
Sector D	Floors 12–16

The ground floor is allocated solely to sector A to emphasise its importance. Sector B above is made the larger of the remaining three sectors, on the basis that people may walk to/from floor 1 and the travel distance to sectors C and D is larger than to sector B.

10.7.3   A fixed sectoring priority timed system (FS-PT)

Static directional sectoring systems can also allocate the lifts on a priority timed basis. The landings in the building zone served by the group of lifts are grouped into up and down sectors. Each sector is timed as soon as a landing call is registered within its limits. The timing is measured in predefined periods of time, designated the priority levels. The time to reach a specific priority level may be different for each sector in order to give some selected sectors an overriding priority. In addition, the six priority levels may not comprise equal time steps. A common situation is to have six priority levels, which, for example, may follow the sequence 10, 15, 20, 30, 40, 60 seconds for a standard floor, but the sequence 30, 50, 70, 90, 105, 120 seconds for a low priority floor. Such arrangements are particularly useful to cater for executive floors, which require high priority service, and for basement floors, which can be given a low priority sector.

The assignment of lifts to the sectors takes into account the number and positions of the available lifts and the sector priority levels. A lift is available for allocation when it has completed its previous assignment, and has dealt with all the car calls that have been registered. The sector with the highest priority is the first to be allocated a lift. If more than one lift is available, it is the nearest lift that is allocated to the sector. After the lift is assigned, it travels without stopping to the sector. However, after leaving the sector, the lift answers landing calls on its way, until the complete set of car calls is served. When the lift stops to answer its last car call, it becomes available, and is allocated to the unattended sector with the highest priority. It may happen, however, that some passengers are waiting at the lift's last call floor. They are discouraged from entering the car, as no landing signal is given as the lift arrives, and the lights in the car are dimmed. In addition, a flashing indicator reading 'do not enter' may be fitted in the rear of the car.

The FS-PT system gives preferential service to the main terminal, when no lift is stationed there, and an available lift is dispatched down, bypassing all but priority levels 5 and 6. The lift stationed at the main terminal floor, with open doors, has an indication that it is the next lift to depart. It is not available to answer demand on other sectors unless no other lift is available, and the lift is stationed for more than 5 s at the main terminal.

Demands at priority level 6 call for an immediate response; the nearest lift travelling towards the top priority sector bypasses all landing calls before that sector. Use is made of this feature to cater (partially) for sudden traffic demands at particular sectors or floors. Should a lift fill up to capacity before completing its sector assignment, this sector is advanced to top priority level, thus applying for immediate service. The next lift to become available is dispatched to the midpoint of this sector. Fully loaded lifts do not stop in response to landing calls.

The FS-PT system is unique among the classical traffic control systems as it considers time when making an assignment; the other algorithms only consider position.

To understand this type of allocation algorithm, consider a group of five lifts serving a zone of 17 floors (including the main terminal). Assume that, at a particular time instant, the position and status of each lift are as indicated in Figure 10.3.

The lift activity can be summarised as:

- Car 1 was allocated to sector S4 and travels express to floor 12, bypassing the up call at floor 9, which is only at priority 1 (P1).
- Car 2 has just become available and two sectors are demanding service, the sector S3 with P1 and sector S8 with P2. The lift is assigned to sector S8 because this sector has the higher priority. It travels express to floor 7, where it reverses its direction of travel and starts answering the traffic demand within its sector, i.e., floor 5.
- Car 3 is the next lift to leave the ground floor, when required.
- Car 4 has been allocated to sector S7, where a landing call at floor 9 has already reached the P4 stage. The lift travels express to floor 9, unless a down call is registered at floor 10 in the meanwhile.
- Car 5 has serviced a down call at floor 16 and is moving downwards to answer a car call at floor 10. It has finished its assignment to sector S5 and will answer the down call at floor 12 on its way down in response to the car call.

10.7.4   A dynamic sectoring system (DS)

The dynamic sectoring group supervisory control system provides a basic algorithm and is suitable to deal with light to heavy balanced interfloor traffic. It is complemented by a number of other control algorithms to cater for unbalanced traffic conditions.

The basic DS algorithm groups landing calls into dynamic sectors, as described in Section 10.3.3. Each lift is allocated to a sector in the sector definition itself and answers the landing calls in the sector according to the directional collective procedure. Usually, a lift defines a single sector, however, if a lift is located at the same floor as a higher reference lift, or if the lift is fully loaded, then the lift does not define a sector. In addition, a lift that remains without a travel command for a specific length of time, typically 5 s, is declared as a free lift and given two sectors, one in each direction.

In parallel with the basic traffic algorithm, another DS algorithm is provided, the free lift algorithm. It operates in a similar way to the basic DS algorithm, but it considers only the free lifts and the heavy duty calls. This free lift control algorithm is intended to insert free lifts ahead of lifts allocated to a high demand sector in order to provide service where the traffic is

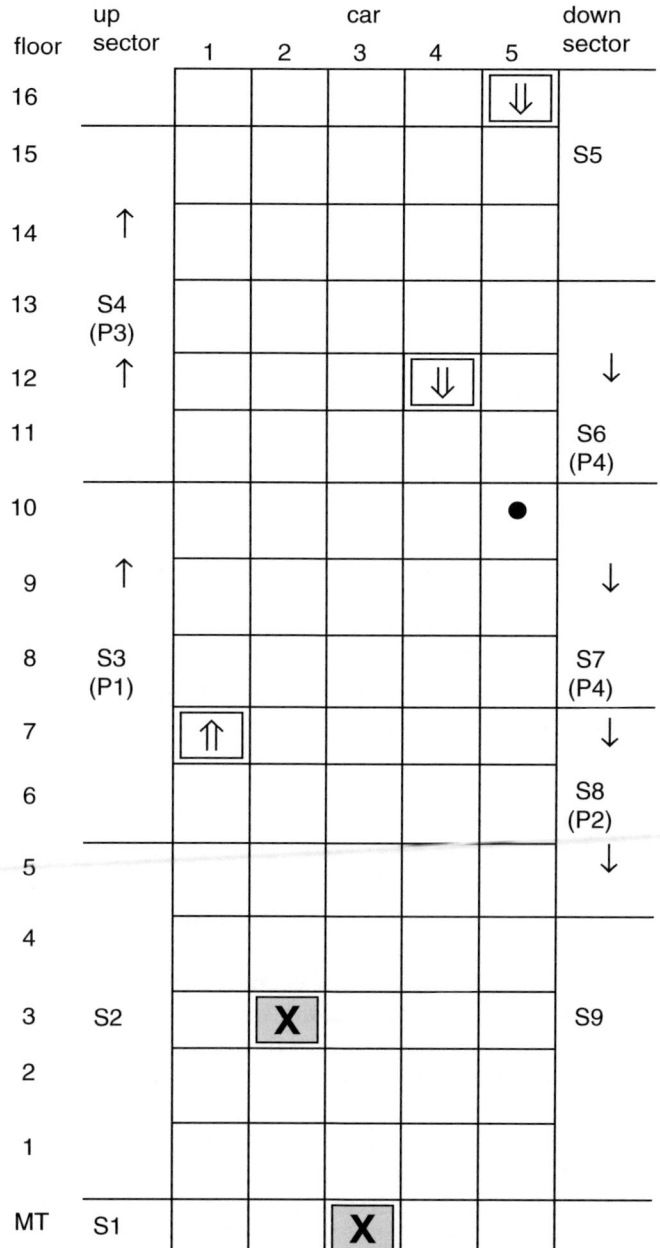

Figure 10.3 Illustration of the assignment of floors to fixed priority timed sectors' car call shown

heavier. The algorithm continuously monitors the number of car and landing calls within the normal dynamic sectors, by counting the calls from the car position. After a predefined number of car and landing calls, 'the optimum counter capacity', the remaining landing calls are registered as heavy duty calls in a free car zone memory. The optimum counter capacity is varied

Table 10.4 Optimum counter capacity

Number of free lifts	Optimum counter capacity
3	2
2	3
1	4

according to the number of free lifts, from a minimum where at least three lifts are free, to a maximum when only one lift is free. As an example, consider Table 10.4.

New symbols: {↑} landing call assigned as heavy duty, hatched car 4 at floor 1 is an up travelling free lift.

In order to express the free lifts to the heavy load centres, a free lift, which is moving in response to a heavy duty call, is taken out of the basic control algorithm.

To illustrate the working of this traffic control algorithm, see Figure 10.4. A group of five lifts serves a local zone of 16 floors above the main terminal floor. The group operates under a dynamic sectoring control system, and the optimum counter capacity is as given in Table 10.4. Where will the sector boundaries be?

Consider first the normal dynamic sectoring procedure:

* Car 1 is dealing with floors 5–12, and the car calls at floors 7, 10, 11, 15.
* Car 2 is dealing with floors 11–5, and the car calls at floors 6, MT.
* Car 3 is dealing with floors 13–16–12, and is a parked free lift.
* Car 4 is a moving free lift.
* Car 5 is dealing with floors 4–MT–4, and the car calls at MT.

In summary, the first, second and fifth lifts are moving in response to car and landing calls. The third and fourth lifts have no call demand placed on them, and therefore are free. Note that the sector of each lift extends from the car position to the position of the next lift ahead that is free or moving in the same direction. When the sector reaches the terminal floor, as for car 5, it continues in the opposite direction. Also note that car 4 is taken out of the normal sectoring scheme because it is a free lift committed to a direction of travel.

The traffic situation considered in the above example is severely unbalanced. The first car has four car calls and four landing calls to answer, which is disproportionate in comparison with the other lifts. The free car algorithm is in control of the situation, and takes action to provide extra service to the sector assigned to car 1. The optimum counter capacity is three, since two lifts are available, so the up landing calls at floors 9 and 11 are declared as heavy duty calls. The two free lifts define two dynamic sectors in the free car zone memory and the two heavy duty calls are within the free car sector of car 4. This is despite car 3 being nearer. The lift starts in response to the calls, moving express to the lowest heavy duty call at floor 9. The insertion of a free lift into the heavy traffic section of the zone, instead of attempting a redistribution of lifts, makes this control algorithm highly efficient.

10.8   Other features of group traffic control systems

The control systems discussed in Section 10.4 illustrate possible methods of dealing with passenger demands (traffic). There are a number of other features that assist this process, and lift

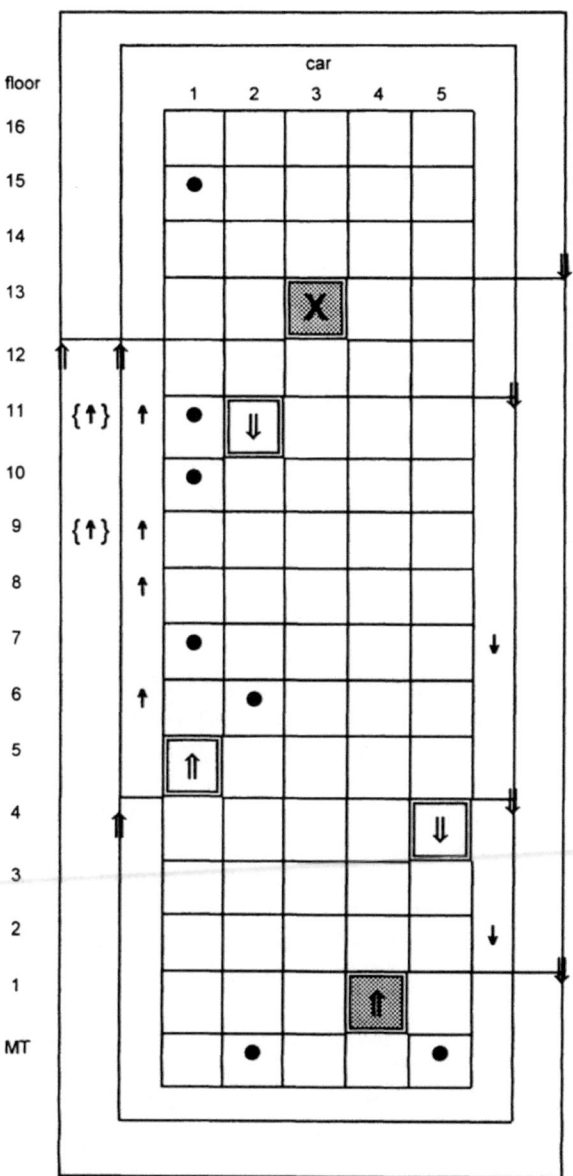

Figure 10.4 Illustration of the dynamic sectoring traffic control algorithm. Inner loop: basic dynamic sectors. Outer loop: free lift sectors

group control systems provide other specialised features, such as basement service, lobby floor preference service, car preference, director/VIP service, heavy demand floors, parking policy, automatic shutdown, energy saving modes, load bypass, maintenance, service, firefighting switches, passenger signals, etc. Whatever the technology, i.e., relay or computer-based, many of the features are important.

10.8.1   Uppeak service

Most lift group control systems detect and take special action for uppeak traffic conditions. While the uppeak condition applies, the lifts answer car calls only, bypassing the landing calls. Thus, as soon as a lift discharges its last passenger on its way upwards, it reverses direction and travels non-stop to the main terminal. If, however, some astute passengers are still in the car and register car calls, the lift will answer the car calls before travelling to the main terminal (Rule 10.1).

There are several uppeak detection mechanisms. A common method is based on weighing devices installed in the car floor, or by measuring the motor load current as an indicator of loading. When heavily or fully loaded lifts leaving the main terminal floor are detected, the uppeak control algorithm is selected for a specific time period. A variation of this method, which is able to cater for slight uppeak situations, detects a lift load at the main terminal in excess of a predefined level, say 50% or 60%. For a certain period of time, a dummy call is set up at the main terminal to ensure that a lift is available there as soon as possible. Another method counts the car calls registered, and when a predetermined number are registered, initiates the uppeak algorithm.

A more reliable and sophisticated approach employs an up/down logic counter that increments when loads are above a predefined level, and decrements for loads below this level. Additionally, the counter decrements on a timed basis, every 60 s or so, to ensure that the uppeak algorithm is switched off quickly as the uppeak traffic diminishes. To prevent instability, hysteresis is built into the mechanism by arranging that the uppeak turn-on level is two or more counts above the turn-off level. In addition, there is a clock inhibit level for the decrementing time pulses, which is at least one unit above the turn-off level. Two important counter levels are defined: the turn-on level, which selects the uppeak state, and the turn-off level, two or three units below the previous level, which switches the uppeak state off.

The automatic uppeak detectors may be substituted or complemented by a time clock or a manual switch in the supervisory controller. These may be a useful complement, but if they constitute the only available means to select the uppeak algorithm, they are poor alternatives. Strakosch (1967) states that in earlier systems, where an attendant was required to change the control program, too often, the lift system was set in one program and not changed, so that when it was time to leave the building the passengers had to fight lifts that had been left on 'uppeak' operation.

An option provided by some uppeak traffic control algorithms is the division of the building into two uppeak sub zones. This division does not concern landing calls, but only the destination floor of the passengers. Once the uppeak state is selected, some of the lifts are allocated to serve the lower sub zone and the remainder to serve the upper sub zone. An indicator is illuminated at the main terminal floor for each lift, stating which floors are served by that lift. Uppeak sub zoning is intended to reduce the number of stops per trip made by the lifts, hence increasing the uppeak handling capacity. It works well, if properly tuned, but it is very sensitive to the sub zone partition. If the uppeak traffic flow is subject to large fluctuations in the destination of passengers, it may be that one subgroup of lifts is saturated and the other is under used. A later chapter examines this and other techniques to improve traffic handling during uppeak as 'uppeak boosters'.

During uppeak, any passenger wishing to travel up from floors other than the main terminal floor should have little difficulty as the lifts frequently stop at the floors while travelling upwards to discharge passengers. However, passengers wishing to travel down may find a restricted service or no service at all during the 10–15 minutes of heavy uppeak demand. In the case of no service, an astute passenger can be serviced by registering an up call and entering the first upwards travelling lift stopping at their floor. With restricted service, down landing calls are served by a single lift at fixed time intervals (30 s to 3 minutes) or, alternatively, when they reach the highest priority level.

Although it has been said, all control algorithms behave in a similar way for uppeak traffic, i.e., lifts fill at the main terminal and distribute to the upper floors, the above are some important control features, which aid efficient handling.

10.8.2   *Down peak service*

During down peak, lifts working on the collective principle, or on a scheduled basis, present a problem: the lifts travel too high up the building, tend to fill up to capacity at the upper floors, and provide a poor service to the lower floors. Therefore, a major feature of a down peak traffic control algorithm should be the provision of a proper cycling of lifts in order to ensure an even service for all the floors. However, group control systems frequently include a means to detect down peak traffic situations, employing similar methods as those used for uppeak detection and considering heavy loaded lift arrivals at the main terminal floor. While the down peak condition applies, these systems will restrict the service provided to any up traffic and cancel the allocation of lifts to the main terminal while the traffic condition lasts.

Fixed sectoring lift allocation systems are particularly adaptable to this traffic pattern as the lower sectors are sure to be allocated a lift, and do not need to rely on the spare capacity of lifts descending from the floors above. Thus, such systems do not need to provide a separate control algorithm for down peak. Priority timed systems, on detection of a fully loaded down travelling lift, might reduce all the up sector priorities for a pre-set time, during which preferential service is given to down calls.

Dynamic sectoring schemes, where the sectors are defined according to the status, position and direction of travel of each lift, do not cope very well with down peak traffic. Most lifts, after becoming available at the main terminal and finding no up landing calls registered, are assigned to the top landing calls. Thus, such systems suffer from the same problems as directional collective systems, namely, very high average reversal floors, and poor service to the lower floors. A separate down peak control algorithm is therefore necessary.

One such system detects heavy down traffic terminating at the main terminal floor. The traffic algorithm then groups the down landing calls into dynamic sectors and assigns lifts to serve call groups in the sectors in turn order. In order to provide a proper cycling of lifts and a systematic and equalised service for down traffic, each time a lift stops in response to a down landing call, the floor is blocked to ensure that each floor is served only once during one service cycle.

10.8.3   *Load bypass*

When a lift fills up to capacity, it should not stop in response to further landing calls as such stops would be useless and particularly annoying to the passengers already in the lift. A device is usually available to provide such a feature. However, landing calls should not be cancelled.

It is important that the car load detection is set correctly. As has already been explained, larger lifts cannot accommodate the rated capacity as indicated in the standards. Consider Example 5.3 again. This refers to a 2,500 kg lift. If the load detection were to be set to 60%, this would equate to 1,500 kg. If an average person weighs 75 kg, this is 20 persons and this is larger than the design capacity of 19.0 persons. Therefore, the lift will frequently stop unnecessarily. The real 60% value would be equivalent to 60% of 23.8 persons (the probable capacity), i.e., 1,080 kg.

10.8.4   Heavy demand floors

Heavy floor demands can occur, for example, at the closing of a meeting or lecture. It is justifiable to bring extra lifts to the floor to deal with such peaks of demand.

In order to cater for local heavy demands, the number of landing calls in each sector can be evaluated by the traffic analyser and compared with the average number of landing calls per sector. A particular sector exceeding the average value by more than a predefined quantity can be set up as a heavy traffic sector. The measure of heavy traffic can be individually adjusted for each sector, allowing preferential service for some selected sectors. A heavy traffic sector is considered to be vacant by the control system, even if one or more lifts are already serving the sector. Extra lifts are brought to this sector, bypassing the landing calls at other sectors.

Another method is to detect, at individual floors or specially designated floors, that a fully loaded lift has left that floor and a new landing call has been registered within 2 s or so, for the same direction of travel. The traffic controller can then send free lifts to the heavy demand floor.

10.8.5   Lobby and preferential floor service

The lobby or main terminal floor in a building is normally of great importance, owing to the steady flow of incoming passengers. Preferential service is usually provided for these passengers by parking a lift at the main terminal prior to any other sector. Lifts are usually parked with their doors closed at the main terminal floor and can be assigned as the 'next' lift to leave. The lobby floor preferential service implies that a slightly poorer service is provided to the remaining floors in the building. The feature is highly undesirable under certain traffic conditions, such as down peak. A good down peak control algorithm must therefore override such a feature. Another arrangement is that when a lift has parked at the main terminal for a period of time, which can vary from 5 seconds to 3 minutes, it becomes available to satisfy demands in other parts of the building.

Various techniques are used to bring a lift to the main terminal whenever it is vacant. The parking algorithm can preferentially park an idle lift at the main terminal rather than at any other vacant sector. A dummy call can also be inserted to force a lift to the main terminal. In a priority timed system, it is the assignment algorithm that allocates an idle lift to the main terminal floor, although it allows the lift to pick up high priority calls on its way down to the main terminal.

A feature related to car preference is the 'director' service, which gives special service to the floors where senior executives or directors are installed. The lift system may be made to recognise landing calls at such floors and to deal with them with higher priority, or, key-operated switches may be available at these preferred landings, which cause a lift to travel direct to the executive floor, bypassing all other landing calls, or a lift may be completely segregated out of the bank of lifts for director service. It is obvious that this sort of preferential treatment can seriously affect the efficiency of the service as a whole, and it should be avoided whenever possible.

10.8.6   Parking policy

Under light to medium traffic conditions, a lift frequently has no calls to answer. The lift is then free for allocation, and if no further demand exists, it might be parked at its current position, or a convenient floor, or sector in the building zone. The parking procedure is mainly intended to distribute the lifts evenly around the building, in anticipation of traffic demand. A proper parking policy is essential for good lift system performance.

The parking sectors can be the same as the demand sectors in fixed sectoring controllers. In the DS algorithm, the free lift control algorithm provides a parking policy by dividing the building into fixed parking sectors to redistribute any free lifts over the whole building zone.

The parking of a lift at the main terminal floor has preference over parking the lift in other sectors, thus providing preferential service to the main terminal floor. If the lift stationed at the main terminal departs in response to a car or landing call, the nearest parked lift could be immediately dispatched to fill its place.

10.8.7   Basement service

Basement service usually requires special consideration. Even a small basement demand under uppeak and down peak conditions can severely degrade the lift system performance. For this reason, most group control systems will seriously restrict basement service. These floors frequently contain services that are of secondary importance during peak periods, and a restricted service may be sufficient. For example, basement floors may be grouped into one sector, which is given very low priority. Some systems will accept a car call, but not a basement landing call; other systems will allow one single designated lift to reach basement level. Some systems may simply provide no basement service at all while a heavy traffic demand prevails. In some cases, it may prove preferable for a building designer to provide alternative vertical transportation systems to cope with basement service, thus allowing the more expensive above ground system to serve the remainder of the building. Section 8.3.3 deals with basement service in more detail.

10.8.8   Car preference

When a lift is taken out of normal passenger control to be exclusively operated from the inside of the lift, it is said to be in-car preference, also referred to as independent service, emergency service or hospital service. The transfer is made by a key-operated switch in the lift, which causes the doors to remain open until a car call is registered for floor destination. All landing calls are bypassed, and car position indicators on the landings for the lift are not illuminated. The removal of the key, when the special operation is complete, returns the lift to normal control.

Car preference may be useful to give a special personal service, or for an attendant to have complete control of the lift whenever it is required. A typical example is in hospital buildings, where lifts for carrying beds and stretchers require a car preference switch.

10.8.9   Automatic shut down

On older systems, when motor generator sets were employed, the traffic control system might shut down some lifts when the traffic demand is very light in order to save wear, tear and power. Today, this feature is not required on modern drives. However, a feature to cause a different lift to answer calls in some predefined sequence during light traffic periods is desirable to avoid one lift servicing all calls. Although automatic shut down of generator sets is no longer extant, the shut down of lifts when demand levels are low, particularly at night will save energy (and wear and tear).

10.8.10   *Other features*

A number of other facilities available in group supervisory control systems include the provision of indicators of lift arrivals at landings, and a number of switches, such as maintenance switches to switch off a lift while maintenance work is in progress, and fire switches whenever required by the fire authority, to enable firefighters to take over the complete control of one or more lifts in a group. Another useful feature is the provision of anti-nuisance devices to ensure that a lift does not answer car calls if it is empty. This avoids unnecessary car trips and stops due to a practical joker who registers car calls by pressing all the car buttons and then leaving the lift.

10.9   Comparison

Figure 10.5 illustrates the performance of a standard collective and the dynamic sectoring algorithms. They follow a similar profile with the classical 'knee' at about 80% car load. Provided the relevant precautions given in Section 10.8 are taken, all the legacy systems offer similar performances.

Down peak traffic demands are often about 50% larger than uppeak. Figure 10.6 illustrates the performance of the three legacy algorithms considered in this chapter.

Performance is similar up to about 140% demand when it starts to deviate. The DS algorithm then provides a poor performance The FS-PT algorithm down peak performance is very good, especially under very heavy traffic conditions. The system not only cycles the lifts to the sectors in a convenient way, but also provides extra service to sectors waiting too long to be assigned a lift. The FS algorithm's down peak performance is poor above 150% demand.

The DS system provides an exceptional performance for interfloor traffic conditions, see Figure 10.7.

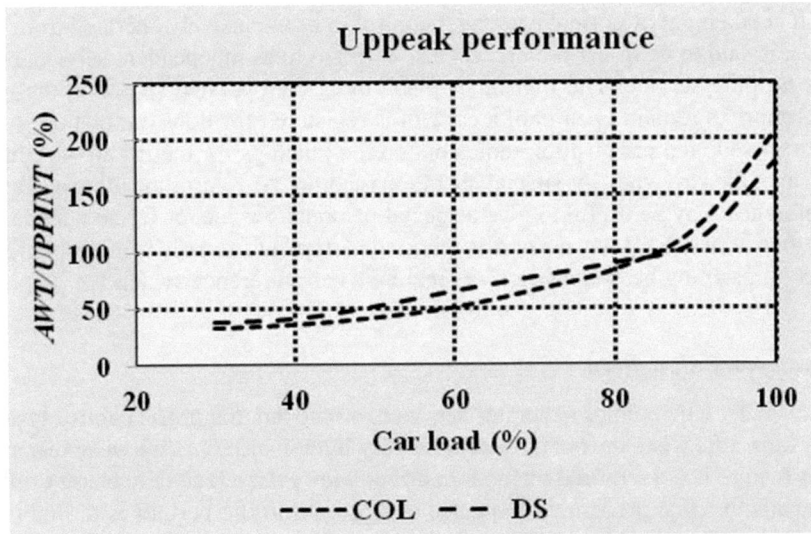

Figure 10.5 Uppeak performance

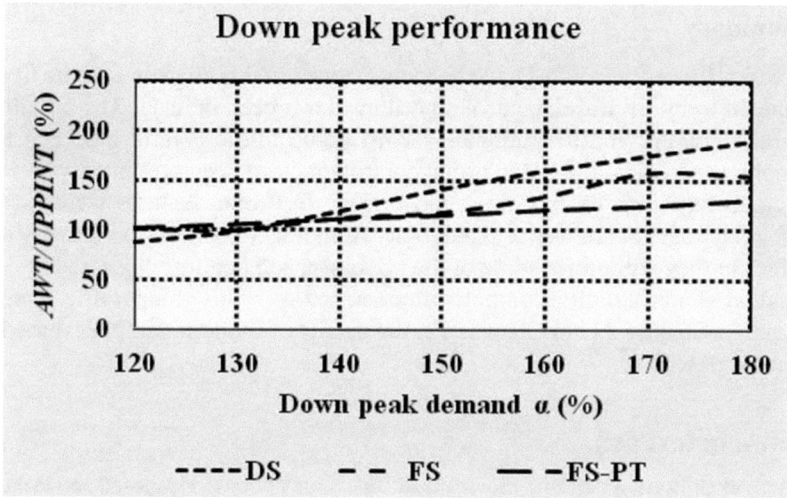

Figure 10.6 Down peak performance

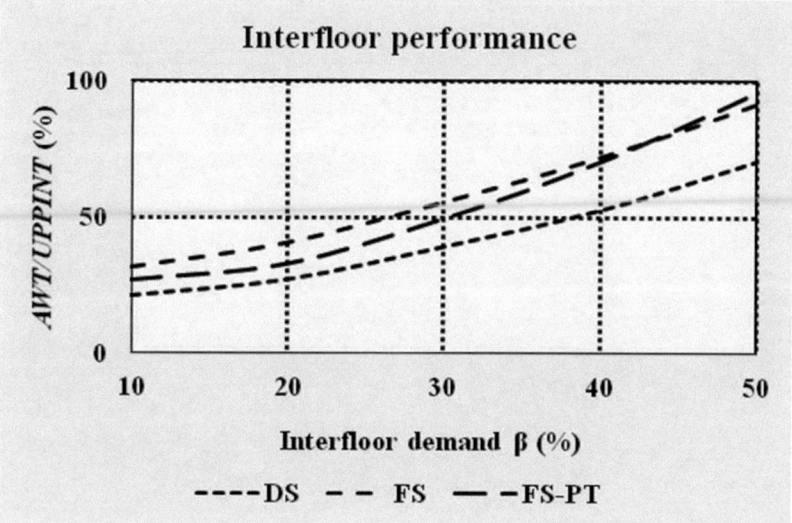

Figure 10.7 Interfloor performance

The FS and FS-PT algorithms provide a fair control algorithm for interfloor traffic, but not as good as can be obtained from dynamic sectoring. The main problem is that the system lacks a parking algorithm to redistribute the lifts in the building. It is true that the allocation procedure distributes the lifts to the sectors, but only after the landing calls are registered; thus, the express jumps necessary for the redistribution are made when the passengers are already waiting for service. Uniform performance is obtained for all floors, but average waiting times tend to become extended.

10.10   Summary

The principles of lift traffic control have been introduced in this chapter, and the five rules that must be adhered to by lift traffic control algorithms have been defined. The possible types of single lift traffic control and the rationale for lift group traffic control has been introduced. Three examples of classical lift traffic group controllers have been presented.

The lift control systems incorporating most of the foregoing features were largely implemented with relay logic or solid state fixed logic. A digital computer is used today to control a group of lifts, but they still utilise some of the principles and features discussed in this chapter. The facts and problems of digital computer implementation of lift group traffic control systems are considered in Chapter 11. In Chapter 12, the details of modern computer-based lift group traffic control is described.

Notes (shown in text as[1])

1 A door interlock prevents a car from moving if the doors are not properly closed and locked.
2 The bars (|) indicate d is to be the modulus of the expression, i.e., independent of sign.

11 Computer-based traffic control

The concept of centralised supervisory control systems for buildings, known as: building auto-mation systems (BAS), building energy management systems (BEMS), etc. using digital com-puters[1] is already well established. It might include lift group supervisory information as part of the comprehensive information system for a whole building, which might also include other facilities such as employee identification, security control, fire control, environmental control, water treatment and data logging. However, it is not possible to include the task of lift traffic control in any centralised building control system. Thus, a lift should have all aspects of its traffic and power control managed independently of other building systems. This does not preclude the monitoring of lift activity.

The opportunity exists with a computer to program complex tasks to assist the call alloca-tion process, which is impossible to achieve with fixed program systems. This chapter first reviews conventional lift traffic control systems and then discusses the general principles of computer-based systems and details of three traffic control systems.

11.1 An analysis of conventional control systems

11.1.1 Two requirements, five rules and four tasks

The two requirements were explained in Section 10.1, and the five rules were laid down in Section 10.4. To these requirements and tasks must be added the four primary tasks of a group traffic control system when serving both car and landing calls:

(a) to provide an even service to every floor in a building
(b) to minimise the time spent by passengers waiting for service at a landing
(c) to minimise the time taken to move passengers from one floor to another
(d) to serve as many passengers as possible in a given time.

There may be conflicts between some of these tasks. For example, an even service may result in some passengers waiting longer.

11.1.2 The nature of passenger demand

Problems are encountered in attempting to achieve the four primary tasks owing to the random nature of the time of call registration, the landing at which passengers arrive and request ser-vice, and their ultimate destinations. Extensive observations of actual buildings tend to dispute the traffic pattern depicted in Figure 4.2. Beebe (1980) performed a study of logged data from two commercial buildings and observed the following:

(a) Both uppeak and down peak traffic periods exist, but last longer than indicated by Figure 4.2 and at lower intensity levels.
(b) During an uppeak period, a considerable volume of down traffic exists.
(c) Similarly, during a down peak period, a significant level of up traffic exists.
(d) Algorithm switching caused by an incorrect detection of a specific traffic condition can seriously degrade performance.

Unlike traffic analysis, which is often based on pure traffic patterns of up traffic only (pure uppeak), down traffic only (pure down peak) and random traffic with no dominant pattern (pure interfloor), traffic controllers have to deal with the real world of mixed traffic patterns. It is probably best to consider all traffic to be an interfloor traffic pattern with various significant overlaying patterns. If this premise is adopted, then it should be possible to deal with the observations (a), (b) and (c). So, an efficient traffic controller should deal well with the pure interfloor traffic pattern, but respond to any dominant trends. The latter point (d) is well known, where time clocks have caused the uppeak traffic algorithm to be operating during (say) balanced interfloor traffic.

The main conclusion to be drawn is that the traffic control algorithm must be able to follow changes in passenger demand at all times in a 'bumpless' (hidden) manner.

11.1.3   *Characteristics of legacy traffic control systems*

Chapter 10 discussed four legacy control algorithms, which have been embodied into proprietary controllers in various ways. They were: nearest car (NC); fixed sectoring, bidirectional sectors (FS); fixed sectoring, priority timed unidirectional sectors (FS-PT); and dynamic sectoring (DS). Although the NC algorithm does possess a limited computational ability (Hirbod, 1975), it is intended for low rise, duplex/triplex installations, and it will not be considered further. The other three traffic control systems exhibit a number of common characteristics.

The systems are 'multi-algorithmic' with different algorithms or sets of rules for the different traffic patterns. The default operating mode is the interfloor algorithm. Uppeak requires a different algorithm in all cases. Down peak is served by the interfloor algorithm in the FS and FS-PT systems, with the minor change of cancelling allocation of cars to the main terminal.

Where a different algorithm is used to handle uppeak or down peak traffic, the changeover is usually abrupt. This can cause disruption of the current traffic, although the severity of the effect depends on the intensity of the current traffic. When the detection is incorrect, the effect is most severe. It would appear that much could be gained by the use of a single control algorithm (Smith and Peters, 2002), which can adapt to different traffic conditions and provide a 'bumpless' transfer.

During uppeak, the actions of the traffic control algorithms are identical (see Section 10.9). All a control system can do is to bring cars, which have completed serving their car calls, back to the main terminal immediately. The preferential service given to the main terminal has the benefit of handling minor up peaks that occur from time to time. Efficient uppeak detection is important as late detection produces a poor uppeak service.

All systems should allow partly loaded cars to collect landing calls in their path until they are loaded to no more than 80% of actual capacity. This is achieved on a directional collective principle. This characteristic is of considerable assistance in conserving handling capacity.

The allocation procedures employed by FS and FS-PT control algorithms allocate free or uncommitted cars to demand sectors. Once the car answers a landing call in its allocated demand sector, it continues to answer all other landing and subsequent car calls in simplex mode, and the group controller loses control of the individual cars. Thus, committed cars work virtually independent of one another, defeating the objectives of group control.

The FS and DS systems have no concept of time, and do not time landing calls. Thus, they cannot give urgent attention to long wait calls. Only the FS-PT system has a concept of time.

11.1.4 *Improving a legacy traffic control algorithm*

The concept of using existing traffic control algorithms, designed to be implemented in relay or solid state logic, is a quick way to obtain a computerised traffic controller. However, is this the correct approach when the computer has much greater computational power available? Lim (1983) decided to test this thesis and selected the DS algorithm as the best 'all round' legacy control system. He rationalised his choice by these observations: that static sectoring algorithms generally allocate a car to a sector without checking whether there is a better placed car; they allow a loaded car to serve the sector ahead of it without checking whether another car already has this duty; and that the DS algorithms focus attention on individual floors.

Having selected the best conventional algorithm, Lim decided to introduce modifications one at a time, and compare the performance to the unmodified system described in Section 11.5. Four poorly performing characteristics were selected, these were optimum counter capacity, free car allocation algorithm, waiting time counter and car journey time. Modifications were then added to the original algorithm and the results observed by simulation techniques (Chapter 17).

Only slight improvements were possible (Barney and Dos Santos, 1985). It would appear that the DS algorithm[2] is an attractive design for a fixed logic controller, and the improvements attempted by Lim were not very effective.

11.1.5 *Summary*

The discussion so far allows a number of comments to be made:

(a) Passenger arrivals at floors and passenger destination floors are random.
(b) During any traffic period, there is never a total trend towards the classical traffic patterns.
(c) It is important that any control algorithm follows the various changes in demand.
(d) A truly optimal solution is not possible.
(e) The modification of any existing control algorithm is not likely to be profitable.

Thus, it has been seen that conventional algorithms possess limitations, and their adoption to computer use restricts the flexibility possible with a computer. Quite rightly, this suggests that an entirely new approach is necessary if the computational, logic, programming and data storage abilities of a computer are to be exploited in full. The design of such an algorithm must meet the four primary tasks given in Section 11.1.1 and also obey the five rules.

11.1.6 *A new approach to computer lift traffic control*

The control of lift systems is not an isolated area for the application of digital computers to control problems. Computer control has been applied widely since the late 1950s in the process and power generation industries. The application of computers to lift traffic group control was an important step in lift technology. For a long time, relay logic had been used to control lift systems. Digital electronic techniques were initially tried in the 1960s by lift manufacturers, unsuccessfully, as at that time, only discrete component circuit implementation was available, with its inherent low reliability, complexity and noise interference problems.

The lift industry moved successfully in the 1970s from relay logic controllers to those employing digital computers. Digital computers use integrated circuits, provide very high reliability and allow considerable versatility in the type of controller that can be implemented. The facilities and features contained in a traffic control algorithm implemented in a computer are only limited by the imagination and ingenuity of the algorithm designer, the speed of operation of the computer and by financial considerations. Their maintenance does require a higher degree of knowledge than the normal lift maintenance fitter possesses in order to diagnose and repair faulty components. This difficulty can be overcome by the use of software diagnostic routines leading to board exchanges on-site with further repairs carried out off-site.

The first approach by the lift industry to the computer control of lift systems was the software implementation of the existing conventional control algorithms. This has been shown not to be the best approach as the features presented by conventional algorithms are naturally limited by their fixed logic implementation.

Such early attempts at lift programming suffered from the major weakness that lift manufacturers were unable to evaluate fully the practicability and performance of new control algorithms without implementing these algorithms and installing them on the lift system. By then, it was too late to make extensive modifications. The availability of simulation programs solved this problem.

11.2   A new approach: the estimated time of arrival (ETA) traffic control system

Another approach was to consider a newly devised control system. By 1973, a lift manufacturer had developed and implemented a computer-based control system where the landing calls were assigned to lift cars according to the time each car was estimated to take to answer the call. A number of other features were also included such as the definition of priority levels, the consideration of high activity floors and a priority service for long waiting calls.

11.2.1   General philosophy

An estimated time of arrival (ETA) digital computer-based traffic control system allocates lifts to landing calls based upon computed car journey times, i.e., how long it will take a lift to arrive. Early systems of this type, developed in the 1970s, substituted relay or solid state fixed logic by a truly programmable computer. This technique was an obvious one to use once programming facilities were available. The ETA technique remains the underlying basis of many computer-based systems on the market. The general philosophy below is that of an actual implementation, and gives the opportunity to indicate the various advantages and disadvantages of such a technique. Proprietary implementations will differ and often offer additional features.

11.2.2   Operational description

A primary characteristic of the ETA system is the fast, continuous collection of lift data and the use of the computational abilities of the computer to process these data in order to make the control decisions. To do this, the system scans the registered car and landing calls, and the position, direction of travel and status of each lift. The scan rate does not need to be faster than every tenth of a second, as little happens in this time. An estimation of the number of passengers in each lift, based on weighing devices, should be obtained. The time elapsed for every landing call since call registration should be recorded, and all the landing call allocations re-evaluated periodically. The frequency of updating can be every second as this allows sufficient time for re-allocations, except if a lift is already in its slow-down sequence.

The procedure for allocation can be as follows. Newly registered landing calls are allocated to the lifts committed to move towards the call in the same direction as the call, and also for any uncommitted lifts. Any other lift is regarded as being 'in service' and is not considered for allocation. To decide the passenger transfer times, a fixed time of (say) 3 s for each stop due to a landing call is assumed. For car calls, the estimated number of passengers in the lift and the number of relevant car calls is considered. Twice the larger of these two numbers would be assumed as the transfer time in seconds. The landing call is assigned to the lift presenting the shorter car journey time (cf. Section 11.3.1). No account is taken of possible new car calls between the current position of the lift and the new landing call.

The storage and processing of performance data is important in order to make decisions. An evaluation (say) every 60 s should be sufficient. Examples of relevant data would include:

(a) the total number of landing calls answered at each floor during the last 60 s
(b) the total time taken to answer these calls at each floor during the last 60 s
(c) the average landing call waiting time at each floor
(d) the maximum landing call waiting time and the floor at which it occurred.

From the data, the overall average waiting time experienced by the landing calls in the last 60 s, and an average landing activity factor based on the number of passengers that required service at each floor, can be evaluated. Any landing call that has been waiting for service more than (say) three times the current average waiting time could then be treated as a priority call, and a lift is assigned to make a special run to deal with such a call. In addition, any floor presenting an activity level at least (say) three times larger than the current average landing activity could be declared as a high activity floor and is given priority for as long as the condition applies. This facility caters for sudden heavy local demands.

It is useful to declare a number of predefined floors as priority floors and give them permanent service priority. Typical examples are the ground floor, restaurant floors and executive floors where heavier than average traffic is expected or special service is required. These floors could be defined as parking floors if these were required. The number of priority (or parking) floors should not exceed the number of lifts.

Lifts are therefore assigned to landing calls according to a priority structure. The system deals:

1 with long wait priority calls
2 with high activity floors
3 with priority levels
4 with the remaining normal landing calls.

In all of these cases, the car journey times to the landing call are evaluated and the call assigned to the lift providing a shorter journey time.

A lift assigned to service a long wait priority call, a high activity floor, or a priority level is only allowed to stop in response to car calls before reaching the calling floor. Thus, a different procedure must be adopted to evaluate the car journey times by considering only car calls to determine the stopping floors and the passenger transfer times. It may happen that a lift, which is assigned to service such priority calls, discharges its last passengers at a floor where some traffic demand exists. In this case, an information signal would be illuminated in the lift to show its special service condition and request passengers not to enter the lift. The landing call is obviously not cancelled.

There are thus three timed levels of data exchange between the real world of the lift and the digital computer. The first is the basic event scanning at some ten times per second. The second is the updating of call to lift allocations at once per second. The third is the determination of priorities once every minute.

11.2.3   Performance

Uppeak traffic situations can be detected using the normal heavy demand feature by detecting when lifts leave the main terminal floor with a specific loading, for example, three times greater than the average floor activity. While the uppeak condition continues, a number of lifts are dispatched to the main terminal, the number of lifts being dependent upon the intensity of the peak demand.

The normal operation algorithm can cope with down peak traffic, although the ETA system is liable to provide a poorer service to the lower floors in the building. Down peak is a heavy traffic situation and, as lifts will not be free for redistribution, 'leap-frogging' is liable to occur. This is due to the allocation of lifts to landing calls being made individually and independently, rather than to sectors, thus causing the lifts to frequently bypass landing calls and overtake other lifts. The allocation of lifts based on computed car journey times gives no assurance that the lower landing calls will be given the same service as the upper landing calls, once calls are assigned. Although some account is taken of the time that landing calls have remained unanswered in the allocation procedure, this is only partially effective.

Although the algorithm has a concept of time, the normal allocation does not consider landing call waiting times; it is only when a landing call has been waiting too long that it is given priority. This is where passengers have to suffer a long wait before being considered for attention.

The system should also include a number of extra features, which are usually provided by good supervisory control systems, namely, a car preference key-operated switch in each car and anti-nuisance devices that prevent an empty lift from answering false car calls.

11.2.4   Conclusions

The ETA control system can be expected to provide a good uppeak performance. By declaring the main terminal floor as a parking and priority floor, cars will be sent down to deal with the incoming traffic.

The system is not, however, particularly suitable for down peak traffic. If a comparison is made with fixed sectoring systems, similar round trip times are to be expected for the two systems, but the ETA system will present a smaller number of stops per trip, thus, it is less efficient.

Under light to medium balanced interfloor traffic conditions, when the free lift parking procedure is active, the system behaviour is very similar to a dynamic sectoring system, and good performance is to be expected. For heavier traffic conditions, the redistribution of free lifts is less efficient, and the system provides no equivalent to the heavy duty call allocation algorithm available in the dynamic sectoring system. Hence, under heavy interfloor demand levels, the ETA system can only be said to be reasonably good, presenting a performance lying between the dynamic sectoring and the fixed sectoring system performances.

It should be emphasised that this control system provides two great advantages, which are inherent to computer control. They are the data logging capability and the possibility of changing the control algorithm parameters by simply reconfiguring the computer control program. For example, the number and location of priority floors can easily be altered on-site and requires no rewiring. A further advantage is that new control programs can be loaded as they become available, obviating the need to take out the old control system as is required with conventional systems. On request, the computer can calculate, record and print out the lift system's performance data.

11.3   Features to be included in a computer-based traffic control system

This section indicates some of the desirable features that a computer-based control system should offer. These features are often impossible to achieve with fixed logic systems.

11.3.1   Computing car journey times

The cornerstone of most computer-based traffic controllers is the requirement to estimate the journey time for each lift to reach each landing call as part of the call allocation process.

> **Definition 11.1:** car journey time[3] is the time it takes a lift to travel to a landing call.

The car journey time is estimated according to certain arithmetical and logical rules. The procedure to evaluate car journey times for a normal landing call considers the car calls, which are already registered and must be honoured, and the landing calls to which the car is currently assigned. These journey times may be calculated on the basis of a direct trip to a landing, or an indirect trip where the car stops at intermediate landings on its journey, in order to service car and/ or landing calls. The number of such calls between the lift and the calling floor is important. A car journey, once assigned to a call, then becomes the estimated system response time for that call.

A car journey time consists of several components:

(1)   Interfloor flight time (including acceleration, deceleration, levelling and travel at rated speed)
(2)   Door operating times (opening and closing)
(3)   Car call and landing call passenger transfer times
(4)   Any door dwell time periods likely to exceed the passenger transfer times.

For component (1), when estimating the car journey time for moving cars, a correction must be made in respect of the acceleration time.

Each time a car stops, it is not known how many passengers will enter or leave the car. This has the effect of making the passenger transfer time component (3) difficult to evaluate. If a lift is picking up landing calls at floors in addition to, or instead of, at the main terminal, it implies it is during an off-peak traffic situation. For reasonable levels of interfloor demand, it will be shown in Chapter 15 (Figure 15.4) that the ratio of the number of car call stops to landing call stops is about 120%. This can be expressed as: at only 20% of the landing stops does more than one passenger enter the lift. Little error would therefore occur in the estimation procedure if one passenger transfer time is assumed for each car or landing stop.

The destination floors of waiting passengers are not known until they enter the car. This leads to the possibility of car calls being registered between that landing call and another landing currently allocated. The discussion regarding passenger transfer times lead to the conclusion that there could be one car stop for each landing call served – but where? Clearly, the call must be between the landing call floor being considered and the last floor of the building in the direction of service. Under balanced interfloor traffic, the average floor number could be half way between the landing floor and the last floor of the building. It would therefore be reasonable to make the assumption of positioning a fictitious car call every time a landing call is assigned to a lift, half way along its possible journey.

It is inevitable that a high degree of uncertainty must exist in the positioning of a fictitious car call, however, the effect is likely to be small. When traffic is light, there is more chance of a larger error, but it can be tolerated, owing to the low demand. Whenever the traffic is heavy, a car may have a large number of allocations making an incorrect deduction less significant. In an uppeak traffic situation, the feature can be turned off as it is of no benefit. It is in down peak where the error would be largest as the destination floor is not half way down the building, but at the main terminal, and the feature should be turned off.

11.3.2   Unbalanced interfloor traffic

The random arrival of passengers requesting service gives rise to the possibility of the traffic arrival profile at all floors becoming momentarily unbalanced, although the average traffic could still be considered as balanced. This situation can arise when more passengers regularly arrive at a particular floor than any other floor. It is desirable to still maintain an even service distribution in order to meet the requirements of the ideal control system. The handling of an unbalanced situation requires two actions: detection and correction.

Detection requires the estimation of the number of passengers entering the car at a particular floor. This can be achieved by load weighing. Assume car loads can be weighed to an accuracy of 10% (i.e., 10 values over the load range). Then, the number of passengers boarding at a particular floor can be detected as the extra load over the original load in the car on arrival at the floor. However, when a car stops at a floor with a landing call as the result of a car call, passengers will leave the car. As has been shown previously, this is likely to be one person, hence, a realistic estimate of extra load can be made. (An exception is when a car stops at its last car call floor, when all passengers are assumed to exit.) It is now possible to estimate the average number of passengers, expressed as a percentage of a fully loaded car, entering any car when a car stops for a landing call. This information can be used to correct the condition.

11.3.3   Reduction in the number of stops

During balanced interfloor traffic, a car serving a landing call will, in general, make at least one further stop (car call) before becoming free. The number of stops made by cars per unit time indicates the level of loading, and any reductions in the number of stops will clearly conserve system capacity. There are two ways in which the total number of stops might be reduced:

(a) By serving passengers wishing to travel to the same floor in the same car.
(b) By serving a landing call where a car call already exists.

The first method is only possible using destination registration stations (see Section 11.6.2). The second method is possible by inserting a special procedure into the basic allocation system. The procedure is activated whenever a landing call being considered coincides with a known car call being served by a particular car. If the expected value of landing call waiting time is not too much longer than the original allocation, then the landing call should be allocated to the particular car even if any other allocations are smaller. This has the effect of conserving lift system capacity to serve other demands better.

11.4   Advanced control techniques

11.4.1   Fundamental limitations

It is important to point out that, although computer-based traffic control systems can allocate lifts more efficiently than the legacy traffic control systems, there is a limit to what can be done. The main limit is the finite handling capacity resource of the underlying equipment to handle the traffic demands. This is more dependent on good equipment, which is properly set up, than advanced control systems. Once the major inefficiencies have been removed, such as stopping full cars, faulty detection of car loads, inefficient door operations, etc., then it is only possible to 'trade' one parameter against another. This means that one passenger's shorter waiting time is another passenger's increased waiting time. The effect on the second passenger could be so small that its deterioration is unnoticed, but the effect on the first passenger could be significant.

This section looks at some advanced control techniques that can be applied individually or in concert.

11.4.2   Minimal cost functions

Calls are often allocated to a suitable car using the concept of minimum cost, i.e., a cost function.[4] This concept operates by performing a trial allocation to all available cars and allocating the call to the car presenting the lowest cost. There are criteria for selecting a suitable cost function. These can, for example, be based on either quantity of service, or quality of service, or both. In general terms, the quantity of service is a measure of the lift capacity consumed to serve a specific set of calls, indicated by the total of the journey times of all the cars. This could be minimised by keeping passengers waiting in a lobby until there were enough passengers to make a trip worthwhile. Airlines apply this principle. The quality of service is indicated by the average value of either the passenger waiting time or the passenger journey time (waiting time plus in-car travel time).

The minimisation of waiting time implies putting passengers into the first lift that arrives. This would result in no change from the usual procedure.

The minimisation of the total car travel time implies using the smallest system capacity, which is equivalent to using the smallest possible number of cars. The result of this policy would be very large passenger waiting times, a result that would not be acceptable. This criterion alone is thus not suitable as a cost function.

The minimisation of a combination of average passenger journey and waiting times is a more acceptable objective. Both times are interrelated, and the minimisation of one might be achieved at the expense of the other. An accurate calculation of passenger journey time can only be achieved if passenger destinations are known at landing call registration time. As conventional two-button signalling systems are being considered in this discussion, only passenger waiting time can be minimised. In practice, of course, passenger waiting time cannot be measured. The period of duration for a particular landing call only represents the time the first passenger at that landing has to wait. All subsequent passengers benefit from the first registration and wait for less time.

Thus, the control algorithm can only (practically) reduce the cost of the system response time to service a landing call. In Section 15.6, the passenger average waiting time (*AWT*) is shown to be very nearly equal to the lift average system response time (*ASRT*) for balanced interfloor traffic. So, for the balanced interfloor traffic condition, this cost function is very suitable. This cost function is less relevant for unbalanced traffic conditions. For example, during uppeak, the *ASRT* is related to the interval (*INT*) by values of car load. For down peak traffic, the *AWT* is more linearly related to demand (Figure 14.3). Special measures must therefore be taken to deal with unbalanced traffic conditions.

There are other cost functions that can be employed, for example, to minimise energy usage (Peters and Mehta, 1998). The hall call allocation strategy has been shown (So and Suen, 2002) to use more energy than its simple predecessor. Peters and Mehta show that an algorithm with a parking policy uses more energy over one that does not. As indicated in Section 11.4.1, it is only possible to 'trade' one parameter against another.

11.4.3   New signalling methods

The computational potential of the digital computer could be further exploited in order to approach optimal computer control. To achieve this, the supervisory system needs to be presented with additional information related to the passenger destinations. Consider four cases, which are illustrated in Figure 11.1.

Case 1. A car travelling towards a new landing call, with a car call for that landing, and one car call after that landing, which the new passenger wishes to travel to.

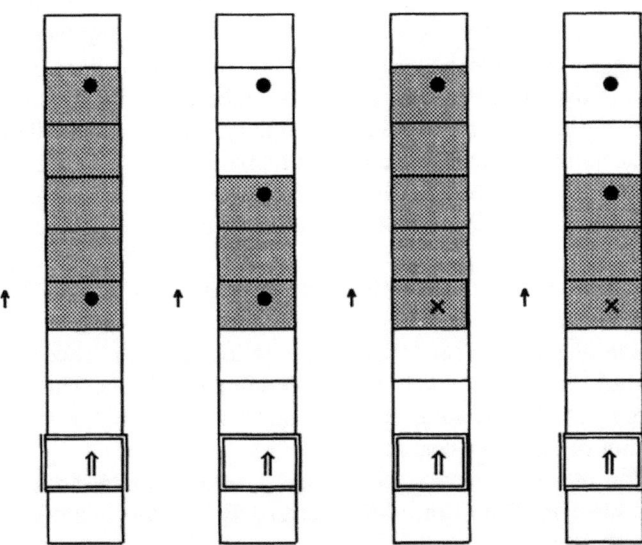

Figure 11.1   Four cases when answering a landing call three floors away from an up moving car. Shaded areas indicate original route allocated to the lift

Table 11.1 Down landing call at floor 5

0	1	2	3	4	5	6	7	8	9
0.51	0.13	0.02	0.23	0.11		n/a	n/a	n/a	n/a

Case 2. A car travelling towards a new landing call, with a car call for that landing, and one car call after that landing, which the new passenger does not wish to travel to.

Case 3. A car travelling towards a new landing call, with a car call after that landing, which the new passenger wishes to travel to.

Case 4. A car travelling towards a new landing call, with a car call after that landing, which the new passenger does not wish to travel to.

Case 1 is most efficient as no extra stops are introduced, but the controller does not know the new passenger destination until boarding. Case 2 is efficient at picking up the landing call, but a new car call is introduced, which the controller does not know about until boarding. Case 3 has similar efficiency to Case 2, as a new landing stop is introduced, but the destination is coincident to the existing car call. The controller did not know this until boarding. Case 4 is the least efficient as the controller introduces a new landing stop to pick up the landing call only to discover on boarding that a new car stop will also be required.

In all these cases, the destination of the passenger making the landing call is unknown, only the direction. A clever traffic supervisor can help with Cases 1 and 2, i.e., by using a car already stopping at the landing. Cases 3 and 4 always consume system capacity.

The usual two up/down pushbuttons only give direction information to the traffic controller, not ultimate destinations. Could the controller have some intelligence? Barney and Imrak (2001b) suggests that a neural network could help. Neural networks are good at analysing a lot of data over a long period. If a log is kept of the destinations that arise from landing stops, then a weighted table can be produced, which the traffic controller could use to determine which is the best car to send to a landing call. To illustrate this, consider Table 11.1.

This shows the likely destinations for a down call at floor 5 in a 10 floor building. It shows that there is a 51% chance of the call being for the main terminal, a 23% chance of the call being for floor 3 and so on. Probably, a fuzzy logic algorithm would be required to produce the weighted table to deal with fuzzy passenger behaviour such as not registering a car call as they enter, not entering a car call at all, entering an incorrect car call, getting out at the wrong floor, etc.

11.5   A new approach: stochastic traffic control systems

11.5.1   General philosophy

A stochastic based[5] traffic control system, named CGC, was developed by Lim in 1983 and published (Barney and Dos Santos) in 1985. Observations of legacy controlled lift systems have indicated that the response times to answer landing calls follow a curved shape similar to the exponential distribution curve of Figure 11.2(a). This distribution curve has a large number of calls answered in zero time or during the first time band. However, there is a long tail to the distribution with some calls waiting for very long periods of time.

The x-axis is units of average system response time. The y-axis is frequency.

The underlying premise of the CGC algorithm design was to bring the tail closer to the average, and to sacrifice the 'instant' collection of some calls by moving the exponential away from

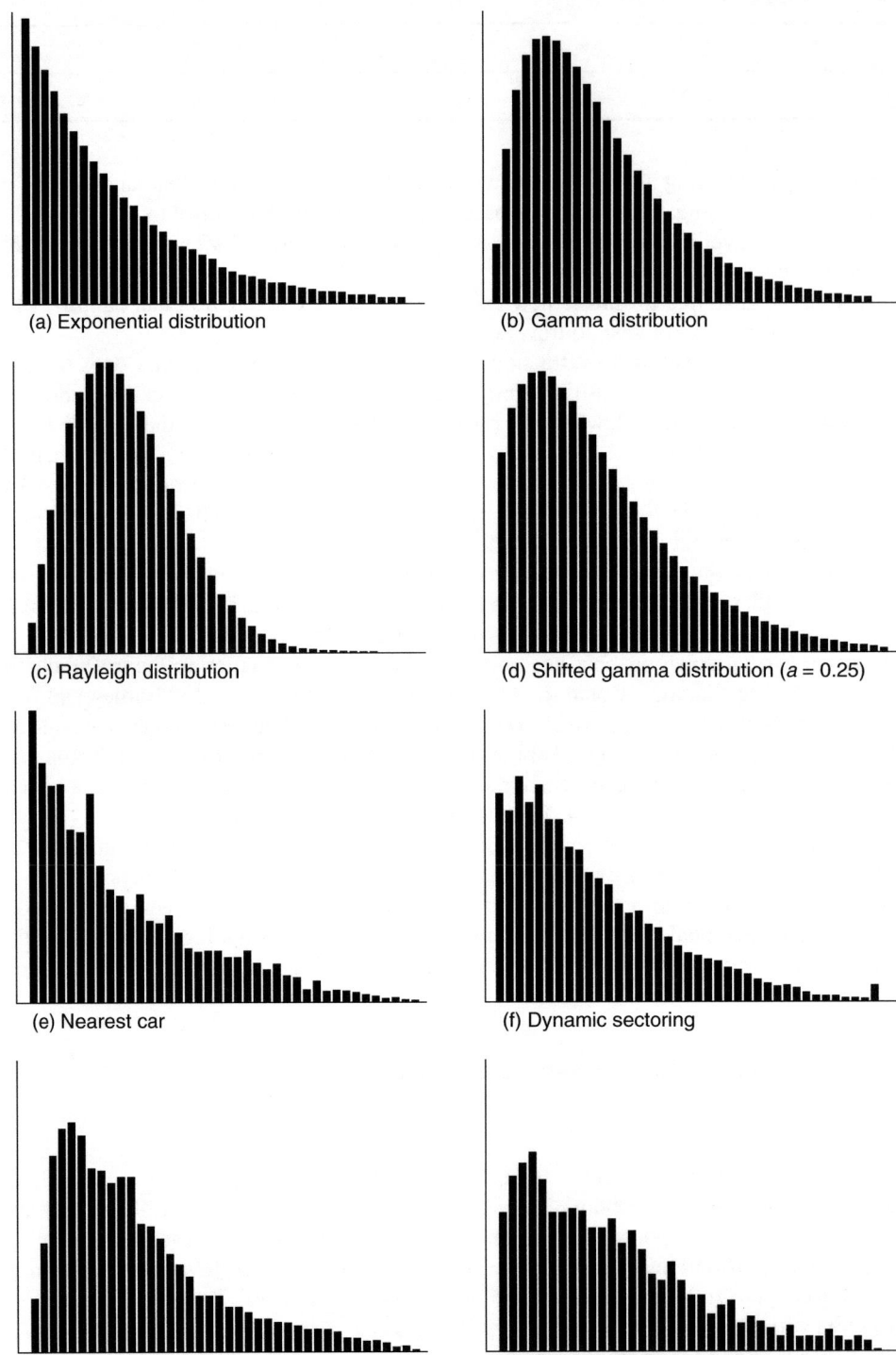

Figure 11.2 Statistical distributions (after Halpern, 1995)

the origin to a Gaussian shape similar to the Rayleigh distribution curve of Figure 11.2(c). The effect would be to give a more even and more consistent service to passengers, by trading the instant response calls for the long tail calls.

What Lim did has subsequently been analysed by Halpern (1992, 1993, 1995). Halpern in his 1992 paper entitled 'Variance analysis' introduced a method of analysis based on a utilisation factor and a variance performance factor. He showed that a legacy traffic control system behaved as a Poisson process. In the 1993 paper, 'Variance analysis of hall call response time', he used a computer-based system where he had control of the traffic algorithm. By altering the cost function, he was able to improve, and worsen, the performance of the traffic control with respect to response times. He was able to trade off response time against variance, i.e., the consistency of the responses. This control, he termed minimum variance.

In Halpern's 1995 paper, 'Statistical analysis . . . of call response times', he analyses two legacy systems: nearest car (NC), termed 'closest call' by Halpern, and dynamic sectoring (DS), and compares them to three computer-based systems: computer group control (CGC) in normal and down peak modes and the minimum variance algorithm (MVA).

It was indicated above that the response time distribution for a legacy traffic control system was likely to be exponential, as illustrated by Figure 11.2(a). The ideal distribution would be as Figure 11.2(c), which inclines to a Gaussian shape, but also shows the inevitable tail exhibited by real world systems. A better distribution shape could be a Gamma distribution as shown in Figure 11.2(b). Here the tail is longer. However, if any computer-based traffic control system were to produce a distribution of response times similar in shape to Figure 11.2(d), which is a shifted Gamma distribution, this would be very acceptable.

The other distributions shown in Figure 11.1 relate to Halpern's analysis of the different computer traffic control systems: (e) the nearest car, (f) dynamic sectoring (DS), (g) stochastic (CGC) algorithm and (h) the minimal variance algorithm (MVA). The nearest car algorithm exhibits the classical exponential shape. The DS algorithm, which uses fixed logic, shows a trend towards the ideal. Both the CGC and MVA algorithms more nearly meet the ideal response distribution. Halpern concludes that the computer-based systems no longer follow a Poisson process, associated with legacy control, but a shifted Gamma process. He also confirms the premise of Section 11.4.1 of a finite (handling capacity) resource.

The intention of a computer group control (CGC) traffic control system is to provide an even service to all floors, where every landing call is given a fair consideration. This means that the landing call that has been waiting the longest should be given the first consideration for service. To achieve this egalitarianism, landing calls are considered to form a queue, and will generally be served in order of their waiting time. The system described[6] below was designed to serve the balanced interfloor traffic condition, with considerations of the other traffic conditions.

11.5.2   *Operational description*

In the CGC algorithm, the landing calls are arranged, using a computer programming technique, into a first-in, first-out (FIFO) queue. Thus, the priority of each call is indicated by its position in the queue, the highest priority being at the head end of the queue (Table 11.2).

A landing call is taken from the head end of the queue and allocated to a suitable lift using the concept of minimum cost. In order to apply the minimisation procedure, it is necessary to calculate the car journey times to travel to a landing call. This is achieved by the procedure given in Section 11.4.1. Part of the procedure is to assume a fictitious car call positioned half way along the possible journey of the lift.

Table 11.2 Landing call queue

Floor	11	10	8	1	15
Direction	UP	DN	UP	UP	DN
Waiting time	25	21	14	9	5
	Head	–	–	–	Tail

The concept of a high threshold time (*HTT*) is introduced. This time is not a constant, but reflects the current demand into the lift system by expressing it in terms of the average system response time (*ASRT*).

$$HTT = KTH \times ASRT \qquad (11.1)$$

where *KTH* is a constant, and *ASRT* is determined over a recent sampling period T_s. Lim (1983) found suitable values of *KTH* to be 3.0 and T_s to be 90 s.

The conservation of system capacity can be improved by reducing the number of stops a lift makes. The method adopted is generally as described in Section 11.3.3. If the expected value of landing call waiting time is calculated to be less than the high threshold time, then the landing call can be allocated to another lift, even if the other allocation was smaller. As the high threshold time varies with demand, an even service is provided.

From time to time, a particular floor may experience increased traffic demands. These can be detected as indicated in Section 11.3.2. Correction of the unbalanced condition is simple to arrange. The heavy duty floor is moved to the head of the landing call queue, where it will receive immediate treatment. For example, suppose the call at floor 10 shown in Table 11.2 were to be given heavy duty status, then the first two head entries in the landing call queue would be reversed.

In some buildings, it is necessary to declare certain floors as high priority floors, e.g., VIP service or executive service. This feature can be achieved by setting a lower and a higher time limit on the waiting time at specified landings. A method is to define the lower and higher limits as proportions of the high threshold time, as in Table 11.3. This ensures the other floors still receive a reasonable service.

The computer group control algorithm can be summarised as follows:

(1) A lift is allocated to a landing call according to least cost except:
(i)   if another landing call has exceeded a specific, but demand varying, threshold value, and is also allocated to the same lift, the call will not be allocated to that lift
(ii) if another lift is scheduled to stop for a car call at the landing being considered, and the landing call is estimated not to wait longer than a specific, but demand varying, threshold value, it will be allocated to the other lift.

Table 11.3 High priority floor limits

Floor	10	5
Lower limit	0.5 *HTT*	0.4 *HTT*
Higher limit	0.8 *HTT*	0.9 *HTT*

(2) Landing calls are considered for allocation by examination of a landing call queue, landing calls at the head of the queue being allocated first.

(3) The landing call queue is normally arranged on the basis that the longest wait call is at the head of the queue except:

(i) a heavy duty floor landing call may be brought to the head of the queue

(ii) a priority floor landing call may be brought to the head of the queue.

(4) The landing call queue is arranged so that the various categories are ordered from the head of the queue as follows:

(i) priority call

(ii) heavy duty call

(iii) long waiting call

(iv) normal call.

11.5.3 Computer group control performance

The CGC algorithm is designed to handle the interfloor traffic condition as it exists for much of the working day. Uppeak traffic and down peak traffic are considered special cases of interfloor traffic. A comparison with the dynamic sectoring (DS), which has been shown to be the best performing fixed logic algorithm, is shown in Figure 11.3, which shows that the CGC algorithm answers less calls immediately and the tail is smaller and smoother.

The CGC algorithm performs in a similar manner to all the other algorithms during uppeak traffic.

The down peak performance of the CGC algorithm is very much poorer than any other algorithm. It would appear that an algorithm designed to perform well under balanced inter-floor traffic does not serve down peak adequately and *vice versa*. It can be concluded that, in its attempt to provide an even service to all floors during the down peak period, it does so by sacrificing performance. Optimal performance is thus not achieved. Lim (1983) suggests '*that lifts be encouraged to serve adjacent calls*'. This would require a modified cost function, which first minimised the average waiting time, and secondly minimised lift journey time between two subsequent landing calls.

Lauer (1984) suggests that once lifts respond to demands in a theoretically optimum manner then '*further invention or investigation in the area of lift supervisor logic would be pointless*'. As has been shown in the considerations of the design of the CGC algorithm, the consideration of a 'bumpless' transfer between different traffic conditions prevents optimality.

The intuitive approach by Lim to the CGC algorithm has subsequently been proved mathematically. The CGC algorithm has been implemented by one lift installer (Godwin, 1986).

11.5.4 Example of operation of stochastic control system

Consider a 16-floor building served by four lifts with lift positions and a landing and car call pattern as given in Figure 11.4, and a call table as given by Table 11.3. How will the lifts serve these calls?

New symbol: X fictitious car call. Shaded area shows route taken by lift.

To consider this problem, a number of assumptions must be made.

Assume only one passenger, with a transfer time of 1.2 s, enters or exits at each stop. The single floor flight time is 5.0 s, the door operating time is 3.8 s, and thus the cycle time is 10.0 s. For each floor jumped, add 1.0 s. Delete 1.0 s if a lift is moving.

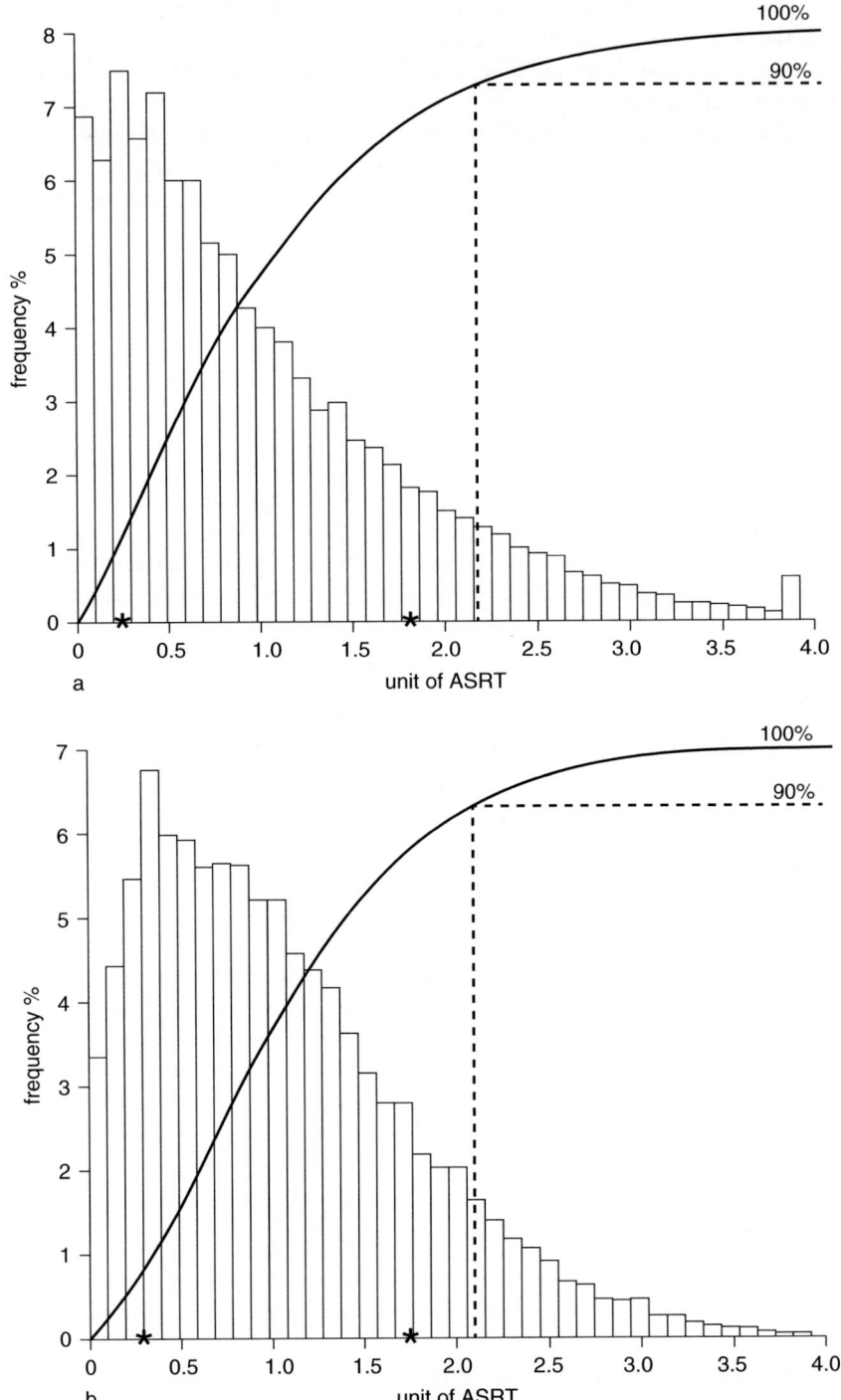

Figure 11.3 Comparison of interfloor performance DS (upper graph) and CGC (lower graph) algorithms

New symbol: ★ fictitious car call. Shaded area shows route taken by lift

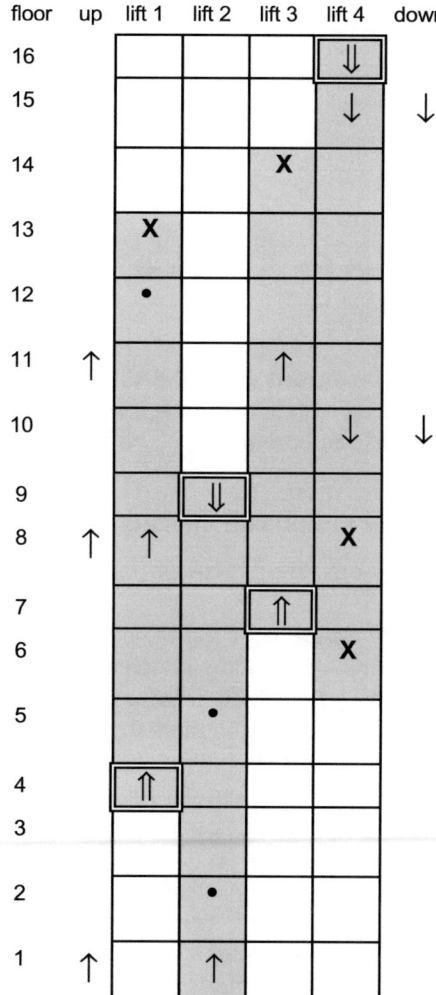

New symbol: **X** fictitious car call. Shaded area shows route taken by lift.

Figure 11.4 Final disposition of lifts and calls for the example

Table 11A

Lift 1:	10 + 6–1 = 15 s	(lift moving)
Lift 2:	(10 + 3–1) + (10 + 2) + (10 + 8) = 42 s	(lift moving)
Lift 3:	10 + 3 = 13 s	(lift stationary)
Lift 4:	10 + 4 = 14 s	(lift stationary)

Assume all calls are being completely reallocated, and calls are taken from the queue shown in Table 11.3.

Taking each landing call in turn, the service times can be calculated. As an example, take the landing call at the head of the queue, i.e., landing call at floor 11-UP.

Table 11.4 Calculation of landing call times and allocation to lifts

Floor	Wait (s)	Lift 1	Lift 2	Lift 3	Lift 4
11 UP	25	15	42	**13 (1)**	14
10 DN	21	27	41	38	**15 (2)**
8 UP	14	**12 (3)**	39	19	39
1 UP	9	56	**34 (4)**	47	42
15 DN	5	46	66	35	**10 (5)**

Fictitious car calls at: (1) floor 14; (2) floor 6: (3) floor 12; (4) floor 9; (5) floor 8.

Table 11.4 can now be formed where the bold entries in the grey cells indicate the allocations made. It is important to note that some later allocations affect earlier ones. For example, the landing call at floor 15 allocated to lift 4 will cause the landing call at floor 10 to wait 25 s.

11.6   A new approach: hall call allocation

11.6.1   *General philosophy*

It would be much more useful if the traffic controller knew the intended destination of each landing call. This information could be obtained by replacing the conventional up/down buttons by a panel of passenger destination buttons at each landing, but this needs different fixtures on the landings and a different attitude by passengers. However, most people today are familiar with the digital keypad on mobile phones, bank teller machines, calculators, etc., so their use should not present a problem. As for the passenger attitude, it has been pointed out that the call registration procedure is not necessarily more difficult to understand than the operation of the usual up/down buttons (Christensen, 1988), and it is probably less of a step in this technological age than when the attendant was made redundant by automatic control and passengers had to operate the lifts by themselves.

The idea of call buttons on the landing was first proposed by Leo Port in 1961 (1961, 1968), but he only had relay logic in which to implement it, and could not provide dynamic allocation, only fixed allocation. Installed in two buildings in Australia, it functioned in one for some 20 years or more. An HCA system was first described by Closs in 1970, and implemented by a major lift company in 1990 (Schroeder, 1990c), when computer technology had caught up with the ideas. Now installed in many buildings, it has gained acceptance across the world as efficient. Most manufacturers have now applied the technique (Hikita *et al.*, 2001; Smith and Peters, 2002).

Hall call allocation gives the opportunity to track every passenger from registration to destination. This has great advantages during uppeak as passengers can be grouped to common destinations, since there are large numbers of them. The individual waiting time may increase, the travel time may decrease, but there would be an overall reduction in journey time. During down peak, there is no advantage as the destination floor is known. During reasonable levels of balanced interfloor traffic, there is little advantage as most landing calls and car calls are not coincident (see Figure 15.4). However, during an uppeak with some down travelling traffic, or a down peak with some up travelling traffic, there are possible benefits. This leads to a conclusion that a hybrid system with a full call registration station at the lobby and other principal floors and two-button stations at all other floors might be beneficial.

This approach can yield cost savings, particularly for modernisations if the existing two-button stations can be interfaced with the new control system. Mixed landing and destination call systems provide the same uppeak boost as HCA. However, they require a car call station in the car, which is susceptible to abuse if passengers at the ground floor join the next car to depart, rather than the car allocated by the traffic controller. This abuse can be discouraged by temporarily disabling (bridling) the car call buttons until the allocated landing calls above the main terminal have been answered.

This traffic control system utilises additional information obtained from the passenger as to their destination, not just their direction. A floor calling station must be placed on each floor to allow passengers to select their destination floor.

11.6.2 *Operational description*

The call allocation algorithm implies a different method of call registration. Instead of the usual two up/down buttons, Figure 11.5 (a), a panel of touch buttons, Figure 11.5 (b), is required at every landing. This can be the familiar telephone keypad. Thus, a passenger arriving, for instance, at floor 3 and wishing to travel to floor 11 touches button '1' twice. Within a very short time (<1 s), the passenger would receive an indication on a display beside the call registration station showing which lift has been allocated to their call. No destination buttons are necessary in the car, but an indicator inside the car shows the floors at which the lift is stopping.

This type of call registration can be adapted to be multifunctional. For example, the * (star) key could be a – (minus) key, which could be used to enter below main terminal floors. The # (hash) key could be the key to indicate that a disabled person wished to use a lift, and an empty lift could be dispatched. Parameter setting keys could be programmed to allow technician access, such as pressing the star and hash key together. There are many other opportunities.

The computer control algorithm obviously requires that every passenger registers their destination calls. The computer algorithm used to allocate a landing call to a lift is simple to implement. Each time a new call is registered, the computer allocates it in turn to each of the lifts, and evaluates the cost of each allocation. The allocation giving the lowest cost is then adopted. Suitable cost functions are (1) passenger average waiting time, (2) passenger average journey time, or (3) a combination of both.

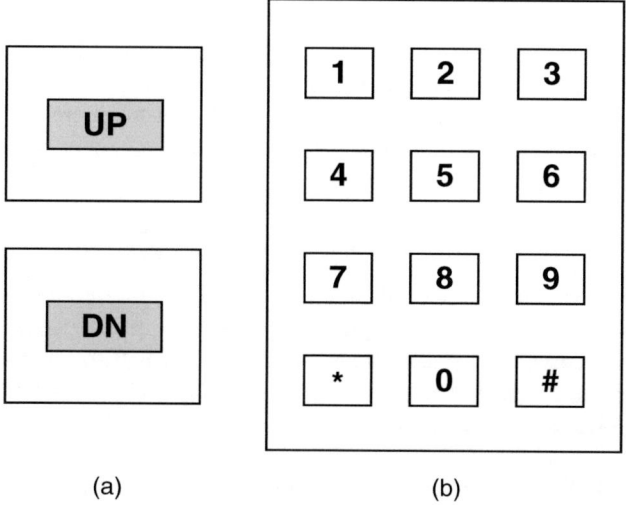

(a)   (b)

Figure 11.5 Registration stations (a) old and (b) new

11.6.2.1   *The simple cost function*

During an uppeak, the obvious cost function to implement with call allocation is journey time (2). This is because a waiting time allocation criterion (1) would do no more than allocate every new call to the first available lift at the main terminal that possessed space capacity, in the same way as the collective-distributive algorithm. If journey time is the cost function, calls terminating at the same floor tend to be allocated to the same lift, hence reducing the number of stops per trip and the round trip time. The system handling capacity is increased, and the main terminal floor more frequently served. However, a waiting passenger may not be allocated to board the first available lift, and this may produce increased waiting times. The overall effect is that better journey times are produced, in comparison to conventional algorithms, for the whole range of traffic intensities, but can result in longer waiting times. It is better to sacrifice some passenger waiting time and use passenger average journey time as the cost function. The mathematics are as follows.

Consider that a new call is to be allocated to a system of L lifts, each lift (I) with $N(I)$ calls to answer, and $JT(I)$ accumulated journey time for the $N(I)$ calls.

Assume that $NJT(K)$ is the new accumulated journey time for $N(K) + 1$ calls, when the new call is allocated to lift K. The average journey time for the complete set of calls is:

$$AJT = \frac{NJT(K) + \sum_{I=1, I \neq K}^{L} JT(I)}{1 + \sum_{I=1}^{L} N(I)} \tag{11.2}$$

This can be written as:

$$AJT = \frac{NJT(K) - JT(K)}{1 + \sum_{I=1}^{L} N(I)} + \frac{\sum_{I=1}^{L} JT(I)}{1 + \sum_{I=1}^{L} N(I)} \tag{11.3}$$

As the two summations in Equation (2) do not depend on the allocation K, the minimisation of AWT only requires the minimisation of the term $NJT(K) - JT(K)$. This simplifies the evaluation of the cost function, as only this incremental cost is to be evaluated, instead of the whole expression for AWT. The quantities $NJT(K)$ and $JT(K)$ are evaluated by simulation.

It should be noted that the incremental cost $NJT(K) - JT(K)$ is made up of several terms. It includes the waiting and journey times for the new call, and the increase in the waiting and journey times of calls already allocated to lift K, the extra passenger transfer time resulting from the new call, and any extra stops to pick up and discharge the new passenger

11.6.2.2   *Average journey time with maximum journey time constraint*

A third type of cost function, proposed by Closs (1970), uses average journey time with a maximum waiting time constraint (3). It operates by costing each allocation against an average journey time cost function, but penalising any solution for which the waiting time of the new call exceeds a predefined value of maximum wait (*MWT*). The adaptive call allocation algorithm operates as follows:

(1) Evaluate cost of allocation of the new landing (hall) call to lift 1:

$$COST(1) = NJT(1) - JT(1) \tag{11.4}$$

(2) Compare the new call waiting time *NCWT(1)* with the predefined value *MWT*. If it is smaller than *MWT*, then *COST(1)* is not altered, but if it is greater, a penalty is added to the cost:

$$COST(1) = COST(1) + \text{penalty} \tag{11.5}$$

The penalty is made up of a fixed value added to a term proportional to the excess of waiting time above *MWT*. For example:

$$\text{penalty} = 300 + 10 \, (NCWT(1) - MWT) \tag{11.6}$$

(3) Repeat the procedure from (1) for all lifts.

The effect of using a penalty is to force the elimination of the allocation to lifts with an existing high number of allocations from receiving another allocation, making it easier to select a more lightly loaded lift.

Of course, if all lifts exceed the limit value, a suitable allocation will always be found in terms of minimising the new call waiting time, as the cost (using the same parameters as above) becomes:

$$COST(K) = COST(K) + 300 + 10NCWT(K) - 10MWT$$

$$= COST(K) + 10NCWT(K) + \text{constant} \tag{11.7}$$

where $10NCWT(K)$ is the most significant term of those depending on the allocation of K.

11.6.2.3   Reduction in number of stops

The 'positive' concept of using a cost function as a performance index can be transposed into a 'negative' concept of penalty functions in order to promote higher efficiency. An example of a penalty function is the rejection of an allocation, which introduces an additional stop.

The call allocation algorithm causes calls requesting the same destination floors to be carried by the same lift. This has the effect of reducing the number of stops. However, in some cases, the cost of allocating a new landing (hall) call to a lift already stopping at the calling landing or destination floor is marginally greater than the cost to allocate the call to another lift not stopping at either floor. Although the allocation is perfectly proper, it might be better not to allocate the new call to the lift with the lowest cost, as by not doing so, capacity is reserved for future calls. To cater for this idea, a penalty $p\%$ is introduced for each extra stop motivated by the new call. To prevent operation of this penalty under low traffic conditions, the penalty is made dependent on the incremental cost of the allocation and is proportional to car load.

$$\text{penalty} = \frac{p}{100} \times \text{incremental cost} \times \frac{\text{load}}{AC} \tag{11.8}$$

where, AC is the actual car capacity and the load is measured as the average value of the number of passengers inside the lift, or queuing for service. The procedure improves performance for values of p up to 10%. For larger values of p the algorithm is self-defeating, as it produces less appropriate allocations.

11.6.2.4   Dynamic uppeak subzoning

During an uppeak, the obvious cost function to implement with call allocation is journey time. This is because a waiting time allocation criterion would do no more than allocate every new call to the first available lift at the main terminal that possessed space capacity, in the same way as the collective-distributive algorithm. If journey time is the cost function, calls terminating at the same floor tend to be allocated to the same lift, hence reducing the number of stops per trip and the round trip time. The system handling capacity is increased, and the main terminal floor more frequently served. However, a waiting passenger may not board the first available lift, and this may produce increased waiting times. The overall effect is that better journey times are produced, in comparison to conventional algorithms, for the whole range of traffic intensities, but under some circumstances, it can result in longer waiting times, mainly for light traffic originating at the main floor.

It will be shown in Section 8.5.3 that uppeak sub zoning is sometimes used by conventional group control systems to improve (boost) the uppeak handling capacity.

Subzoning is very sensitive to where the zone partition is fixed and should ideally be adjusted for every traffic situation. As, in practice, a fixed partition is implemented, it cannot respond to the wide fluctuations found in arrival traffic patterns. Knowing the advantages of uppeak sub zoning, and the adaptability of a computer implemented algorithm in coping with input traffic variations, a dynamic sub zoning concept can be implemented in the ACA system. The building is divided into three sub zones, as shown in Figure 11.6.

The lifts are divided into two subgroups, one for the lower sector, and the other for the upper sector. No indication of this partition is given to the passengers. A newly registered landing (hall) call is allocated to a lift in the usual way, by evaluating the costs of the allocation of the call to every lift, and choosing the allocation giving the lowest cost. However, during the evaluation of the cost, the allocation of a call registered for the lower sub zone to a lift allocated to the upper sub zone is penalised, and so is the allocation of a call with a destination in the upper sub zone to a lift in the subgroup serving the lower sub zone. The penalty, which is added to the cost of the allocation, is a function of the load of the two subgroups of lifts, and can be expressed as:

$$\text{penalty} = \left(1 + \frac{b}{100}\right)M \tag{11.9}$$

where, M is a constant value and b measures the imbalance of lift loads between the upper and lower subgroups as a percentage of the highest subgroup lift load.

The fact that the loads of the two subgroups of lifts are taken into account contributes to equalise these loads. For example, the allocation of a call terminating at a floor in the lower sub zone to a lift assigned to the upper sub zone can be penalised by a quantity ranging from zero, if all the upper sub zone lifts are idle, to $2M$, if the lower sub zone lifts are idle.

A call registered to the median sub zone can be allocated to either subgroup of lifts, with preference for the subgroup with the smallest load. The allocations to the lifts assigned to the heavier loaded subgroup are penalised by a quantity that equals the absolute value of b multiplied by M.

A correction mechanism allows this technique to deal with extremely unbalanced traffic destinations, as if excessive unbalance between the subgroup loads is detected, the sub zone limits are automatically adjusted.

11.6.2.5   Walking time

A further feature is necessary in the call allocation control algorithm. After registering the required destination floor and receiving a reply as to which lift will service the landing (hall)

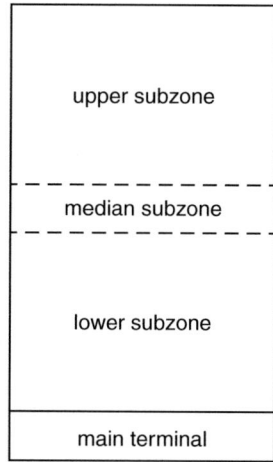

Figure 11.6 Uppeak sub zoning for the ACA traffic control system

call, a passenger must walk to the lift. Thus, the allocation procedure must allow sufficient walking time for the passenger to reach the lift from the call station when allocating the call to a lift.

11.6.2.6   Look ahead

Although the mathematics suggest allocations to all lifts in groups of K lifts, in practice, a 'look ahead' (k) of 2–4 only is practical. This implies groups of more than six cars.

11.6.3   ACA performance

Waiting time has proved to be the most suitable cost function to implement in the ACA control system under various traffic patterns, with the exception of uppeak, for which journey time is used.

The ACA system is able to detect a traffic situation where most traffic originates at the ground floor, independently of the level of traffic intensity, and can automatically switch the cost function from waiting time to journey time. This will provide very good journey time performances, although under light up traffic situations, the waiting times may be considered unacceptable, because sometimes a call is not allocated to the first lift to depart. However, it can be considered that waiting times are psychologically more important than journey times, and that under light traffic conditions, it is not important to reduce the number of stops, and therefore waiting time is the appropriate cost function. To cater for these considerations, the ACA system can use waiting time as cost function until heavy uppeak traffic is detected, when it switches to journey time.

A lift system under adaptive call allocation control presents increased uppeak handling capacities. As a result, it can cope with uppeak traffic intensities that would saturate a similar lift system under conventional control. Average journey times are improved over the whole range of traffic intensities, and average waiting times are improved for the heavy traffic demands.

The performance of ACA under interfloor traffic is substantially better than for conventional systems.

A similar situation occurs for medium to heavy down peak traffic. Under heavy down peak traffic, where the knowledge of the passenger destination floor is no advantage as most passengers have destinations at the main terminal, the average waiting times presented by the ACA algorithm may exceed those presented by a fixed sectoring algorithm. This reflects the handicap resulting from the necessity of fixing the allocation of a new call at call registration time.

The ACA system presents a powerful capability to adapt to unbalanced and changing traffic circumstances. It also possesses those advantages inherent to computer control, namely, the flexibility and ease of algorithm modification without any rewiring being performed.

The HCA system is often used to boost (Russett, 2002) an under lifted building during the uppeak period. If the boosted handling capacity is temporarily insufficient, some HCA systems may refuse to make an allocation requiring the passenger to try again later. These refusals are very frustrating for passengers and should be avoided if at all possible. If refusals are disallowed, there are some instances where transporting passengers in the wrong direction (Rule 4) cannot be avoided. This is a rare occurrence in a well-designed system unless the building is under lifted.

11.7 Comparison

Figure 11.7 compares HCA and ACA algorithms to the COL and DS legacy systems for uppeak performance. The benefit of a switched, i.e., adaptive cost function, can be seen when comparing ACA and HCA curves. Although at low loads the waiting times are longer, the journey will be smaller.

Figure 11.8 illustrates well the even performance of the FS-PT legacy algorithm during down peak. This can be attributed to the use of time stamping waiting calls. The performance of CGC is unacceptable. The ACA algorithm performs badly at heavy loads as the destination is always known under down peak, i.e., the main entrance (exit) floor.

Interfloor traffic levels are rarely above a 30% demand, which is about one-third of the occupants using a lift during any one hour. At these demands, there is little difference (to the passenger) in service times whichever traffic control is used. Figure 11.9 shows that at higher demands. the legacy algorithms FS and FS-PT perfom badly compared to ACA, CGC and DS algorithms.

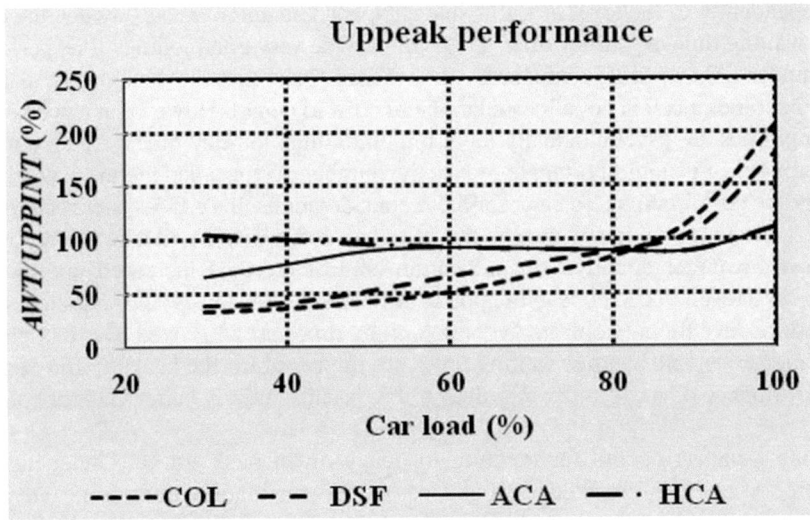

Figure 11.7 Uppeak perfomance

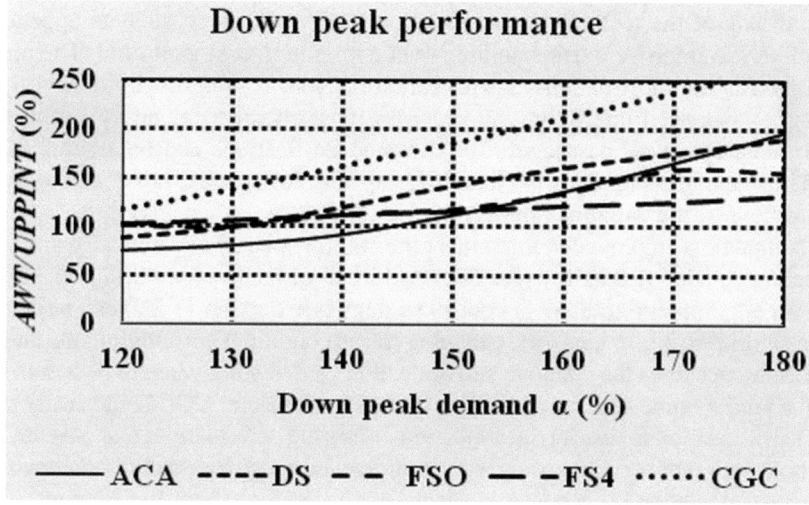

Figure 11.8 Down peak perfomance

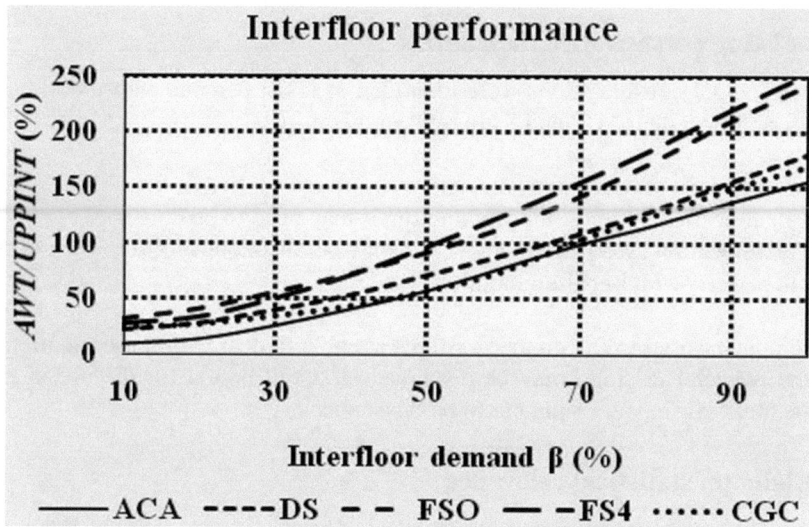

Figure 11.9 Interfloor performance

11.8   Conclusions on advanced control techniques

Many of the advanced control techniques employ complex mathematics and involved programming, which makes the practical implementation of the traffic controllers difficult. In addition, the proper understanding and correct adjustment on-site by installation and service persons is doubtful, and there is also an increased risk of system unreliability. Powell (2001) states '. . . *the added complexity involved in creating these (neural) networks and putting them into production could not be justified on the (slightly) expected gains in dispatching performance . . . over less complicated techniques*'.

The use of any of the techniques during a dominant traffic flow, such as uppeak or down peak, is unlikely to improve traffic handling over a minimum cost algorithm. The provision of additional destination information, as with call allocation, is unnecessary during light traffic conditions, i.e., balanced interfloor, and becomes most effective for heavy traffic situations, particularly uppeak. Then, passengers for common destinations can be assembled to travel together. The technique improves the handling capacity for uppeak, but does not assist down peak or interfloor traffic handling (Barney 2000a, 2000b).

Once a computer is employed to implement the control strategy, the final algorithm is limited only by the imagination and ability (see Section 11.1.6) of the program designer. For example, the search for a 'bumpless' transfer of control strategy (see Section 11.2.3) can be dealt with by having one algorithm able to adapt the changing traffic conditions. In addition, the hall call allocation algorithm becomes the adaptive call allocation by detecting when to switch from a waiting time to a journey time cost function. The stochastic algorithm CGC could easily be married to the hall call allocation to restrict the allocation of landing calls to those that have been waiting for a threshold period of time. Learning algorithms can be added to 'predict' outcomes and learn to improve the calculation processes, such as the estimated time to reach a landing call.

All these techniques allow the use of the underlying resource (handling capacity) more effectively for the benefit of all passengers. An added advantage is that the systems become more consistent in their response to passenger demands.

11.9   Assisting persons with disabilities

In Section 8.7, eight considerations were identified to assist persons with mobility problems and other disabilities; items 6, 7 and 8 are relevant here:

6   signage may need to be larger and clearer
7   with call allocation systems, the walking times may need to be longer
8   with call allocation systems, an empty car may be taken out of group control to be provided to persons with impaired mobility.

Many computer-based traffic control systems are able to deal with these requirements. For example the 'wheelchair' logo may be displayed on a call button for the use of persons in wheelchairs, thus causing an empty car to be dispatched.

11.10   Endnote: statistical relevance

There are a number of advanced techniques, which can be classified as artificial intelligence (AI) techniques and which can be applied to the conventional two-button signalling system. These include: expert systems (Qun *et al.*, 2001); fuzzy logic (Ho and Robertson, 1994); dynamic programming (Chan and So, 1996); genetic algorithms (Siikonen *et al.*, 2001; Miravete, 1999); knowledge-based systems (Prowse *et al.*, 1992); neural networks (Barney and Imrak, 2001b); and optimal control (Closs, 1970). The references are representative of many published papers on this subject, and each gives a background to the technique.

All these techniques depend on large numbers of events in order to operate satisfactorily. To put these techniques into perspective, consider two buildings, where the uppeak provision is satisfactory.

Building 1 has eight lifts, serving 16 floors, with a total population of 1,600 persons. The lift system can handle 15% uppeak arrivals, i.e., 240 persons in 5 minutes. This is a significant building, probably prestigious, with a large scale of activity. Statistically, an arrival rate of 0.8 persons per second at one floor (the main terminal) is sufficient justification to claim that the calculations of H and S, in the determination of the handling capacity and interval, are reasonable.

Now consider that 25% of the building population, i.e., 400 persons, uses the lifts in a balanced interfloor fashion over a period of one hour. This is an arrival rate of 25 persons, per hour, per floor, or an arrival rate of 0.007 persons per second. This is less than one hundredth of the uppeak arrival rate at the main terminal floor. This level of arrival activity is not sufficient to allow any statistical consideration. Even if it is assumed that 50% of the population use the lifts every hour, i.e., an arrival rate of 0.014 persons per floor per hour, it is still not statistically significant.

Now consider Building 2, which has three lifts, serving eight floors with a total population of 400 persons. The lift system can handle 15% uppeak arrivals, i.e., 60 persons in 5 minutes. This is a small office building, at the low end of the market place, with an adequate handling capacity. The arrival rate of 0.2 persons per second at the main terminal floor is still sufficient to claim justification for the calculations of H and S.

If 25% of the building population, i.e., 100 persons, uses the lifts per hour in a balanced interfloor fashion, this is an arrival rate of 12.5 persons per hour per floor or an arrival rate of 0.003 persons per second. This is just over one hundredth of the uppeak arrival rate at the main terminal floor. Once again, this level of arrivals is not sufficient to allow any statistical consideration. At 50% of the population using the lifts every hour, i.e., an arrival rate of 0.007 persons per floor per hour, again, is not statistically significant.

The above discussion indicates that a statistical approach to a balanced interfloor traffic pattern is not sustainable, simply, there are not enough passenger events. The two examples, from each end of the building quality spectrum, indicate that if the uppeak sizing is satisfactory, then the balanced interfloor activity should provide a similar satisfaction. As the balanced interfloor activity moves into some dominant traffic situation, i.e., more traffic at some floors, then the statistical justification improves. In addition, the demand on the traffic controller lessens as it is more efficient to serve fewer floors with more intending passengers than most floors with one passenger per stop.

Some of the techniques use data gathered over time to predict traffic demand. This would be satisfactory if the demands imposed on a lift system could be relied on to occur reliably each day at the same time. As the number of events is so low, statistically, these techniques are unlikely to improve traffic performance.

Chapter 12 discusses a number of advanced techniques.

Notes (shown in text as[1])

1 Throughout this chapter the word 'computer' will mean digital computer.
2 The proprietary implementation of this algorithm was said to have been designed by the use of simulation.
3 This is *car* journey time, not *passenger* journey time.
4 'Cost function' is optimal control theory terminology. Its inverse is the 'performance index'.
5 Usually, the term 'statistically based', meaning 'facts systematically collected', is used. In this case, it seems more appropriate to use the term 'stochastic based', meaning 'aim at a mark, guess'.
6 A fuller description is given in Barney and Dos Santos (1985).

12 Overview of modern lift group control systems

12.1 Introduction

This chapter provides an overview of the modern lift group control systems. Modern lift group control systems have two types of functions.

Main function:

a) The main function of the lift group controller is the allocation of the landing calls to the lift cars in the group. This is also referred to as dispatching. This function is discussed in detail in Section 12.2.

Supporting functions:

b) These are functions that provide information that is necessary to enable the successful execution of the main function. A number of these functions are listed and discussed in detail in Section 12.3.

12.2 Main function of lift group control: dispatching

The function of allocating the landing calls to the lift cars can be generally classified into four types of methods. These are rules-set-based systems, objective function optimisation systems, sectoring systems, adaptive systems. Many practical implementations are hybrids of two or more of the systems above.

12.2.1 Rule-set-based systems

These types of systems prevailed in the seventies and eighties. They are based on using a set of rules that suit the prevailing mode of traffic in the building (e.g., uppeak traffic, down peak traffic . . .) (Hikita *et al.* 2001; Tsuji *et al.*, 1989; Prowse *et al.*, 1992; Chenais & Weinberger, 1992). Once a dominant mode of traffic has been detected, a set of rules become effective. For example, once a dominant uppeak traffic mode has been detected, then lifts are generally dispatched back to the main entrance as soon as they have delivered the passengers to their destinations. One lift opens its doors and keeps them open until it fills up and departs. Once departed, another lift opens its doors and so on.

It has become customary to link sets of rules to certain types of traffic. Once a traffic mode has been detected or selected, the corresponding set of rules is activated. There are three means by which the controller activates the set of rules corresponding to a certain traffic type/mode:

a) Automatic detection of the presence of a certain traffic mode

There are many rules that have been developed over the years for this. The different methods of detecting traffic modes are discussed in more detail in Section 10.8.1 and later in Section 12.3.1.

b) Manual switching

An operator in the building would switch the controller into different modes, depending on their perception of the prevailing traffic.

c)   Timer switching

A timer is used to switch the controller into a traffic mode based on the predicted passenger movements in the building.

The main advantage of this method is that the rules are easy to implement, and do not require excessive processing power in the controller. This made them very suitable for implementation in conventional relay-based systems, prior to the advent of microprocessor-based systems. However, they suffer from many disadvantages (Chenais & Weinberger, 1992):

a)   There is generally no clear cut prevailing traffic mode. In any real-life building, the prevailing traffic mix at any one point in time is a mixture of the three traffic modes: incoming traffic; outgoing traffic; and interfloor traffic.

b)   The traffic mode sometimes changes very quickly, so by the time that controller has detected the prevailing traffic mode, it is too late, and the traffic mode has changed again.

c)   Sometimes, there is heavy demand from one floor (e.g., end of a function or lecture . . .), which is difficult to classify into any one of the known types of traffic. So *et al.* call this mode of traffic four-way-traffic (So *et al.*, 1996).

d)   The rules have to be extracted from an expert, and it is not always easy to extract and express these rules in a clear format suitable for programming.

e)   It is not always clear what the expert is trying to optimise in a set of rules (i.e., the rules are sometimes very objective and depend on the expert's point of view) (Chenais & Weinberger, 1992).

Two examples of rules (Chenais & Weinberger, 1992):

a) Coincident call rule

Try to allocate the landing call to a car that already has a car call to the same floor.

b) Heavy traffic

In cases of heavy down traffic, give lower priority to up landing calls, and in cases of heavy up traffic, give lower priority to down landing calls.

Although fuzzy logic has been used in a number of different contexts in lift group control, in some cases, it is also applied to code rule sets (Umeda *et al.*, 1989; Danapalasingam *et al.*, 2005; Gudwin *et al.*, 1998; Kim at al., 1998).

12.2.2   Objective function optimisation systems

This method relies on the optimisation of an objective function. There are a number of objective functions that have been used, in varying degrees of importance.

It is important to remember when discussing the objective function, that most parameters of interest are random variables. It is possible to use the average (expected) value of the random variable, the variance of the random variable or the maximum value of the random variable or a combination thereof to build the objective function.

Examples of possible objective functions are listed below:

1   **Landing call system response time**

Landing call system response time (the average, the longest, or the variance or weighted combinations thereof). In conventional control systems, the number of passengers behind a call is not known, and the system is only aware that a landing hall request has been made. In these cases, it is meaningless to talk about passenger waiting time, and it is more realistic to talk about system response time.

2   **Passenger waiting time**

Passenger waiting time (the average, the longest or the variance or weighted combinations thereof). In the case of hall call systems, the lift group controller has full knowledge of the origin and destination of each passenger once he/she has registered their call. In such a case, it is meaningful to use as the optimisation function the passenger waiting time. Where passengers have different priorities or statuses, then it is also possible to optimise a weighted average of the total passenger waiting time.

3   **Car call response time**

Car call response time (the average, the longest, or the variance or weighted combinations thereof). In conventional control systems, the number of passengers behind a car call is not known to the control system (although it could be inferred by using feedback from the door photocell signals and the car load signals). It is thus more meaningful to use the car call response time, rather than the passenger travelling time, as an objective function.

4   **Passenger travelling time**

In the case of hall call allocation control systems, the destination of every passenger is known. It is thus possible to use the passenger travelling time as part of the objective function. As with the previous three functions, it is possible to minimise the average, longest or variance of the objective function or weighted combinations thereof.

5   **The handling capacity**

There are certain situations where the lift system is overloaded, such as when the original system was under designed, or where the population or function of the building change after it has been commissioned. In these situations, the highest priority is the enhancement of the system handling capacity, even where this is at the expense of the waiting time. The objective becomes, in this case, maximising the handling capacity ($HC\%$) or effectively minimising the round trip time. An effective method is thus needed to detect that the system is overloaded, and this could be based on detecting the car loads and the immediate re-registration of the landing calls following the departure of a lift car from that landing.

6   **The round trip time**

Under certain conditions, the objective function is the round trip time. More specifically, the objective could be to minimise the average round trip time of all the lifts in the group, to minimise the longest round trip time or to minimise the variance of the round trip times for the lifts in the group.

7   **Energy consumption**

It is rare to use energy consumption as the only component of the objective function. It is usually combined with one of the passenger performance measures. It is also possible in some cases to use the maximum power demand on the electrical supply.

8   **The car loading**

In order to meet heavy passenger demand, it is also possible to use the car loading as an objective function in order to ensure that the cars are fully loaded.

It is also possible to use a weighted sum of a number of these functions as an objective function. Typical examples include:

a) **A weighted average**

A weighted average of waiting time and travelling time (Smith & Peters, 2002; Smith & Peters, 2004; Peters, 2006).

b) **Another weighted average**

A weighted average of waiting time and energy consumption. It is, for example, possible to increase the weight of the waiting time during periods of heavy passenger activity, and increase the weight of the energy consumption part during periods of light passenger activity (Smith & Peters, 2004; Zhang & Zong, 2014).

c) **A further weighted average**

Using a weighted average of the average system response time and the longest waiting time, it is possible to reduce the mean as well as the variance of the system response time (Halpern, 1993).

Any optimisation system requires three vital elements (Qun *et al.*, 2001):

a) **Objective function**

A number of examples of objective function have been listed earlier in this section.

b) **Optimisation tool**

This is the tool used to optimise the objective function. Examples of such tools are listed later in this section.

c) **Model**

The model is important because it is used by the optimisation tool to evaluate the feasibility/viability of a suggested allocation. This could be a set of equations, discrete event simulation models, or other types of models, e.g., Petri net models, (Qun *et al.*, 2001; Cho *et al.*, 1999a).

Examples of optimisation tools are listed below:

1 **Conventional search techniques**

Examples include fixed step size and random step random search techniques, simulated annealing and threshold accepting (Markon & Nishikawa, 1991).

2 **Linear programming**

Linear programming can be used to place constraints and obtain an optimal solution (Guifeng, 2001; Ruokokoski *et al.*, 2008). The Hungarian method is the most widely used method for solving the linear programming problem.

3 **Genetic algorithms**

A lot of research has been carried out in applying genetic algorithms in the area of lift group dispatching (Tyni & Yilnen, 1999; Siikonen, 2000; Eguchi *et al.*, 2004; Cortes *et al.*, 2004; Bolat & Imrak, 2006), and some systems are already running using such a feature. Genetic algorithms allow the lift group controller to arrive at an optimal or near-optimal solution in short periods of time that are suitable for application in real time control.

4 **Neural networks**

When using neural networks for lift group dispatching, it is necessary to train them offline first, and then they could carry on training and learning online (Markon *et al.*, 1994; Tolosana *et al.*, 1998).

5   Particle swarm optimisation

An example applied to destination lift group control systems can be found in (Li *et al.*, 2007).

6   Optimisation techniques

Optimisation techniques (exact re-optimisation (Hiller *et al.*, 2013a; Hiller *et al.*, 2013b); robust optimisation (Zhang & Zong, 2014); bi-level optimisation problem (Sorsa *et al.*, year unknown).

7   Multiple agents and blackboard architecture

An example of reinforcement learning using multiple agents can be found in (Crites & Barto, 1998); another example on multiple agents can be found in (Koehler & Ottiger, 2002) and on blackboard architecture in (Brandin *et al.*, 1989) and (Pang & Nandy, 1989).

8   Threshold-based policies

An example can be found in (Pepyne & Cassandras, 1997) where, under conventional up-peak control, the point in time where the car door is closed achieves the optimal car loading.

The main advantages of optimisation methods over rule-based methods are (Chenais & Weinberger, 1992):

1   The optimisation method automatically adapts to the changes in the prevailing traffic mix.
2   There is no need to transform the optimisation function into rules; this ensures that the optimisation function is, in fact, optimised.
3   The optimisation function can be easily modified in response to the client's requests (e.g., the sales engineer can independently change the function to be optimised without the need to refer back to the software engineer).

Constraints can be automatically incorporated in the optimisation tool. Examples include linear programming and genetic algorithms (a good example on power maintenance scheduling can be found in Negnevitsky (2005)). It is not evident from the available research whether there is a universal optimisation function that can be applied under all traffic conditions.

12.2.3 Sectoring systems

Under this mode of control, the building is split into a group of sectors. Each sector comprises a group of floors. The floors comprising a sector could be contiguous or non-contiguous. The allocation of lifts to each sector could be fixed or variable. The size of each sector and the floor it contains could also be fixed or variable.

It is important first to distinguish between the two terms sectoring and zoning. In much of the literature, the terms are used interchangeably. However, in this chapter, they refer to two different concepts. Sectoring is a *soft* feature that is programmed within the lift system controller and is activated under the specified condition, by which each lift only serves a group of floors during a round trip. Zoning on the other hand is a *hard* feature that refers to a fixed arrangement by which a lift will always only serve a set group of floors. This is fixed in the hardware of the building (e.g., shafts, lobbies; landing doors, etc.). Sectoring can be easily changed during the lifetime of the lift system. Zoning is very difficult to change and has to be decided from day one. The terms soft and hard are very relevant in this context. Several optimisation studies have been carried out regarding the optimal zoning arrangement of buildings (Powell, 1971; Powell, 1975; Mitric, 1975a; Mitric, 1975b; Browne & Kelly, 1968). Zoning is also referred to as banking. Zoning is beyond the scope of this chapter.

Depending on the size and boundaries of sectors, sectoring can be further subdivided into the two following types:

a) **Static sectoring**

In the case of static sectoring, the size and delineation of the sectors is fixed. More details on static sectoring can be found in Section 10.6.21.

b) **Dynamic sectoring**

In this case, the size and delineation of the sectors changes continuously depending on the passenger movements within the building. More details about dynamic sectoring can be found in Section 10.6.2.2.

Moreover, the allocation of lifts to the sectors could also be static or dynamic allocation, as follows (more details about the allocation of cars to sectors can be found in Section 10.6.3):

a) **Static allocation**

Under this method of allocation, each lift car is always allocated to the same sector. The main advantage of this method is that it is more convenient for passengers as they congregate in front of the lift that is always allocated to the sector that contains their destination floor. It thus causes less confusion for the passengers. However, in order to equalise the handling capacity of the different sectors, the size of sectors must be made unequal in order to equalise the round trip time of the different lifts in the group (lower sectors have a larger population, while upper sectors have a smaller population).

b) **Dynamic allocation**

This is referred to as sequencing or rotating the lift cars to the sectors. It is also referred to as 'round-robin' allocation by Otis (Bittar & Thangavelu, 1988; Bittar & Thangavelu, 1989). Each lift car can be allocated to a different sector in each round trip. This has the effect of equalising the effective round trip time for the different sectors and hence the handling capacity of the sectors, with the convenience of still having equal size sectors. The main disadvantage of this method is the confusion it can cause to passengers waiting in the lobby, and the difficulty passengers find in moving through a crowded lobby to the car serving their desired sector (Fortune, 2002). Area control (Sakai & Kurosawa, 1984) is a variation on the concept of dynamic sectoring.

Sectoring is usually applied (and is most effective) in the cases of uppeak traffic with a single entrance, and down peak traffic with a single exit. It is usually applied when the system is undergoing heavy uppeak or down peak traffic. It boosts the handling capacity. A general overview of sectoring systems can be found in (Christy, 2014).

It Otis's system (Powell, 1992; Bittar & Thangavelu, 1988; Bittar & Thangavelu, 1989; Thangavelu, 1989; Bahjat & Bittar, 1992; Kameli, 1994; Kameli, 1996), it can be seen that the system evolved over time, starting with static sectoring, adding dynamic sectoring, and then culminating in a feature that allows automatic detection of the traffic type and intensity, thus switching the controller into uppeak, static sectoring and then dynamic sectoring as the intensity of the traffic increases (Kameli, 1994; Kameli, 1996). The system learns the historical and real time movements of passengers in the building, and tries to predict and adapt to the movement of passengers in real time (Kameli, 1994).

Sectoring has the advantage of being very simple to apply and requiring little processing power or time. However, as it is only suitable for uppeak or down peak traffic, it does suffer from the disadvantage that it needs to detect the presence of the type and intensity of the traffic mode in order to be activated or deactivated.

Sectoring has also been applied in destination lift group control systems where it produces an even larger boost in handling capacity under uppeak or down peak systems with a single entrance (Siikonen & Ylinen, 2006; Siikonen, 2008; Sorsa *et al.*, 2005).

12.3   Supporting functions of lift group control

In addition to the main function of allocating landing calls to the lift cars in the group, there are several supporting functions that lift group controllers use in order to enhance performance and provide supporting data and information. These are discussed in the following subsections.

12.3.1   Automatic detection of prevailing traffic mix

Some group control algorithms rely on the knowledge of the current traffic mix in order to trigger a different set of rules or a different optimisation objective function. For this reason, group control software could include automatic detection of the prevailing traffic mix. For example, the controller could detect that the dominant traffic mode is uppeak (incoming traffic), down peak (outgoing traffic) or interfloor traffic. In more advanced cases, the controller could even produce an estimate of the mix of traffic (e.g., 40%:40%:20% incoming: outgoing: interfloor, respectively). In some cases, the system would also detect the magnitude of the traffic or each component of the traffic. This could even be detailed on a floor by floor basis.

Three examples of automatic detection of traffic patterns in a building are described below.

Rules of thumb have been developed to detect the presence of uppeak and down peak traffic (Beebe, 1980). When the number of up stops divided by down stops is larger than unity, this can be used as an indicator of the presence of an uppeak traffic condition. When the number of down stops divided by up stops is larger than unity, this can be used as an indicator of the presence of down peak traffic conditions. An alternative indicator of down peak is when the ratio of all landing calls divided by all stops approaches unity.

Neural networks have been used to detect the type of traffic mix; So *et al.* used a neural network to estimate the mix of prevailing traffic (So *et al.*, 1996). The neural network undergoes supervised learning first, and then unsupervised learning. The inputs to the neural network include number of up stops, number of down stops and the history of stopping times, among others. Five modes of traffic patterns are detected, namely: uppeak traffic; down peak traffic; off-peak traffic; two-way traffic; four-way traffic.

Otis presents an algorithm for detecting the presence of uppeak traffic when two consecutive fully loaded cars depart from the main terminal within a preset time period (Bahjat & Bittar, 1992).

12.3.2   Generation of virtual passenger traffic

Historically, it has been possible to revoke a landing call assignment to a lift car if a better assignment is found later. Revoking such an assignment is acceptable in systems that do not provide early call announcement to passengers. However, it is not possible to revoke a landing call assignment in early call assignment systems as this causes confusion to passengers, whereby the passenger has to move to the newly assigned lift car, probably by moving though a crowded lobby.

There are two cases where early call announcement is adopted:

1   Conventional lift group control system

In this case, the announcement relates the landing call at that floor (could amount to two announcements on any floor, assuming an up landing call and a down landing call). This offers the advantage of reducing traffic conflict at the landing, and reducing boarding time by encouraging passengers to move to the announced lift car that will serve their landing call.

2   Hall call allocation lift group control

By definition, hall call allocation group control systems are early call announcement systems. Each passenger receives an immediate allocation as soon as he/she enters the destination, and will start moving towards their assigned lift car.

The inability to revoke a landing call assignment and replace it with a better assignment in early call announcement systems causes a serious problem. The landing call assignment is usually made based on minimising the system response time to the landing call (or the passenger waiting time in hall call group control systems). The estimation of the system response time is based on the information available at the time of making the landing call assignment. However, with the passage of time, further landing calls and car calls are registered, and the estimated system response time becomes incorrect. The actual system response time turns out to be larger than the original estimate, with detrimental effects on the performance of the group controller.

For this reason, in systems that operate early call announcement, a method is needed in order to provide an accurate estimate of the system response time to the landing call at the time of making the early assignment of the landing call to the lift car. The problem above has been solved using two approaches: the use of neural networks or statistical approximation to provide a better estimate of the system response time; and the generation of virtual traffic to represent future calls.

An artificial neural network is used in order to estimate system response time to the landing call, which Otis denotes as remaining response time (*RRT*) (Powell *et al.*, 2000). It uses a linear perceptron with 47 inputs (that include number of registered car calls, number of passengers and distance to the floor). The estimation of the remaining response time is compared to the conventional estimation of the *RRT*, based on information available at the time of making the assignment, and shows an improvement in absolute error ranging from 7.4% to 20.8%.

Statistical approximation has been used in a similar way to provide a better estimation of the system response time (Cho *et al.*, 1999b).

Another possible solution to address the problem of accurately estimating the system response time to the landing calls in early call announcement systems used by KONE (Siikonen, 2002b) and Fujitec (Fujitec, 2014) is to generate virtual traffic. The generated virtual traffic represents the future possible car and landing call registrations, and allows the system to produce a better estimate of the system response time. Obviously, these virtual calls are cancelled from the system with the progress of time, and as real calls are registered.

12.3.3   *Estimating passenger traffic in the building: origin–destination matrix (OD matrix)*

It is vital for a lift group controller to have a good measurement of the historical and real time passenger movements within the building. Historical data are data gathered over previous days. Real time data are the data gathered over the latest 5 to 10 minutes. The passenger demand can be compactly represented in the so-called original-destination matrix. Such data can be used in two applications:

1   **Generation of virtual passenger traffic**
    The OD matrix is vital for generating the virtual passenger traffic (discussed in subsection 12.3.2).
2   **Operation of dynamic sectoring system (Kameli, 1994)**
    The controller uses the historical data and real time data in order to dynamically sector the building under the dynamic sectoring group control algorithm.

The passenger movements within the building vary over the time of the day. It is usually quantified over a predefined period of time, typically over 5 minutes. The passenger movements can be captured as a set of probabilities within a matrix form, referred to as the origin–destination (*OD*) matrix. The *OD* matrix can be measured in real time by observing the lift movements within the building. Signals representing the registration and cancellation of car and landing calls are complemented by car load signals and door light beam signals in order to estimate the *OD* matrix. Two main streams of work have been carried out in this area by Kuusinen *et al.* (Kuusinen *et al.*, 2014; Kuusinen & Malapert, 2014) and Basagoiti *et al.* (Basagoiti *et al.*, 2012; Basagoiti *et al.*, 2013).

12.3.4   *Identification of the number of waiting passengers*

A conventional lift group control algorithm can improve performance if the number of passengers behind the landing call is known. Computer vision has been used in order to identify the number of waiting passengers at the landing in (Alani & Mehta, 1996; So *et al.*, 1992; and So *et al.*, 1993a).

12.3.5   *Prior standby control*

This feature is more of a supporting function, rather than an information provision function. It has been shown that improvements in the system response time can be achieved if the lifts are dispatched in advance to the expected floors, based on past historical data (Yamashita *et al.*, 2000; Tam & Chan, 1996). This is in effect similar to the parking policy that has been traditionally used in uppeak conventional systems where lift cars are returned back to the main entrance as soon as they have delivered the uppeak travelling passengers.

12.3.6   *Time series forecasting*

This is basically another variation on the learning of the historical passenger movement in the building. It typically uses neural networks for its operation (Zong *et al.*, 2000), although other methods could also be used.

12.4   Notes on destination call allocation control systems

Several of the features used for conventional lift group control have also been applied to hall call allocation (destination group control) systems (Schroeder, 1990d; Siikonen & Ylinen, 2006; Siikonen, 2008; Sorsa *et al.*, 2005; Yoshikawa *et al.*, 2013; Stanley *et al.*, 2011; Gerstenmeyer & Peters, 2014; Lauener, 2007). The most important of these are the sectoring tools (static/dynamic sectoring; static/dynamic allocation). Most of these systems detect the presence of uppeak or down peak traffic in order to operate the sectoring function (Siikonen & Ylinen, 2006). Outside the uppeak or down peak, they revert to objective function optimisation algorithms.

Of particular interest is a modern feature that never arose in conventional systems, called reverse journeys. One of the five rules that the single lift controller and the lift group controller should adhere to is that the lift car should not reverse its direction of travel with passengers still in the lift car. With hall call allocation lift group control systems, the destination of each passenger is known in advance, and there are situations where it is beneficial to violate this rule, e.g., reverse journey calls as shown in Gerstenmeyer & Peters, (2014).

Part D

Advanced analysis

D1 Introduction

Uppeak

Uppeak traffic sizing defines the underlying capability of the lift installation. This size should be decided by the requirements of the target building, whether speculative or high prestige. Once the uppeak capability has been decided the quantity and quality of the service in all the other traffic conditions is also decided. Chapter 13 discusses some considerations and techniques that can be applied to the uppeak traffic condition normally encountered during the morning incoming traffic. Some of the sections answer pragmatic questions, but others are simply for curiosity.

Down peak

At the end of a working day the occupants of a building wish to leave as quickly as possible. Most, if not all, of the lift traffic during this outgoing or down peak traffic period terminates at the main terminal and the demand for lift service is usually very high. Surprisingly very little attention is devoted to down peak traffic, and the evaluation of a lift system for this traffic pattern is far from common practice. The conventional lift traffic designer is happy to conclude that if a system is properly designed to cope with the uppeak traffic then it will be able to move the down peak traffic satisfactorily and this is generally so. The hypothesis is that a lift system will be designed to serve the requirements of the uppeak traffic according to the type of building, whether it is high prestige or low end/speculative. The assumption then is that if the lift system provides the necessary Quantity and Quality of Service required in uppeak, then a matching Quantity and Quality of Service will be obtained during down peak. This conclusion will be shown to be a criterion based on Quantity of Service rather than on Quality of Service.

The questions are 'Can down peak be analysed in a similar way to uppeak?'. And 'Can a concept of a round trip time be used?'. The answer is yes to both questions as can be seen by examining Figure 4.8, which shows a spatial plot of car movements during down peak traffic. There is a similarity to the uppeak traffic patterns in two respects. The first is that all trips pass through the main terminal floor. The second is the regular staircase stopping pattern. This pattern has a reverse slope to the uppeak, but with less stops. It suggests that an analysis similar to the uppeak analysis is possible. Chapter 14 addresses these questions

Interfloor

Once a building is occupied, its population will move around the building about their normal business randomly (in the statistical sense). A small percentage of occupants will leave the building and be replaced by new arrivals. A typical time to observe such an interfloor traffic pattern is during the mid-morning or the mid-afternoon periods. The traffic demand is frequently balanced and there is no dominant pattern of arrival at any floor or any dominant destination floor. Passengers using the lift system at this time eventually return to their original departure floor, hence the use of the term balanced interfloor traffic. This is illustrated in Figure 4.9, which shows a typical spatial plot of lift movements around a building during an interfloor traffic period. The plot can be viewed upside down, or back to front, and still displays a random pattern! At other times there may be some unbalanced interfloor movements for example to/from refreshment floors, hotel lobbies, etc. These traffic flows are not as dominant as those found during pure uppeak or down peak traffic, and not as random as those found during balanced interfloor traffic. A more significant traffic pattern occurs at midday and this will be considered in Section 13.5.

The analysis of lift systems under interfloor traffic conditions is rare and it is very seldom mentioned by lift writers. It is however the traffic pattern that exists for most of the working day. A few authors have developed methods to evaluate the round trip time for interfloor traffic. Strakosch (1967),[1] based on lift engineering experience, considered interfloor traffic and the expected number of stops by intuitive means and provided a step by step method to calculate a round trip time. Tregenza (1972) presents a formula for the evaluation of the round trip times and deduces relationships for the expectancies of the high reversal floor and the expected number of stops per trip. The cars are assumed to circulate uniformly around the building, but always returning to the main terminal floor.

The mathematical evaluation of interfloor traffic is even rarer than an intuitive analysis. Alexandras (1977) developed formulae for the most general traffic condition of balanced interfloor traffic using statistical analysis. He considered a Poisson probability distribution function and produced equations for both H and S. They are very complex and are not reproduced here. He proved their validity by reducing them to the classical formulae for H and S for uppeak traffic. Barney and Dos Santos (1985) report these results. Peters (1997a) provides other formulae of equal complexity as his General Analysis formulae. These formulae are summarised in CIBSE Guide D: 2000. Both the Alexandris generalised formulae and the Peters General Analysis formulae only apply when there is a dominant traffic flow, i.e.: when lifts pass through the main terminal every trip.

In practice, during balanced interfloor traffic, lifts do not make a round trip as defined for both the uppeak and down peak traffic patterns. Examining Figure 4.9, where there are three cars, shows 251 reversals of direction of which only 87 occurred at the main terminal. Each round trip contains two reversals in the circuit. This would account for 174 of the reversals and therefore indicates nearly one third of all trips do not include the main terminal. Thus calculations of round trip times using values of H and S derived from the Alexandris/Peters formulae will provide results which have no meaning and this approach is not recommended.

The methods of evaluation proposed by Strakosch and Tregenza, and the mathematical formulae proposed by Alexandris and Peters, could all be suitable for the now discarded scheduling systems, where cars are dispatched at regular time intervals from the terminal floors, but they are not applicable to modern non-scheduling (on demand) lift systems. Modern systems work on an allocation basis, and the cars are frequently taken out of their normal cycles to make an express run to particular floors or sectors, either as a parking strategy or to provide extra service to heavy traffic demand sectors. Chapter 15 elaborates on the interfloor traffic condition.

Midday

In the previous sub sections, uppeak, down peak and interfloor traffic patterns have been analysed and discussed. These were all considered as pure patterns, i.e.: with NO mixed mode traffic. The midday traffic pattern is defined in Section 4.4.3 and other traffic patterns were touched upon in Section 4.4.5. This former pattern can be analysed mathematically. Other patterns, especially where there are mixed modes of demand, are best examined using simulation techniques (see Chapter 15).

D2 Review

The primary traffic condition is pure uppeak. This traffic condition is usually the condition to be satisfied in the traffic design. It is analytic and formula can be derived. The pure down peak and mixed mode, midday traffic patterns are usually satisfied by the sizing carried out for the uppeak traffic pattern. Both of these patterns are analytic if assumptions are made. The only traffic pattern that defies a mathematical examination is the interfloor traffic pattern. Fortunately a lift system subjected to this pattern is never called upon to deliver its full capability. Interfloor demands are modest compared to the other three traffic patterns. Table D.1 indicates the formulae derived.

All the analysis presented in this book is mainly based on the classical traffic pattern, depicted in Figure 4.2. It has been used to describe how people use lifts in a building and to size lift systems successfully for over half a century. But it is largely a figment of the imagination, as it probably has never existed. Siikonen (1998, 2002), Peters and Sung (2000), (Powell, 2001) and many others all report considerable differences to Figure 4.2. Just because Figure 4.2 has been "discredited" it does not mean it should be abandoned as a valuable tool. As a "benchmark" it is generally accepted worldwide. Countless buildings have been designed to its "illusion" and the designs work! Why does it work? The answer is that the uppeak design provides an underlying capacity, which sets the performance of the three other major traffic conditions of down peak, interfloor and midday traffic.

The following chapters look at the ratio of inherent handling capacity for three traffic patterns compared to uppeak handling capacity. With uppeak considered to be unity, the ratios are:

uppeak	1.0
down peak	1.6
midday	1.3
interfloor	1.4

Table D.1 Review of traffic patterns

Traffic pattern	Round trip time equation	Handling capacity equation	Passenger average waiting time equation
Uppeak	$RTT = 2Ht_v + (S+1)t_s + 2Pt_p$ Equation (4.11)	$UPPHC = 300P/UPPINT$ Equation (4.8)	$AWT = [0.4 + (1.8P/RC - 0.77)^2]/INT$ Equation (6.1)
Down peak	$RTT_D = Nt_v + (0.5S+1) + 2Pt_p$ Equation (14.9)	$DNPHC = 300P/DNPINT$ Equation (14.10)	$DNPAWT = 0.85aUPPINT$ Equation (14.5)
Interfloor	No equation available	No equation available	$IFAWT = UPPINT (0.22 + 1.78\beta/100)$ Equation (14.2)
Midday	$RTT_M = 2Ht_v + 2St_s + 4Pt_p$ Equation (16.1)	$MDHC = 300 \times 2P/MDINT$ Equation (16.2)	$MDAWT = 0.85MDINT$ Equation (16.3)

These ratios assume that all the handling capacity can be utilised.

The interfloor handling capacity is never utilised, as the demand is about one fifth of the uppeak demand. In fact, lift system performance under interfloor traffic demand has an inherent capability as high as 1.4 times the underlying uppeak capability (Siikonen, 2002a):

All the formulae given in Table D.1 relate to pure traffic patterns, except the midday formulae. The pure patterns are unlikely to be seen in practice. However, the tradition of applying the classical approach to sizing has generally proved satisfactory. Chapter 17 deals with the more complex traffic patterns by using the simulation technique. This technique allows a full understanding of a lift system's capability. Chapter 18 looks at practice.

Note (shown in text as[1])

1   Page 87.

13 Advanced analysis of uppeak

13.1 Uppeak formulae

Two major parameters in the evaluation of the round trip time equation are the probable number of stops (S) and the highest reversal floor (H). Both these parameters are dependent on the number of floors served (N) and the number of passengers in the lift (P). The number of passengers is dependent on their arrival rate, and the assumed probability distribution function (pdf) of the passenger arrival process determines how they arrive.

Consider the simplest arrival process, which is the rectangular pdf. This is a pdf where there is a simple linear relationship defining the times of passenger arrivals. The following formulae have been derived and are shown in Table 13.1 for reference.

For equal demand: S (see Equation (5.7), Section 5.7)

H (see Equation (5.14), Section 5.8)

For unequal demand: S (see Equation (7.7), Section 7.3.1)

H (see Equation (7.14), Section 7.3.3)

These formulae, derived using the rectangular pdf, consider the probability of a particular floor having a specific attraction.

Consider the Poisson pdf. This is a pdf where there is an exponential relationship defining the times of passenger arrivals. The following formulae have been derived, and are shown in Table 13.2 for reference.

Table 13.1 Formulae for S and H for rectangular pdf

Demand	Probable number of stops (S)	Highest reversal floor (H)
Equal	$S = N\left[1 - \left[\dfrac{N-1}{N}\right]^P\right]$	$H = N - \displaystyle\sum_{i=1}^{N-1}\left[\dfrac{i}{N}\right]^P$
Unequal	$S = N\left[1 - \dfrac{1}{N}\displaystyle\sum_{i=1}^{N}\left[1 - \dfrac{U_i}{U}\right]^P\right]$	$H = N - \displaystyle\sum_{j=1}^{N-1}\left[\displaystyle\sum_{i=1}^{j}\dfrac{U_i}{U}\right]^P$

Table 13.2 Formulae for *S* and *H* for Poisson pdf

Demand	Probable number of stops (S)	Highest reversal floor (H)
Equal	$$S = N\left(1 - e^{-\frac{\lambda INT}{N}}\right)$$	$$H = N - \sum_{i=1}^{N}\left(e^{\frac{-\lambda INT}{N}}\right)^i$$
Unequal	$$S = N - \sum_{i=1}^{N} e^{-\lambda INT\frac{U_i}{U}}$$	$$H = N - \sum_{j=1}^{N}\prod_{i=N-j+1}^{N} e^{-\lambda INT\frac{U_i}{U}}$$

For equal demand: *S* (see Equation (5.17), Section 5.13.2)

H (see Equation (5.18), Section 5.13.2)

For unequal demand: *S* (see Equation (7.18), Section 7.7.3)

H (see Equation (7.19), Section 7.7.3)

These formulae, derived using the Poisson pdf, consider a time-related exponential distribution.

Notice that Equations (5.17) and (5.18) are not presented exactly as shown in Section 5.13.2. The identifying subscript p is omitted (trivial). More importantly, the term λT has been replaced by λINT as the time interval T used by Tregenza (1972) and Alexandris (1976) can be related to the lift system interval INT. This allows another substitution as λINT is P, as defined by Equation (5.19). This brings all the equations into the same form, and some further mathematical approximations can be carried out (not shown here) to show their close equivalences.

13.2   Low call express floor

When evaluating the round trip time Equation (4.11), various parameters are required such as *H, S* and *P*. The parameter *H* is the high call reversal (HCR) floor, which the lift reaches on an average journey during a round trip. It is important to derive a value for *H* accurately (Section 5.4) and not assume it to be *N* for a low to mid rise building zone or *N*–1 for a tall building zone as some designers do. A theoretician will then question:

'Where is the first stopping floor that the lift serves during an uppeak trip?'

It might be assumed to be the floor next to the main terminal. But it may not be. The reason for the question is that Section 7.4 indicated that a lift may not reach its rated speed after a single floor run, particularly for high speed lifts. Thus, an error in the round trip calculation may arise. The question has no great relevance where the served zone is not immediately adjacent to the main terminal, as the time to travel the express zone will be large. But what is the magnitude of the error?

Kavounas (1992a, 1993a, 1993b) examined this question, terming the first stopping floor the 'Lowest Call Express (LCE)' floor. The first paper briefly reported his findings and the latter two give the theory more fully. The problem Kavounas faced is illustrated in Figure 13.1. Here, LCE is given by the symbol h. For equal floor populations, he first 'rephrases' Equation (5.14) as:

$$H = N - X \tag{5.12bis}$$

and then says:

$$h = 1 + X \tag{13.1}$$

By following the principles applied in Section 5.8, and after a great deal of mathematics, Kavounas shows that X is:

$$X = \sum_{i=1}^{N-1} \left[\frac{i}{N} \right]^P \tag{13.2}$$

Kavounas comments that the term '1' represents the lowest possible floor that can be served, i.e., the first one above the main terminal, and considers that the term × represents the reciprocal nature of the problem. For unequal populations, Kavounas derives a similar expression:

$$h = 1 + Y \tag{13.3}$$

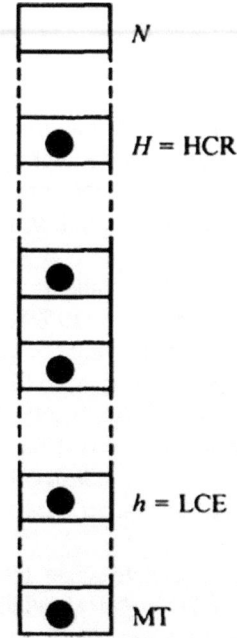

Figure 13.1 Illustration of Lowest Call Express floor

where:

$$Y = \sum_{j=1}^{N-1} \left[\sum_{i=j+1}^{N} \frac{U_i}{U} \right]^P \tag{13.4}$$

It should be noted that the inner summation range of the equation for Y differs from the corresponding part of Equation (7.14) given below. This is due to the inner summation for H being concerned with the influence of the populations of the lower floors, whereas the inner summation for h is concerned with the influence of the populations of the upper floors. The numerical difference is small.

$$H = N - \sum_{j=1}^{N-1} \left[\sum_{i=1}^{j} \frac{U_i}{U} \right]^P \tag{7.14}$$

The result indicates that the first stopping floor is as far from the main terminal as the high reversal floor is from the last floor in a building zone. This gives symmetry to the car call pattern, which is of little consequence, except for small lifts serving many floors. For example, a 10-person lift serving 22 floors would have a stopping range from floor 2 to floor 20 (see Table 5.11).

Kavounas (1993b) concludes:

> *Although the LCE consideration always affects some of the components of the RTT summation, it is often in ways that neutralise each other. Therefore, the RTT may often turn out to be virtually unaffected by the LCE consideration. This is why the LCE principle was originally termed (Kavounas 1992a) a third order consideration.*

The theoretician should now be satisfied.

13.3 The 80% car loading factor

Conventional traffic design calculations assume that lifts only load, on average, to 80% of their probable capacity (Definition 5.1). Note that the actual capacity is not the simplistic rated load (kg) divided by 75 (see footnotes to Table 5.8), which ensures safe operation, but the available space divided by 0.21 m² (see column 5, Table 5.8). But why has 80% been chosen (see column 6, Table 5.8)?

Some workers (Forwood and Gero, 1971; Tregenza, 1972; Zimmermann, 1973) say it is normal practice based on experience. Strakosch (1967)[A] states that passengers left to themselves '*will seldom fill an elevator to more than that (80%) during uppeak*' and in later editions (1983; 1998) offers the advice: '*Actual loading is less than the weight allowed since people will not crowd that close together*'. Loading a service facility to 80% is a well-known classical statistical technique used in bulk-queue multi-server facilities when a good service is desired. It could be possible that someone in the lift industry, long ago, with statistical experience, whose name is lost to posterity, suggested that 80% be used. However, can the 80% figure be proved?

Barney and Dos Santos (1977) set out to investigate this assumption by computer simulation. Figure 13.2 was plotted, as the result of over 400 simulations for groups with from four to six lifts. Here, the vertical axis indicates the uppeak performance figure as *AWT/INT* and the horizontal axis shows the corresponding percentage car load. The value for the passenger average waiting time (*AWT*) was obtained from the simulation. The value for the interval (*INT*) and the percentage car load was calculated using the procedure described in Section 6.11. This procedure matches the actual arrival rate to the lift system handling capacity. It thus provides values for the interval and the percentage car load when this match is made. The relationship for the uppeak performance figure *AWT/INT* is chosen in order to normalise the results for all lift systems regardless of their size, floors served, dynamics, etc. Each point plotted (■) represents one simulation.

Some interesting facts emerge at once from Figure 13.2. First, there is a spread of points, due to the randomness of the passenger arrivals and destinations. For example, some simulations approach the 100% performance figure at only a 60% value of car load. Second, there is a definite pattern. A weighted average performance figure line can be superimposed on the graph, as shown. It indicates the increase of *AWT/INT* with an increase of car loading. A large number of the simulations present a performance figure smaller than 50% at low car loadings. As the car loads increase, this excellent performance gradually vanishes and, at a 60% percentage load, the average performance figure is approximately 50%.

For the car loads above 70%, there are some runs showing a performance figure in excess of 100%. The number of such runs for percentage loads up to 80% is small and probably reflects the effect of car bunching (see Section 13.5). The average performance is still good in this range. The number of poor runs increases for loads above 80%, and for 90% loads, the average performance figure exceeds 100%. Above 90% percentage loads, the situation deteriorates very rapidly, and only a few runs give an acceptable performance. Such runs will correspond to the lifts running with a constant headway and large queues building up. Thus, the uppeak lift system performance depends on the car loading, and for heavy percentage loads, say above 90%, average waiting times in excess of an interval time must be expected. Figure 13.2 is presented in an idealised form as Figure 6.1, where the spread of values is indicated by dashed lines.

Computer simulation is an empirical modelling method; another method of analysis is by mathematical modelling. A notable feature of Figure 13.2 is a 'knee' at a system utilisation of 80% (as characterised by percentage car load), above which the performance deteriorates rapidly, leading eventually to intolerable queuing situations. The form of this curve is by no means unique and, as stated above, occurs with bulk-queue multi-server facilities. This leads to the proposition that a theoretical derivation of Figure 13.2 might be possible.

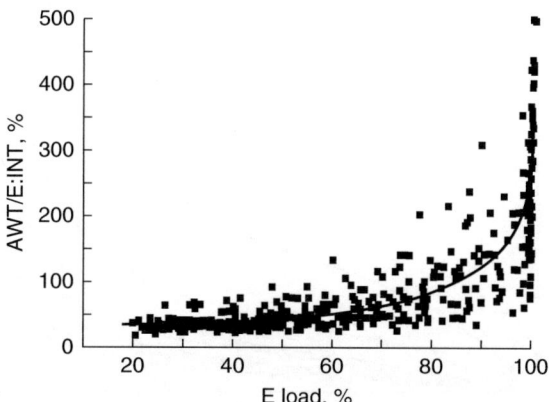

Figure 13.2 Uppeak performance from simulations

Alexandris *et al.* (1979a) considered a multi-car lift system as a bulk service queuing problem. To perform the mathematical analysis, certain model assumptions were made:

(1) Only the uppeak traffic pattern is to be considered.
(2) Lifts do not leave without passengers.
(3) Lifts return to the main terminal, even when there are no calls.
(4) There is no limit to the length of the queue of passengers waiting for service.
(5) Passenger arrivals obey the Poisson process.
(6) There are no priorities, passengers use whichever lift becomes free.
(7) The queue service discipline is first-in, first-out.
(8) Service is by batches of size no greater than the rated capacity of the lifts.
(9) The service time for each batch is exponentially distributed.

All of these assumptions are reasonable as most proprietary controllers operate on the basis of assumptions (2), (3), (6) and (8), and passengers often behave according to assumptions (5) and (7). Complex mathematical processes (Alexandris, 1977) resulted in the curves that are shown in Figure 13.3 being produced. This figure shows a facility utilisation factor R (horizontal axis) plotted against mean passenger waiting time (vertical axis). The number of lifts in the groups ranged from one to eight. The mean passenger waiting time was normalised by dividing it by the lift system interval time. The graphs are similar in shape to Figure 13.2.

The simulation runs from which Figure 13.2 was constructed assumed groups of four to six lifts. So, by eliminating those curves for lifts ranging from one to three in Figure 13.3 and averaging the rest, Figure 13.4 can be drawn. The shape of this curve is very similar to that of Figure 6.1. Thus, justification for Figure 13.2, which was obtained empirically by simulation, is confirmed by mathematical analysis.

13.4 Traffic analysis – the inverse *S-P* method

Often, there is a need to know the number of passengers in a lift, for example when deciding whether the sizing is correct, or when an estimate of bunching is needed, or in order to estimate the likely passenger waiting time. Several methods exist to determine the number of passengers in a lift, including: load weighing, photocell signals, sensitive pads, imaging systems and observers. Al-Sharif (1992a, 1992b) suggests another method.

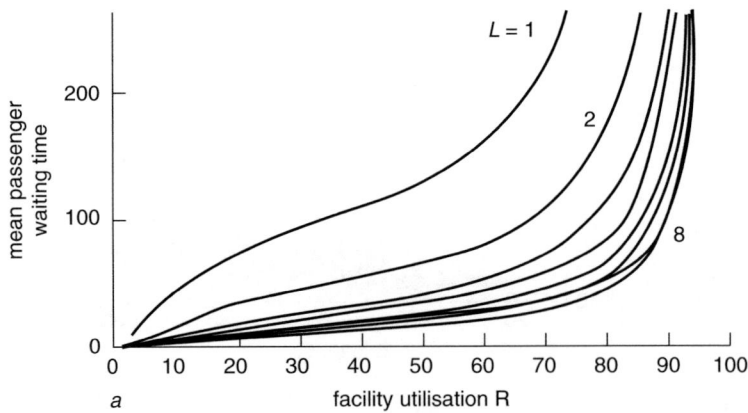

Figure 13.3 Mathematical model of uppeak performance

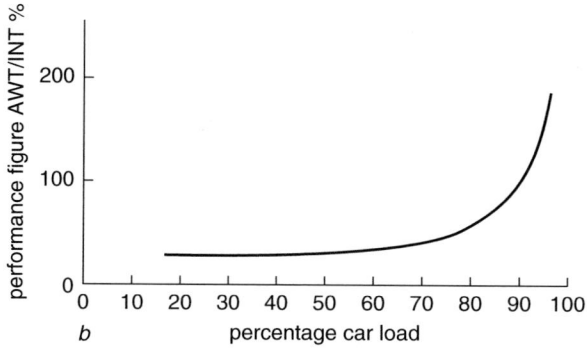

Figure 13.4 Simplification of Figure 12.3

In the design of lift systems, the traditional design method is to derive the round trip time from various parameters. One of these parameters is the number of stops (S), which for equal floor populations is given by:

$$S = N\left[1 - \left[\frac{N-1}{N}\right]^P\right] \tag{5.5}$$

Equation (5.5) can be inverted to produce a formula for P:

$$P = \frac{\ln\left[\dfrac{N-S}{N}\right]}{\ln\left[\dfrac{N-1}{N}\right]} \tag{13.5}$$

Al-Sharif validates the method thoroughly in his paper, and shows it has a close correspondence to real systems. Thus, a very simple method of 'load weighing' is obtained, which could be easily incorporated into traffic supervisors, enabling them to switch algorithms to meet traffic demands. It is interesting to note that Basset Jones (1923) used this formula for another purpose in order to determine the variance of S from the expected value $E(S)$.

13.5   Bunching

Uppeak calculations are based on average values for the various parameters. Deviations from these average values in the actual lift installation will cause deviations from the calculated performance. One phenomenon that prevents optimum performance is 'bunching'. Ideally, lifts arrive to transport passengers at the main terminal with a separation time equal to an average interval. In practice, this does not happen, and lifts arrive with an irregular interval. In an extreme case, all lifts in a group could arrive and leave simultaneously, like a huge single lift with a capacity equal to the sum of their individual capacities.

Early traffic controllers dispatched lifts from terminal floors with a fixed headway. This was satisfactory, provided that the lifts returned before the headway time had expired. A disadvantage was that the introduction of a headway larger than the underlying average interval reduced the underlying handling capacity. The technique of using a lobby dwell time during intense uppeak traffic works well as it ensures lifts leave the main terminal substantially filled to full capacity. There is no degradation in handling capacity, as long as the lobby dwell time is shorter than the time to load a lift. Modern hall call allocation systems, where passengers register their destinations and are then allocated a car, will not, by definition, suffer from bunching.

Al-Sharif (1993) illustrates bunching by considering two lifts with a round trip time of 50 s serving a passenger arrival rate of one person per second. If the lifts arrive uniformly in time, i.e., every 25 s, then the average waiting time might, simplistically, be half the interval, i.e., 12.5 s, and 25 persons will leave in each lift. If the first lift now arrives at 10 s and the second at 50 s, then the first lift transports 10 people ($AWT = 5$ s) and the second transports 40 people ($AWT = 20$ s), assuming it is big enough. The total average waiting time will be the weighted average, i.e.:

$$AWT = (10/50 \times 5) + (40/50 \times 20) = 17 \text{ s}$$

which is longer than the theoretical 12.5 s.

Suppose now, that the lifts do not have infinite capacity, but can only accommodate 25 persons, then 15 persons will be left to wait another 10 s for the next lift. They will be joined by a further ten persons ($AWT = 5$ s). The weighted average AWT will then be:

$$AWT = (10/60 \times 5) + (25/60 \times 20) + (15/60 \times (20 + 10)) + (10/60 \times 5) = 17.5 \text{ s}$$

Thus bunching does not matter too much if all waiting passengers can board the lifts to reach their destinations. The average interval may still be as designed, but the average passenger waiting time will increase. Provided all waiting passengers are served, the handling capacity of the lift installation is not affected. However, if passengers are left behind, then the handling capacity has been compromised.

Schroeder (1990a) considers bunching to be one of the reasons for the rapid increase in passenger average waiting times, when cars load to greater than 50% of their capacity. He suggests estimating bunching by forming the ratio: actual passenger average waiting time divided by the average waiting time that would be obtained if the linear relationship $AWT = INT/2$ is applied, i.e., $2AWT/INT$. This ratio is unity for low car loadings (<50%) and a large number as car loadings increase to 100%. It would be better to invert the ratio, i.e., $0.5INT/AWT$, so that unity represents no bunching and zero represents total bunching. This method requires the acquisition of values for interval and passenger average waiting time.

Figure 6.1 illustrates uppeak performance with the value of INT on the vertical axis being the actual value the lift provides in response to each level of car loading. Table 6.4 tabulates Figure 6.1. Considering all car loadings below 50% to give a value of 0.5, Table 6.4 can be converted to give the inverted Schroeder table, see Table 13.3.

Table 13.3 indicates that, at the usually assumed car loading of 80%, the bunching factor is 0.6. This would appear to be the limit of passenger tolerance under the Schroeder method. Schroeder suggests that as the number of lifts in a group increases, then the bunching ratio will worsen. However, the irregularities will be less noticeable as, generally, the interval will be smaller. A single lift can also be considered to have a bunching factor simply manifested as an irregular round trip time.

Al-Sharif (1993) proposed a formula based measure of bunching. If the time difference between lift number i departing and lift number $i + 1$, is defined as $t_{i, i+1}$, then the difference between this time and the ideal time can be taken as a measure of how much bunching exists. The ideal time is the average interval. Thus, the time difference is:

$$(t_{i,i+1} - RTT/L) \tag{13.6}$$

Table 13.3 Bunching factor (after Schroeder, 1990a)

Car loading	50%	60%	70%	80%	85%	90%	95%
AWT/INT (Table 6.4)	–	0.50	0.65	0.85	1.01	1.30	1.65
Bunching ratio	1.0	1.0	0.8	0.6	0.5	0.4	0.3

All figures are rounded

This time difference can be either positive or negative, as an early arrival of a lift is as bad as a late arrival. To penalise large deviations from the interval, this time difference formula was squared, a technique which also takes care of the positive and negative values. Using this method, a value of zero represents no bunching and unity, total bunching. Because of the reversed valuation, when applying the Schroeder criteria in the method above, a tolerable acceptable value may be 0.4. This method relies on obtaining the times of each lift departure. The full formula is not given here.

Kavounas (1992b) asked if bunching is (a) a self-correcting situation, i.e., the non-bunched state is one of lesser energy, or (b) a self-aggravating situation (like wildfire), which needs early detection and correction, or (c) neither. The foregoing suggests it is (b).

13.6 Conclusions on uppeak

Uppeak traffic design depends on how many passengers (*P*) a lift can transport when it is serving the uppeak traffic condition. The procedure is the same regardless of whether the design is for a new lift or an existing lift. The design procedure is to match the number of arriving passengers with the handling capacity of the installed lift system. This chapter has indicated some factors the designer may like to consider.

Note (shown in text as[1])

1 1st edition, page 64; 2nd edition, page 74; 3rd edition, page 84.

14 Advanced analysis of down peak

14.1 Early work on down peak analysis and concepts

Two authors give some indication of the down peak traffic intensities (quantity criterion), which need to be served. Strakosch in 1967[1] established a requirement that a lift system should be able to evacuate the population of a building in 15–30 minutes, which he later refined in 1983[2] to 25–40 minutes as an update to modern practice. He says that for office buildings, the 5-minute down peak of traffic '*may exceed any other traffic peak by 40–50%*'. Zimmermann (1973) says that, from his experience, the evening 'crush' lasts 7–9 minutes in most buildings, moving 40–50% of the population. He also suggests a peak demand of 25% of the population for 5 minutes.

What are the main characteristics of the down peak traffic condition? Lift cars discharge passengers at the main terminal floor, travel back up the building to the floors above, fill up at a number of stops and express back down to the main terminal. This is almost the reverse of the uppeak traffic pattern, but there are exceptions. Where there is a suitable traffic control system ensuring a reasonably equal service to all floors, the average reversal (from up direction to down direction) occurs at a lower floor than the uppeak high call reversal floor. See Figure 4.8 for an illustration of this effect. So (H_D) is smaller. The number of stops is observed to be smaller than during uppeak as cars fill at three, four or five floors. This is owing to the intensity of the passenger demand at 'going home time', even in flexitime workplaces, which fills the lifts at a smaller number of floors. Hence, a much lower number of stops (S_D) than during uppeak. There is also a tendency for the cars to fill nearer to the probable car capacity (PC, Table 5.8), whenever there is still space available (the 'no touch' syndrome apparently being abandoned at 'going home time'). This could lead to more efficient loading and unloading of passengers and shorter passenger transfer times (t_p). So there will be higher car loads, which are more quickly transferred. Thus, the round trip time will be smaller than for other traffic patterns and, consequently, the down peak handling capacity will be inherently greater than for uppeak, but by how much?

If the expected number of stops and the expected highest reversal floor were known, and if cars are assumed to fill to maximum capacity, then the down peak round trip time could be calculated using a modified version of Equation (4.8), *viz*:

$$RTT_D = 2H_D t_v + (S_D + 1)t_s + 2PCt_p \qquad (14.1)$$

Both Strakosch and Zimmermann describe empirical rules for down peak round trip time evaluations. Strakosch (1967) presents a method where he considers that the lift cars nearly always reach the top floor, as 'a conservative measure', and the number of stops may be determined from the knowledge of the population of each floor and the nature of that population, for example, whether the cars could fill at a single floor. If the exact nature of the occupancy is not known, then he estimates that the down peak expected number of stops per trip (S_D) is approximately 75% of the probable number of stops in the uppeak situation (S).

Zimmermann (1973) uses 75% of the number of floors (N) serviced above the main floor as the highest reversal floor (H_D), and he assumes that the expected number of stops is equal to the ratio of the number of floors and the number of cars (L). Table 14.1 summarises these ideas of down peak evaluation.

Example 14.1

Consider a system of four, 16-person lifts serving 16 floors above the main terminal floor with lift dynamics of $t_v = 1.6$ s, $t_s = 8.4$s, ($T = 10$ s), $t_p = 1.0$ s. The building population is 647 persons.

Using the ideas of Table 14.1 and Equation (14.1), it is possible to evaluate the down peak interval of cars at the main terminal floor, the down peak handling capacity, and the down peak to uppeak handling capacity ratio. Table 14.2 summarises the results.

Table 14.2 indicates a wide disparity between the two authors. They do, however, indicate that the down peak handling capacity is larger than the uppeak handling capacity.

There is another important aspect of lift behaviour during a down peak traffic situation that must be considered, which is the effect of the traffic control system. If the high call reversal technique is applied, then a lift reaching the top of the building may fill to capacity in a small number of stops, owing to the heavy down peak traffic demands. It will then travel to the main terminal floor, bypassing a number of landing calls at the lowest floors. Thus, if each car, after discharging the passengers at the main terminal, is allowed to answer landing calls registered from the top of the building, it is possible that the lowest floors will never obtain service. To avoid this possibility, the supervisory control algorithm applying to the down peak traffic

Table 14.1 Evaluation of down peak parameters

Parameter	Strakosch	Zimmermann
Expected number of stops S_D	$S_D = 0.755$	$S_D = NIL$
Expected highest reversal floor H_D	$H_D = N$	$H_D = 0.75N$

These ideas are applied in Example 14.1

Table 14.2 Illustrative example

Parameter	Uppeak	Strakosch	Zimmermann
P	12.8	16	16
H	15.3	16	12
S	9.0	6.75	4.0
INT	39.6	37.1	28.1
HC	97	129	171
DNPHC/UPPHC ratio	n/a	1.3	1.8

All figures are rounded

pattern must allocate the cars in some 'round robin' fashion to guarantee a balanced service to all floors in the building. This is best achieved by dividing the building into sectors (Definition 10.5). All modern group supervisory control systems provide such a facility.

The evaluations of Table 14.2 only consider the quantity of service. Bearing in mind the effect of the traffic control system cycling the cars to groups of floors, it is now possible to consider the quality of service. Strakosch (1967)[3] states that:

Service should be available at every floor at intervals no longer than 60 s.

This statement emphasises a very common confusion between the interval of cars at the main terminal floor and the service interval at a particular floor. Indeed, if the lift cars fill up at a few floors, they cannot call at all floors during each trip, and the average time interval between successive services at a particular floor (*FINT*) is longer. One way to estimate this floor interval is:

$$FINT = N/S_D \times DNPINT \tag{14.2}$$

Considering the lift system given in Table 14.2, the frequency of service at a particular floor, according to Strakosch's postulation is:

$$FINT = 16/6.75 \times 37.1 = 88 \text{ s}$$

and from Zimmermann's postulation gives:

$$FINT = 16/4 \times 28.1 = 112 \text{ s}$$

Another way to calculate the floor interval is to divide the building into the same number of sectors as there are lifts. This gives:

$$FINT = L \times DNPINT \tag{14.3}$$

Strakosch's values become:

$$FINT = 4 \times 37.1 = 148 \text{ s}$$

and Zimmermann's values become:

$$FINT = 4 \times 28.1 = 112 \text{ s}$$

All the floor service intervals above break Strakosch's rule of a service time of less than 60 s.

A conclusion can be drawn from Table 14.2 and the results for *FINT*. The values obtained from the two methods and the two authors are so different that they offer no degree of design confidence. Thus, what values should be used for the various parameters?

14.2   Definitions for down peak

Barney and Dos Santos (1977) simulated over 2,000 lift installations under down peak conditions covering a wide range of variables and different traffic control strategies. They defined a number of parameters.

14.2.1 Down peak demand

> **Definition 14.1:** down peak demand (α) is expressed as a percentage of the number of potential passengers (λ) arriving at a lift system and requiring service during a 5-minute peak period, with respect to a reference value (*UPPHC*).

The reference value is the uppeak handling capacity (*UPPHC*), which is calculated for the same lift system under consideration for a probable car loading of 80%.

Definition 14.1 is given as Equation (14.4):

$$\alpha = \frac{\lambda}{UPPHC} \times 100\% \qquad (14.4)$$

For a particular lift configuration, which is properly designed for uppeak (from a conventional point of view), the uppeak rate of passenger arrivals matches the uppeak handling capacity *UPPHC*. A reasonable range of values for the down peak demand placed on the lift system in terms of passenger arrival rates could be from a value just below *UPPHC*, say 80% of *UPPHC*, through 100%, where uppeak and down peak handling capacities are equal, to a value well above *UPPHC*. If the information on peak demand values during down peak provided by some authors (e.g., Strakosch, 1967; Zimmermann, 1973) is used, then it is necessary to consider demand levels exceeding *UPPHC* by 40–50%. This parameter, which acts as an independent variable, should therefore cover a range of values from 80% to over 150% of *UPPHC*.

14.4.2 Down peak number of stops

> **Definition 14.2:** the down peak percentage number of stops is the average number of stops per trip (*DNPSTPS*) during a down peak traffic situation, measured as a percentage of a reference value (*UPPSTPS*).

The reference value (*UPPSTPS*) is the uppeak expected number of stops per trip for 80% probable car loadings during uppeak.

Figure 14.1[4] shows the ratio of number of down peak stops to uppeak stops against passenger demand. The average curve (solid line) shows a ratio of 40% at a low demand, and at the more likely demand level of 150%, the ratio is 50%. Thus, the number of down peak stops is about half the uppeak stops. This is lower than Strakosch, and higher than Zimmermann indicated.

14.2.3 Down peak interval

> **Definition 14.3:** down peak percentage interval (*DNPINT*) is the interval of cars at the main terminal during a down peak traffic situation, measured as a percentage of a reference value (*UPPINT*).

The reference value (*UPPINT*) is the uppeak interval for cars with an 80% probable car loading.

Figure 14.2 shows the ratio of down peak interval to uppeak interval against passenger demand. The average curve (solid line) shows a value of about 66% at a demand of 150%.

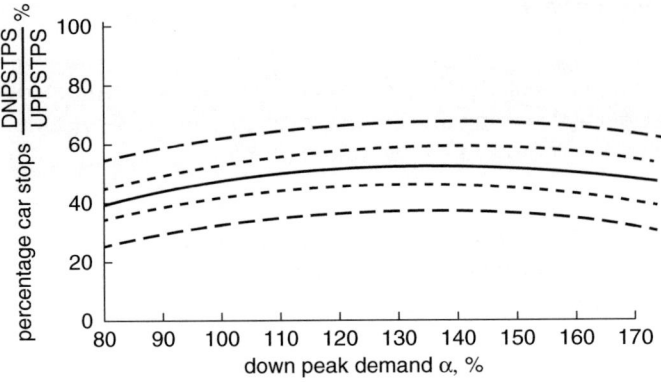

Figure 14.1 Down peak stops compared to uppeak stops

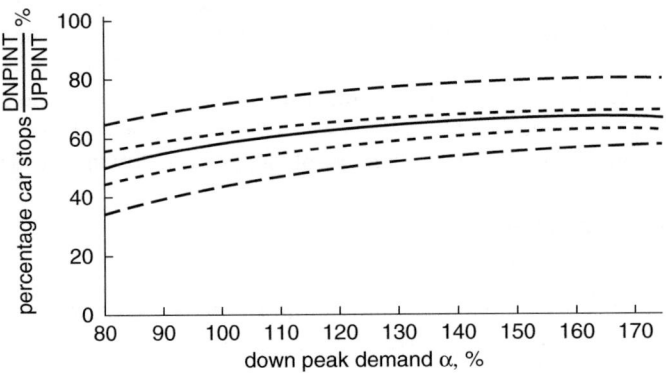

Figure 14.2 Down peak interval compared to uppeak interval

Thus, the down peak interval is about two-thirds of the uppeak interval. The conclusion is that the down peak handling capacity will exceed the uppeak handling capacity, in a similar ratio, i.e., by approximately 60%.

14.2.4   Down peak performance

Definition 14.4: the down peak performance figure is the measure of the quality of service provided by a lift system during a down peak traffic situation, expressed as a percentage of average down peak passenger waiting time (*AWT*) and of the reference value *UPPINT*.

Figure 14.3 shows the performance figure of average down peak passenger waiting time (*AWT*) normalised by dividing it by the uppeak interval (*UPPINT*).

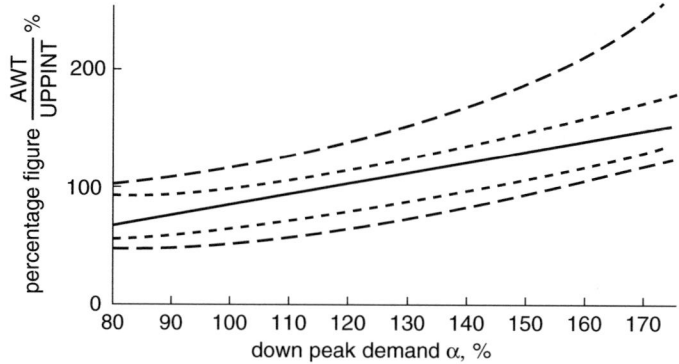

Figure 14.3 Down peak performance compared to uppeak performance

A straight line approximation can be drawn on Figure 14.3, and results in the equation:

$$DNPAWT = 0.85\alpha UPPINT \tag{14.5}$$

but from solving Equation (6.1): $UPPAWT = 0.85\ UPPINT$, and so:

$$DNPAWT = \alpha UPPAWT \tag{14.6}$$

If a down peak handling capacity of 150% of uppeak is assumed, then:

$$DNPAWT = 1.5 \times UPPAWT \tag{14.7}$$

Thus, although the down peak handling capacity is improved, for example in this case, by 1.5 times, the passenger average waiting time deteriorates by the same value, i.e., by becoming 1.5 times longer.

In terms of the uppeak interval and using Equation (14.5):

$$DNPAWT = 1.5 \times 0.85 \times UPPINT = 1.275 \times UPPINT$$

This means the typical down peak average passenger waiting time, in this case, is 27.5% longer than the calculated uppeak interval.

Example 14.2

Suppose a lift system has an uppeak handling capacity of 100 persons/5 minutes at an uppeak interval of 30 s. What will the average passenger waiting time be for a down peak arrival rate of 140 persons/5 minutes?

$$\alpha = 140/100 = 1.4.$$

Using Equation (14.5):

$$DNPAWT = 1.4 \times 0.85 \times 30 = 35.7\ s.$$

14.3   Evaluating equation (14.1)

14.3.1   *Average number of stops (S_D)*

What is the value for the down peak number of stops S_D?

Schroeder (1984), after making some assumptions, formed a mathematical expression for the optimum number of stops for a range of lifts and rated capacities. He concluded '*the optimum number of stops is rather low*', but did not quantify 'low'. The value for S_D, which has been found by simulation in Figure 14.1 as 0.5S, concurs with this statement.

14.3.2   *Average highest reversal floor (H_D)*

What is the value for the down peak highest reversal floor H_D?

Schroeder (1984) suggests H_D might be $0.5N + 0.5S_D$. However, the highest reversal floor will depend largely on the traffic controller. If it operates in such a way as to cycle cars around the building into 'sectors', in order to ensure an even service, then H_D will be somewhere near to the central floor of the served zone. It could be 0.5N. However, as the traffic controller will require the lifts to reverse direction at the highest call in each sector, H_D may well be higher, say $0.5N + 0.5S_m$, where S_m is the number of floors in a service sector. If a lift has not filled to capacity in its assigned sector, a good traffic controller will permit the lift to stop for calls registered in the next sector encountered. This will be particularly true where sectors contain a small number of floors. Thus, there will be a tendency for the highest floors in a sector to have already been serviced, except, of course, for the highest sector. This may well cause the value for H_D to be lower. Perhaps a best estimate for H_D is 0.5N.

14.3.3   *The average number of passengers*

What is the value for the down peak number of passengers P_D?

If lifts tend to fill nearer to their actual capacity during down peak, then the transfers will be more efficient. Although the value of P will increase, the value for t_p will decrease. Thus, the value for $P_D t_p$ could remain constant at Pt_p, i.e., the same as during uppeak.

14.3.4   *The down peak round trip equation*

Equation (14.1) can now be written as:

$$RTT_D = 2 \times 0.5Nt_v + (0.5S + 1)t_s + 2Pt_p \tag{14.8}$$

$$RTT_D = Nt_v + (0.5S + 1)t_s + 2Pt_p \tag{14.9}$$

The down peak interval (*DNPINT*) can be obtained by dividing RTT_D by the number of lifts. The down peak handling capacity (*DNPHC*) will then be:

$$DNPHC = 300P/DNPINT \tag{14.10}$$

Reconsider Example 14.1. Using Equation (14.9), an extra column can now be added to Table 14.2 to give Table 14.3.

Examination of Table 14.3 shows that the interval is 24.3 s, the handling capacity is 157.7 persons/5 minutes and the DNP/UPP ratio is 1.63. This is approximately half way between Strakosch (pessimistic) and Zimmermann (optimistic), and indicates the down peak handling capacity is some 63% larger than during uppeak.

Is the improvement in down peak handling capacity consistent for a different number of floors and rated car capacities? Applying Equation (14.8) to the 'four corners of the lift world and its centre of gravity' (see Section 5.12.3) results in Table 14.4.

Table 14.3 confirms the hypothesis that an installation during down peak traffic inherently has at least 60% more capacity than in uppeak. As a rule of thumb, traffic designers can assume the ratio *DNPHCIUPPHC* is 1.6.

14.3.5   *Quality of service*

How will the quality of service change? What will the floor interval be, and what will the passenger average waiting time be?

Again, considering the installation in Example 14.1, assume with the four lifts that there are four sectors of four floors. If a sector is served once every four (main terminal) intervals, then the floor interval from Equation (14.4) is:

$$L \times DNPINT = 4 \times 24.3 = 97.2 \text{ s}$$

Using Equation (14.5), the down peak passenger average waiting time will be:

$$0.85\alpha UPPINT = 0.85 \times 1.63 \times 39.6 = 54.9 \text{ s}$$

Table 14.3 Illustrative example – Table 14.2 extended

Parameter	Uppeak	Strakosch	Zimmermann	Barney et al.
P	12.8	16	16	12.8
H	15.3	16	12	8
S	9.0	6.75	4.0	4.5
INT	39.6	37.1	28.1	24.3
HC	97	129	171	157.7
DNPHC/UPPHC ratio	n/a	1.33	1.76	1.63

All figures are rounded

Table 14.4 Comparison of down peak to uppeak handling capacity

Number of floors	Rated car capacity	DNPHC/UPPHC ratio
10	10	1.60
10	24	1.54
16	16	1.62
20	10	1.62
20	24	1.61

This is much better than the floor interval would suggest, because it is an average value, whereas the floor interval is a maximum value. In addition, because the traffic controller can be programmed to pick up landings calls ahead, while the lift still has available capacity, some high floors in a sector will have already been served before a lift is assigned to the sector. If Strakosch's rule of service within 60 s is applied to passenger waiting time and not floor interval, his criterion is met.

To achieve this quality of service requires the careful setting up of the load weighing system and the number of floors in a sector. If the load weighing is set at say 60% of actual car capacity (see Table 7.2), then a lift with just over 60% loading will not stop for further landing calls ahead. As it is considered that passengers will squash closer together when going home, the load weighing could be set to about 75% during down peak. This may mean that some lifts may become loaded above the conventional 80% value on occasions.

It is important that the number of floors in each sector is large enough for cars to fill in a sector. It is also important to ensure that the floor interval is not too long. The design range for floors per sector is probably from three to five floors.

Consider Example 14.1 again. The down peak handling capacity is:

158 persons/5 minutes.

This is an arrival rate for all floors of:

158/300 = 0.53 persons/second.

At a single floor this is an arrival rate of:

0.53/16 = 0.033 persons/second.

If the floor interval is 97.2 s, then, as each lift arrives, the number of passengers waiting, on average, will be:

0.033 × 97.2 = 3.2 passengers.

To fill the car will require only four stops. For this system, dividing the building into four sectors of four floors is ideal. If there were more cars available, six say, and four sectors were retained, then the floor interval would improve by 33% to 64.8 s, and the passenger average waiting time would similarly improve.

14.3.6   *Estimating the down peak handling capacity and average passenger waiting time*

The traffic design procedure should always satisfy the uppeak requirements. Equation (14.9) should enable a lift designer to estimate the down peak round trip time (and hence the handling capacity) in relation to an uppeak design.

For example, using a *DNPHC/UPPHC* ratio of 1.6, a 12% system may have a 5-minute down peak handling capacity of 19.2% and a 17% system may have a down peak handling capacity of 27.2%. Over a 10-minute outgoing traffic period, this represents from 38% to 54% of the zone population leaving the building. These figures relate well to the experience expressed by Strakosch and Zimmermann given in Section 14.1 of this chapter.

Using Equation (14.5), the designer can also estimate the down peak average passenger waiting time.

Example 14.3

Consider Example 14.1, which has an uppeak handling capacity of 15%, i.e., 97 persons/5 minutes, serving a building population of 647 persons, with an uppeak interval of 40 s. What is the down peak passenger average waiting time, the down peak handling capacity in persons/5 minutes, and the percentage of the building population that can leave in 10 minutes?

Assuming the cars are loaded to 80% probable capacity, the uppeak passenger average waiting time (see Table 6.4) is:

$$UPPAWT = 0.85 \times 40 = 34 \text{ s.}$$

From Table 14.5, the *DNPHC/UPPHC* ratio is 1.63, and using Equation (14.5), the down peak average passenger time is:

$$DNPAWT = 1.63 \times 34 = 55 \text{ s.}$$

The uppeak handling capacity is 97 persons/5 minutes.
The down peak handling capacity is $97 \times 1.63 = 158$ persons/5 minutes.
In 10 minutes, 316 persons can leave, which is 49% of the population.

14.4   Down peak conclusions

A detailed examination of the results of a large number of simulations allowed two objectives of the down peak analysis to be fulfilled, namely, the evaluation of the quantity of service provided by a lift configuration under down peak traffic conditions, and an estimation of the corresponding quality of service. The examination, like that for the uppeak examination, has been for a pure down peak traffic demand with no demand elsewhere.

The average interval of cars at the main terminal is slightly below two-thirds of *UPPINT*. Thus, the down peak handling capacity exceeds the uppeak handling capacity by over 60%. If it could be assumed that the lift cars can be loaded to 100% of probable car capacity during down peak, then a down peak handling capacity as high as 200% of *UPPHC* could be obtained.

However, the interval of cars at the main terminal does not mean that a particular floor is served at the same frequency since each car only serves a few floors on each trip. The simulation analysis shows that the down peak lift performance is heavily dependent on the control policy in this case. The conclusion is that, for most lift systems, in order to make full use of the down peak handling capacity, it is necessary to allow for higher average passenger waiting times than those encountered during uppeak traffic. These waiting times are generally longer by the same ratio as the handling capacity is improved. That is, if the down peak handling capacity is 1.6 times larger than the uppeak handling capacity, then the passenger average waiting time during down peak will also be 1.6 times longer than those endured during uppeak.

It may happen that the designer does not know the lift supervisory control system that will be installed and how the floors are served. Consequently, there is a difficulty in deciding what the performance might be. It is interesting to note that the best down peak algorithm is similar to the uppeak sectoring 'booster' algorithm (Section 8.5). In down peak, the landing calls are collected in a sector, and the passengers are taken to the main terminal, i.e., the opposite action to that of the uppeak operation.

It should also be noted that the figures presented are average figures, and that particular configurations will deviate from the average, as illustrated in Figures 14.1, 14.2 and 14.3. In addition, the situations considered are for equal floor demands. Thus, care must be taken in the interpretation of any evaluations where this is not so. Simulations will be necessary in complex situations.

Down peak traffic handling is always dependent on the underlying uppeak handling capability.

All the down peak equations discussed can be easily programmed into a spreadsheet.

14.5   A warning

The methods used to improve or boost the uppeak handling capacity do little to improve down peak handling capacity. This is still dependent on the underlying installation performance.

To understand this dependence, consider Example 14.1 again. The underlying handling capacity is 97 persons/5 minutes, the original building population is 647 persons, and the percentage population handled is 15%. As Example 14.3 shows, in down peak this becomes 24.5%. Now, suppose that the population increases to 1,000 persons. The original system now has an uppeak percentage population of 9.7% (97/1,000). An uppeak booster is applied, which brings the uppeak handling capacity back to 15%. However, the down peak handling capacity will still only be 1.63 times the underlying uppeak handling capacity, i.e., 15.8%. Under down peak conditions, only 31.6% of the increased building population will still be able to leave over 10 minutes, i.e., 31.6% of the population.

Thus, the uppeak handling capacity can be improved, but the down peak handling capacity remains unchanged. If a system were to be designed with uppeak boosters, to meet the uppeak design criteria, the down peak criteria will not be met. Siikonen (2000) suggests that if uppeak boosters are employed, that the uppeak criterion for handling capacity should be increased by 20–30%. This only goes part of the way. In the example above, the uppeak handling capacity criterion would need to increase by 50%.

Notes (shown in text as[1])

1 Page 102.
2 Page 112.
3 Page 103.
4 Figures 14.1, 14.2 and 14.3 are a summary of the detailed results shown in Barney and Dos Santos (1977). The curves show the overall probable limit of values (long dashed line) and the possible range of controller influence (short dashed line).

15 Advanced analysis of interfloor traffic

15.1 Balanced interfloor demand

A characteristic of balanced interfloor traffic is that there are larger demands made on the lift system in terms of the number of stops made by each lift. Whereas during uppeak and down peak traffic, there are a number of passengers boarding or alighting together, during balanced interfloor traffic, there is a stronger likelihood that each individual passenger requires one stop to be picked up and one stop to be set down. A good indicator of overall activity during interfloor traffic may well be the number of stops a lift is making.

With the exception of Tregenza's (1971, 1972, 1976) work, no other researchers have given mathematical consideration to the intensity of traffic flow. It is simply predicted that passenger average waiting times will deteriorate with an increase in the passenger demand.

Strakosch in his books defines three categories of demand: light, medium and heavy. Light traffic is where the number of passengers requesting service is no more than two to three times the number of lifts in service. Medium traffic is defined when lifts fill to less than 50% of their capacity and heavy demand is when lifts fill to over 50% of their capacity. Heavy demand is unlikely as cars will not fill to 50% of their capacity during balanced interfloor activity. Lift car loadings at this level would generally indicate a dominant flow of some sort. Bedford (1966), analysing a fixed sectoring proprietary supervisory control system similar to that described in Section 15.5, considers a system busy, if the number of stops made per car per minute is about 2.25.

15.2 Quality of service: performance figure

In the balanced interfloor traffic situation, the handling capacity of a lift system would be a less important parameter than for the uppeak and down peak traffic conditions, even if it had a meaning. This is because the number of passengers being transported is generally smaller. It is most unlikely that all the occupants of a building would want to use a lift during each hour of the working day. However, the lower demand is offset by a greater diversity in destination requirements. Thus, quality of service is more significant under interfloor traffic conditions as the lift system is expected to provide an excellent service in order to save occupants' time.

The quality of service may be measured in terms of passenger average waiting times, or in terms of passenger average journey times. From an employer's point of view, journey times are more important as they measure the total time spent by an employee obtaining lift service. From the user's point of view, waiting times are psychologically more important. It is considered that, once a passenger enters a lift car, they will not mind if the car takes a certain time to reach their destination floor, as long as that time is not too long. However, passengers will become impatient if they have to wait too long for service. This is particularly so at the main terminal floor, which may receive special treatment during interfloor traffic by, for example, parking a free lift there whenever possible.

15.3   Interfloor passenger demand

The lift system performance during interfloor traffic is certainly dependent on the rate of passenger arrivals. A traffic flow can be measured by the number of passengers requiring service per unit time and would represent different demand levels for different lift configurations. It would, however, lack normalisation. Thus, in order to provide normalisation and also to relate balanced traffic performance with the uppeak design, it is sensible to present the balanced interfloor passenger demand as a percentage of the uppeak handling capacity (*UPPHC*). In Chapter 13, down peak demand was related to *UPPHC*, and this enabled the down peak handling capacity to be directly compared to the underlying uppeak handling capacity. Following the same principle, a definition of passenger demand could be as follows.

> **Definition 15.1:** balanced interfloor 5-minute demand (β) is expressed as the number of passengers requiring service (λ) from the lift system during a 5-minute period, expressed as a percentage of a reference value, *UPPHC*.

The reference value (*UPPHC*) is the uppeak handling capacity of the lift configuration for a passenger loading to 80% of probable car capacity.

$$\beta = \frac{\lambda}{UPPHC} \times 100\% \tag{15.1}$$

Thus, each lift configuration should be considered under a variety of passenger arrival rates, from very low traffic intensities to demand levels that saturate the lift system. No precise indication could be found in the lift literature of when a demand level was considered as a light or a heavy demand load, except those by Strakosch indicated earlier in this chapter. Thus, a wide range of demand levels is used in the analysis, varying from 10% to 100%.

Again, as for the down peak study, the passenger average waiting time will be presented as a percentage of *UPPINT* in order to define normalised performance figures, which makes them less dependent on particular lift configurations, and also allows this traffic condition to be related to the uppeak design. A definition can now be provided for balanced interfloor performance.

> **Definition 15.2:** the balanced interfloor performance figure is the measure of the quality of service provided by a lift system under balanced interfloor traffic conditions, expressed as a percentage of the passenger average waiting time to a reference value, *UPPINT*. The reference value (*UPPINT*) is the uppeak interval for a passenger loading of 80% of probable car capacity.

As with the analysis of uppeak traffic (Chapters 6 and 13) and down peak traffic (Chapter 14), simulation techniques are best employed to study interfloor traffic. Barney and Dos Santos (1977) together with Lim (1983) conducted over 2,000 simulations covering a wide range of variables and different traffic strategies. The balanced interfloor traffic performance figures plotted in Figure 15.1 show performance deteriorating with demand. The performance is also dependent on the supervisory control policy over the whole range of demands. The performance figures for the different control algorithms at low demands are very similar, except for the adaptive call allocation (ACA) algorithm (11.6), which will be discussed later.

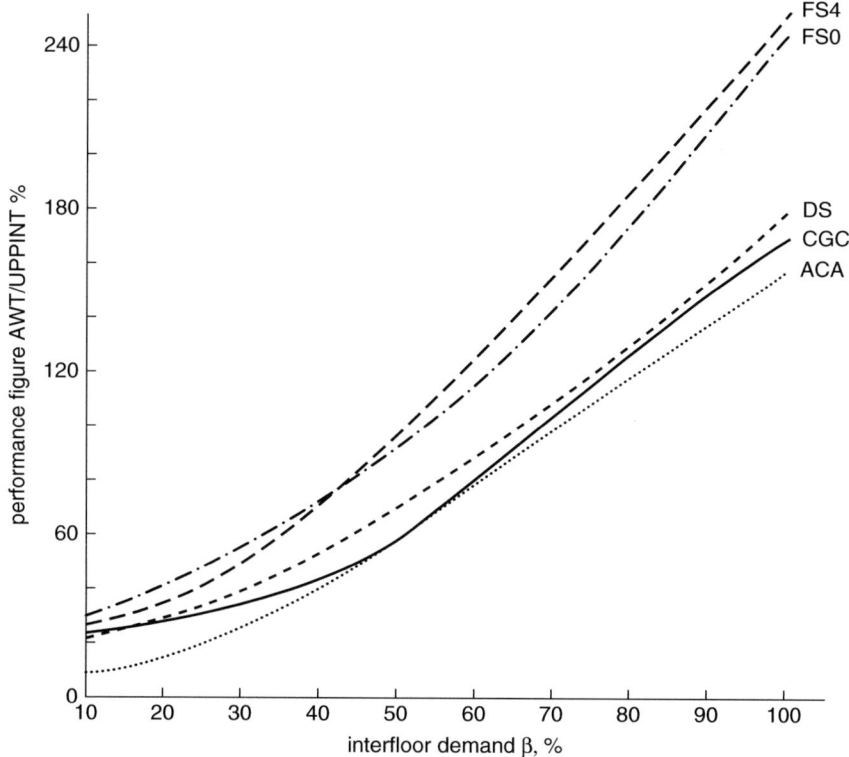

Figure 15.1 Performance -v- interfloor demand (by algorithm)

At the 10% very low demand level, they present average waiting times around 30% of *UPPINT*. With an increase in demand, the performance degrades at an increasing rate, and two groups of traffic control algorithms appear on Figure 15.1. The fixed sectoring algorithms FS4 (Section 10.7.3) and FSO (Section 10.7.2) perform badly compared to the dynamic sectoring (Section 10.7.4) and the computer based algorithms (Section 11.5 and 11.6). The similarity of performances for the FS4 and FSO algorithms shows that the allocation of cars on a priority timed basis presents no advantage over a policy of redistribution of free cars around the building, complemented by a sector heavy load detection mechanism as used in the DS dynamic sectoring algorithm. This observation confirms the manufacturer's statement that the dynamic sectoring algorithm was designed specifically for the interfloor traffic condition.

The ACA algorithm, as depicted in Figure 15.1, appears to provide a superb performance at low traffic levels. This is due to a slightly different method of representation. As passengers in an ACA system register calls at a landing destination station, and then walk to an appropriate car, the ACA system only starts to register passenger waiting time after some 5 s of walking time has elapsed. Hence, the ACA curve is a little overstated. A recalculation would show, that up to mid-range values of β (40% < β < 60%), the ACA system and the CGC system give virtually identical results, after which, the CGC algorithm is worse than the ACA algorithm.

Figure 15.1 clearly shows that fixed sectoring systems are less effective for balanced interfloor traffic. At very high demands, all the lift systems saturate, and at 100% demand, the passenger average waiting time is 1.4 to 2.5 times the uppeak interval (*UPPINT*), dependent on the traffic algorithm used.

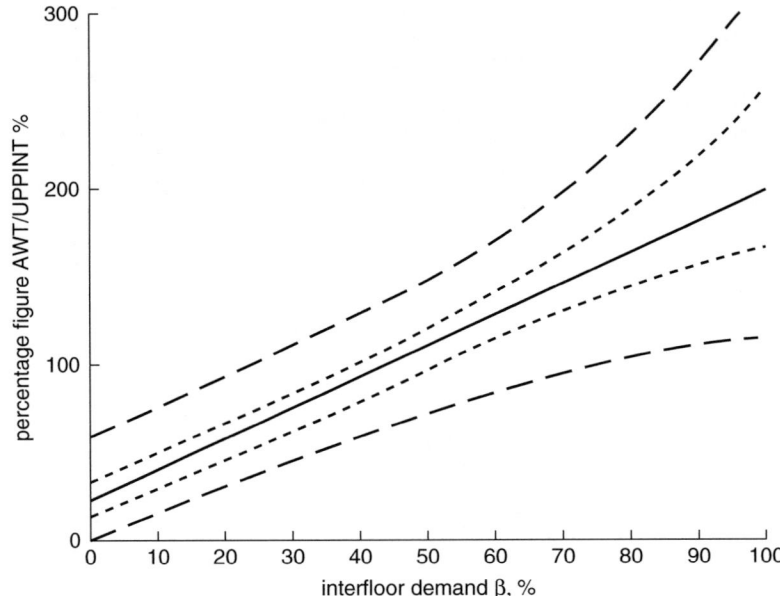

Figure 15.2 Performance -v- interfloor demand (normalised)

Figure 15.1 can be simplified to Figure 15.2 by drawing a weighted average line through all the different control algorithms. The long dashed lines indicate the overall probable limit of possibilities, and the short dashed lines represent the possible limit caused by the influence of the control algorithms.

An equation can be derived for the straight line in Figure 15.2 to allow an estimation of interfloor performance to be obtained, as:

$$IFAWT = UPPINT(0.22 + 1.78\beta/100) \tag{15.2}$$

Example 15.1

Consider an interfloor demand of 30% made on a lift system with an uppeak interval of 34 s. What is the likely passenger average waiting time?

$$IFAWT = 34[0.22 + 1.78 \times 30/100] = 25.6 \text{ s}.$$

15.4   Actual passenger demand

The demand levels used to derive Figure 15.1 range from 10% to 100% as defined by Definition 15.2 and Equation (15.1). How do these demand levels relate to actual demand?

In uppeak and down peak traffic situations, there are short periods of 5–10-minute peaks of demand for the lift system to handle. Lift system sizing is determined by the 5-minute handling capacity during uppeak, which if correctly chosen, generally means that the lift system also meets the longer periods of down peak demand. During balanced interfloor traffic, the passenger demand is continuous over longer periods of time, i.e., in excess of one hour. The demand levels are thus lower, but the activity of the lifts is not necessarily smaller as the number of stops is higher. There is a need to relate the demand indicated in Equation (15.1) with a longer-term passenger activity.

Assume that each morning period is four hours long between occupant arrival (uppeak) and the midday break, and that the afternoon period is four hours long between the midday break and the departure time (down peak). Then, in the morning and afternoon periods, there will be about a three-hour period when the building occupants are working. The occupants will be at their desks or in meetings, and will be much less active. It might be reasonable to assume that each occupant uses the lift system once every morning period and once in every afternoon period. The demand on the lift system would thus be about one third of the building population using the lift system every hour. This demand parameter will be defined.

Definition 15.3: balanced hourly interfloor demand (γ) is the number of passengers requiring service from the lift system during a one hour period, expressed as a percentage (x) of the building population (U).

$$\gamma = \frac{xU}{100} \tag{15.3}$$

Relating this demand to the uppeak demand, if 30% of the building population uses the lifts during one hour of interfloor activity, this is equivalent of only 2.5% demand over 5 minutes. Similarly if 36% of the population uses the lifts during one hour of interfloor activity, this is the equivalent of 3.0% demand over 5 minutes. These interfloor demand levels are considerably smaller than the uppeak values, which are typically from 12–17%, i.e., interfloor demand is about one fifth of the uppeak demand.

It is convenient to relate Definitions 15.1 and 15.3 by considering the following.

From Definition 15.3, the number of passengers handled in 5 minutes is:

$$\gamma = \frac{xU}{100 \times 12} \tag{15.4}$$

Substitute into Equation (15.1):

$$\beta = \frac{xU}{100 \times 12} \times \frac{100}{UPPHC} = \frac{xU}{12UPPHC} \tag{15.5}$$

but:

$$\beta = \frac{xU}{100 \times 12} \times \frac{100}{UPPHC} = \frac{xU}{12UPPHC} \tag{15.6}$$

Rearranging Equation (15.6) and substituting in Equation (15.5):

$$\beta = \frac{xU}{12} \times \frac{100}{U\%POP} = \frac{x100}{12 \times \%POP} \tag{15.7}$$

To illustrate the relationship, suppose a lift system has an uppeak %*POP* of 12% and a 36% interfloor one hour demand, then the value for β will be:

$$\beta = \frac{36 \times 100}{12 \times 12} = 25\%$$

Similarly, a lift system with an uppeak %*POP* of 15% and a 30% interfloor one hour demand will have an interfloor demand (β) of:

$$\beta = \frac{30 \times 100}{12 \times 15} = 17\%$$

Thus, the likely range of usage of, say, one third of the building occupants using a lift every hour, is represented by the low end of interfloor demand β at less than (say) 25%. At this range of usage, the performance figures shown in Figures 15.1 and 15.2 are smaller, and the passenger average waiting times are unlikely to exceed 50% of the calculated uppeak interval.

Example 15.2

For a lift system designed to handle a 15% uppeak traffic demand (%*POP*), with an uppeak interval (*UPPINT*) of 34 s, determine the likely balanced interfloor performance when 50% of the building population use the lifts each hour. Compare the fixed sectoring and computer based systems, stating approximate performance times.

Using Equation (15.7):

$$\beta = \frac{50 \times 100}{12 \times 15} = 28\%$$

Using Figure 15.1 for fixed sectoring systems:

performance figure = 50%
AWT = 0.5 × 34 = 17 s.

Using Figure 15.1 for computer based systems:

performance figure = 30%
AWT = 03 × 34 = 10 s.

15.5   Number of stops

Part of the actuality of interfloor traffic patterns is that lifts make more stops, while handling fewer passengers. Thus, it may be of interest to record the number of stops as evaluated from the simulation runs. Figure 15.3 shows the number of stops made for an increasing interfloor demand. The number of stops made by each car shown in the figure increases with demand until a saturation level is reached. The saturation value, in the order of 4.5 stops/car/minute, is not very dependent on the lift configuration or the control algorithm. This is a predictable result, because a saturated system causes the cars to stop at almost every floor, and the number of stops per minute is consequently limited by the floor cycle time.

The number of stops can be estimated for the lift systems simulated using Equation (15.8). The limit to number of stops car/minute is given by:

$$\frac{60}{T + \left(P_l \times t_l \right) + \left(P_u \times t_u \right)} \tag{15.8}$$

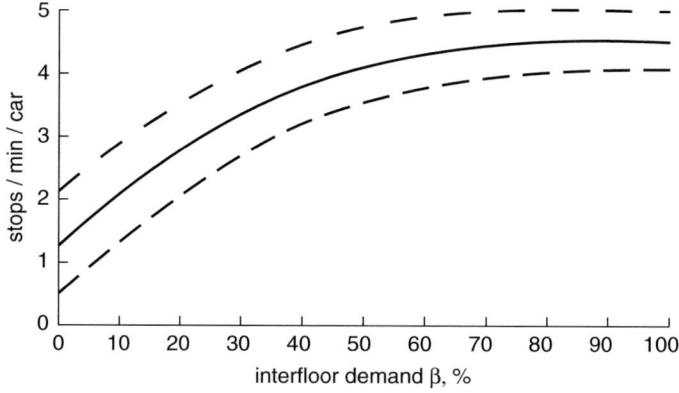

Figure 15.3 Stops/min/car -v- interfloor demand (normalised)

where:

T is the single floor performance time

P_l and P_u are the number of passengers entering and leaving the car at each floor

t_l and t_u are the single passenger loading and unloading times.

This formula assumes that the passenger transfer times exceed the door dwell times. To evaluate the formula, a number of assumptions can be made. The single floor performance time can be taken as 10.0 s and the individual passenger transfer times taken as 1.0 s. At saturation, assume two passengers leave and two passengers enter at each stop. This determines the denominator terms as 14.0 s. Then, the number of stops is 4.3, a similar result to that obtained by simulation.

Siikonen (2002a) has indicated that the limiting number of stops may be even lower, as any systems that employ long door dwell times, without passenger detectors, increase the cycle time as these times are longer than the passengers require to board or alight. She puts the limiting value at 3.3 stops/car/minute.

It is interesting to note that Bedford (1966) said a lift system was busy with 2.25 stops per car per minute. This suggests that a busy system is below that of a saturated system.

Lim (1983), as part of his work to compare the dynamic sectoring (DS) and computer group control (CGC) algorithms, plotted the different types of car stops against interfloor demand. This is shown as Figure 15.4.

Figure 15.4 shows that at low levels of demand, there are very few common landing and car call stops (CLSTP). Even at high demands, common landing and car call stops only amount to about 20% of the total car stops (CSTP). This observation is confirmed by examining the curve showing the ratio of car call stops to landing call stops (CC/LCSTP). This is almost constant at a figure of 1.2. This supports the earlier suggestion that each prospective passenger causes a car stop to be picked up and introduces a new call to be set down. There are very few algorithmically induced movements.

Applying the Bedford 'busy system' of 2.25 stops per car (CSTP) to Figure 15.4 indicates an interfloor demand range of 17% to 25%, which corresponds to the 30% to 36% values of passenger activity mentioned in Section 15.4.

During uppeak and down peak traffic, the usually applied performance measure is passenger average waiting time, except this parameter cannot be measured. Therefore, the obvious and measurable parameter is the interval. During interfloor traffic, the concept of the round trip and an interval is not viable. So, what can give an indication of performance?

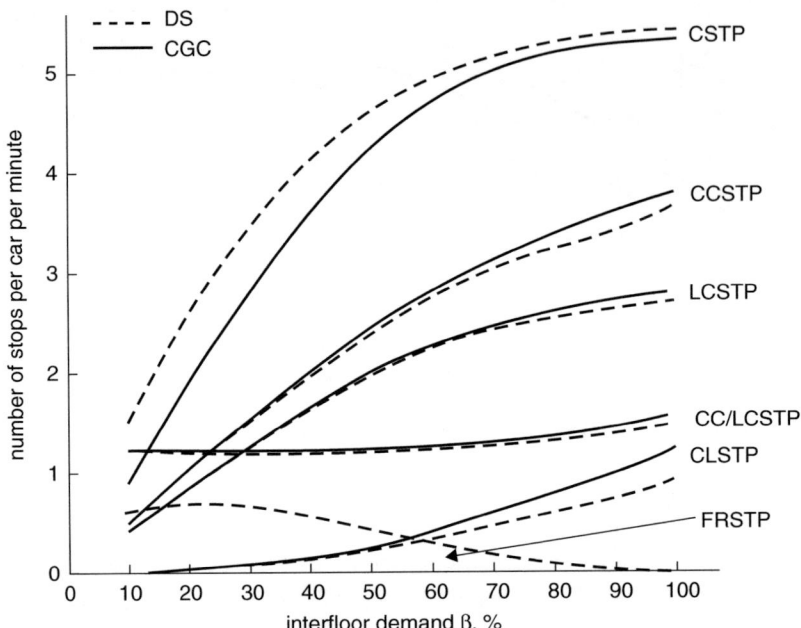

Figure 15.4 Stops/min/car -v- interfloor demand (by algorithm)

Figure 15.2 gives a theoretical relationship between the passenger average waiting time and the underlying uppeak interval. Lim (1983) carried out a number of investigations to find a better performance criteria for interfloor traffic. This was to use ninety percentile (90%) values of passenger waiting time. This idea fits well with the one hour view that must be taken of interfloor traffic. This moves the debate away from averages, which only satisfy half the passengers, towards a 90% value, where only 10% of the passengers are disadvantaged.

Figure 15.5 gives a comparison between average waiting time and a ninety percentile waiting time for the range of interfloor demand. If these two parameters are expressed as a ratio, the graph of Figure 15.6 is obtained. This graph shows an almost constant ratio of 2.2 for interfloor demands over 30%. However, this ratio is not useful as passenger average waiting time cannot be easily measured.

Because Lim was simulating the lift systems he was able to plot, in Figure 15.7, the ratio of the ninety percentile to the average system response time (see Definition 6.7), which can be measured.

Figure 15.7 shows a ratio of approximately 2.1 between the ninety percentile and the average system response time for interfloor demands greater than 30%. It can therefore be concluded that 90% of passengers wait no longer than twice the average system response time. It should be noted that at the more likely demand levels of less than 30%, the ratio becomes smaller, to approximately 1.8.

From Figure 15.6, $NPER = 2.2AWT$.
From Figure 15.7, $NPER = 2.1ASRT$.
So, $AWT = 0.96 \times ASRT$.

Thus AWT and $ASRT$ are almost equal.

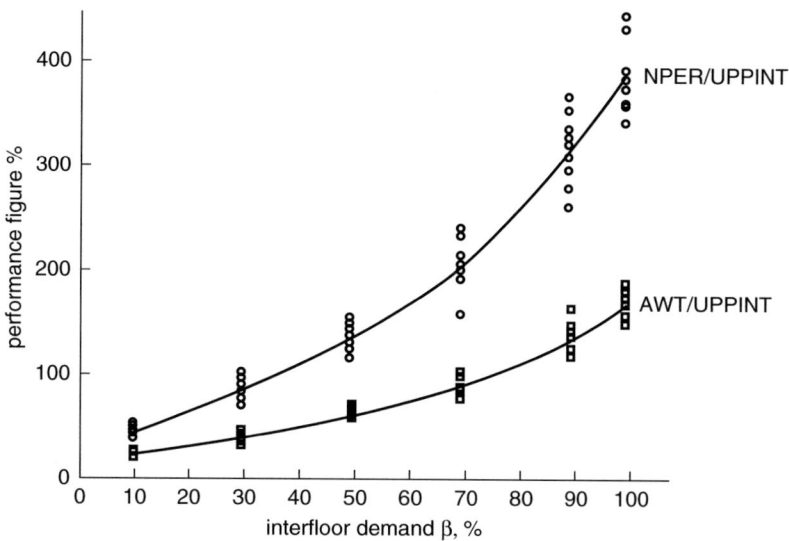

Figure 15.5 Performance -v- interfloor demand (relative to uppeak)

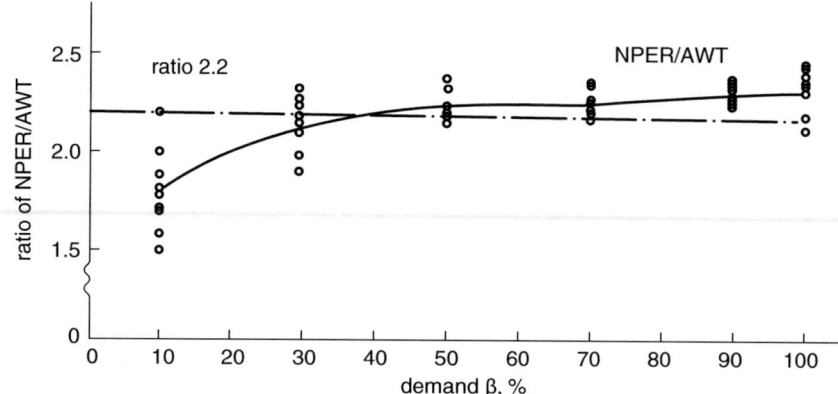

Figure 15.6 Interfloor ratio (AWT)

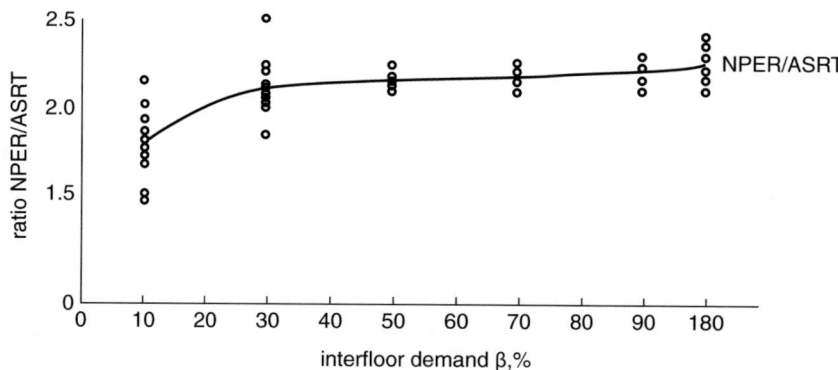

Figure 15.7 Interfloor ratio (ASRT)

15.6 Conclusions on balanced interfloor traffic

A large number of simulation runs for a variety of lift configurations, demand levels, number of floors and supervisory control systems have allowed the assessment of the quality of service provided by a lift system under balanced interfloor traffic conditions.

The lift performance degrades severely with demand, and is dependent on the type of supervisory control system implemented. A dynamic sectoring system is more efficient than the fixed sectoring systems, but the adaptive call allocation computer control algorithm presents the best performance.

A 'busy' lift system can be defined as one where cars make 2.25 stops/minute and handle about one third of the building population in one hour. Interfloor demands higher than 50% tend to saturate the lift system, causing the cars to stop at most floors, thus degrading the quality of service. At the saturation level of demand, lifts stop over twice as often as during a busy period.

The system response time provides a measure of quality of service as it is related to ninety percentile of all passengers not waiting longer than twice the average system response time.

The concept of a round trip time has no meaning during balanced interfloor traffic. Once interfloor traffic becomes unbalanced, then it may transform into a dominant traffic pattern such as uppeak or down peak, which are analytical. The traffic pattern could mutate into an unbalanced traffic pattern such as that experienced at lunchtime. This will be considered in the next chapter. The underlying handling capacity of the lift system is never utilised during balanced interfloor traffic. Siikonen (2002a) suggests this underlying handling capacity might be 40% larger than during uppeak. If all this handling capacity were to be used, each person in the building would be using the lift twice every hour, and waiting times would be very large. Because only some 25–30% of the underlying capacity is used, the passenger waiting times are generally excellent.

15.7 A warning

The information presented in this chapter enables a lift designer to estimate the balanced interfloor lift performance in relation to an uppeak design. However, it should be noted that the figures given here refer to a balanced demand per floor. If a building is unequally populated, it is known from practical simulations carried out that waiting times can become extended.

Most methods employed to improve, or boost, the uppeak handling capacity do not improve balanced interfloor performance as the capability of the underlying lift installation remains the same. The effect on the performance during interfloor traffic of a 'boosted' system will be less obvious, as the system utilisation is much lower. The only traffic control technique that could be used to boost interfloor performance would be a call allocation system. This system could serve the calls more efficiently as both the calling floor and destination floors are known to the controller. But as the passenger demand is much lower than the system capability, it is unlikely that the expense is justified over a two-button system.

16 Midday traffic

16.1 Preamble

In the previous chapters, uppeak, down peak, and interfloor traffic patterns have been analysed and discussed. These were all considered as pure patterns, i.e., with *no* mixed mode traffic. The midday traffic pattern is defined in Section 4.4.3. Other patterns, especially where there are mixed modes of demand, are best examined using simulation techniques (see Chapter 17). Midday traffic, by its nature, is mixed mode.

16.2 Midday or lunchtime traffic

16.2.1 Background to the traffic condition

The midday or lunchtime traffic is today regarded as a most severe test of the capability of a lift system to handle passenger demands. It is a traffic condition that makes use of all of a lift system's capability. This period of time can last over two hours, depending on the business arrangements in a building. During this period, passenger demand builds up as passengers depart from their 'home' floors and then return some time later. A simple case would be a building that does not contain refreshment or other facilities, and the traffic generated would include passengers:

(a) leaving their home floor to travel to the main terminal
(b) returning from the main terminal to their home floor
(c) moving between floors (interfloor traffic).

In a more complex case, where there are refreshment or other attractions, these may be placed on a facilities floor. If the facilities floor were the same floor as the main terminal, its effect would be similar to the simple case described above. More likely, the facilities floor will not be the main terminal, and might simply be a restaurant. It may, however, include other facilities such a travel agent, banking facilities, shops, leisure centre, etc. In addition, it may not be situated in the lower part of the building and could be the highest floor.

Figure 7.5 shows a midday peak scenario in an office building where there are restaurant facilities on the top two floors. The peak traffic is a combination of passengers:

- travelling from their offices to the restaurant for lunch
- travelling back to their offices after lunch
- travelling to the ground floor to leave the building
- returning to the building.

The facilities floor can therefore be very attractive to the building occupants. Besides traffic types (a), (b) and (c) listed above, other traffic generated would include passengers:

(d) leaving their home floor to travel to the facilities floor
(e) leaving the facilities floor to travel to their home floor
(f) leaving the facilities floor to travel to the main terminal
(g) leaving the main terminal to travel to the facilities floor.

The character of the traffic changes throughout the midday period. Initially, in the first 30 minute period, persons begin to leave their home floors, i.e., traffic types (a) and (b) in the simple case, and traffic types (a), (b) and (d) in the more complex case. These persons eventually travel away from the main terminal floor, creating traffic type (c) in the simple case, and traffic types (e), (f) and (g) where there is a facilities floor. But at the time that the first period persons begin to travel, new persons will start their midday break. This counter traffic flow can create a very high demand on the lifts. The lower demand at the beginning of the midday break thus becomes a complex high demand in the middle period of the midday break, which reduces to a lower demand, similar to the first period, at the end of the midday period. The most severe demand of the whole midday period is the central portion, and the lifts can load to their full capacity.

How can this demand be modelled, and what should its value be? In uppeak traffic, there is a benchmark of a pure up demand, in down peak, there is a benchmark of a pure down demand and in interfloor traffic, there is a benchmark of a completely random demand. Can a benchmark be suggested?

Peters and Sung (2000) suggest that if the uppeak handling capacity were to be (say) 15%, then there could be a 5% uppeak traffic, a 5% down peak traffic and a 5% interfloor traffic. This suggests the modal split of traffic would be three ways, and would be a value related to the uppeak handling capability of the lift system.

Siikonen (1998), based on some traffic measurements across the world, quantifies the values at 40% incoming, 40% outgoing and 20% interfloor traffic. She does not specify the demand level, but indicates that whatever it is, the proportions are as stated above.

These two authors use different terminology: Peters uses uppeak/down peak traffic, and Siikonen uses incoming/outgoing traffic. Both these terminologies suggest the simple case rather than the complex case. It might be better to term the traffic as up traffic and down traffic as this will deal with both cases. The Siikonen scenario would seem more likely for the middle part of the midday period, where the traffic would divide as 40% up traffic, 40% down traffic and 20% interfloor traffic. A further step can be taken by assuming that the interfloor traffic is divided equally in the up and down directions. The best benchmark could therefore be 50% up traffic and 50% down traffic.

16.2.2   *Calculation of midday traffic*

Can the midday traffic pattern be calculated? To establish a formula for the round trip time during the midday period, a number of assumptions need to be made:

(a) The restaurant floor and main terminal floor are near together, i.e., in the lower part of the building.
(b) Lifts leave the lower part of the building with a car occupancy of 80% of the probable capacity (Table 5.8), i.e., with P passengers.

(c) Lifts stop S times on the way up, and reverse at the highest reversal floor H.

(d) Lifts leave the highest reversal floor, stop S times on the way down, and arrive at the lower part of the building, loaded to 80% of the probable capacity, i.e., with P passengers.

(e) The lifts exhibit a round trip, i.e., they call at the main terminal every trip.

Initially, the main terminal will be an extra stop, but as the traffic pattern establishes itself, the lifts will make $2S$ stops every round trip. The time consumed for stopping is therefore $2ST$ (where T is the performance time).

The number of passengers that board and exit on both the way up and the way down the building is P passengers. The time consumed for passenger transfers is therefore $4Pt_p$.

The lift will pass through $(H–S)$ floors both on the way up and the way down the building (see Equation (4.7)). The time consumed to travel past these floors is therefore $2(H–S)t_v$.

$$RTT_M = 2(H–S)t_v + 2ST + 4Pt_p$$

$$RTT_M = 2Ht_v + 2St_s + 4Pt_p \qquad (16.1)$$

Where, from Equation (4.11): $T–t_v = t_s$ the handling capacity can be determined by considering the number of passengers transported over the whole round trip. This is $2P$ passengers, giving the formula:

$$MIDHC = 300 \times 2 \times L \times P/RTT \qquad (16.2)$$

No account has been taken of interfloor traffic. It might be reasonable to assume that under this heavy demand situation, any interfloor traffic boards and alights at the stops serve the main traffic flow, and that the interfloor traffic does not introduce any extra stops.

It is also assumed that the traffic controller can deal with a restaurant floor adjacent to and below the main terminal floor, or conversely, the restaurant floor adjacent to and above the main terminal floor, when lifts might fill to capacity at the first floor visited and therefore be unable to accommodate any further passengers at the next adjacent floor. The controller would need to detect skipped floors and send lifts to service them. The traffic controller should also employ load weighing to avoid stopping when there is no capacity available for new passengers to board.

Where the restaurant floor is at the top of the building, it would be assumed that the cars fill to 80% probable capacity prior to departure. Similarly, cars would be assumed to fill to 80% probable capacity prior to departure from the main terminal floor.

Table 16.1 shows the results of applying Equation (16.1) to the 'four corners of the lift world and its centre of gravity' (see Section 5.12.3).

Table 16.1 Comparison of midday to uppeak handling capacity

Number of floors	Rated car capacity	MID/UPP ratio
10	10	1.34
10	24	1.23
16	16	1.24
20	10	1.29
20	24	1.17

Rated car capacity is derived from rated load divided by 75

The table shows a range of values during the midday period. The average handling capacity is some 1.3 times the underlying uppeak handling capacity. These calculated figures do not take into account any incidental interfloor traffic handled. Siikonen (2002a) has shown, by the simulation of a small number of systems, that the midday to uppeak ratio is 1.2–1.4. Using the same technique, she also states that she considers the down peak to uppeak ratio is 1.6–1.8, compared to the value of 1.6 derived using Equation (14.10). The work reported in Chapter 14 relies on the extensive simulation of over 2,000 systems, and should encourage a designer to use the lower value of 1.6. The differences between the Siikonen values and those reported here could be accounted for by the different calculation and simulation techniques, the number of systems considered and the traffic control systems employed. However, the values are very similar.

16.2.3   *Midday demand*

Peters and Sung (2000) related their modal split of traffic to the underlying uppeak handling capacity. Table 16.1 also relates demand to uppeak capacity. But what is the demand? Can this demand be modelled and simulated? Again, no benchmark of likely demand exists. Factors to be considered are:

(a)   What is the length of the midday period?
(b)   What is the length of each person's midday period?
(c)   What percentage of floor occupants use the lifts during the midday period?
(d)   How long do they stay away from their home floor?
(e)   In a restaurant, how long do they stay (residence time)?
(f)   What is the departure/arrival profile of passengers?
(g)   What is the modal split between a main terminal and a facilities floor (if any)?

These are broad assumptions, and will have a number of other factors that affect them. For example, how many covers are there in the restaurant, how efficient is the service in the restaurant, are there snack/self-service kitchen facilities on, or close to, each floor, etc.? The following are practical suggestions based on experience:

(a)   2 hours
(b)   30 minutes
(c)   One-third per hour; two-thirds per midday period
(d)   30 minutes
(e)   30 minutes
(f)   In a 30-minute period, there are six 5-minute periods. Assume the departure profile is similar to down peak, and that all the demand is over 10 minutes, i.e., two 5-minute periods. Also assume that the return profile occurs at the start of the next 30-minute period and also the last 10 minutes. Note the coincident demand.
(g)   Two-thirds to the facilities floor, one-third to the main terminal.

Assumption (f) relates well to evening down peak passenger behaviour. Then, building occupants wish to leave the building as quickly as possible. The heavy demand lasts for 10 minutes and uses all the underlying capacity of the lift system. The underlying capacity of the lift system was shown in Chapter 14 to be 1.6 times the underlying uppeak handling capacity. Building occupants are in a similar situation during the midday period as they are in their own time, and want refreshment, or to leave the building; therefore, it is reasonable to assume that they make similar demands on the lift system.

The scenario indicated above may be more severe if the midday period is shorter, or if the numbers of persons using the lifts is larger. The demand may, however, be smaller if the demands are more random and do not occur simultaneously, as indicated in the scenario.

It is possible to relate assumption (f) to a demand level. Each floor has a 100% population. There are four 30-minute midday periods, thus, 25% of the population is available to use the lifts every 30-minute period. Only two-thirds of them use the lifts (16.7%) and then the demand is spread over two 5-minute periods. The demand in each of the 5-minute periods is thus 8.3%. This is a demand well below any ever considered for an uppeak design. There is a difference, however, as when the down traffic demand coincides with the up traffic demand, then the lifts must provide twice this requirement, i.e., 16.7%. Using the midday/uppeak ratio of 1.3, the underlying uppeak handling capacity must be at least 13%.

Example 16.1

Consider a building with 11 floors above the main terminal. Floor 1 is a restaurant floor. The single floor transit time is 2.0 s, the performance time is 10 s, and the passenger transfer time is 1.0 s. There is a population of 180 persons on floors 2–11. What is a suitable system to serve a 14% uppeak arrival rate (i.e., 252 persons/5 minutes)?

With eight lifts of a rated load of 1,800 kg, the probable capacity is 18.6 persons (Table 5.8), with an 80% occupancy of 14.9 persons.

A spreadsheet calculation gives the following data.

Uppeak arrival demand 13.8 %
Number of persons: 248 persons
Interval: 18.0 s
Other data: $H = 10.7$, $S = 8.3$, $t_v = 1.6$ s, $T = 10.2$ s, $t_p = 1.0$ s.

What is the midday traffic handling capacity?
Using Equation (16.1):

$$RTT = 2 \times 10.7 \times 1.6 + 2 \times 8.3 \times 8.6 + 4 \times 14.9 \times 1$$

$$= 236.6 \text{ s}$$

The midday interval *MDINT* is:

$$MDINT = 236.6/8 = 29.6 \text{ s}$$

Using Equation (16.2):

$$HC = 300 \times 2 \times 8 \times 14.9/236.6$$

$$= 302 \text{ persons/5 minutes.}$$

The uppeak handling capacity was 248 persons/5 minutes, giving a midday to uppeak ratio of 1.22.

Table 16.2 Representation of passenger departures from each floor during a midday period of 2H: 10m

Fl	First midday hour												Second midday hour												Third	
	1	2	3	4	5	6	7	8	9	10	H	12	13	14	15	16	17	18	19	20	21	22	23	24	25	26
10	15	15	0	0	0	0	15	15	0	0	0	0	15	15	0	0	0	0	15	15	0	0	0	0	15	15
9	15	15	0	0	0	0	15	15	0	0	0	0	15	15	0	0	0	0	15	15	0	0	0	0	15	15
8	15	15	0	0	0	0	15	15	0	0	0	0	15	15	0	0	0	0	15	15	0	0	0	0	15	15
7	15	15	0	0	0	0	15	15	0	0	0	0	15	15	0	0	0	0	15	15	0	0	0	0	15	15
6	15	15	0	0	0	0	15	15	0	0	0	0	15	15	0	0	0	0	15	15	0	0	0	0	15	15
5	15	15	0	0	0	0	15	15	0	0	0	0	15	15	0	0	0	0	15	15	0	0	0	0	15	15
4	15	15	0	0	0	0	15	15	0	0	0	0	15	15	0	0	0	0	15	15	0	0	0	0	15	15
3	15	15	0	0	0	0	15	15	0	0	0	0	15	15	0	0	0	0	15	15	0	0	0	0	15	15
2	15	15	0	0	0	0	15	15	0	0	0	0	15	15	0	0	0	0	15	15	0	0	0	0	15	15
Rest	0	0	0	0	0	0	100	100	0	0	0	0	100	100	0	0	0	0	100	100	0	0	0	0	100	100
MT	0	0	0	0	.0	0	50	50	0	0	0	0	50	50	0	0	0	.0	50	50	0	0	0	0	50	50

Example 16.2

What would be the worst case scenario? Following the suggestions in Section 16.2.3.

Population of each floor	= 180 persons
Proportion who use lifts in midday period	= 120 persons
Floor population who use lifts in one hour	= 60 persons
Floor population who travel to the main terminal in one hour	= 20 persons
Floor population who travel to the facilities floor in one hour	= 40 persons
The departure profile from each floor over two hours	= 120 persons
The departure profile from the main terminal over two hours	= 400 persons
The departure profile from the facilities floor over two hours	= 800 persons
The departure pattern is shown diagrammatically in Table 16.2.	

Table 16.2 shows that the maximum demand occurs in periods 7–8, 13–14, 19–20 and 25–26. In each of these 5-minute periods, the total up plus down demand is 300 persons. As the lifts can handle 300 persons/5 minutes, this means the midday period can be satisfied in terms of handling capacity.

16.3   Midday conclusions

The lift system may be able to handle the demand, but what will the passenger waiting times be? The uppeak interval for the underlying lift system in Example 16.1 was 18.0 s, and the midday interval was 29.6 s. The lifts were assumed to load to the same occupancy level in both cases. This means that the passenger average waiting time would be longer during the midday period by the ratio of the interval values. The ratio in this case is 1.6, which is almost twice as long. The likely formula for the midday passenger average waiting time would therefore be:

$$MIDAWT = 0.85MIDINT \qquad (16.3)$$

In practice, the midday demand may not be as large as discussed above. Examine Section 17.7.5, and Example 17.4, which shows an all-day simulation for a typical demand pattern.

Whatever the assumptions that are made, the procedure above gives a possibility of a calculation analysis. A simulation, however, allows a more realistic judgement to be made.

Part E
Techniques and trends

17 Simulation and computer aided design (CAD)

17.1 Introduction

Chapters 4–8 have used calculation methods to design and evaluate the performance of lift installations. These calculation methods are based on mathematical models of the lift system. Other methods that have been discussed in Chapters 13–16 are based on the computer modelling of the lift systems to provide empirical formulae. This computer modelling was carried out using a simulation program operating in batch mode.[1] All calculation methods rely on simplifications and assumptions in order to make the calculations possible. Such simplifications and assumptions are discussed in Chapter 7. Other simplifications need to be made to deal with special situations, see Chapter 8.

Where the traffic design requirements are more complicated, computer simulations may need to be carried out. Simulation allows all traffic conditions to be examined more exactly. They allow a check to be made of the underlying basic traffic design and a comparison to be made with the more realistic representation obtained by simulation. The realism of the simulation depends on the richness of the model used. The simulated results may be quite different from those obtained by calculation. This is because all calculations are based on statistics, and generally deliver an average answer to a design, not a specific one.

Most lift companies and many consultants have calculation programs, and some have simulation programs. The information regarding their internal structure is not generally available. This chapter looks at the traffic design of lift systems using simulation within a computer aided design (CAD) process.

17.2 Conclusions[2]

Simulation allows a greater understanding of complex situations, but is not necessary for straightforward designs. Because simulation can follow each passenger from arrival to departure, a vast amount of data can be collected. Graphical figures give a close feel of what is happening, and confirmation of the underlying exact values can be obtained from tabulated print outs. The ability to plot graphs for any period of time, for any floor, or all floors is important to the appreciation of the design. It can be concluded from the examination of the results presented here that digital modelling can be very realistic.

The question that can be posed is will the calculation results be the same as those obtained by simulation?

It is possible, but less likely for several reasons; here are some:

1   The calculation method uses a rectangular probability distribution function to describe passenger arrivals, whereas a simulation often uses a Poisson probability distribution function as it is simple to generate in a digital computer and closer to the real world.

2   As the calculation can use non-integer values for the variables, e.g., $P = 6.7$, a simulation can only use an integer value for each passenger (although the results can be averaged to non-integer values).

3   The destinations in a calculation are statistically derived, whereas in a simulation they may be generated by a pseudo random binary sequence (PRBS), of a specified length and feedback, which is easy to generate in a digital computer, and closer to the real world.

4   Because of statistical generation methods, it is difficult for a simulation to place exactly the specified demand into the lift system, e.g., in a calculation, the arrival may be defined as 100 persons/5 minutes, whereas the simulation might present a range from 95 to 105 persons/5 minutes, so the comparison is not always like for like.

5   A simulation can deal with the conflict between the number of passenger transfer times and the effect of the door dwell times exactly, whereas in a calculation, an assumption has to be made.

6   A simulation can deal with dwell times almost exactly; calculation can only estimate their effect.

In general, but not always, the use of the calculation method to size a lift system will v in a slightly over lifted installation.

17.3   History of lift simulation

The first major steps in lift simulation application programs in the public domain occurred in 1972 with the Lift Simulation and Design (LSD) program, which was developed by Dos Santos (1972) at the University of Manchester Institute of Science and Technology. Programmed in FORTRAN IV for a mainframe (now called a server) time-sharing computer (Digital Equipment PDP10), it was later transferred to a DEC minicomputer (PDP11), and in 1985, to the first IBM compatible personal computers, running under the DOS operating system, and renamed PC-LSD (Personal Computer – Lift System Design). After Dos Santos, the design suite was further enhanced by others (Moussallati, 1974; Hirbod, 1975; Swindells, 1975; Lim, 1983). All development, except bug fixes, ceased in 1985.

The programming facilities available in 1972 were such that PC-LSD is modular in design and suffers from a very poor user interface for the input of data. It could, however, simulate a wide range of legacy traffic algorithms, which included: two simple collective algorithms (COL, MCO), one duplex/triplex algorithm (THV), three fixed sectoring algorithms (FSW, FS4, FSO), two dynamic sectoring algorithms (DS, SCH) and four modern computer control algorithms (CGC, ACA, CCU, MAS).

PC-LSD was very rich in graphical displays of passenger waiting and journey times, percentile plots, call response graphs, car spatial movements, etc. Full numerical printouts were also available.

PC-LSD was supplied to over 20 enterprises in the UK, Europe, North America and Asia, including lift companies, standards bodies, consultants and educational establishments. Difficult to use, it never became an industry standard.

ELEVATE[3] was first available in October 1997, and has the look and feel of most modern computer applications, with a simple interface for the input of data and easy navigation around the suite. Initially, it had only three in-built traffic control algorithms (collective, ETA and HCA/ACA). These three generic types were, at that time, sufficient.

The latest (2015) version is rich in traffic control algorithms. Elevate also provides a facility for users to incorporate their own algorithms. As the large amounts of data are collected into a spreadsheet, users can design their own graphs, tables and reports.

The simulation phase is illustrated by a display of lift movements as the simulation proceeds, and is much appreciated by architects. The ELEVATE suite also includes calculation programs, which range from the simple methods, such those based on Equation (4.8), through those based on the generalised analysis procedure (Peters, 1990).

Thousands of copies of ELEVATE are in operation in at least 60 countries worldwide. It is under continuous development, and has become the industry de-facto standard for the 21st century.

17.4   Simulation and CAD defined

The computer simulation of engineering processes is particularly appropriate where the study of the actual process is difficult or dangerous, too costly, would take too much time, or would be inconvenient. Existing lift systems fall into this category. In the case of a new lift, the installation does not even exist.

> **Definition 17.1:** simulation is the development and use of models to aid in the evaluation of ideas, and the study of dynamic systems or situations.

Digital computers are most suitable for the simulation of discrete systems (but not continuous systems)[4] that can be described by sets of logical equations. A lift system is a discrete system:

*   Each individual passenger arrival is a discrete event.
*   Each individual passenger departure is a discrete event.
*   Each lift is a discrete unit.
*   Each floor is a discrete entity.
*   Each car movement is a discrete occurrence.
*   Each door operation is a discrete occurrence, etc.

Digital computer simulation programs can be either event based, i.e., the model is updated every time something happens, or time-based, i.e., the model is updated at regular intervals. A lift system is sparse in the number of events that occur (see Section 11.10), compared to some engineering systems. Most events also do not require immediate action, and some events initiate identical actions, making it efficient to service them at the same time. It is therefore sensible to select the time-based method, with an update interval chosen to service all events in a reasonable time.

Engineering design involves the appreciation of shape, form and relative values, thus, the graphical presentation of data allows the designer to appreciate a design quickly. The designer should be able to submit the input data, receive the results back quickly, appreciate them and resubmit the design, if required. The process of computer aided design is to input data, carry out an application, e.g., a simulation, receive output data to consider, and if necessary, repeat the process with new input data.

> **Definition 17.2:** computer aided design is where a designer interacts directly with a complex computer process, and in so doing, closes the design loop.

17.5   Underlying structure of a digital computer simulation program for lift traffic design

The necessity for the computer aided design of lifts was foreseen by Jackson (1970) who wrote:

> *a real need . . . is a computer program to simulate the likely performance of proposed lift systems . . . Different numbers, speeds and groups of lifts should be considered, as well as different control systems . . . the results would show designers the performance of several proposals . . . [and allow] . . . rational decisions.*

17.5.1   Time slices

Section 17.4 explained that it is sensible to select the time-based method of digital simulation. Thus, an update interval must be chosen to service all events in a reasonable time. The choice of the update interval, or time slice, is important if the accuracy of the numerical values obtained is to be maintained.

One method is to have a relationship with the speed of a lift. The simulation program would automatically select the update interval. If (say) the update interval was one tenth of the single floor transit time, then most events would be accurately represented. For a lift with a high rated speed (say 5 m/s) travelling a short distance (say 3.0 m), the update interval would be small (0.06 s). However, a slow speed lift (say 0.25 m/s) with a long travel distance between floors (say 10 m), would have a long update interval (4.0 s), and this could mean a loss of accuracy. This loss of accuracy can be overcome by keeping all times as an integer number of update intervals plus a remainder.

Another method is to select a fixed update period (say 0.1 s), and to keep real values of all numbers. The maximum error with this method would then be no more than 0.1 s.

17.5.2   Display update

The display presented during the simulation phase could be updated at every time slice. This is unnecessary as the simulation will be running many times faster than real time. A suitable display update could be every ten update intervals. If the user does not want this detail, then the display update could be made a variable and altered to suit the user's requirements.

17.5.3   Random number generation: arrivals and destinations

The arrivals and destinations must be selected. A common approach is to use a random number generator, which on a digital computer will not be produced from 'white noise', but from a PRBS subroutine. PRBS subroutines need a starting point known as a 'seed'. Sometimes, this seed is fixed or can be different for each design run.

Some programs use a PRBS number generator to randomly introduce passengers into the simulation and to select their destination. Other programs use a PRBS number generator for destinations, but a Poisson process for arrivals.

17.5.4   Number of simulations

A digital computer always gives the same answer in response to the same input (and seed). So, the more simulations that are run with a different seed, the more accurate will be the results.

Good results can generally be obtained with ten simulations. If the time to compute a problem is too long (and simulations are slow), and the user is not prepared to wait for ten simulations to run, it is suggested that one simulation is run until a result close to the design criteria are achieved. Then, up to ten simulations could be run to obtain more authoritative data.

17.5.5 Simulation period

Lift systems have to serve the four main traffic conditions: uppeak, down peak, interfloor and midday (lunchtime). These can be studied as distinct time bands of (say) one hour of activity. If different time bands for study were desired, then the start and finish time of each period would need to be defined. It has been established (Section 4.6) that 5 minutes is a realistic minimum time period in which to analyse data. Longer periods can be considered for some traffic conditions. It therefore follows that in any simulation period, the data that are accumulated over 5-minute periods can be analysed over one or more 5-minute periods. These 5-minute periods can be called data collection periods.

17.6 Simulation design data required

Chapter 5 indicated that three sets of data were required to carry out a design using a calculation method. For a simulation, these data may need to be extended. The sets are: building data, lift system data and passenger data. The data sets that follow are a basic set as some calculations and simulations require additional data.

17.6.1 Building data

The basic set of building data for a calculation are:

- number of floors
- average interfloor distance
- express jump
- express additional time

For a simulation, these data may need to be extended to:

- floor names
- individual floor heights (elevation)
- definition of entrance floor(s)

17.6.2 Lift data

The basic set of lift data for a calculation are:

- number of lifts
- rated load
- car capacity (number of passengers by area)
- rated speed
- average flight times between floors
- door opening times
- door closing times
- advance door opening time
- start delay

For a simulation these data may be extended to:

- car floor area
- individual flight times

- door dwell time (landing call)
- door dwell time (car call)
- door dwell time (main terminal)
- differential door opening time
- number of door re-openings
- traffic control system
- acceleration (m/s²)
- jerk (m/s³)
- floors served.

17.6.3   *Passenger data*

The basic set of passenger data for a calculation are:

- building population
- floor population
- passenger transfer time.

For a simulation, these data may need to be extended to:

- passenger arrival rates at specified floors in specified periods
- floor bias.

17.6.4   *Management*

The basic set of data for a calculation are:

- project name
- designer name
- date.

For a simulation, these data may need to be extended to:

- simulation period
- time slices between calculations
- display update frequency
- random number seed
- number of simulations
- units used
- energy model used
- data output format (tables, graphs, etc.).

17.7   Simulation case studies

17.7.1   *Introduction*

What follows is based on the ELEVATE lift traffic simulation package as this program is widely used. The authors do not necessarily recommend ELEVATE as there are other simulation programs available, particularly from manufacturers. These latter programs are usually on restricted access, and their innermost workings commercially confidential. However, Day and Barney (1993) proposed a checklist to evaluate computer programs used for traffic design, which Peters refined in 2000.

The ELEVATE simulation program has an extensive range of facilities not explored here. Readers who might contemplate using the program need more knowledge and practical experience than this book can provide. This chapter forms an introduction to simulation, and Chapter 18 offers more advanced analysis.

The use of simulation is best examined by examples. The examples presented here are compared, where possible, with the results obtained from the mathematical model. Advanced examples are given in Chapter 18.

Three 'pure' traffic situations, uppeak, down peak and interfloor traffic will be considered by simulation in these case studies. Another appropriate case study is a simulation for the office day from 07.00 to 19.00. These case studies use data acquired from an actual building.

In Example 6.7, a lift system was designed using the iterative balance method. The data used are given in Table 6.10.

The case studies enable the designer to more closely examine the quality of the design.

Table 6.10 Summary of Example 6.7

Input data	Value
Number of floors	10
Rated load	1275
Actual car capacity	13.8
Number of passengers	9.4
Number of lifts	5
Rated speed	1.6
Building population	800
Interfloor distance	3.3
Express jump	0
Express additional time	0
Single floor flight time	5
Door close time	3
Door open time	1.8
Advance door opening	0.5
Start delay	0.5
Passenger transfer time	1.2

Results	Value
Number of passengers	9.4
Highest reversal floor	9.5
Number of stops	6.3
Performance time	9.8
Round trip time	118.7
Interval	23.7
Handling capacity	119
Percentage population	14.9
Capacity factor (%)	68

Table 17.1 Basic data used in Example 17.1

Feature	Data
Traffic control system	Uppeak (uppeak 2)
Traffic arrival pattern (template)	Barney uppeak (one hour)
Peak traffic value	15%
Floor populations	80
Floor bias (attraction)	Equal

17.7.2 Example 17.1: uppeak simulation case study

To carry out a simulation of the uppeak traffic pattern, the simulator requires more information. Various input screens (or design fields) will need to be filled in (not shown here). The data required, shown in Table 17.1, includes:

In an uppeak calculation, the designer wishes to confirm the handling capacity (*%POP*) and the interval (*UPPINT*). In a simulation, a different approach is necessary.

The arrival pattern is as shown in Figure 17.1. It will be noted that the profile closely resembles Figure 4.4, and shows one hour of arrivals divided into twelve 5-minute periods. Table 17.2 tabulates the 5-minute arrivals by period. The peak arrivals are at period 8 with the peak set at 15%, i.e., 120 persons.

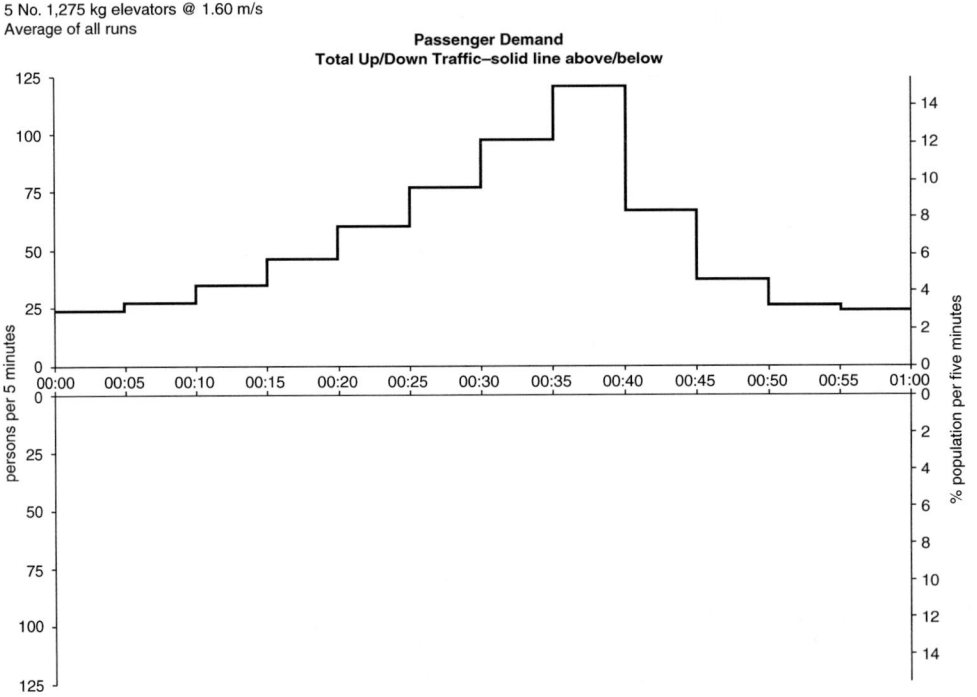

Figure 17.1 Uppeak arrival pattern, template: Barney uppeak (one hour)

Table 17.2 Arrivals by period during uppeak

Period	Time	Arrival rate (persons/5-minutes)
1	0–05	29
2	5–10	38
3	10–15	50
4	15–20	63
5	20–25	76
6	25–30	91
7	30–35	105
8	35–40	120
9	40–45	81
10	45–50	52
11	50–55	35
12	55–60	28

The peak arrival rate at period 8 is the same as that for the calculation method.

But what is the interval at the main floor during the peak five minutes? This is shown in Figure 17.2.

At the peak interval (period 8), the average interval is 25.0 s. This compares closely with the calculated interval, which is 23.7 s.

Of interest to the designer is the car loading in period 8 as each car leaves the main floor. Figure 17.3 shows the car loading by period.

The average car loading in period 8 is 72.4%. This compares to 68% by calculation.

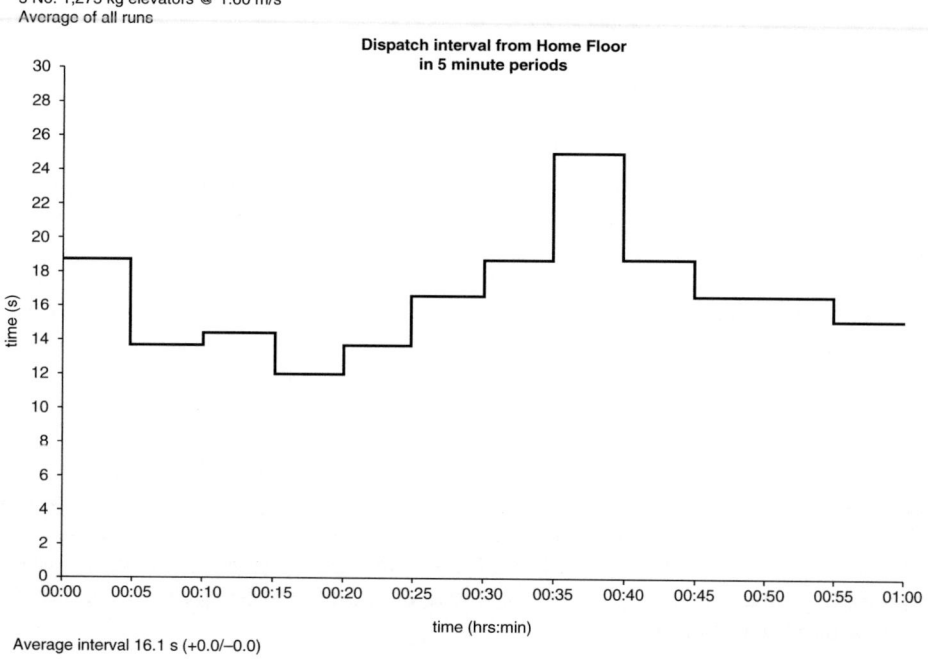

5 No. 1,275 kg elevators @ 1.60 m/s
Average of all runs

Dispatch interval from Home Floor in 5 minute periods

Average interval 16.1 s (+0.0/–0.0)

Figure 17.2 Interval by period

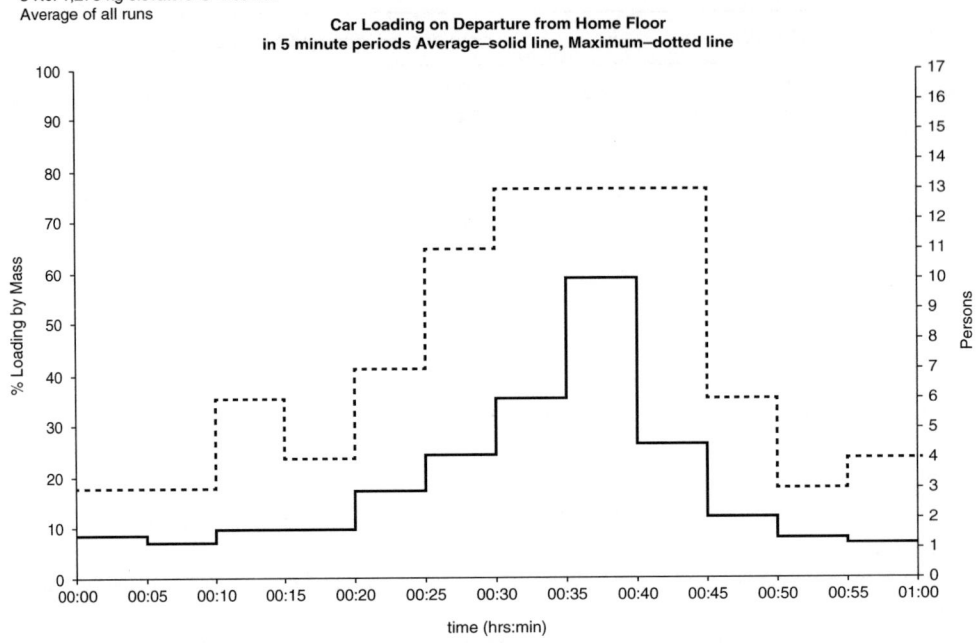

Figure 17.3 Car loading on departure from the main floor by period

Having established a suitable correlation with the calculation method, the designer is now interested in the queue length as a criterion for passenger satisfaction.

The maximum queue is 14 persons, and occurs for two brief periods. This would be acceptable as it is about one car capacity, and it clears quickly.

With a simulation, the designer can obtain more accurate information on passenger average waiting times (*AWT*) and passenger average times to destination (*ATTD*). Figure 17.5 shows these two criteria.

In period 8, the passenger average waiting time is 6.4 s, but is larger at 7.8 s in period 7. These data can be calculated. Section 6.7 gives a formula for passenger average waiting time and a value is given in Table 6.11 of 14 seconds. The correlation is thus poor. The calculation method relies on statistical data for hundreds of simulations. Example 17.1 is a specific design and may not conform to the average. The value obtained by simulation is likely to be the most accurate value.

The passenger average travel time can be calculated using the formula in Section 6.8.3.

This gives a calculated value of 50.4 s. Adding this to the passenger average waiting time of 14 seconds gives a passenger average time to destination of 64.4 seconds. The simulated passenger average time to destination is 71.7 s. For the reasons given in regard to passenger average waiting time, the difference may be due to the simulation being for a specific installation, not an average installation.

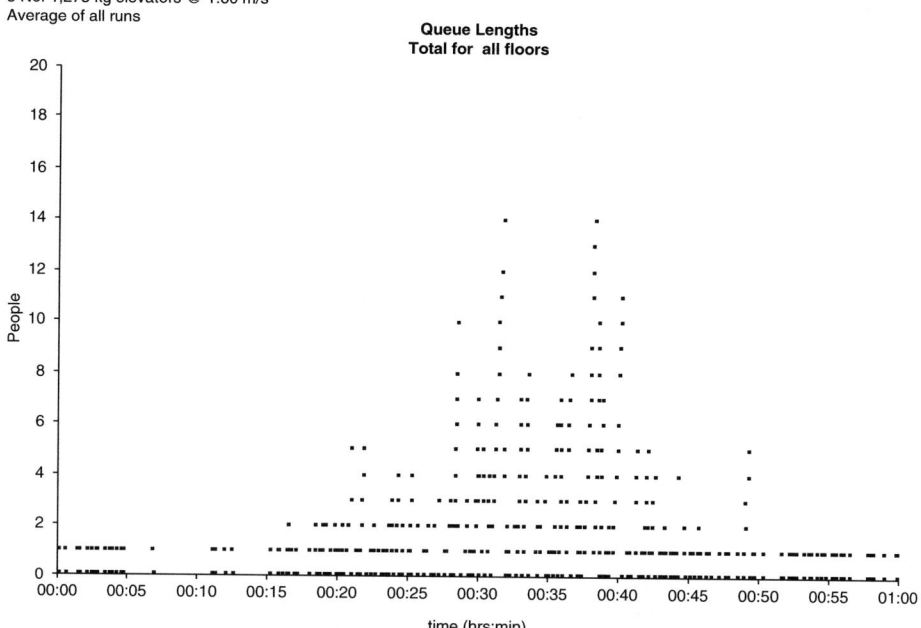

Figure 17.4 Queue lengths

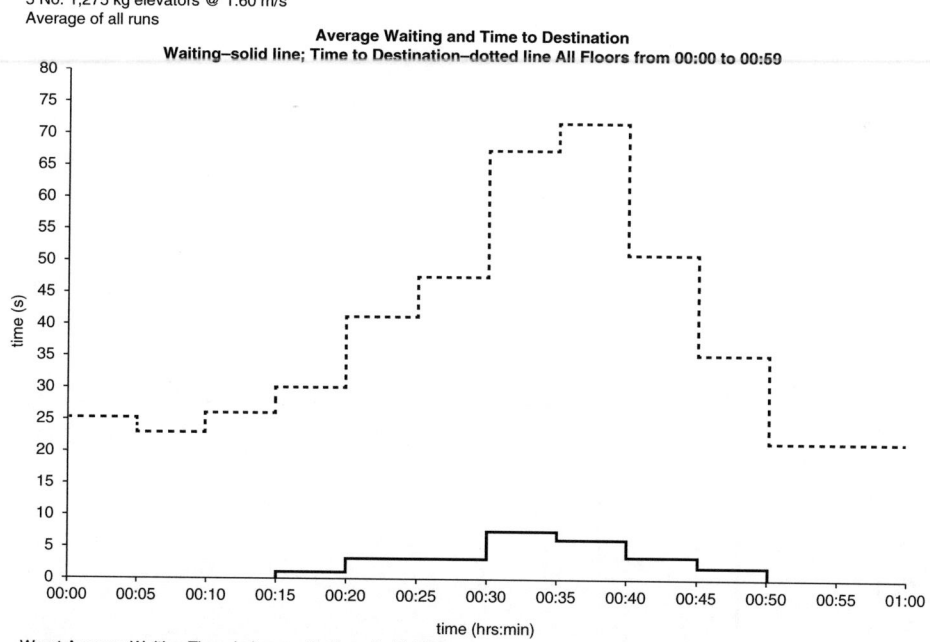

Figure 17.5 Passenger waiting and time to destination

17.7.3 Example 17.2: down peak simulation case study

To carry out a simulation of the down peak traffic pattern, the simulator requires more information. Various input screens (or design fields) will need to be filled in (not shown here). The data required, shown in Table 17.3, includes:

The arrival pattern is as shown in Figure 17.6. It will be noted that the profile is similar to Figure 4.7, and shows one hour of arrivals divided into twelve 5-minute periods. Table 17.4 tabulates the 5-minute arrivals by period.

The peak arrivals are at period 3, with the peak set at 22.5%, i.e., 180 persons (1.5 × 120).

The ELEVATE program provides a comprehensive detail of calls made, i.e., where from, where to, time registered, the lift taken, etc. (not shown here). Examination of this data shows that 180 passengers entered the system in the peak five minutes at period 3. This confirms that the simulation represents the correct passenger demand.

Table 17.3 Basic data used in Example 17.2

Feature	Data
Traffic control system	Group collective (down peak)
Traffic arrival pattern (template)	Barney down peak (one hour)
Peak traffic value	22.5%
Floor populations	80
Floor bias (attraction)	Equal

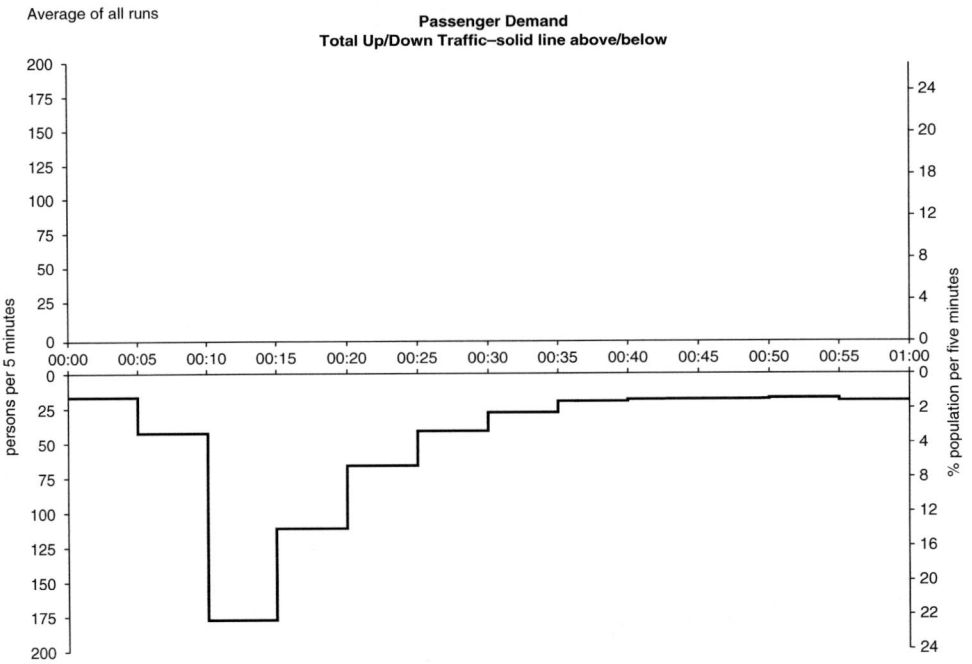

Figure 17.6 Down peak arrival pattern, template: Barney down peak (one hour)

Table 17.4 Arrivals during down peak

Period	Time	Arrival rate (persons/5 minutes)
1	0–05	22
2	5–10	47
3	10–15	180
4	15–20	115
5	20–25	71
6	25–30	46
7	30–35	33
8	35–40	26
9	40–45	24
10	45–50	24
11	50–55	23
12	55–60	25

As for uppeak, the interval at the main floor can be obtained, see Figure 17.7.

At the peak (period 3), the average interval is 14.3 s. This interval is about 60% of the uppeak interval (see Section 14.2.3).

Of interest to the designer is the car loading in period 3 as the car arrives at the main floor (passengers are leaving the building). Figure 17.8 shows the car loading by period. This graph was constructed using a spreadsheet tool from the raw data provided by ELEVATE.

The average car loading (by area) in Period 3 is 53%. This is satisfactory.

The designer is now interested in the queue length as a criterion for passenger satisfaction. Figure 17.9 shows the total of the queue lengths at all ten floors. The worst case in Period 3 with 25 persons, i.e., an average per floor of 2.5 persons.

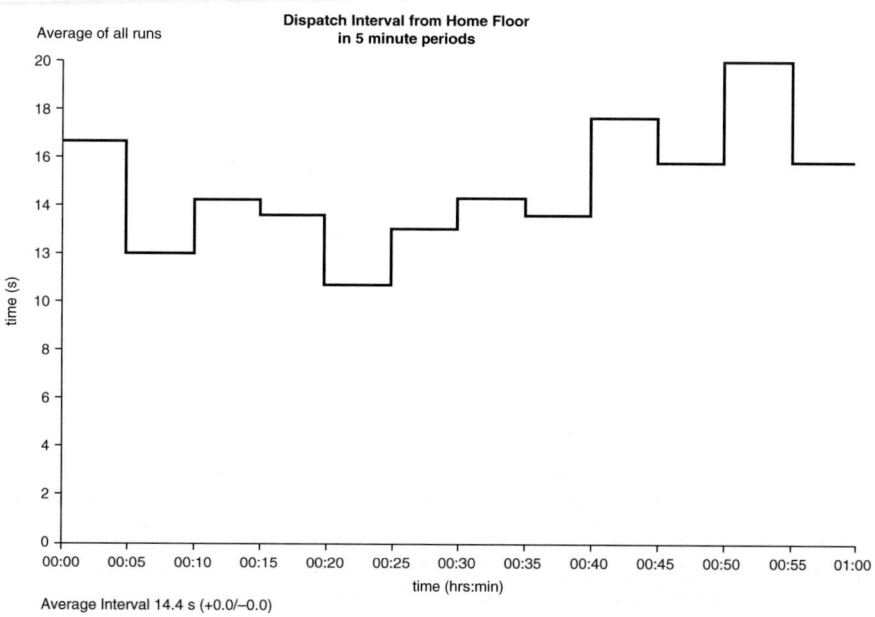

Figure 17.7 Interval by period

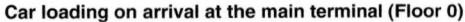

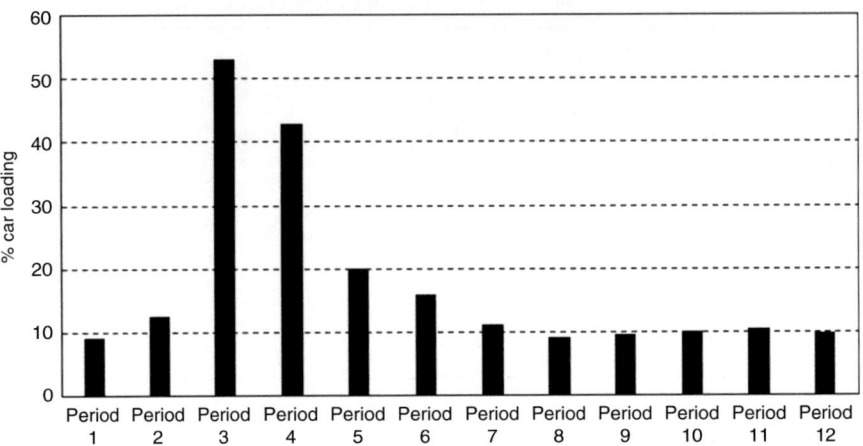

Figure 17.8 Car loading on arrival at the main floor by period

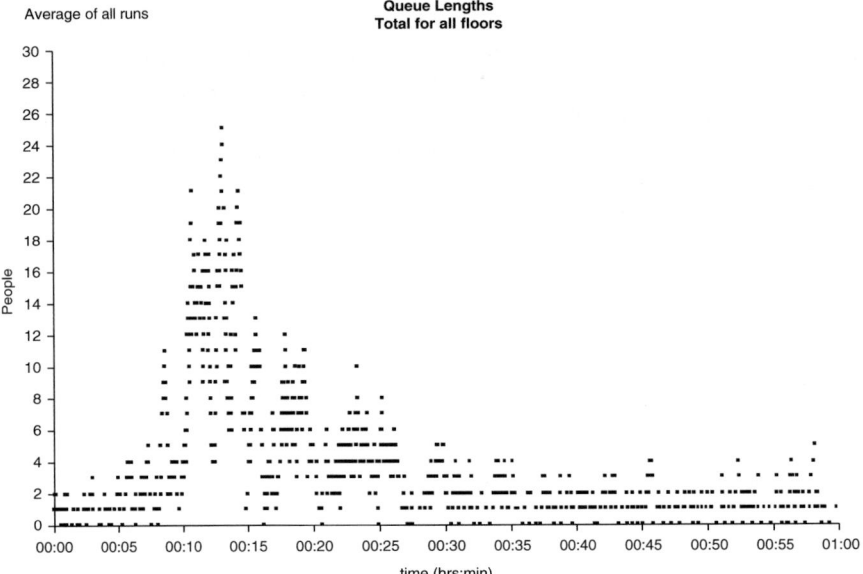

Figure 17.9 Queue lengths

With a simulation, the designer can obtain more accurate information on passenger average waiting times (*AWT*) and passenger average times to destination (*ATTD*). Figure 17.10 shows these two criteria for all floors.

The passenger average waiting time for all floors is 20.1 s. However, some floors may be disadvantaged. For example, floor 9 as shown in Figure 17.11. Here, passengers have to wait 37 seconds in period 9.

The ability of a simulation to examine an individual floor enables the traffic algorithm to be tuned.

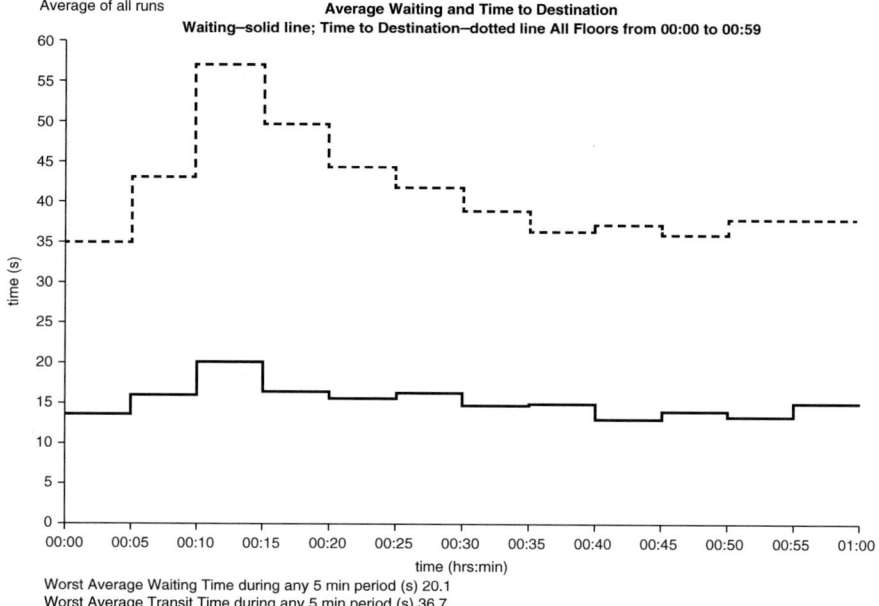

Figure 17.10 Passenger waiting and time to destination – all floors

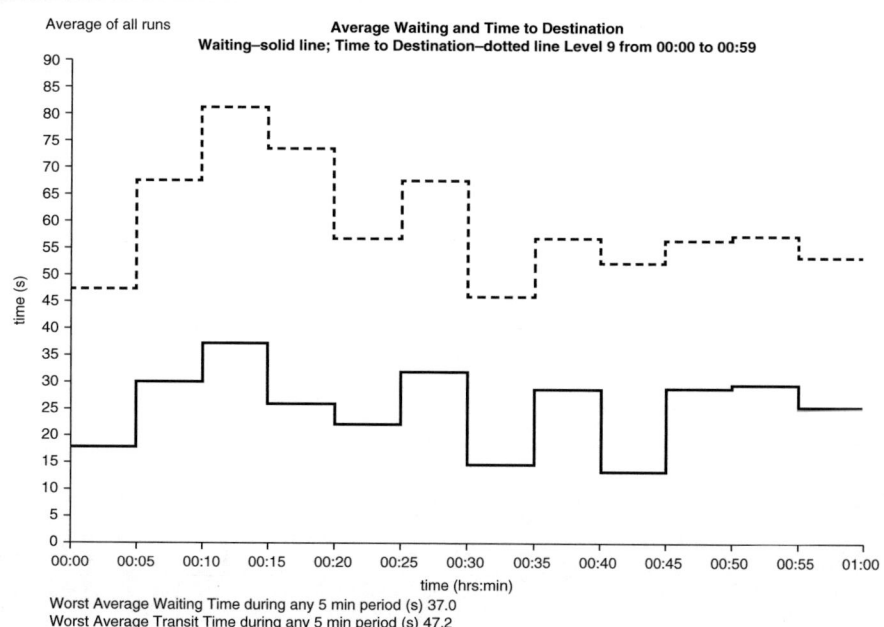

Figure 17.11 Passenger waiting and time to destination – floor 9

17.7.4   Example 17.3: interfloor simulation case study

To carry out a simulation of the interfloor traffic pattern, the simulator requires more information. Various input screens (or design fields) will need to be filled in (not shown here). The data required, shown in Table 17.5, include:

Table 17.5 Basic data used in Example 17.3

Feature	Data
Traffic control system	Group collective (auto)
Traffic arrival pattern (template)	Barney interfloor (one hour)
Peak traffic value	n/a
Floor populations	80
Floor bias (attraction)	Equal

The arrival pattern as shown in Figure 17.11. Table 17.6 tabulates the 5-minute arrivals by period. The arrivals steadily increase through the hour. This range of passenger demands tests the ability of the installation to service different levels of demand.

The total demand over one hour is 36% of the population of 800 persons (see Chapter 15). This is a likely demand in an office building.

The interval at the main floor has no meaning in interfloor traffic as lifts do not always call there. Similarly, car loading is of no interest. However, passenger waiting times are of interest. With a simulation, the designer can obtain more accurate information on passenger average

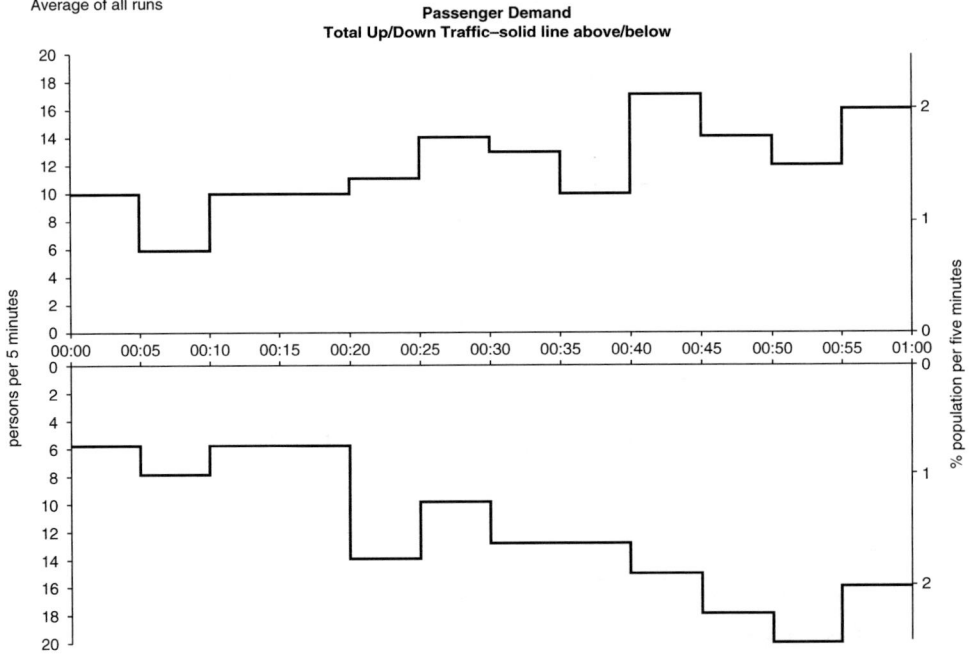

Figure 17.12 Interfloor arrival pattern, template: Barney interfloor (one hour)

Table 17.6 Arrivals during interfloor demand (persons/5 minutes)

Period	Up demand	Down demand	Total demand
00:00	10	6	16
00:05	6	8	14
00:10	10	6	16
00:15	10	6	16
00:20	11	14	25
00:25	14	10	24
00:30	13	13	26
00:35	10	13	23
00:40	17	15	32
00:45	14	18	32
00:50	12	20	32
00:55	16	16	32
Totals	143	145	288

waiting times (*AWT*) and passenger average times to destination (*ATTD*). Figure 17.13 shows these two criteria for all floors.

The passenger average waiting time for all floors is 9.7 s. However, some floors may be disadvantaged. This may be judged by considering the system response time at all floors. Figure 17.14 shows this. The figure was constructed using a spreadsheet graphing tool.

For example, floor 9 as shown in Figure 17.14 has the longest system response time of 21 seconds. The ability of a simulation to examine individual floor response times enables the

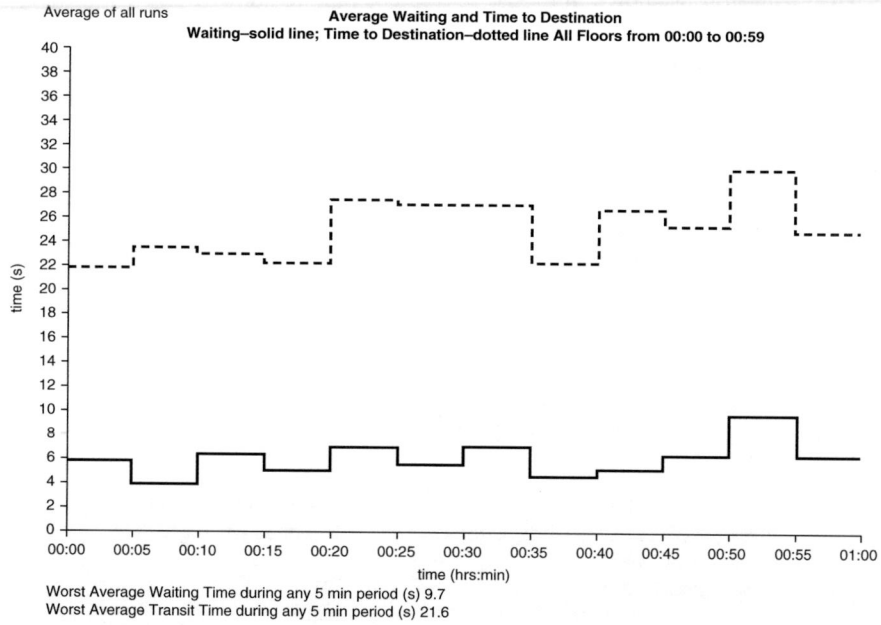

Figure 17.13 Passenger waiting and time to destination – all floors

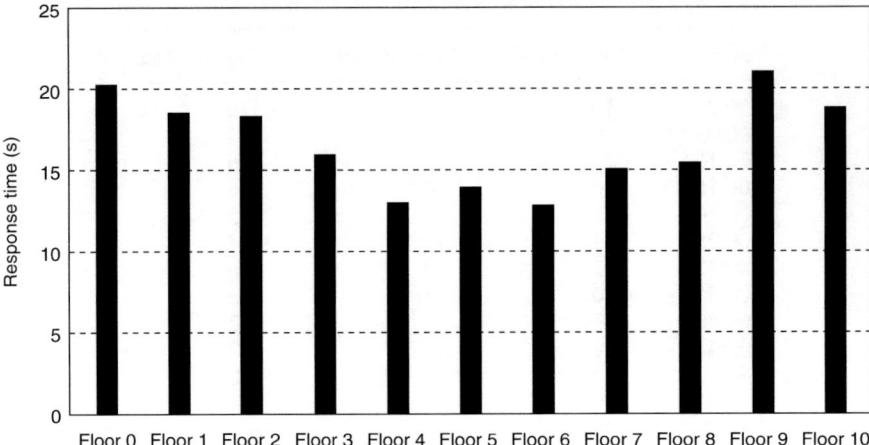

Figure 17.14 Average system response times by floor

traffic algorithm to be tuned. For example the main floor (floor 0) and floor 9 might need to be given a better service.

17.7.5   Example 17.4: all day simulation case study

An all day simulation gives a very detailed account of what can happen in an office building during the working day. This case study uses data collected by Siikonen, and the basic data are given in Table 17.7.

The arrival pattern is as shown in Figure 17.15. The pattern shows many characteristic passenger demands. The uppeak can be seen to occur at around 09.00, the midday period from around 12.00 to 14.00 and the down peak from around 17.30. At no time does the individual up or down peak demand exceed 8%, and is easily accommodated in the inherent underlying 15% uppeak capability designed in Example 6.7.

Figure 17.16 shows the queuing pattern across the day. It is generally below the average car capacity of 13.8 persons, except at around 09.00 when it reaches 21 persons momentarily. This is acceptable performance.

Figure 17.17 shows the car loading over the working day. The worst average is 51.5%, with some peaks extending up to full car capacity (by area). It is this capacity that allows the lift installation to respond to peak demands.

Table 17.7 Basic data used in Example 17.4

Feature	Data
Traffic control system	Group collective (auto)
Traffic arrival pattern (template)	Siikonen full day office
Peak traffic value	n/a
Floor populations	80
Floor bias (attraction)	Equal

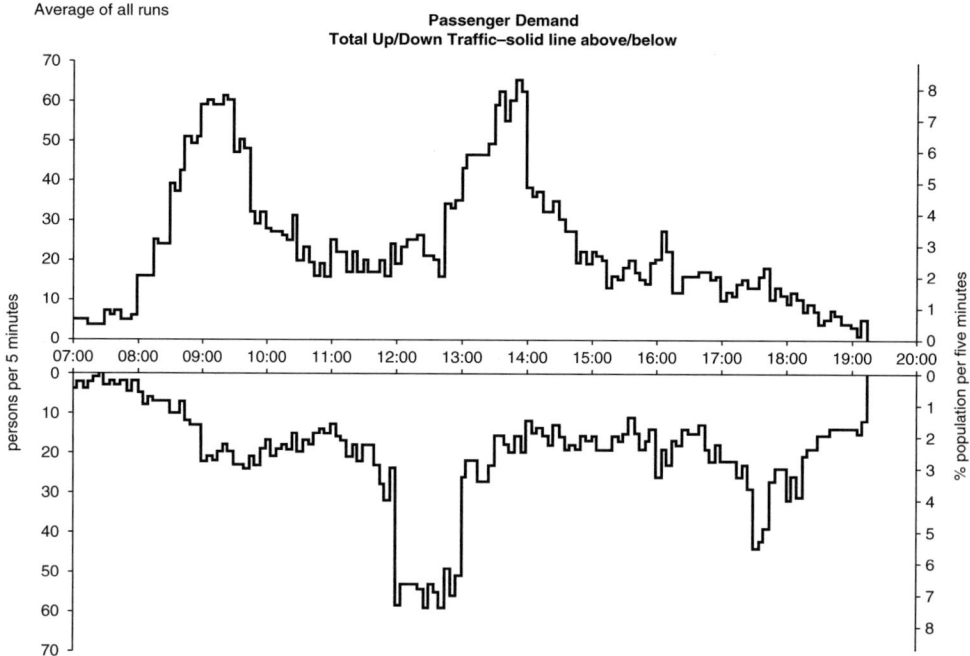

Figure 17.15 Traffic demand during a working day in an office

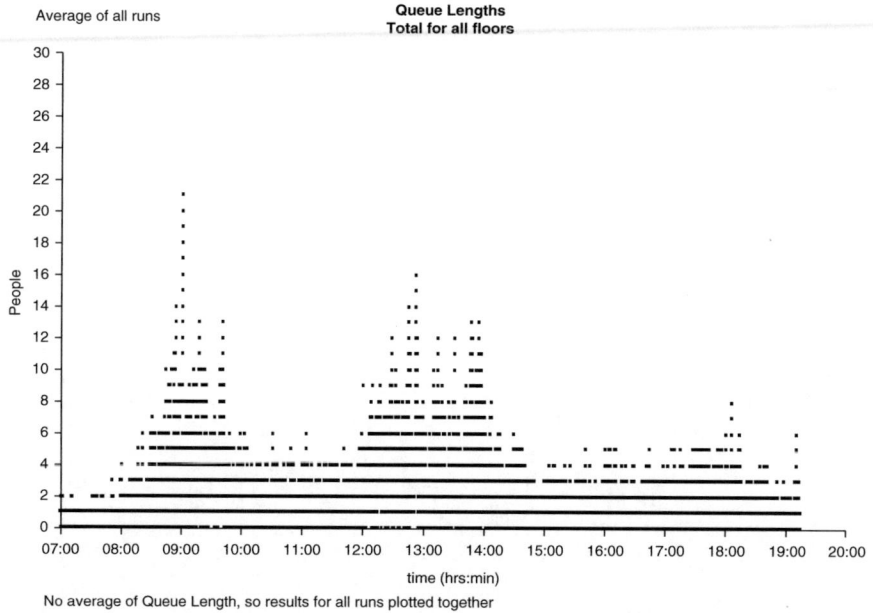

No average of Queue Length, so results for all runs plotted together

Figure 17.16 Queues over the working day

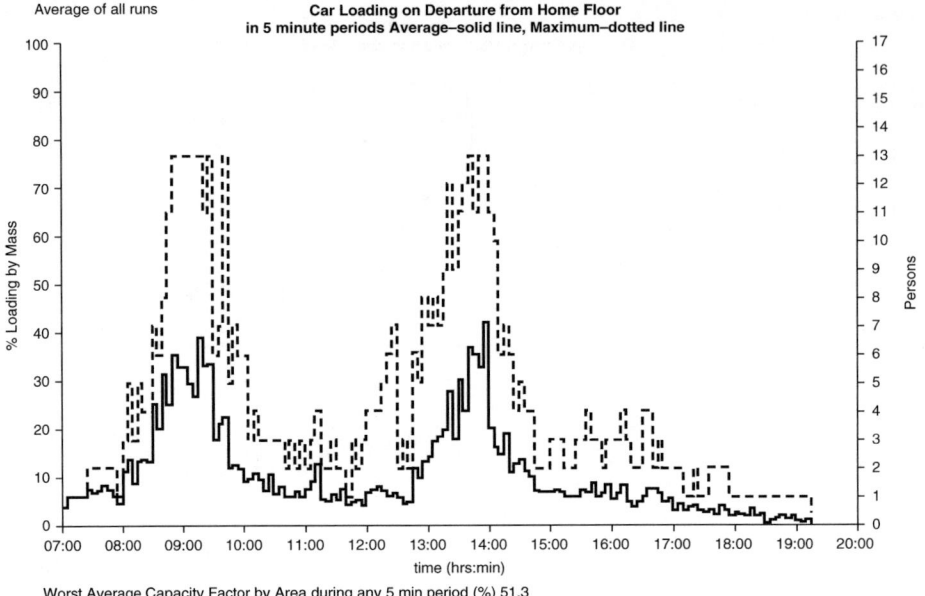

Figure 17.17 Car loading at floor 0

Figure 17.18 shows the interval. It averages 29.3 seconds.

The passenger average waiting time and passenger average time to destination are shown for the simulated period in Figure 17.19. It can be seen that the longer times occur at the peak periods in close correlation with the demand graph of Figure 17.15.

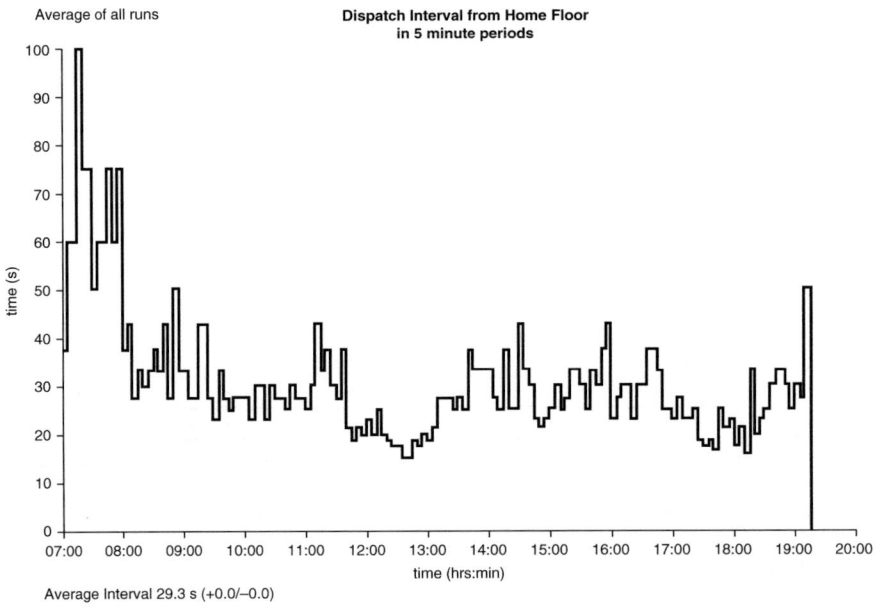

Figure 17.18 Interval

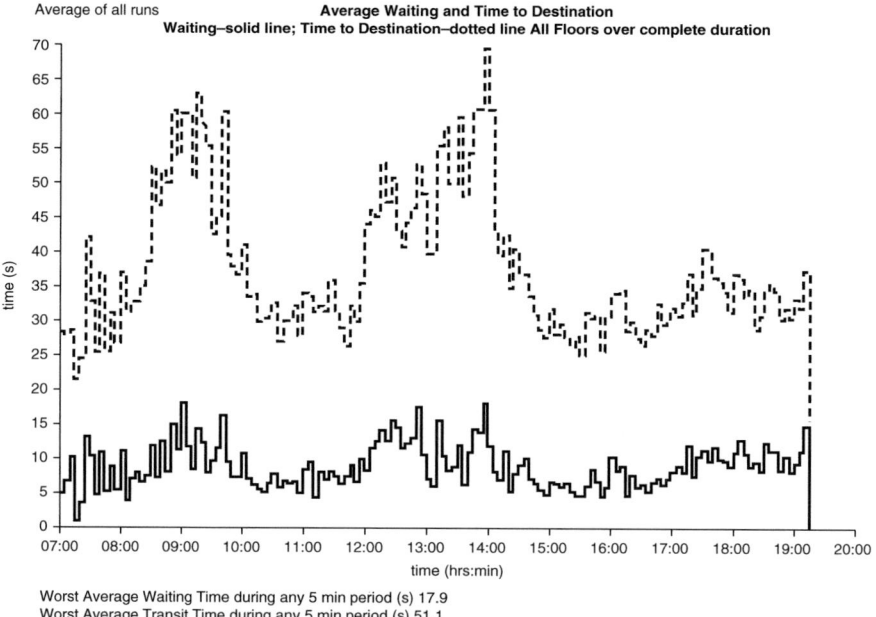

Figure 17.19 Passenger average waiting time and passenger average time to destination

In a simulation, distributions of passenger waiting times, passenger transit times and passenger times to destination are given. Examples are shown in Figures 17.20, 17.21 and 17.22. All the graphs provide: number of calls answered in specified time bands, percentile information, indications of average, longest and ninety percentile values. All the values indicate a satisfactory design.

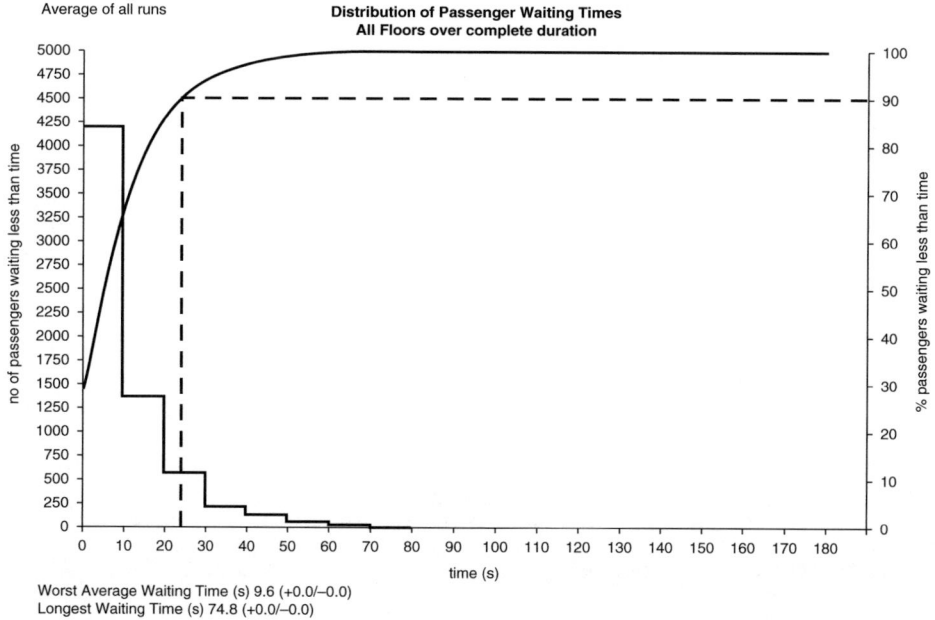

Figure 17.20 Distributions of passenger waiting times

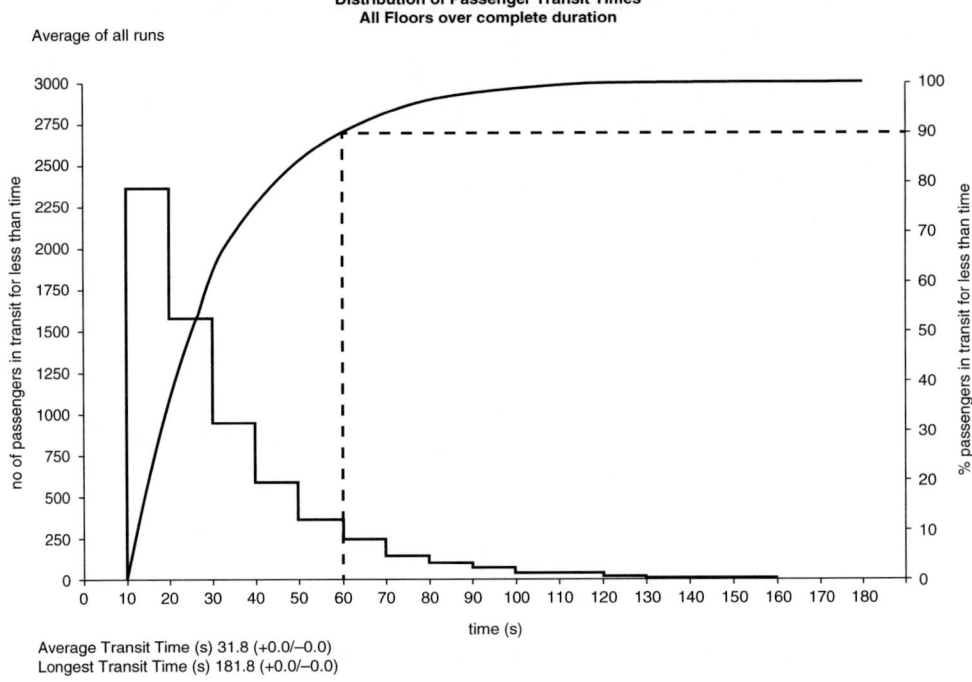

Average Transit Time (s) 31.8 (+0.0/–0.0)
Longest Transit Time (s) 181.8 (+0.0/–0.0)

Figure 17.21 Distributions of passenger transit times

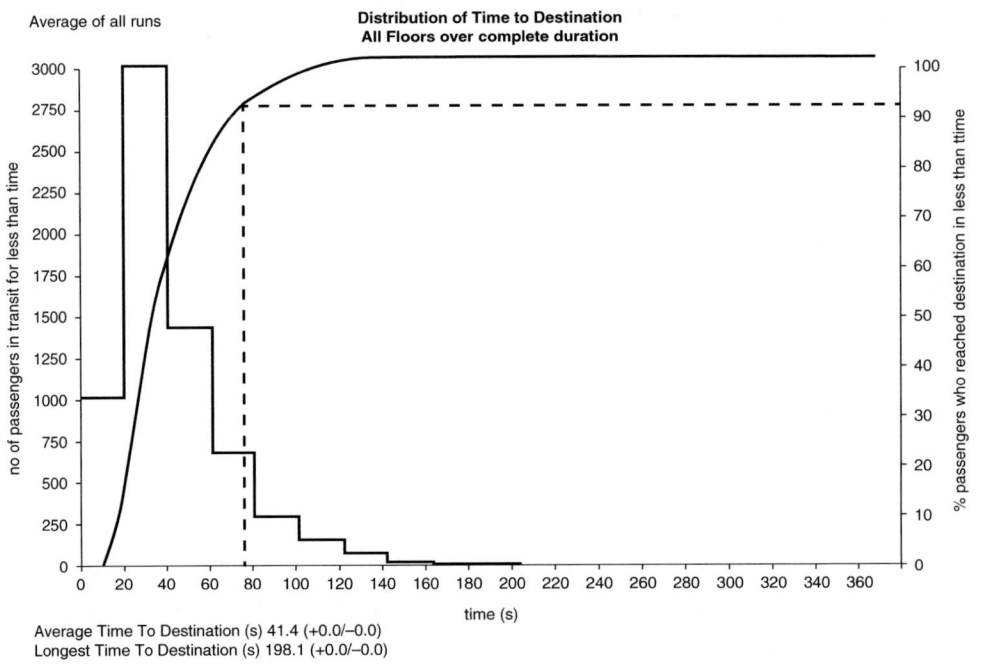

Average Time To Destination (s) 41.4 (+0.0/–0.0)
Longest Time To Destination (s) 198.1 (+0.0/–0.0)

Figure 17.22 Distributions of passenger time to destination

17.8 Defining a simulation – advanced approaches

Most simulation programs have extensive data input facilities in order to define, as precisely as possible, the lift model to be simulated. It is not possible to consider all of these here, and each simulation program is different, but here are three: destination bias, passenger demand and templates.

17.8.1 Destination bias

In Chapter 4, values for H and S were calculated for all floors having equal populations. An assumption was made that, because they had equal populations, then they would be equally attractive as destinations. Such as situation is illustrated in Table 17.8. This table shows the floor biases (in percentages) for a simulation period. The period is usually five minutes, but can be any length. Note the bias on a floor to floor basis must add to 100%. The table illustrates a pure uppeak traffic pattern.

It is possible that the population on a floor does not relate to the attraction of the floor as a destination. It may be that a floor has a restaurant facility, so it would attract more calls. Table 17.9 illustrates a case where floor 3 is more attractive than other floors, with about 25% of the traffic.

Table 17.8 Bias between floors in an 11-storey building (10 floors above the main terminal) for a specified simulation period

Floor	0	1	2	3	4	5	6	7	8	9	10	Total
0	0	10	10	10	10	10	10	10	10	10	10	100
1	10	0	10	10	10	10	10	10	10	10	10	100
2	10	10	0	10	10	10	10	10	10	10	10	100
3	10	10	10	0	10	10	10	10	10	10	10	100
4	10	10	10	10	0	10	10	10	10	10	10	100
5	10	10	10	10	10	0	10	10	10	10	10	100
6	10	10	10	10	10	10	0	10	10	10	10	100
7	10	10	10	10	10	10	10	0	10	10	10	100
8	10	10	10	10	10	10	10	10	0	10	10	100
9	10	10	10	10	10	10	10	10	10	0	10	100
10	10	10	10	10	10	10	10	10	10	10	0	100

Note the use of the BS ISO 4190–5 floor identification recommendation.

Table 17.9 Unequal bias floor to floor

Floor	0	1	2	3	4	5	6	7	8	9	10	Total
0	0.0	7.4	10.7	26.8	3.7	18.5	11.1	3.0	2.7	5.4	10.7	100
1	14.2	0.0	9.9	24.9	3.4	17.1	10.3	2.8	2.5	5.0	9.9	100
2	14.3	7.1	0.0	25.8	3.5	17.7	10.6	2.9	2.6	5.2	10.3	100
3	15.0	8.6	12.5	0.0	4.3	21.4	12.9	3.5	3.1	6.2	12.5	100
4	14.1	6.6	9.6	23.9	0.0	16.5	9.9	2.7	2.4	4.8	9.6	100
5	14.6	7.7	11.2	28.1	3.9	0.0	11.6	3.2	2.8	5.6	11.2	100
6	14.3	7.1	10.3	25.9	3.6	17.8	0.0	2.9	2.6	5.2	10.3	100
7	14.1	6.5	9.5	23.8	3.3	16.4	9.8	0.0	2.4	4.8	9.5	100
8	14.1	6.5	9.5	23.7	3.3	16.3	9.8	2.7	0.0	4.7	9.5	100
9	14.2	6.7	9.7	24.4	3.3	16.7	10.0	2.7	2.4	0.0	9.7	100
10	14.3	7.1	10.3	25.8	3.5	17.7	10.6	2.9	2.6	5.2	0.0	100

So far, the discussion has been concerned with one simulation period. No two simulation periods will be the same. For example, there may be surges in activity. An example is in a university that operates with a 50-minute lecture followed by a 10-minute changeover. Thus, for ten of the 5-minute periods, there is little or no activity.

17.8.2 *Passenger demand (arrival rate)*

It does not follow that the passenger arrival rate at a floor will correspond with the population of that floor. The arrival rate will vary from simulation period to simulation period as the university example suggests. Table 17.10 shows three adjacent simulation periods: A, B and C. The arrival rate at the 11 floors varies, except for floor 4, which is consistent.

17.8.3 *Templates*

Predefined traffic demand templates enable faster data entry into a simulation program. A number of basic templates can be defined. These include:

Uppeak (see Example 17.1)
Down peak (see Example 17.2)
Interfloor (see Example 17.3)
All day (see Example 17.4)

In each category, there can be variations. An example is the Barney uppeak template (Figure 17.23) and the CIBSE modern office template (Figure 17.24). The two figures show a different demand pattern. Figure 17.23 represents the classical pure uppeak demand, and Figure 17.24 shows the demand likely in a modern office.

17.9 Conclusions on simulation case studies

The four case studies show the powerful facility that a simulation program can provide.

The selection of five 1,275 kg lifts with a rated speed on 1.6 m/s carried out by calculation is confirmed for uppeak performance, with many calculated values being similar to the simulated values.

Table 17.10 Passenger arrival rates through three different simulation periods

	Period		
Floor	*A*	*B*	*C*
0	70	14	54
1	6	40	15
2	9	6	7
3	20	7	18
4	23	21	22
5	16	10	12
6	9	14	7
7	3	21	2
8	2	8	2
9	5	9	4
10	9	1	2

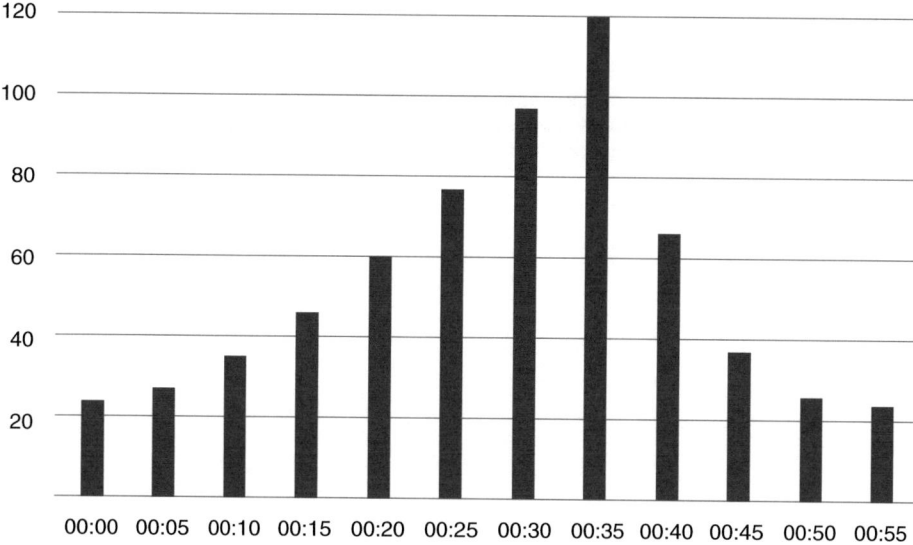

Figure 17.23 Barney uppeak template

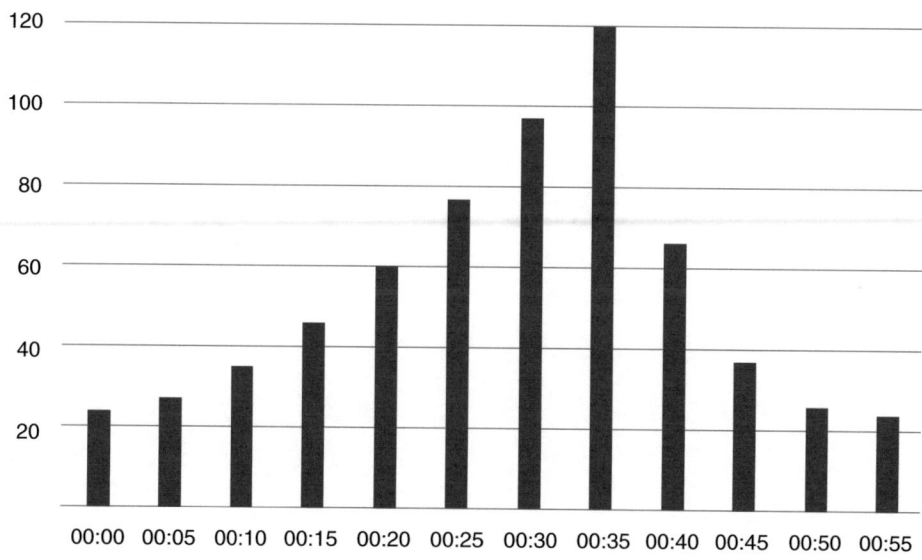

Figure 17.24 CIBSE Modern office template

Simulation has also shown that the system provides a good performance for down peak and interfloor traffic demands.

The all day performance did not show any poor performance.

This chapter is a simple examination of how simulation may be used. It is important that a calculation is carried out first to establish a possible design. Then, simulation can be used to provide more detailed data.

Section 17.8 gives an insight to the facilities that may be available.

Conclusions were also drawn in Section 17.2

Notes (shown in text as[1])

1 Batch mode is where a large number of simulations are carried out, without human interaction, against an input file of data, writing the data to output files for later analysis.
2 The conclusions are placed here as the final part of this chapter illustrates simulation by a number of case studies with their own conclusions.
3 www.peters-research.com.
4 Continuous systems are usually analogue systems and are described by differential equations.

18 Advanced techniques

18.1 Introduction

The traditional methods for lift traffic system design have been based on calculation or simulation. This chapter looks at some of the advanced techniques used in calculation and simulation, and carries out a comparison between them. It also presents a typical blueprint for the lift traffic system design process. Queuing theory is presented as an alternative tool for analysing lift traffic systems. It is also shown how queuing theory can be used to provide a reconciliation tool between calculation and simulation.

Attempts have been made to model lift passenger traffic and lift movements as chaotic systems, and this is discussed at the end of the chapter.

18.2 Calculation methods in lift traffic design

Calculation methods rely on the evaluation of the round trip time, which still forms the basis for classical lift traffic design. Once the round trip time has been evaluated, the target interval can be used to find the required number of lifts in the group. Methods for compliant lift traffic designs can then be followed (Al-Sharif *et al.*, 2013b). The round trip time can evaluated using analytical or numerical methods, which are outlined in the next two subsections.

18.2.1 Analytical equation-based calculation

The analytical equation-based design method relies on a number of equations that have been derived with a number of assumptions. It is important to stress the fact that any equation would have been derived based on simplifying assumptions. It is extremely important to check the validity of these assumptions prior to applying an equation.

At the most basic level, the three equations for the round trip time (RTT), the probable number of stops (S) and expected highest reversal floor (H) are based on simplifications under a number of special conditions (CIBSE 2010, Barney 2003a). These equations implicitly make the following ten assumptions:

1 Equal floor populations.
2 Equal floor heights.
3 Rated speed attained in one floor journey.
4 Incoming traffic conditions.
5 Single entrance.
6 Passenger choices of destinations are independent.

7   Perfect door timing is assumed (as soon as the last passenger enters or leaves the car, the doors start to close).

8   One passenger leaves or enters the car at any time.

9   Constant passenger arrival model.

10   No door re-openings.

If any of the assumptions above are not true, then the equations can be modified in order to cope with that. For example, the equations for the probable number of stops, and the highest reversal floor for the case of unequal floor population have been derived and are available (CIBSE 2010, Barney 2003a). However, as conditions become more complicated, deriving the relevant equations becomes more involved (Peters, 1990; Al-Sharif *et al.*, 2013c).

18.2.2   *Numerical calculation methods*

In order to overcome the difficulties encountered in deriving new equations for the more general cases, numerical methods can be used to find the value of the round trip time. Examples include Monte Carlo simulation (*MCS*) (Al-Sharif *et al.*, 2012) and the Markov Chain Monte Carlo (*MCMC*) method (Al-Sharif & Hammoudeh, 2014d).

In Monte Carlo simulation (*MCS*), a probability density function (*PDF*) and a cumulative distribution function (*CDF*) are generated for the entrance floor arrival percentages. Another *PDF* and *CDF* are generated for the floor population percentages. Random sampling is used to generate random origin and destination floors for each passenger using the arrival percentages *CDF* and the population percentages *CDF*, respectively. This is repeated for each of the *P* passengers. The individual journeys are then combined to form the full round trip, and the time taken to complete the full round trip is calculated. The scenario is repeated for a large number of trials (e.g., 100,000) and the average of all the values taken.

In the Markov Chain Monte Carlo (*MCMC*) method, a transition probability matrix is first developed. This is an *N* by *N* matrix, each element of which represents the probability of the lift moving to the jth floor, given that it is currently at the ith floor. The derivation of each element of the matrix is a function of the percentage arrivals, population percentages and the number of passengers carried in a round trip. A Monte Carlo simulation is then carried out by moving the lift between floors using random sampling of the next floor, based on the transition probability matrix (*TPM*) for a large number of scenarios. Then, the total time is divided by the number of round trips completed, resulting in the value of the round trip time.

The reason why the use of these numerical methods is easier than deriving new equations is the fact that the random sampling used with the cumulative distribution function (*CDF*) embeds the probabilistic nature in the result without the need to derive a probabilistic complex equation.

18.3   Simulation in lift traffic design

Simulation is currently widely used as a complement to calculation for the design of lift traffic systems. The two types of simulation tools are outlined in the next two subsections.

18.3.1   *Discrete event simulation*

Discrete event simulation (Siikonen et al., 2000; Al-Adhem, 2012) exploits the fact that lift traffic systems are discrete by their very nature. For example, when a passenger arrives in the lobby for service, the only times of interest from a traffic design point of view are: the passenger's arrival time; the passenger's boarding time into the lift car; and the passenger's alighting time from the lift car at his/her destination. What happens between these events is of no interest from the passenger's point of view.

Thus, discrete event simulation generates the times at which significant events take place within the system. This fact has a substantial effect on the calculation time, and the complexity of the software. This is the reason why a discrete event simulator can simulate lift traffic activity in a fraction of real time.

18.3.2   *Time slice simulation*

In order to enable the designer to simulate continuous, as well as discrete, events taking place within the lift system, it is necessary to employ time slice simulation (Matlab, Peters Research 2014) as opposed to discrete event simulation. Time slice simulation uses a fixed period of time (called the time slice) as the basis for advancing the simulation. The status of the system is evaluated and updated every time slice, regardless of whether any significant events have taken place or not. This has two disadvantages:

1   It places a heavier processing burden on the software, which means that the simulation takes longer to complete. As the value of the time slice is reduced, this effect becomes more pronounced.
2   Where a large value for the time slice is selected (i.e., in order to reduce the processing burden mentioned in the previous point), it is no longer possible to record the exact time at which a discrete event takes place.

The selection of the suitable value of the time slice is a compromise between the two points above. Selecting a large time slice will reduce the processing burden but increase the error in detecting the exact time that a discrete event has taken place. Reducing the time slice will increase the processing burden but will improve the accuracy of the time recorded for the time at which discrete events have taken place.

18.4   Comparison of calculation and simulation methods

Before comparing the different traffic design methods, it is worth pointing out a major difference between calculation and simulation. Comparisons have been carried between calculation and simulation (Christy, 2012; Jappsen & Rieke, 2012). One of the obvious comparisons is the use of the interval as the basis for design in calculation methods, whereas passenger waiting and passenger transit times are used as the basis for the design under simulation. Within calculation, it is difficult to calculate the values of the passenger waiting time and the passenger transit time. The interval is a much more obvious parameter for the quality of service, and is very easy to calculate in the case of a single entrance under incoming traffic conditions. A good comparison between the calculation of the round trip time and the simulation can be found in section 4.10 of CIBSE Guide D 2010 (CIBSE, 2010).

Although this is true, it does not have to be the case. The main reason why average passenger waiting time and average passenger transit time are not used as the basis of the design in calculation is the absence of suitable formulae. It has become possible to develop analytical-based equations for calculating the value of the average passenger waiting time and average passenger transit time (So & Suen, 2002), or to use numerical methods to calculate them (Al-Sharif *et al.*, 2013a). Moreover, it is also possible to use simulation to find the value of the interval. Definitions of passenger waiting and transit times can be found in Barney (2005).

The use of SimEvents to simulate queue server systems can be considered a form of discrete event simulation. The four criteria that will be used to compare the different methods are listed below (Al-Sharif & Al-Adhem, 2013d):

1   Simplicity/complexity.
2   Calculation time.
3   Convergence to a final value.
4   Ability to evaluate group control algorithms.

These are discussed in more detail in the following four subsections, which compare the four methods: analytical calculation; numerical calculation; discrete event simulation; and time slice simulation.

18.4.1   Simplicity/complexity

This criterion looks at the simplicity of the application of each of the four methods to lift traffic design. It is obvious that the simplest of the four methods is the analytical equation-based method, as it simply requires the substitution of the values of the variables in the analytical equations to arrive at the value of the round trip time. However, this simplicity does not apply to the case where new equations are being developed for new situations. Such derivation of new equations has always proven to be very complicated and difficult as it requires the development of new equations to reflect more complex scenarios. A good example is the case where a new round trip time equation has to be derived for the case where a mixture of traffic is prevailing, or where an equation is needed to calculate the average passenger waiting time under a new group control algorithm. Nevertheless, work has been done in this area that can deal with these special conditions (Peters, 1990).

The time slice simulation is the most complex of the four methods as it takes advanced processing to update the state of the system at every time slice.

18.4.2   Calculation time

The fastest method is the equation-based method, whereby the answer is obtained instantly. The Monte Carlo simulation method can provide an answer in short periods of time (around 10 seconds). Simulation-based tools take much longer times to provide answers (in the range of 1 minute to 5 minutes, depending on the complexity of the building, the group control algorithm, the simulated time and the number of simulations).

18.4.3   Convergence

The analytical equation-based methods are instantaneous and thus the problem of convergence is non-existent in this case. The numerical methods usually take between 1 to 30 seconds to converge. The convergence is difficult to achieve in simulation methods. As the number of simulations is increased, and as the simulation time is increased, the average values start approaching the true value. In practice, designers will repeat ten simulations, and take the average values. It is generally accepted that an approximation is sufficient to guide the designer.

18.4.4 Evaluation of group control algorithms

The calculation methods are very limited in terms of dealing with and evaluating group control algorithms. The same problem is encountered when using numerical methods. Both can only be used for evaluating uppeak group control algorithms.

Simulation methods are ideal for evaluating group control algorithms, but repeatability is a problem (due to the difficulty of convergence). The comparison is summarised in Table 18.1 below.

An additional, more esoteric, point is worth mentioning here. It is important to distinguish between *transient* behaviour and *steady state* behaviour. Simulation-based tools are ideal for capturing this transient behaviour, whereby the lift system increases its throughput in response to an increase in the passenger arrival rate. By definition, calculation tools are more representative of the steady state behaviour of the system once it has settled down following a change in input demand.

Much work has been done recently to ensure consistency in the results from simulation (Finschi, 2010; Hakonen & Siikonen, 2008). This mainly depends on the assumptions in the simulation model and on the used passenger traffic conditions and templates.

In order to improve convergence, the simulation is repeated a number of times (using different random number generator seeds), and the average value is taken (Jappsen & Rieke, 2012 and section 4.7 of CIBSE 2010). This removes the effects of the simulation parameters and passenger artefacts.

In order to allow comparison of different proposals during the detailed design phase, standard design templates are used, such as morning uppeak and lunchtime uppeak, as well as all day templates for offices and residential use (section 4.6.4 of CIBSE 2010).

Table 18.1 Comparison of the four methods

Criterion	Simplicity	Calculation Time	Ability to deal with continuous models	Convergence to a final value	Ability to evaluate group control algorithms
Method Equation-based analytical	Very simple	Very fast	Yes (if equation is available)	Instant	Not practical, although some equations have been developed (Barney: 1992, 1998)
Monte Carlo Simulation	Simple	Fast	Yes	Practical (but is a compromise between accuracy and processing time)	Limited to uppeak group control algorithms
Discrete Event Simulation	Medium level of complexity	Medium	Impractical	Requires the increase of the simulation time and the number of simulations	Ideal
Time slice Simulation	Medium level of complexity	Slow	Ideal	Requires the increase of the simulation time and the number of simulations	Ideal

18.5   Current practice in traffic design

Thus far, an overview of the different methods used within calculation and simulation has been presented. In practice, designers rarely treat calculation and simulation as mutually exclusive. Most lift traffic designers use a hybrid of the two methods.

A designer usually starts with the user requirement stated as the value of the target interval and required handling capacity. The speed of the lifts is then found by calculating the preferred speed that allows the lifts to travel between terminal floors in around 20 seconds. An assumption is made regarding the capacity of the lift car, and it is assumed that the number of passengers in a round trip will be equal to 80% of the rated car capacity. One of the two calculation methods are then used to evaluate the value of the round trip time (usually under incoming passenger arrival conditions). Dividing the value of the round trip time by the target interval and rounding the answer to the next whole number provides the number of required lifts. This suggested arrangement is taken as a first pass solution.

The designer then usually proceeds to the next stage, whereby simulation is used to evaluate the average passenger waiting time and the average passenger travelling time of the passengers, using the arrangement suggested by the calculation method. Based on the outcome, the designer might alter the number, speed and capacity of the lifts.

More importantly, the designer would also attempt to find the effect of the following two parameters on the average passenger waiting time and the average passenger transit time.

1   Mixed traffic conditions: the round trip time from the calculation method is usually based on pure incoming traffic conditions (also known as uppeak traffic). With simulation, the designer can understand the effect of mixed traffic conditions on the passenger waiting and transit times. For example, it has become common to simulate the suggested configuration under preset mixed traffic conditions. Examples of traffic mix conditions include: 40%:40%:20% (Barney 2003a); 45%:45%:10% (CIBSE 2010, British Council of Offices 2014); and 42%, 42%, 16% (British Council for Offices, 2009) incoming, outgoing and inter-floor traffic, respectively. Such traffic mix ratios are believed to be representative of the lunchtime peak traffic conditions in many modern office buildings.

2   Group control algorithm: the round trip time from the calculation methods usually assumes that no group controller is used to allocate landing calls to the various lifts in the group. In order to assess the effect of the group control algorithm on the average passenger waiting time and average passenger transit time, simulation has to be employed. The designer would assess the impact of the different group control algorithms (e.g., destination group control, static sectoring, and dynamic sectoring) on the average passenger waiting time and average passenger transit time. It is not unknown for designers to reduce the number of lifts in the group by one lift for systems that employ destination group control algorithms. However, where this is done, great care has to be taken to ensure that the system is still capable of meeting the lunchtime peak traffic conditions.

So, the simulation process would allow the designer to *fine tune* the design based on the resulting average passenger waiting time and the average passenger transit time.

18.6   Application of queuing theory in lift traffic analysis

An alternative tool that can provide a macroscopic view of the lift traffic system is queuing theory. This section provides an overview of the method and some of the results that can be obtained.

18.6.1 Introduction to queuing theory

Lift traffic systems provide an ideal arena for the application of queuing theory. A lift traffic system is a typical example of a queue server system, whereby the passengers represent customers arriving for service, the lobby is a first-in, first-out (*FIFO*) queue, and the lifts represent a multi-server system. Applying queuing theory to lift traffic systems can be very insightful (Braun, 2003), and provides the designer with a macroscopic view of the operation of the system and the interrelationship between different performance parameters such as the queue length, the waiting time in the queue, and the service time.

Queuing theory has been applied to lifts in order to assess the values of the waiting time and the queue length in the lobby. The most important piece of work that has been carried out in this area is that done by Alexandris *et al.* (Alexandris *et al.*, 1979a; Alexandris *et al.*, 1979b; Alexandris *et al.*, 1981; Alexandris, 1986). The work in (Alexandris *et al.*, 1979a) derives a set of equations that model the lift system under uppeak traffic conditions in order to evaluate four parameters that are considered representative of the system performance. The four parameters are: average passenger waiting time in the main terminal (i.e., lobby); the average queue length of passengers waiting in the lobby; the percentage of busy lift cars in the group; and the passengers delay time. In Alexandris *et al.*, (1979b), formulae for the highest reversal floor in a round trip (*H*) and the average number of stops in a round trip (*S*) are presented, based on an assumed Poisson arrival process. The work assumes steady state operation, and thus assumes that the system manifests this performance continuously, once subjected to the specified level of passenger arrivals. There are two consequences to this critical assumption:

1 The lift traffic system must be designed for an assumed infinitely continuous arrival rate, and must be able to infinitely sustain such an arrival rate without excessive queues forming and without excessive passenger waiting times.
2 The equation cannot deal with the case where the system is subjected to a passenger arrival rate that is sustained for a finite period of time (e.g., 5 minutes, 10 minutes; 15 minutes) whereby the arrival rate is equal to 100% of the design capacity (i.e., system loading is 100%) or even exceeds the system design capacity for short periods of time.

The work shown later in this section attempts to use queuing theory and openly available software to explore the macroscopic view of the performance of a lift traffic system under finite workspaces (e.g., 5 minutes, 10 minutes, and 15 minutes) at system loading rates equal to or exceeding 100%. The motivation is that most lift traffic simulation studies that are carried out today are usually run for a finite period of time at system loading conditions of 100%. Moreover, it is useful, in some cases, to assess the performance of a system when it is overloaded and subjected to system loading in excess of 100% (e.g., understanding the effect of the increase in a building population due to change of function; or assessing an under-designed lift traffic system).

As mentioned in (Markon *et al.*, 2006), queuing theory modelling of lift traffic systems allows a *macroscopic* view of the performance to be formed. It can be used to provide insights into the operation of the system, and to allow general conclusions to be drawn. The lift system details are ignored and the whole lift system is treated as a black box that processes passengers.

Discrete event simulation software, SimEvents, which is part of Simulink/Matlab can be used in order to evaluate the passenger average waiting time, and the average queue length in lifts. It is used to illustrate the effect of the workspace (i.e., simulation time) under the combined effect of the following two conditions:

1   Finite workspace (i.e., transient conditions as opposed to steady state conditions).
2   System loading values equal to or more than 100%.

18.6.2   *The concept of a workspace*

A new term will be introduced in this chapter, that of the *workspace*. A workspace is the period of time over which a variable is calculated (e.g., the round trip time) or the period of time over which a simulation is run. In the classical design of lift systems, it is customary to calculate the value of the round trip time over one round trip. In such cases, it can be assumed that the workspace is one round trip. In cases where the lift system is simulated for finite periods of time, then the simulation will be equivalent to the workspace. It is, for example, customary to simulate the lift system for a period of 5, 10 or 15 minutes (300, 600 or 900 seconds, respectively). Passengers are generated for the duration of the simulation time, but the simulation continues until the last passenger has been *processed* (i.e., has arrived at his/her target floor and alighted). In queuing theory, equations are derived that assume steady state conditions are achieved (e.g., Little's formula, Little, 1961), which is equivalent to setting the workspace (i.e., simulation time) to infinity. As discussed earlier, work in (Alexandris *et al.*, 1979a; Alexandris *et al.*, 1979b; Alexandris *et al.*, 1981; Alexandris, 1986) make such an assumption and derive steady state probabilities of the variables of interest. Three cases of workspace values are listed in Table 18.2 below.

It is unusual in lift traffic simulation to use long simulation workspaces that lead to steady state conditions, as this would lead to excessive queue lengths and excessive waiting, unless the system has been over designed in order to cope.

Once the concept of an infinite workspace has been abandoned, it is important to note that using different workspace values will lead to different results. For example, using a workspace of 15 minutes would result in a longer passenger waiting time and a larger value for the average queue length compared to using a workspace of 5 minutes. It is thus meaningless to enquire about the average passenger waiting time or the average queue length without specifying the workspace (i.e., the simulation time).

18.6.3   *System loading*

As discussed earlier, the system loading is an important concept in the design process. By varying the system loading, the designer can test the system performance. In cases where the design is carried out using a finite workspace, the system loading can be set to exceed the capacity of the system for finite periods in time.

The corresponding term usually used in queuing theory is system utilisation. However, by definition, the system utilisation cannot exceed 100% as it is meaningless to say that a server

Table 18.2 Various values of the workspace

Case	The value of the workspace
Calculation of the round trip time	One round trip
Simulation (transient conditions)	The simulation time which is expected to be larger than the round trip time (e.g., 300 s, 600 s, 900 s)
Steady state conditions	Infinity

is occupied for more than 100% of the time. For these reasons, when attempting to subject the system to loading in excess of 100%, it is more appropriate to use the term system loading.

The emphasis on the arrival rate as an input to the system (in order to vary the system loading) is shown diagrammatically in Figure 18.1. The system has been designed for a passenger processing rate of μ passengers per second (or $c\mu$ in the case of multiple servers). The system is subjected to a passenger arrival rate of λ passengers per second. Once the system has been simulated for a period of time equal to the workspace (*WS*), the performance figures of average queue length and average waiting time can be extracted. By varying the value of λ above and below the value of μ, the system loading can be set to values below and above 100%, respectively.

The general equation for system utilisation (or system loading as will be referred to in this chapter) can be calculated as shown below:

$$\rho = \frac{\lambda}{c \cdot \mu} \tag{18.1}$$

where:
λ is the passenger arrival rate in passengers per second
μ is each lift's passenger processing rate in passengers per second
ρ is the system loading (dimensionless)
c is the number of servers in the system.

But the number of servers in this case is equal to the number of lifts:

$$c = L \tag{18.2}$$

The inter-service time can be related to the lift parameters as shown below (assuming bulk servers (Bailey, 1954)):

$$\frac{1}{\mu} = \frac{\tau}{P_{max}} \tag{18.3}$$

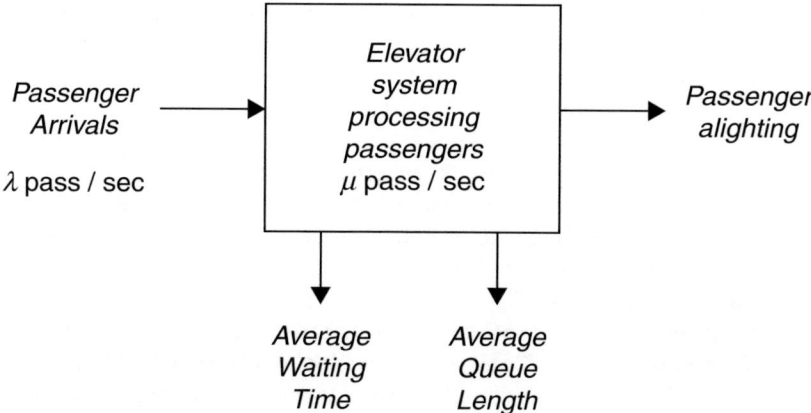

Figure 18.1 System loading diagram

where

τ is the round trip time in seconds when it fills up with P_{max} passengers

P_{max} is the maximum number of passengers that can board the lift car

μ is the passenger service rate for each lift in passengers per second

Substituting (2) and (3) in (1) gives:

$$\rho = \frac{\lambda \cdot \tau}{L \cdot P_{max}} \tag{18.4}$$

Noting that:

$$P = \frac{\lambda \cdot \tau}{L} \tag{18.5}$$

where:

P is the average number of passengers boarding the lift car.

Then, substituting (5) in (4) gives:

$$\rho = \frac{P}{P_{max}} \tag{18.6}$$

... which is only valid for system loading values smaller than or equal to 100%.

This shows that the average car loading will, in fact, be equal to system loading. The work in (Jarrar *et al.*, 2013) was based on a system loading evaluated as the ratio of the car loading. This was valid as the scope of the graph did not cover system loadings exceeding 100%. However, in cases where the system loading exceeds 100%, this measure becomes inadequate. It is then better to use the ratio of the arrival rate divided by effective service rate, shown in equation (7) below.

$$\rho = \frac{\lambda}{L \cdot \mu} \tag{18.7}$$

18.6.4   *The use of SimEvents*

SimEvents is a graphical and modular discrete event simulation tool that is part of Matlab (specifically it is a toolbox within Simulink). It allows the user to build and simulate discrete event systems, and to extract meaningful data about the performance of the system such as queue lengths, waiting times and system utilisation (King and Bouketir, 2008).

A simple example of a straightforward queue/server system is shown in Figure 18.2. It contains an entity generator (the term entity is the generic term used in SimEvents to represent customers/jobs/passengers . . . etc.); a first-in, first-out queue; a multiple server system; and an entity sink (entities do not just disappear, they have to be disposed of somewhere).

A more detailed model has been built in order to represent a lift traffic system, operating in uppeak conditions, as shown in Figure 18.3. The model is an *M/M/c* model (Taha, 2011), whereby the arrival process has an exponentially distributed inter-arrival time, the server has an exponentially distributed inter-service time, and there are c servers (i.e., lifts).

The model allows the workspace time to be varied, whereby passengers will be generated for the duration of the workspace, and the servers are then allowed to continue processing any passengers remaining in the system.

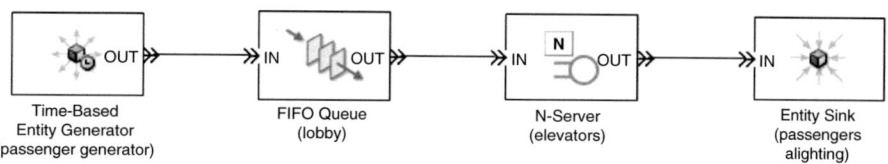

Figure 18.2 Simplified block diagram in SimEvents

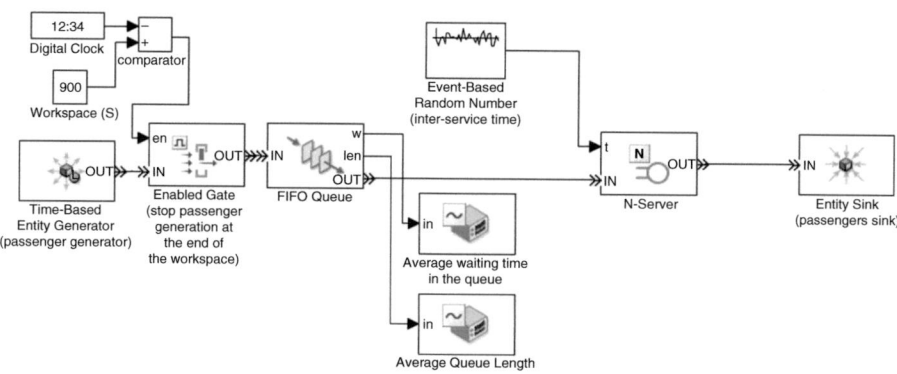

Figure 18.3 Detailed block diagram used in SimEvents in order to simulate the lift system

It is important to remember that nominal simulation time (the workspace) will usually be different from the actual simulation time. To illustrate such a difference, a queuing system has been simulated in SimEvents. The simulation time was set to 900 seconds. However, at the 900-second point, passengers were still present in the queue, as shown in Figure 18.4. The system continues processing the passengers until all have been processed through the servers (i.e., lifts), and discharged, which takes place at the point in time of 985.6 s. Although the nominal simulation time is 900 seconds, the actual full simulation time is 985.6 s in order to allow the system to clear all the passengers waiting in the queue.

18.6.5   Results

In order to verify the results obtained from SimEvents, a Matlab code was written that also generated the average queue length (AQL) and the average waiting time (AWT) for passengers.
   a) Effect of system loading on the average waiting time (AWT) and the average queue length (AQL)
   In the analysis below, system loading values have been varied from 0.1 to 2 (i.e., 10% to 200%). As the workspace in this case is finite (i.e., not steady state condition), the queue lengths and waiting times at system loading values of 100% will not be infinite, but will depend on the workspace duration. The results from the SimEvents package were compared to the results from the Matlab code, and are shown in Figure 18.5 extracted from (Jarrar et al., 2013). Good agreement can be seen in general.
   b) The effect of the workspace on the AWT and the AQL
   The Matlab code was used in order to understand the effect of the workspace duration on the average waiting time and average queue length. Workspace values ranging from 300 seconds

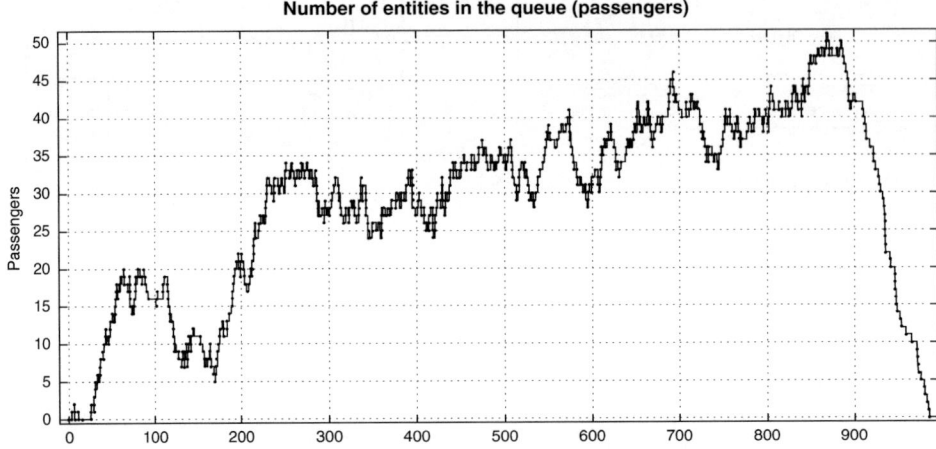

Figure 18.4 Number of passengers in the queue during the simulation

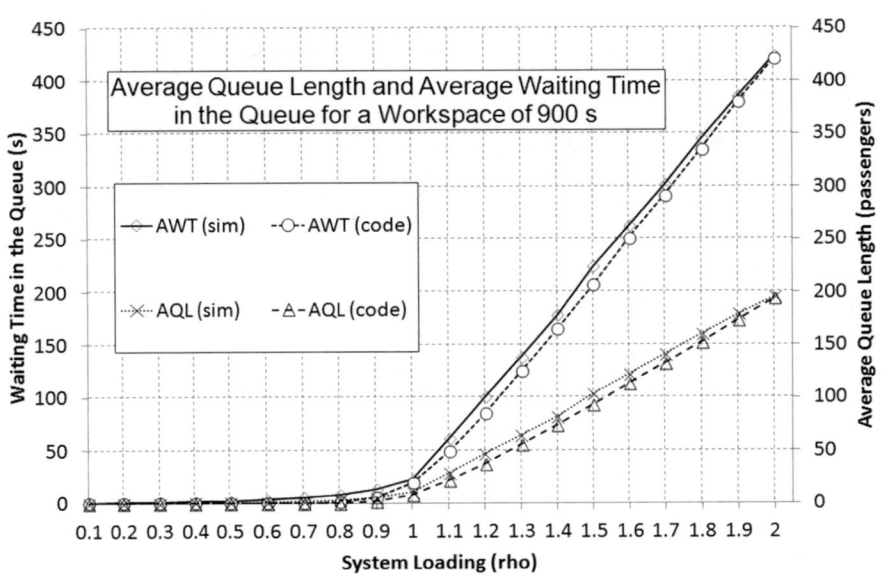

Figure 18.5 Comparison of the results from the SimEvents software and the Matlab code

to 2,700 seconds in steps of 300 seconds were used at a system loading of 100% (i.e., $\rho = 1$). In each case, the average waiting time and the average queue length were recorded. As can be seen in Figure 18.6, the added trend lines show that there is a near linear relationship between the system loading and the average waiting time and the average queue length, at a system loading of 100%.

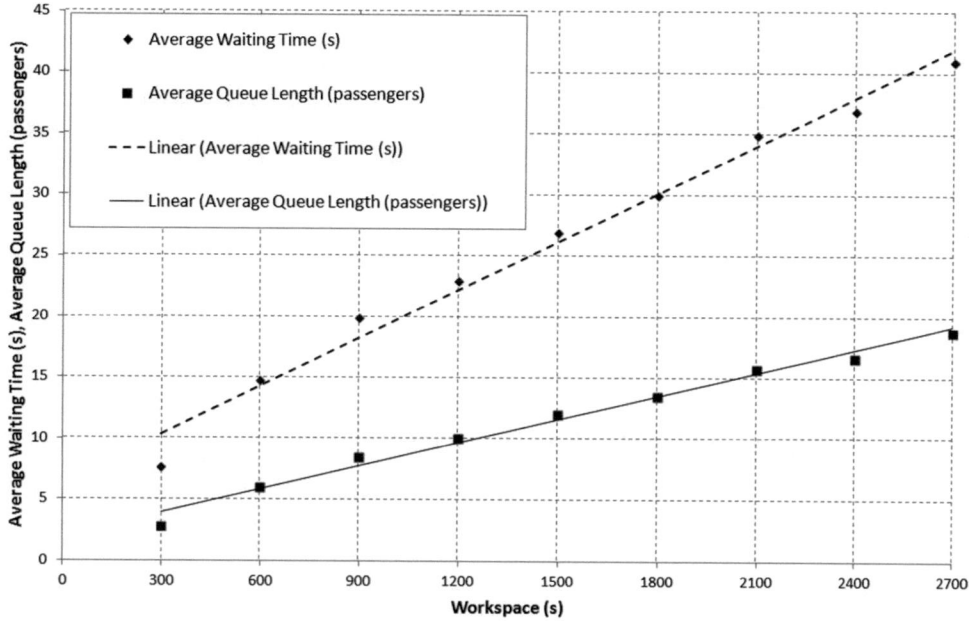

Figure 18.6 The effect of the workspace on the average waiting time and the average queue length at 100% system loading ($\rho=1$)

18.6.6 *Discussion of queuing theory*

The concept of a workspace, as used in lift traffic simulation, has been introduced. It is the length of time over which a variable is calculated, or over which a simulation is run. The concept of a workspace is closely linked to lift traffic simulation. Traditional queuing theory analysis assumes steady state conditions when deriving the equations for the average waiting time and the average queue length. The concept of a workspace, however, assumes finite running time and thus transient conditions. Such an assumption allows the designer to subject the lift traffic system to system loading values in excess of 100%. This could represent real life conditions such as an increase in the population of a building, or an under-designed lift traffic design.

A model in SimEvents has been developed that can be used to provide a macroscopic view of the operation of a lift traffic system. It has been used to show the effect of system loading values in excess of 100% on the average waiting time of passengers and the average queue lengths. The results have also been compared to results from a Matlab code, and good agreement has been found.

The effect of the duration of the workspace was also investigated at system loading values of 100%. At 100% loading, a conventional queue server system under steady state conditions would be overloaded, and waiting time and queue length would approach infinity. However, at finite values of the workspace, finite values of the waiting time and queue length will be attained. It was shown that there is a near linear relationship between the workspace duration, the average waiting time, and the average queue length.

18.6.7 Using queuing theory to reconcile calculation with simulation

Queuing theory can also be used to explain any discrepancy between the results from calculation and simulation. Calculation assumes a constant arrival rate and a constant service rate. Simulation assumes a random arrival rate (i.e., Poisson passenger arrival process), and a random service rate (the value of the round trip time is a random variable that changes in each trip depending on the passenger destinations). Queuing theory shows that if the system is loaded to 100% loading, queues will start growing, and the passenger waiting time will become excessive. It is for this reason that the calculation process must assume a car loading of less than 100%. So, a value of 80% car load can be assumed in the calculation, so that the system will show acceptable performance when run under simulation with random arrival rate and random service rate.

18.7 Modelling of lift traffic systems as chaotic systems

Chaotic systems are examples of non-linear dynamical systems. They are deterministic, rather than random, but are very sensitive to initial conditions. Lift traffic systems are non-linear by virtue of the fact that if a passenger is delayed by a small period of time, they might miss their lift and will have to wait for the next lift. This results in the lift that was missed carrying one fewer passenger, and thus making fewer stops and returning earlier than expected (i.e., had it carried the full number of passengers). For the passenger who missed the lift, they will suffer a longer waiting time. When the next lift arrives, it would have to increase its actual load by one passenger, or displace a waiting passenger, thus causing further disruption for later journeys, and so on. This is a good example, where a small change in initial conditions will cause a significant change later in time, which is characteristic of non-linear dynamical systems (chaotic systems).

A number of researchers have attempted to model the lift passenger traffic as a chaotic system, or the lift movements in response to passenger arrivals as a chaotic system (Nagatani, 2003; Li *et al.*, 2011; Ding *et al.*, 2013). The aim of such modelling is to try to improve the group control algorithm, or to predict future traffic based on current and past traffic. There is no evidence that such a technique has been applied in practice to lift traffic control systems.

18.8 Conclusions

This chapter started with a general overview of the different techniques used for calculation and simulation as applied to lift traffic analysis. This was then followed by an overview of some advanced methods used to further analyse the performance of a lift traffic system, such as modelling using queuing theory and chaotic systems. It is envisaged that these techniques could, in the future, complement the conventional techniques of calculation and simulation in order to provide a better understanding of the selected traffic designs.

19 Traffic surveys

19.1 Introduction

Integral to the lift traffic analysis and design process are traffic surveys. Traffic surveys are an important tool in the lift traffic designer's armoury. They allow the designer or the user to assess demand in a building in order to use this value in future designs, or to verify the performance of an existing building in order to confirm the adequacy of a lift traffic design or to carry out remedial action if necessary.

Research involving extensive surveys of different buildings has shown that traffic patterns in buildings are repeatable over different days, despite small variations (Peters *et al.*, 2011). The traffic pattern in a building can be thought of as a *signature* of that building.

This chapter examines the types of traffic surveys that can be carried out in a building, describes the methodology of carrying them out, and lists the reasons for doing them.

19.2 Types of lift traffic surveys

There are two main categories of lift traffic surveys: a group of surveys that are directly related to passengers; and a group of surveys that are directly related to the lift doors and the lift kinematics, but that has a direct impact on lift traffic performance.

Passenger related surveys

1   Assessing the passenger demand (e.g., the arrival rate) in an existing building (Al Zubi *et al.*, 2011). This type of survey aims to assess the passenger demand in a building. In addition to attempting to quantify the magnitude of the demand, it is also necessary in some cases to understand the split of traffic into different modes of traffic (i.e., incoming traffic; outgoing traffic; and interfloor traffic). Assessing the passenger demand in a building is necessary in two cases. The first case is carried out once a building has been commissioned and occupied to confirm whether the actual passenger demand is equal to that used in the design. The case study presented later in this chapter (section 19.7) was carried out for this former reason (i.e., checking that the performance in the new buildings meets the design requirements). The second case is where the lift system will be refurbished or upgraded, and a survey is carried out in order to ensure that the new design will meet the existing passenger demand. The level of detail in which demand can be quantified varies as follows:
   a)   The simplest form of representation of demand is simply one figure representing the percentage of passenger arrivals in five minutes (λ).

b) As an intermediate representation, the demand can be represented as a mixture of traffic in percentages of incoming traffic, outgoing traffic and interfloor traffic with the arrival rate (λ).

c) In its most complex form, it is represented as a full two-dimensional matrix called the transition matrix that shows the probability of a passenger travelling from one floor to another floor in a round trip, along with the overall arrival rate (λ).

2   Measuring the performance of a lift traffic system in an existing building. This type of survey is more complicated and involved compared to the one presented in point 1 above (Saadeh *et al.*, 2012). Not only does it assess the demand, it also measures the lift system performance under such passenger demand. Such a survey could be carried out, for example, following the commissioning of a new lift system in order to verify the adequacy of a traffic system design, or to understand the reason for the poor performance of a lift system (Malet *et al.*, 1980; McKay, 1980). The types of data that can be gathered within this type of survey are:

a) **Actual round trip time:** the elapsed time between the arrival of a lift at the main entrance and the next arrival of the same lift at the main entrance is the actual round trip time for that lift. It is well accepted that this is a random variable, and the average value is calculated by finding the mean value of these round trips during the period of the survey. Under periods of no passenger activity, the lift might stand still, and this is referred to as *idle time* and must be removed from the value of the round trip for comparison purposes with the calculated value of the round trip time.

b) **Actual interval:** the elapsed time between the arrival of one lift in the group at the main entrance and the next lift in the group is the actual measured interval. By measuring all such values, and taking their average value, the average interval can be calculated. This value can be compared with the calculated value, which is representative of the quality of service.

c) **System response time:** the system response time is the elapsed time between the registration of the landing call and the arrival of a lift in response to the call. This can be easily recorded by an observer located in the lobby. See Definition 6.7.

d) **Maximum car load and average car load:** an example of which can be found in (Peters & Evans, 2008).

e) **Number of door re-openings** (due to a late-arriving passenger, for example): This parameter, for example, is critical to reconciling the evaluation of the round trip time by calculation and simulation.

f) **Stair usage:** This is discussed in Peters & Evans (2008) and in Green & Smith (1977). As mentioned for the last point, this parameter is also critical for reconciling the results of calculation and simulation.

3   Understanding and modelling the nature of the passenger arrival process. An improved understanding of the nature and the model for the passenger arrival process can be used to run more representative simulations for lift traffic. Three examples are shown below:

a) Surveys have been carried out to confirm that the arrival process can be represented as a Poisson arrival process (Alexandris & Barney, 1976; Aqel, 2011; Sorsa & Siikonen, 2014; Lea, 1990). These types of surveys aim to show that the hypothesis that the arrival process follows a Poisson distribution process cannot be rejected (a procedure that is called in statistics *hypothesis testing*). This type of survey is described in more detail later in this chapter (Section 19.6).

b) A survey has been carried out in order to better understand the process of batch arrivals (Kuusinen *et al.*, 2012). Historically, it has been assumed in lift traffic analysis that passenger arrivals are independent, and that the destinations that passengers select are independent. The research by Kuusinen *et al.* (2012) showed that a better representation of passenger arrivals is to assume that the passengers arriving could sometimes arrive in groups of two, three or more passengers in accordance with specific probability density function (pdf).

c)   Surveys are sometimes carried out in order to understand relationships between various lift traffic variables. Examples include trying to confirm the formula describing the relationship between the number of uppeak passengers boarding the lift car, and the number of stops generated by the calls that they register (Al-Sharif, 1992a); or trying to reconcile the values of the round trip time found by calculation and by simulation with the actual values recorded in practice (Al Masri & Rishmawi, 2011).

Lift related surveys

4   Extracting the lift kinematic parameters (distance, velocity, acceleration and jerk). These parameters allow a simulation to be carried out of an existing installation. As mentioned in Peters (2012), the aim of these measurements is to simulate performance, rather than assess passenger comfort. Measuring the kinematic parameters can easily be carried out using an accelerometer sensor placed on the floor of the car and connected to a signal conditioning and processing device. The vertical axis of the accelerometer provides the acceleration in the vertical direction. Integrating this measurement once will produce the velocity, and twice will provide the vertical displacement (from which actual floor to floor heights can be extracted). The acceleration signal can also be differentiated in order to obtain the value of jerk. In systems that still have a levelling speed phase prior to stopping, the measurement could also include the levelling delay time (Figure 19.1 shows an example of such a levelling delay). The processing of the raw data signal extracted must be carried out in accordance with ISO 18738 (ISO, 2012).

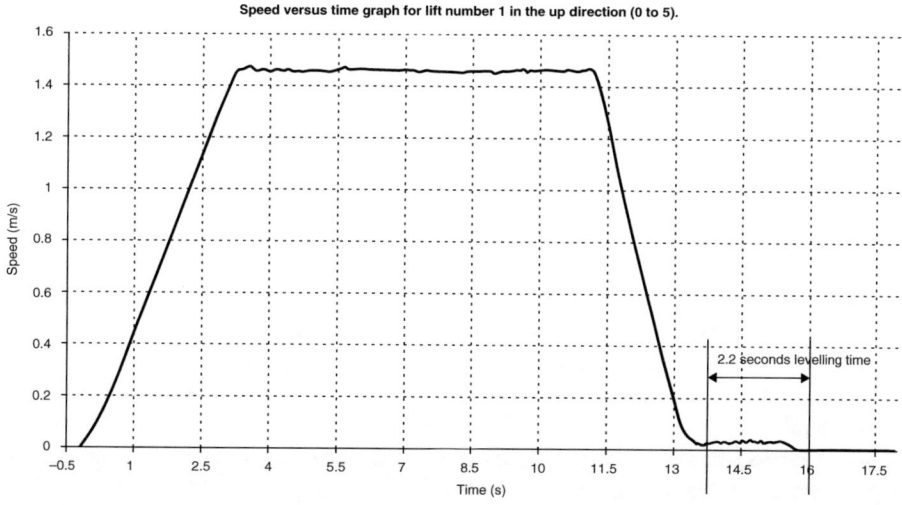

Figure 19.1 Velocity time profile

Table 19.1 Template for on-site data

Item	Unit	Value	Comment
Interfloor distance	m	—	Measure one stair riser and multiply by number of steps
Rated speed up	m/s	—	Use tachometer
Rated speed down	m/s	—	Use tachometer
Car depth	mm	—	Check with rated capacity table
Car width	mm	—	Check with rated capacity table
Car shape		—	Note ease of passenger circulation
Rated capacity	kg/person	—	Check with rated capacity table
Door width	mm	—	Clear opening width
Door type		—	Side, centre, swing, single panel, two panel, etc.
Door closing time	s	—	See Definition 4.15
Door opening time	s	—	See Definition 4.16
Single floor flight time up	s	—	See Definition 4.20
Single floor flight time down	s	—	See Definition 4.20
Performance time up	s	—	See Definition 4.23
Performance time down	s	—	See Definition 4.23
Cycle time up	s	—	See Definition 4.24
Cycle time down	s	—	See Definition 4.24
Car call dwell time	s	—	See Definition 4.18
Landing call dwell time	s	—	See Definition 4.19

Note: The Value column would be filled in on site.

5   Automated measurement of door timing performance: this includes door opening time, door closing time, door advance opening and door start delay. A data logger, combined with an accelerometer, and activated with a pushbutton can be used to record these times. A good discussion on this item is contained in Peters (2012).

It is recommended that the parameters measured in 4 and 5 above are documented using a template such as Table 19.1.

19.3   Tools for carrying out surveys

The different types of passenger related surveys can be carried out using one of the following tools:

1   **Manual paper-based surveys**: manual paper-based surveys are carried out by human observers who locate themselves in the lobbies (for lobby surveys), or in the lift cars (for in-car surveys). Filling in these paper forms is tedious and requires synchronising the form data to a watch. Building tenants (e.g., office employees) are sometimes asked to help by filling in so-called 'diaries' in order to cover interfloor journeys that they travel (within 10 minutes periods) (Courtney & Davidson, 1974), although their effectiveness is questionable.

2   **Manual laptop-based surveys:** although manual surveys are usually based on filling in paper forms, the use of laptops and tablets is becoming more prevalent in order to automate the process (Peters & Evans, 2008). Using a laptop/tablet offers three important advantages:

a) As opposed to the paper form data which require manual entry into the laptop/tablet, this tool obviates the need for manual data entry into a laptop/tablet. By the end of the survey, the data from the survey are populated within a spreadsheet.

b) The person carrying out the survey does not need to synchronise the entries to a watch, as this is done automatically by the laptop or the tablet. This relieves the surveyor from this burden and allows them to concentrate on observing the arriving passengers.

c) The exact arrival time of the passengers to the nearest second can be automatically time-stamped by the laptop/tablet. This allows for more sophisticated analysis of the passenger arrivals later, or for processing them in a different way. For example, when the exact arrival times of the passengers are available to the nearest second, an analysis of the nature of the passenger arrival process can be later carried out (e.g., Poisson arrival process).

3 **Video recording:** it is also possible to place one or more cameras in the lobby in order to record the detailed passenger arrivals and departures. The video can then be later analysed and survey data extracted by human observers. Using video recording to obtain survey data has three advantages

a) The data can be analysed in great detail later at the surveyor's leisure. This removes the risk of data being missed during the survey.

b) The surveyor can revisit certain points in the survey to check any discrepancies or to better understand the events that took place at a certain point in the survey.

c) The presence of human observers can have a disturbing effect on the behaviour of passengers. Using cameras makes the passengers oblivious to the fact that a survey is being carried out, and this removes the effect that the surveyors have on the data (i.e., the 'white-coat effect').[1]

Cameras are usually placed in the lobby. Although possible, there is no evidence that cameras have been placed inside the car for the purpose of in-car surveys.

4 Automatic image processing of videos to infer passenger loadings and unloading. Image processing has become very powerful now that passenger loadings and unloadings from lift cars can be extracted from video data (So & Kuok, 1992; Schofield *et al.*, 1995).

5 Fully automated data extracted from the lift controllers, data loggers/analysers or the building management system. These are listed below:

a) The lift controller data can be communicated via serial ports or Ethernet. The lift controller could record a wealth of data that can be used to extract data about passenger movements (such as weight sensing device and door light beam data). It can also record system response times, call registration and cancellation times, and lift car movements.

b) Infrared beams installed on lift doors. By having two beams, the boarding and alighting of the passenger can be reliably inferred (Peters *et al.*, 1996a), (Kaakinen & Roschier, 1991), (Siikonen, 1991).

c) Dedicated data loggers and data analysers that are independent from the lift controller (Bril & Marsh, 1984).

d) Security access systems connected to building management systems. For example, swipe card data from employees could indicate the passenger arrival and departure times and their destination, as well as their movement (i.e., interfloor) within the building during the day.

19.4 Traffic demand surveys steps

This section examines lobby surveys and in-car surveys in more detail. A comprehensive systematic survey procedure for lobby surveys and in-car surveys can be found in Peters & Evans (2008), and readers are advised to refer to it for more detail.

When carrying out lobby surveys in order to assess the passenger arrival rate as a percentage of the building population in the busiest five minutes (λ), it is necessary to divide the surveyed number of passengers arriving in the busiest five minutes by the building population. There are two options for the value of the population to be used: the *nominal population* or the *observed population*. The nominal population is the population calculated based on floor area and the expected occupant density. The observed population is the total number of passenger arrivals during the day, obtained from a survey (Peters & Evans, 2008). The nominal population will always be larger than the observed population, and using one or the other will make a significant difference to the final arrival rate obtained.

Thus, when an arrival rate (λ) is extracted from the data, it must be clearly stated whether this figure is based on the observed population or the nominal population. For example, Green and Smith (1977) used the observed population, rather than the nominal population.

19.4.1 Lobby survey

It is assumed that the lobby survey is, in effect, a passenger arrival/departure survey. In cases where the building has multiple entrances, it is recommended that an observer be stationed at each of the entrances. Heavy arrival entrances might require two observers.

The steps in this survey procedure are listed:

1 One or more observers (depending on the number of lifts and how busy the lobby is), stand(s) in the lobby 15 minutes before the start of the peak.
2 Each observer has a copy of the form shown in Table 19.2. Each observer must also have a watch that shows seconds (e.g., a digital watch).
3 If more than one observer is present, they must ensure that their watches are synchronised to the nearest second, if possible.
4 Every time a lift arrives at the lobby, fills up and leaves, an entry must be made in the table.
5 The entry has four elements: the time (in seconds) that the doors are fully closed, the lift ID/number, the number of passengers that leave the lift upon arrival (if any) and the number of passengers entering the lift.
6 It is suggested that the survey is done for the duration of the morning uppeak (e.g, 07:55 to 09:35).
7 It is advisable to repeat this over five working days of the same week to capture any variations.
8 Special comments regarding an entry can be entered in the last comments column (e.g., not all passengers could enter the lift due to crowding, or lift doors reversed three times).
9 It will be useful to have some commentary from the building management as to how busy that week was compared to other weeks.

19.4.2 In-car surveys

The observer enters the car and starts observing the passenger movement into and out of the car. The form shown below in Table 19.3 can be used for the survey. Special marks can be used

Table 19.2 Form for a lobby survey

Time doors closed to depart from main lobby	Lift number	Number of people leaving that lift upon arrival	Number of people boarding the lift	Comments
08:03:42	4	1	1	
08:04:18	1	0	4	
08:05:21	2	0	3	r/o
08:05:35	1	0	1	
08:06:14	4	0	1	
08:06:41	1	0	1	
08:07:05	6	0	5	
08:07:34	3	0	3	
08:08:16	5	0	4	
08:08:16	4	1	0	r/o
08:08:47	2	0	1	
08:09:44	4	0	2	
08:10:47	5	0	2	
08:11:46	3	0	2	
08:12:18	2	0	2	
08:12:25	5	0	1	
08:13:06	3	0	2	
08:13:20	1	0	2	r/o
08:13:34	2	0	1	
08:13:52	5	0	3	
08:14:14	3	0	2	r/o
08:14:44	2	0	2	

Table 19.3 Survey form for interfloor traffic

09:10	/Gnd 4 in/2nd 1 out 1 in/ 5th 2 out/ 11th 2 out/ 10th door did not open/ 3rd 1 in/ 0th 10 in 1 out/ 2nd 2 out/ 6th 7 out/ 11th 1 out// Gnd 5 in/
09:15	1st 1 out/ 4th 1 out 1 in/ 6th 2 out 1 in/ 10th 2 out/ 11th 1 out 4 in// 5th 2 out/ 4th 2 out 1 in/ Gnd 1 out 6 in// 2nd 1 out/ 4th 4 out/ 10th 1 out/

to separate journeys (/) and to denote direction reversal (//). The rows are filled in periods of 5 minutes in order to synchronise them later with the lobby surveys.

Legend
/   floor to floor journey
// reversal
1 in   one passenger leaves the car
3 out three passengers leave the car

Thorough examination of Table 19.3 needs then to be carried out in order to separate the interfloor passengers from the incoming and the outgoing passengers. Once the total number of interfloor passengers has been extracted for each five minutes, the total number of interfloor passengers can be calculated by multiplying this number by the number of lifts in the group (assuming that the one lift survey is representative of the other lifts).

The total traffic arrival rate (λ) can then be found by adding the interfloor traffic to the incoming traffic and the outgoing traffic compiled from the lobby surveys.

19.5   Examples of lobby arrival surveys

It is possible to assess the passenger demand in an existing building by carrying out a survey to measure the arrival rate.

A chart showing typical passenger arrivals in a lobby is shown in Figure 19.2. It shows a minute-based survey of passengers arriving in the lobby. The x-axis (abscissa) has units of time, gradated in minutes. The y-axis (ordinate) is the number of passenger arrivals. Each bar represents the number of the passengers arriving in each minute. In order to find the peak arrival in five minutes, a five-minute-wide window is slid horizontally over the data. The window is stopped at the position that produces the maximum total number of passengers in five minutes. The window is shown in Figure 19.2, placed in the position that produces the largest number of passengers arriving in five minutes.

Based on the survey results, it is possible to quantify the arrival rate (λ). Once the maximum number of passengers arriving in five minutes has been found using the sliding window tool, this yields the number of passengers per five minutes. In order to normalise it, this number must be divided by the nominal/observed building population, U.

$$\lambda = \frac{P_{5\min}}{U} \tag{19.1}$$

where:

λ is the arrival rate expressed as a percentage of the building population arriving in the busiest five minutes

P_{5min} is the number passengers arriving in the busiest five minutes expressed in units of passengers per five minutes

U is the total building population in persons.

Demand can be assessed independently of supply. It is important to note that the passenger arrivals are independent of the lift system. Passenger demand can be assessed and quantified regardless of the presence or otherwise of a lift system that is capable of serving it.

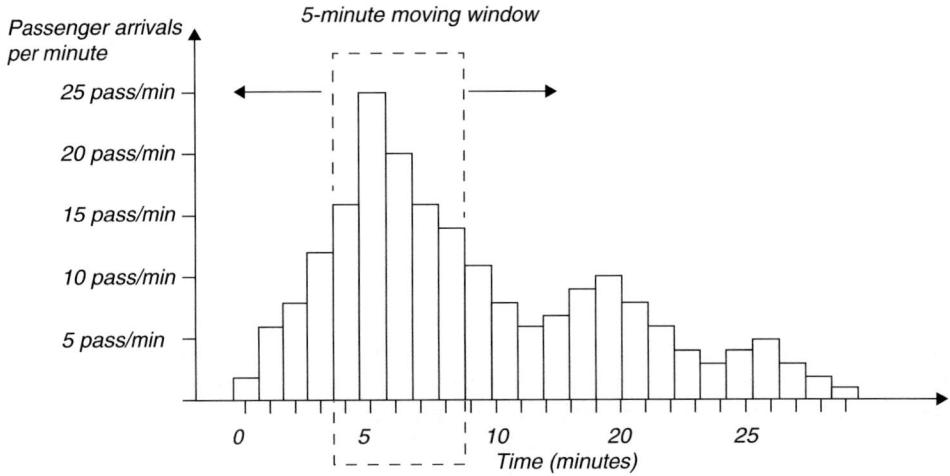

Figure 19.2 A minute by minute survey of passengers arriving in the lobby

The following two examples reinforce the concept of finding the arrival rate from the five-minute survey. It is worth noting that the arrival rate (λ) extracted from the data was based on the nominal building population, rather than the observed building population. The nominal building population is inferred from the floor areas and the expected floor density based on the building usage function (e.g., offices). The observed building population is the total number of passengers who are counted in a full day survey (which is obviously not always easy to carry out).

Example 19.1

The results from an arrival survey carried out at the reception entrance of an office building are shown Table 19.4. Assuming that the nominal building population is 500 persons, assess the passenger demand by evaluating the arrival rate (λ).

Using a five-minute wide sliding window and moving it over the data in the table above, it can be found that the maximum number of passengers arriving in a consecutive five-minute period occurs between 08:05 and 08:09, with a total number of passengers arriving totalling 78 passengers (10+14+24+18+12).

Dividing this number by the nominal building population of 500 persons gives an arrival rate (λ) of 15.6%.

Example 19.2

Results from an arrival survey in an office building are shown in Table 19.5. If the nominal population of the building is 1,500 persons, find the peak arrival rate in the percentage population per 5 minutes (λ).

The arrivals per minute have been plotted in the form of a bar chart in Figure 19.3. It can be seen that using a sliding window shows that the peak five-minute demand occurs between 07:37 and 07:41. The total number of passengers arriving during this five-minute period is 183 passengers per five minutes. Dividing this number by the total building population of 1,500 persons gives an arrival rate (λ) of 12.2%.

Table 19.4 Survey data for Example 19.1

Time (hh:mm)	Passengers/minute
08:01	3
08:02	5
08:03	6
08:04	12
08:05	10
08:06	14
08:07	24
08:08	18
08:09	12
08:10	4
08:11	10
08:12	6

Table 19.5 Survey data for Example 19.2

Time (hh:mm)	Passengers/minute
07:35	12
07:36	15
07:37	30
07:38	45
07:39	60
07:40	30
07:41	18
07:42	15
07:43	9
07:44	2
07:45	3
07:46	6
07:47	7
07:48	4
07:49	2
07:50	1

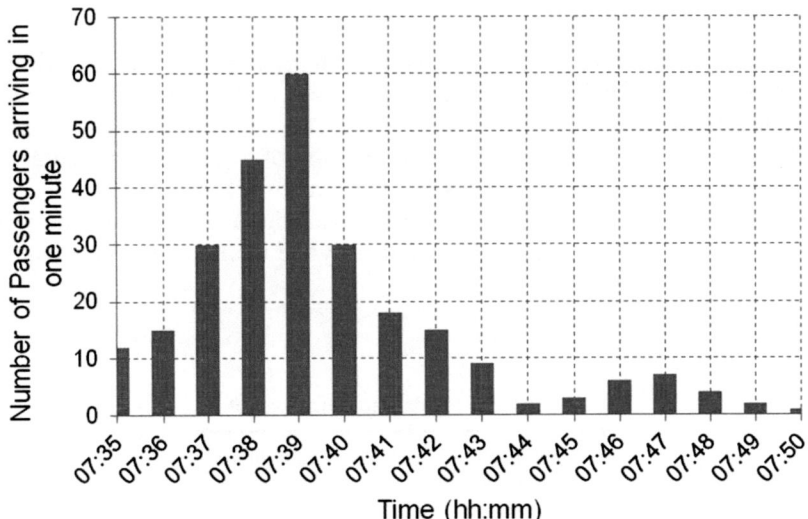

Figure 19.3 A bar chart plot of the passenger arrivals for Example 19.2

19.6   Example of a Poisson process arrival survey

As discussed earlier in this chapter, one of the reasons for carrying out surveys is to better understand the nature of the passenger arrival process. It has been long believed that the passenger arrival process is a Poisson process (where the random variable is the number of passengers arriving in a period of time (t)).

This section presents an example of a survey carried out to check the probability distribution function of the arrival for service. It illustrates the methodology used to collect the data and compile it.

A survey was carried out for one complete hour. The number of passengers arriving in each minute is recorded. In total, there are 60 minutes, and the number of minutes in which zero passengers arrived is found, as well as the number of minutes in which only one passenger arrived, and in which two passengers arrived, etc. This results in Table 19.6.

The data from Table 19.6 have been plotted as a frequency distribution in Figure 19.4, and as a probability density function (pdf) in Figure 19.5.

The steps that have to be followed in order to check whether a process follows a Poisson arrival process are listed below:

1 Decide on the process that needs to be checked.
2 Decide on a suitable survey time (T). This is the total time of the survey (e.g., 1 hour).
3 Decide on a suitable period of time (t) during which the passenger arrivals will be surveyed (for example, 1 minute or 5 minutes).
4 Carry out the survey (i.e., record the number of passenger arrivals during each period of time (t).
5 Develop the frequency table.
6 Convert the frequency table to a probability density function (pdf) by dividing by the total number of readings.
7 Plot the pdf and the cumulative distribution function (cdf).
8 Find the effective arrival rate of passengers in units of passengers per period, usually denoted as λ.
9 Generate the theoretical pdf and the cdf that would be expected under a true Poisson process using the same t and the same λ.
10 Carry out a chi-squared test on the pdf of the observed data and the pdf of theoretical arrival process.

Table 19.6 Processed survey data

n	Frequency	Number of passengers (n)	Probability $P(n)$
0	15	0	0.250
1	11	11	0.183
2	9	18	0.150
3	3	9	0.050
4	4	16	0.067
5	2	10	0.033
6	4	24	0.067
7	4	28	0.067
8	2	16	0.033
9	4	36	0.067
10	0	0	0.000
11	1	11	0.017
12	0	0	0.000
13	1	13	0.017
14	0	0	0.000

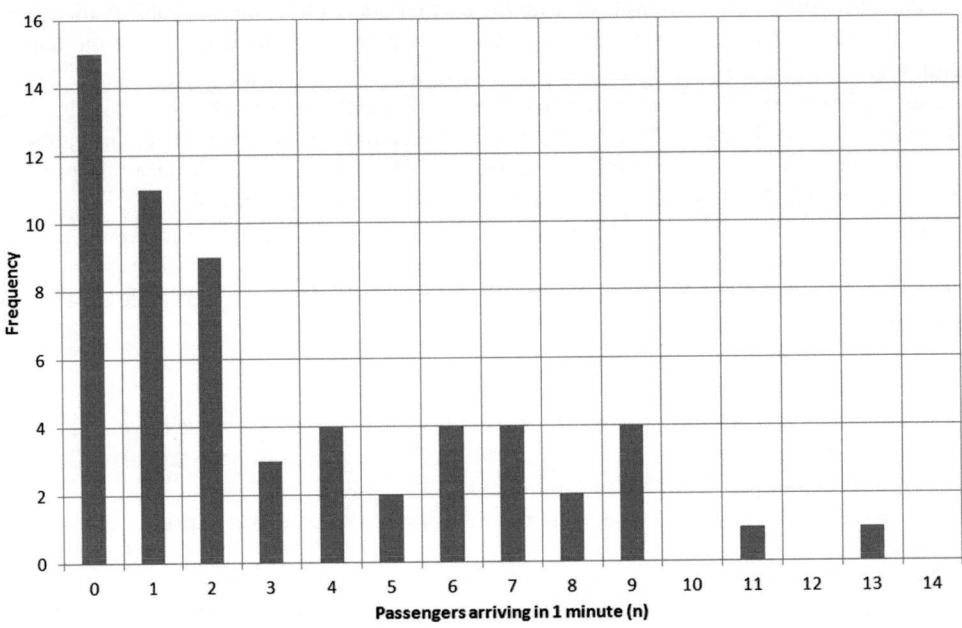

Figure 19.4 Frequency of observations during the survey

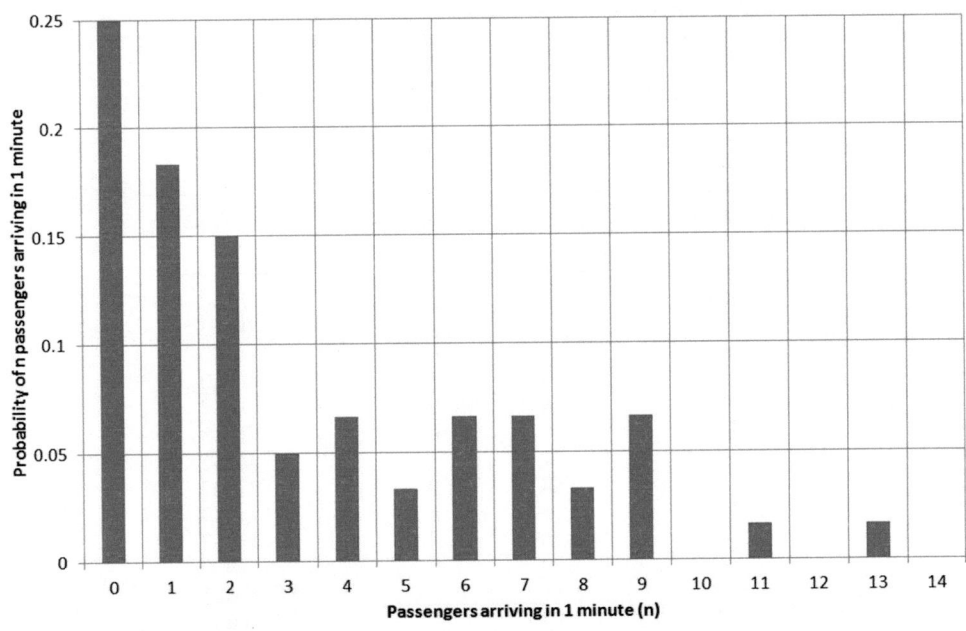

Figure 19.5 Probability distribution function (pdf) of the survey data

19.7   Survey case study

This section describes a demand assessment survey carried out in a new building after it had been occupied by the tenants.

19.7.1   Description of the building

The building comprises 12 floors above ground. It has two basement car parks denoted as −2 and −1. The car parks contain 30 car park spaces. Access to the lifts from the basement is achieved by the use of swipe cards.

The building has 12 lifts of rated load of 1,600 kg and running at 1.6 m/s. The design of the building assumes a split into two zones: one group of six lifts serving floors 1 to 6; and the other group of six serving floors 7 to 12. The original study concluded that the expected interval was in the range of 18 to 21 seconds. The studies and the survey carried out while the main tenant was occupying the previous building concluded that an excellent level of service would be achieved at this building, provided that the same traffic patterns that were observed in the old building persisted in the new building.

When the main tenant first moved into the building, all 12 lifts were allowed to serve all floors. This resulted in a poor performance and long waiting times. They then reverted to the zoned arrangement where the lifts are zoned for floors 1 to 6 and 7 to 12, with both groups serving common floors (3, 10 and 11).

The main tenant in the building occupies floors 3 to 11. Floors 1 and 2, as well as 12, are sublet to other tenants. The floor 10 contains meeting rooms for the main tenant as well as offices and the main reception. Floor 11 contains the restaurant as well as offices for the main tenant. Floors 1 to 9 are identical in the plan. Floors 10, 11 and 12 are stepped in (i.e., reduced area). There is an ATM on the floor 3. Hence, floors 3, 10 and 11 are common floors served by both groups of lifts.

The net area in the building is 49,943 m² net. Based on an area per person of 14 m², this works out at a total design population of 3,138 persons. It is worth noting that 14 m² with no absenteeism is equivalent to a density of 12 m² with 10% of the population absent (90% of the population present on any one day).

One group of lifts serves the lower zone (floors G, 1 to 6, in addition to 10 and 11). These are called the south six lifts, numbered 1 to 6. The other six lifts, the north six (numbered 7 to 12), serve the upper zone (floors G, 3, 7 to 12).

19.7.2   Background information

This section presents some definitions that are essential to understanding the surveys carried out and the system performance.

There are three main types of traffic: up traffic, down traffic and interfloor traffic. These are defined in Definitions 4.1, 4.2 and 4.5.

The passenger performance parameters used: average waiting time (AWT), average travel time (ATT) and average time to destination (ATTD) are defined in Table 6.9.

19.7.3   Description of surveys

Two surveys were carried out: a lobby survey on the first day and an in-car survey on the following day.

The aim of the lobby survey was to understand the arrival pattern at the lobby, obtain an estimate of down traffic from the building, and assess typical car loading. The lobby survey involved counting the number of people leaving each car arriving at the lobby/ground, and counting the number of people boarding each car leaving the lobby/ground. The number of people boarding the cars in a five-minute period represents the magnitude of the up (incoming) traffic and the number of people leaving the cars in a five-minute period represents the magnitude of the down (outgoing) traffic. The maximum number of people boarding the car at any one point in time represents that maximum possible loading of the car.

The aim of the in-car survey is to understand the interfloor traffic patterns. An in-car survey involves an observer standing in the car, recording the number of people boarding the car at a specific floor, and the number of people alighting from the car at a specific floor. The number of people boarding the car from the lobby/ground in a five-minute period represents the up (incoming) traffic. The number of passengers alighting from the car at the lobby/ground floor in a 5-minute period represents the down (outgoing) traffic. The number of passengers moving between floors (i.e., excluding these going to or coming from the lobby/ground floor) represents the interfloor traffic.

A kinematics and timing survey was also carried out. This measured the speed, acceleration, jerk of the lifts and the door timings.

19.7.4   Survey results

The following was noticed from the survey:

1   The peak of the morning traffic took place in the five-minute period between 09:20 and 09:25. Expressed as a percentage of the design building population, this amounted to a peak of 7.0%. This figure comprised 4.8% incoming traffic, 1.2% interfloor traffic and 1.0% outgoing traffic. When expressed as the percentage of the observed building population, this works out at around 12% in total. A plot of the traffic expressed as a percentage of the building population present is shown in Figure 19.6.

2   There was a large number of door re-openings during the survey (nearly one of every two boarding events at the ground floor involved at least one re-opening).

3   Passengers waiting in the lobby were willing to let a lift go and wait for the next one, despite the fact that they could have squeezed in. This is an indication of the fact that they felt confident another lift would be down soon.

4   The cars filled up to around 11 passengers, after which, passengers waited in the lobby and would not board them (i.e., perceiving them as full). This is much lower than the theoretical capacity of 21 persons and even the capacity used in the design (80%, 16.8 persons). In effect, the actual capacity of the lifts is 53% of the theoretical maximum. See Table 5.8.

5   There was a significant amount of traffic to and from the restaurant on the floor 11 in the morning peak (people coming in and travelling to floor 11, and then to their floor; or people coming in, heading for their desk, leaving their briefcase and then going to floor 11 and back).

6   The door dwell time was measured as 2.5 s. The door closing time and the start delay were measured as 5.8 seconds. Assuming the door closing time is 2.9 seconds, then the start delay is 2.9 seconds. The door opening time was measured as 1.5 sec.

7   The speed was measured as 1.6 m/s. The acceleration was measured as 1.0 m/s^2. The jerk was measured as 1.2 m/s^3.

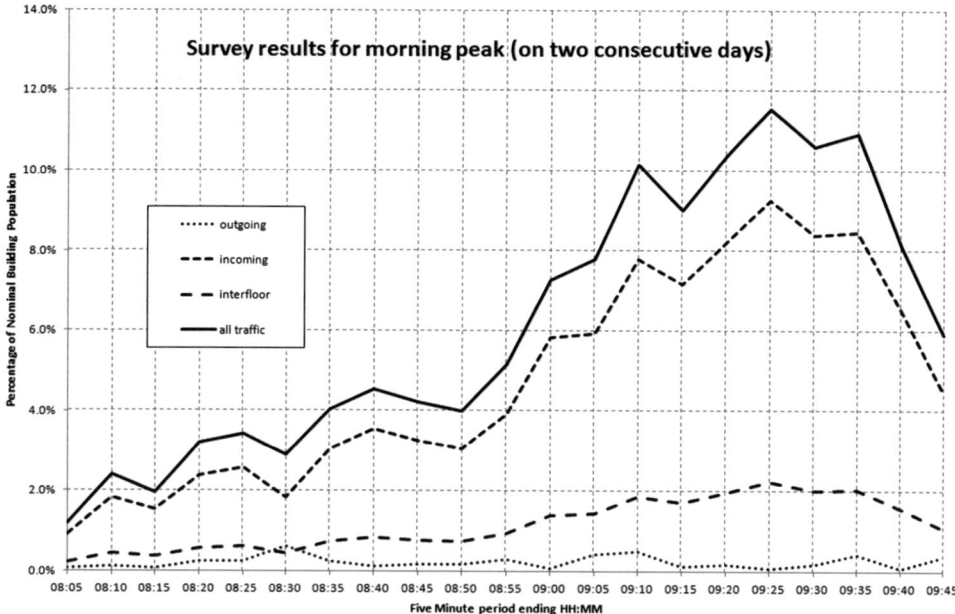

Figure 19.6 Results of traffic survey

Using the building parameters and all the results from the survey, a simulation of building morning peak was run. This showed that current passenger performance parameters are as follows:

- Passenger average waiting time: 8.4 seconds.
- Passenger average travel time: 53.3 seconds.
- Passenger average time to destination: 61.8 seconds.

This is an excellent level of service, and is well within the industry de-facto standard of 30:60:90 (waiting, transit, time to destination, respectively). In fact, under the current zoned arrangement, the lift system can cope with an arrival of 11% in the morning peak.

However, the simulation also shows that under an unzoned arrangement (i.e., all lifts serving all floors), the system can only cope with an arrival rate of 7.5% (providing passenger performance figures of 16.3 s, 75.2 s and 91.5 s). This explains why the system performance was poor when the system was unzoned at the start of the tenancy of the main tenant in the building.

19.7.5 Conclusions

The current lift arrangement in the building provides an excellent level of service. This is mainly due to the low level of arrival in the morning peak (around 7.0% of the observed population) compared to the assumed design arrival (15%). The current arrival rate of 7.0% comprises 4.8% incoming traffic, 1.2% interfloor traffic and 1% outgoing traffic.

It is clear that the current zoned arrangement is critical to the success of the lift traffic system in the building. Moving to an unzoned arrangement would effectively place the system on the limit of its performance, and any increase in demand would lead to deterioration in performance.

The survey has also shown a number of discrepancies between the design parameters assumed and the actual values measured on-site. The most significant of these parameters is the actual car loading, measured at around 53%, whereas the design value used is 80%. The other significant parameter is the start delay measured at 2.9 seconds, whereas the assumed design parameter is 0.5 seconds. These and other discrepancies explain the difference between the assumed level of performance and the actual level of performance of the system.

Note (shown in text as[1])

1 The term 'white-coat effect' originates from blood pressure measurements where the reading taken by the physician in the clinic is elevated compared to the normal value due to the nervousness of the patient during the measurement process, and is thus not representative of the true value of the blood pressure.

20 Current trends and future developments

20.1 Introduction

This chapter examines the current trends in lift traffic system analysis and design, and how they are driving the future of the industry. It is expected that these future developments will decide the shape in the area of lift traffic analysis and design.

Five different areas are discussed in the next five sections. Despite the fact that some areas relate to the electrical and mechanical engineering design of the lift system, they have been selected due to the significant impact they have on the traffic engineering design.

20.2 Complex traffic arrangements requiring more complex design and control

As the traffic arrangements become more complicated in buildings, this is expected to place a heavier burden on the lift group controller. The conventional lift system arrangement comprises one lift car travelling in one vertical shaft. Such an arrangement is developing, and becoming complex in two aspects:

1   Multiple lifts in the same vertical shaft, where the movement is bidirectional. A good example of such complex arrangements is the installation of two roped lift cars in the same vertical shaft, referred to as the twin lift system, which was introduced by ThyssenKrupp of Germany (Thumm, 2004). The twin system, as introduced by ThyssenKrupp, has four safety levels, listed below.

   a)   The first safety level involves the allocation of landing calls to the individual lifts in the group. The lift group controller has to try to allocate the landings calls to the different cars in order to reduce the probability of one car causing an obstruction to another, or in extreme cases, causing a deadlock or lockup (i.e., a 'Catch 22' situation). For these reasons, a hall call allocation group controller is necessary in order to make the system viable (Smith & Peters, 2004). Such requirements place a heavy burden on the single car and the group controller.

   b)   The second safety level involves the control of the speed and position of each car to preserve a minimum safety distance between them. For safety reasons, this is achieved by a completely independent electronic controller that senses the speed and position of the individual cars, and sends commands to the individual controllers to slow down in order to preserve the stipulated minimum safety distance.

   c)   The third level involves the mechanical detection of the distance between adjacent cars, tripping a safety device that applies the operational brakes on both cars.

d)　The fourth, and ultimate, safety level involves the application of the safety gear on both cars using purely mechanical devices.

Another related challenge is the sizing of the twin traffic system to meet the expected arrival rate. Calculation cannot be realistically used in this case, and the designer has to fully rely on simulation.

2　Multiple lifts in the same shaft with horizontal as well as vertical movements: such systems are referred to as 2D (two-dimensional) and 3D (three-dimensional systems). As the cost, practicality and safety of linear motor drive systems improve, such systems will eventually become available on the market. A basic overview of some of the possible 2D systems is presented by So (So *et al.*, 2014). Three such systems are shown in Figure 20.1, Figure 20.2 and

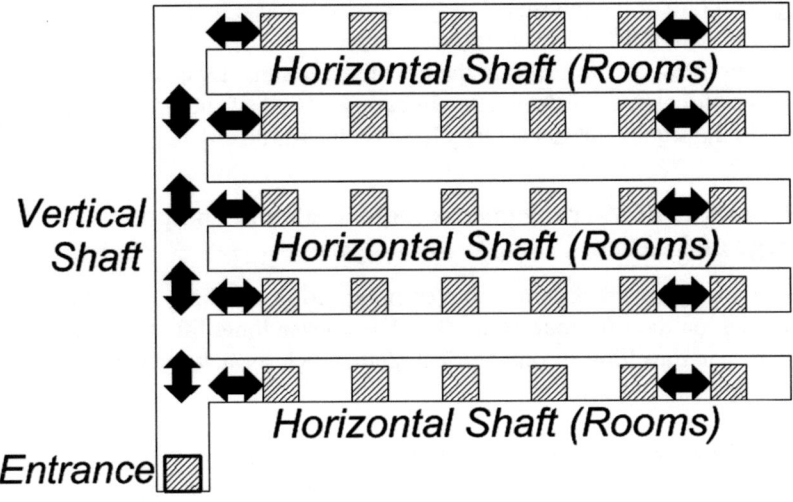

Figure 20.1　L shaped layout

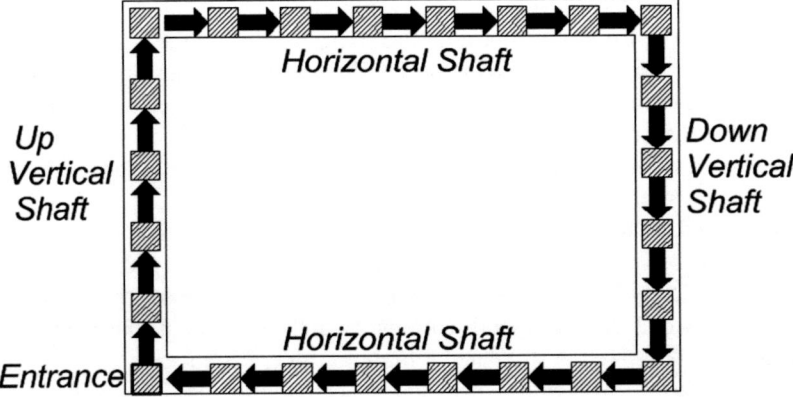

Figure 20.2　Ring shaped layout

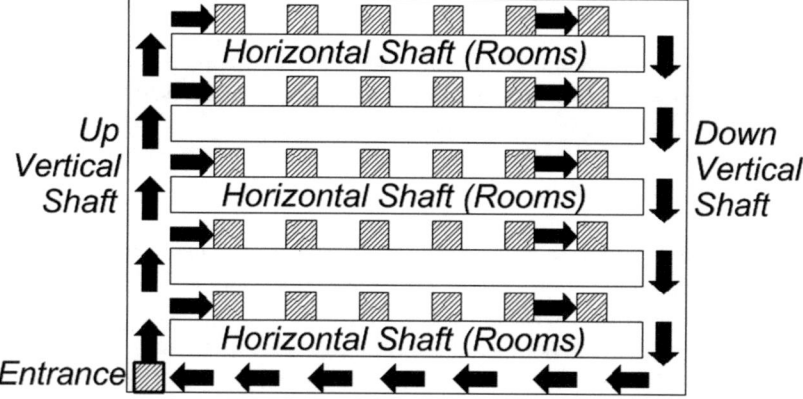

Figure 20.3 Rectangular shaped layout

Figure 20.3 for three different possible arrangements. Two 2D prototype systems have been developed; one by Hitachi and the other by ThyssenKrupp (called the multi, Thyssenkrupp Elevators 2014). The most attractive arrangement is the one whereby two adjacent vertical shafts are used, one for the up direction, and the other for the down direction (very similar to the Paternoster of old). The two vertical shafts are linked by two horizontal sections to allow the lift cars to transfer from one to the other. It is expected that such systems could offer a 50% reduction in the required core space.

20.3   Will hall call allocation group control spell the end of sectoring?

It has been long accepted that lift traffic systems installed in buildings with more than 20 floors should be sectored or zoned. The main reason for such a requirement is to prevent excessive passenger travelling times. Another requirement for sectoring or zoning is to segregate the different modes of traffic in the building for reasons of privacy and/or security (e.g., a tall building that comprises a hotel, residential floors and offices). It is nevertheless recognised that having a dedicated bank of lifts for each section of the building can prove to be wasteful, especially if traffic peaks in different sections of the building do not coincide.

Hall call allocation group control systems are becoming more widely used in new lift installations. One of the main advantages of using them is the fact that they are able to group passengers so that the number of stops are reduced, and hence, the passenger travelling time is kept below specified values.

Some studies have thus suggested that the lifts within different groups be combined into one group and controlled by hall call allocation group control. Wit (2014) explores this concept of exploiting the spare capacity. In a series of simulation for a number of buildings, he shows that a saving of around 25% to 35% in the number of lifts and the core space can be realised if the capacity is shared by combining the lifts into common groups serving sectors or zones. However, special measures need to be introduced in order to preserve the privacy/security between the different types of floors (e.g., hotel, residential, offices). One possible suggested solution for preserving privacy is to use through-lift-cars (i.e., front and rear doors).

20.4   Reconciling calculation and simulation

One of the perennial problems that the lift industry has long grappled with is the reconciliation between the design of a lift system using calculation, and the corresponding results obtained from such a system under simulation. It has always been disconcerting to find that systems designed using the conventional calculation techniques do not perform very well under simulation. It has become customary that the designer would tweak some of the parameters of the lift traffic design in order to achieve acceptable performance under simulation. This is usually carried out by trial and error, rather than following a clear systematic procedure.

Some work has been done in order to reconcile the design produced by calculation, and the results from a simulation of the system. Peters (2013) applied a gradually increasing passenger arrival rate to a lift system, and compared the time response of arrival rate and service rate in a way that was analogous to techniques in systems engineering and system dynamics. He also compared the value of the round trip time under calculation and simulation, applying a loss percentage value in order to account for inefficiencies during simulation.

Another piece of work was carried out in order to use queuing theory to explain why the system design by calculation does not perform adequately under simulation (Al-Sharif *et al.*, 2014b).

There is currently a lot of pressure from the lift industry for more research to be carried out in this area in order to provide a systematic, clear explanation for the differences between calculation and simulation. This will also result in clear procedures for designing lift traffic systems, using a combination of calculation and simulation.

20.5   Evacuation of buildings by lifts

It has been accepted for a long time that lifts are not to be used for evacuation, especially in case of fire. Stairs are relied on as the means of evacuation from buildings.

However, as buildings progressively increase in height, it is obvious that this is not a viable and safe option for evacuating passengers from buildings in the case of emergency. Evacuating by stairs takes an excessively long time, since the walking speed down stairs becomes very low as the stairs become increasingly crowded when the slower walkers block those coming from behind. Moreover, the 11 September 2001 incident placed more pressure on the lift industry to reconsider the *status quo*. It has become obvious that evacuation by stairs is putting building residents at an increased risk compared to the perceived risk of using lifts for evacuation.

An earlier piece of work by Klote (1993) presented a methodology for calculating the evacuation time from the building using lifts. It derived a set of equations from first principles, which are very similar to those used to calculate the round trip time for the basic lift traffic design process.

A ground-swell of opinion has been forming with the view that lifts must be used for evacuation of buildings, and that most lifts, if not all, must be designed to be able to operate for a specified period of time following an emergency (Barney, 2002; Barney, 2005b; Fortune 2010). The view is that total evacuation time should be in the range of 15 to 20 minutes. Evacuation by stairs will still remain preferable for low rise buildings (Barney, 2002).

The Council on Tall Buildings and Urban Habitat (CTBUH) identified the following top five priorities for research in the area of 'circulation, vertical transportation and evacuation' (Oldfield *et al.*, 2014). It is worth noting that all of the top five priorities listed below are related to the topic of *evacuation* (shown in italics).

1   Research on the planning, design and implications of using lifts for *evacuation* in tall buildings.
2   Research on appropriate *evacuation* and egress strategies for disabled persons (including emergency planning, the use of safety zones, etc.).
3   Research on strategies and technologies to deliver information to occupants in *evacuation/ emergency* scenarios (including dynamic route guidance systems, integrated audio and video technology, wireless systems, occupants' attitude to such systems, and conformance to legislation).
4   Research on the use of lifts for *evacuation* post extreme events, e.g., after an earthquake.
5   Research on real-time tall building *evacuation* management strategies and technologies.

It is expected that as numbers of high rise buildings increase, and as the lift hardware technology advances in order to allow the lift to fully operate during emergency conditions, that full or partial evacuation by lifts will become a building standard requirement.

This will have a threefold effect on the traffic engineering of lifts:

a)   Evaluating the total evacuation time out of the building (denoted as *egress time*) will become one of the user requirements for any traffic system design in tall buildings.
b)   The traffic system designer will have to ensure that the vertical transportation system meets the egress time requirement, as well as the traditional ingress requirements. It has long been recognised that the lift system can carry as much as 50% extra handling capacity during down peak, as much as uppeak traffic, and this is a factor that will ensure that lift traffic systems will not have to be enlarged to meet this requirement.
c)   Control system strategies will have to be developed in order to deal with the most appropriate, optimal method of partial or total evacuation by lifts.

Siikonen and Hakonen (2002) carried out a simulation into the use of stairs only, lifts only, and both stairs and lifts in the evacuation of tall buildings. A number of important conclusions are drawn from the study:

1   The study concludes that for buildings with fewer than 15 floors, the use of stairs only for evacuation is faster than using lifts only. Thus, it is only worth considering the use of lifts for evacuation for building with 15 floors or more.
2   The study also shows that it is not necessary to artificially zone or sector the building during evacuation in order to efficiently evacuate passengers out of buildings. The modern group control systems automatically optimise the operation, and thus minimise the evacuation time. It is also not necessary to encourage passengers to gather on every other floor or every third floor (in refuges) in order to reduce the number of stops in the round trip and thus reduce the evacuation time.
3.   The third and most important conclusion relates to the expected target evacuation time. '*If lifts are used in mega high-rise buildings during an emergency situation, evacuation times can drop to 15–30 minutes instead of [the current value of] 2–3 hours. In these buildings, shuttle lifts may become a bottle-neck during the evacuation and down-peak. Handling capacity of a shuttle lift group with only two stops can be considerably increased with double-deck or triple-deck lifts.*' (Siikonen and Hakonen, 2002).

Fortune (2010) presents the concept of the *lifeboat lift*, whereby the lifts and the lift shafts are fully equipped to continue operating for the full evacuation of the building within a specified period of time (e.g., less than 60 minutes). In order for this to be possible, sufficient emergency power must be available in order to operate the designated lifeboat lifts. A new phase 3 would be added to the standard phase 1 and phase 2 of the emergency operation. Passengers will congregate on designated refuge floors, and the lifeboat lifts will shuttle between each refuge floor and the building entrance in order to complete the evacuation of the building tenants. Wit (2010) presents an overview of the work being carried out in the Netherlands to incorporate evacuation by lifts in the national building code.

Guidelines have been provided in a CTBUH (2004) study on the use of lifts for building evacuation.

To sum up, the effect of such a requirement (i.e., the use of lifts for the partial or full evacuation of a building using lifts) would be to transform the lift traffic design process for medium and high rise buildings so that the egress time becomes an additional user requirement. Moreover, it is also expected that the lift group control system would evolve in order to meet certain evacuation strategies.

20.6   The concept of an idealised optimal benchmark

Although the term has not been previously used in the lift context, idealised optimal benchmarks (IOB) are widely used in the lift traffic analysis and design field. An idealised optimal benchmark is the optimal (minimum or maximum) possible value of a critical design variable that attains its value under idealised conditions. It can then subsequently be used as a benchmark to assess the efficiency of a lift traffic design. Thus, the value of the IOB is attained under idealised conditions, and not under practical operating conditions. This is understood and accepted as it allows an objective outcome that is not dependent on the randomness of lift movements, or the randomness of passenger decisions. By definition, they are average values. In addition, they are optimum values that can only be attained under the most favourable conditions. They are used as benchmarks allowing designers to understand how far their designs deviate from optimum performance.

In reality, the allocation of the landing calls to the cars in the group would have to be done in real time, and the look-ahead capability of the system (Barney 1992) might have to be restricted to a smaller number of cars (e.g., two to three cars). Otherwise, the waiting time for passengers will become excessive. This has been introduced as the factor k in the formula suggested by Barney (1992) reproduced below:

$$S_{\text{des}} = \frac{N}{k} \cdot \left(1 - \left(1 - \frac{L}{N} \right)^{k \cdot P} \right) \tag{20.1}$$

The parameter k can take on values of 2 or 3 or 4, up to the maximum value of K (the number of lifts in the group). If k is set to 1, the equation reverts to the classical equation in which the passengers are served on a first-come, first-served basis. The use of the product $k \cdot P$ envisages a large car with all the passengers who will eventually board the cars within the span of the allocation.

Smith suggested the presence of an ultimate performance figure for a group control algorithm that had perfect knowledge, which he called the *Divine Algorithm* (Smith, 2003). The IOB assumes perfect knowledge of the future destinations of passengers, and then compares the performance of the idealised system with the performance of a real-time allocation algorithm that makes an irrevocable allocation of each call as it arrives. This is similar to the concept of competitive analysis presented by Krumke (2001).

Some work has been carried in the area of idealised optimal benchmarks in order to assess the maximum possible performance of a hall call allocation group control under uppeak traffic conditions (Al-Sharif *et al.*, 2014c). This work resulted in a new set of formulae for the evaluation of the round trip time under idealised optimal conditions. At the core of the calculation of the value of the round trip time, are the new equations for the expected number of stops (*S*) and the highest reversal floor (*H*) shown below in equations (20.2) and (20.3), respectively.

$$S_{des} = \frac{N}{L} \cdot \left(1 - \left(1 - \frac{L}{N}\right)^P\right) \tag{20.2}$$

$$H_{des} = \frac{N}{2} \cdot \frac{L+1}{L} - \sum_{j=1}^{\frac{N}{L}-1} \left(\frac{j \cdot L}{N}\right)^P \tag{20.3}$$

where:
N is the number of floors above the main terminal
L is the number of lifts in the group
P is the number of passengers boarding the lift car in each round trip.

It is expected that as the lift traffic systems become more complicated, further research will be carried out into relevant idealised optimal benchmarks for different parameters. The *IOB* presents a powerful tool at the disposal of the designer, which can provide an assessment of the optimal possible performance that a design can attain.

20.7   Epilogue: future developments

It is always tempting and dangerous to try to predict the future as humans tend to overestimate the short term and underestimate the long term. Barney & dos Santos (1985) attempted to do that, to varying degrees of success, in the epilogue to the second edition of *Elevator Traffic Analysis Design and Control* (1985):

> *Research on the theoretical front should continue, preferably initiated and guided by the lift industry (manufacturers and consultants), in the areas of traffic design, engineering design and construction, controller algorithms and lift management. Thus the promise of the high technology of today will in time become the reality of the 'nuts and bolts' of tomorrow. In this vein it would be interesting to move into the future and speculate on the lifts for the next century.*
>
> *As each passenger approaches a lift system the new arrival will be identified on motion detectors and by voice synthesis the lift system will say 'Give me your destination floor'. The passenger will reply in any style, but include something like '. . . floor 9 please. . .', and the lift system by voice recognition techniques will accept the request.*
>
> *Using an (artificial) intelligent traffic control system such as call allocation the passenger will be told: 'Please take car C'. When car C arrives it will announce its destination to all intending passengers and count the number of passengers in the car, alighting and boarding.*
>
> *Thus there will be no call registration stations on the landings or in the car and exact car loads will be determined. Should a passenger feel mischievous and ask for another floor, before entering a lift, the lift will respond: 'You have selected destination floor 9, do*

you wish to cancel your first request!'. This response is possible by the voice recognition system identifying each voice precisely. In fact, the lift system could be programmed to allow access to specific floors only for certain passengers as an aid to security.

The drive system will be by means of magnetic levitation controlled by a solid state power controller operated directly from the mains supply. Thus the lift will be rope-less, gearless and machine-less! Safety standards will be maintained and improved by use of passive regenerative braking. Over-speed governors and safety clamps will not be required.

The landing and car doors will probably be driven by linear motor techniques, locked magnetically and interlocked electronically, hence reducing further the number of moving parts. The complete system will be subject to computer surveillance and management, allowing fault prediction and safety aspects to be continually monitored and maintained.

Lift atheists will not believe any of the above is possible but the agnostics, whilst repressing their scepticism, will competently say they do not know. Fifteen years ago many industry practitioners sneered at solid state drives and computer control. In fifteen years a new generation will turn speculation into fact.

Hall call allocation systems are becoming very widely used; a trend that is expected to continue further into the future, to eventually become a standard feature of all lift systems. Lift traffic calculations using equations such as Equations (20.2) and (20.3) would become standard components of lift traffic analysis.

With the launch of the first prototype of a ropeless linear motor in 2015, it is expected that the linear motor lift systems will initially complement the roped lift systems for the next 25 years, ultimately fully replacing them in the following 15 years, as the technology matures and the costs come down (especially for 'rare-earth magnets').

Lift traffic calculations to date have been concerned with a linear up and down movement. In the future, the movement will be rectangular. Thus, traffic calculations will need to be adapted for multiple lifts in the same rectangle (or multiple rectangles depending on the arrangement).

With the strong demand from consultants and developers, it is expected that there will be pressure on manufacturers to provide clearly defined, fully transparent software simulations. These may be specified more closely by an international standard for lift traffic simulations to ensure all simulators operate in the same way, e.g., whether to describe incoming traffic by a Poisson process or by a rectangular arrival process. Any traffic study that is supported by simulation would then become fully repeatable and reproducible, thanks to well defined lift traffic simulation open standards.

It is expected that lift controllers will become IP (internet protocol) addressable as part of the 'internet of things'. Large amounts of lift data as well as passenger data will become available for analysis. Beebe (2015) has suggested a standard coding interface (XML interface) for the description of lift data. This will eventually obviate the need for manual surveys, and provide large amounts of data for software packages for the verification of performance.

Appendix
Lift kinematics

A.1　Introduction

As seen throughout chapters 4 to 9, the round trip time forms the basis for assessing the handling capacity of the elevator system in order to ensure that it meets the demand. At the heart of the derivation of the round trip time equation lie the assumptions that are made regarding the distance covered, the rated speed, rated acceleration and the rated jerk of the elevator.

> *Kinematics is a branch of mechanics that is concerned with the motion of bodies without consideration for the forces that cause the motion.*

This appendix examines the kinematic principles used in lift systems, specifically concentrating on their use in lift traffic analysis. It presents an overview of the assumptions made and the equations used to calculate the time taken by the lift to traverse the specified distances. This appendix originated as the result of industry collaboration by Barney (2003a),[1] and it has been expanded using material from Al-Sharif (2014e). A more detailed treatment of the subject can be found in (Motz, 1976; Motz, 1991a; Motz, 1991b; Peters, 1995b; Peters, 1996c). The use of lift kinematics in the generalised calculation of the round trip time can be found in (Al-Sharif *et al.*, 2012; Al-Sharif & Hammoudeh, 2014d; Al-Sharif *et al.*, 2014f).

A.2　Nomenclature

a is the rated acceleration in m/s^2
a_{max} is the maximum attained acceleration in m/s^2
d_f is the typical height of one floor in m
j is the rated jerk in m/s^3
j_{max} is the maximum attained jerk in m/s^3
$t(d_f)$ is the time taken by the lift to travel one floor, starting and finishing at standstill in s
t_{acc} is the time taken to accelerate up to the top speed from standstill in s
t_{dec} is the time taken to decelerate from the top speed down to standstill in s
t_f is the time taken to complete a one floor journey in s assuming that the lift attains the top speed v
t_v is the time taken by the lift to travel one floor while travelling at rated speed in s
v is the rated speed in m/s
v_{max} is the top attained maximum speed in m/s

A.3   The rationale for speed profiling

A lift can be thought of as a variable speed drive system. Most variable speed drive systems are closed loop feedback systems. They monitor the speed of the motor. They compare the actual value of the speed to the desired value of the speed (sometimes called the reference value), and then drive or brake the motor according to the difference (error) between these two signals.

In order to know what the speed *should be* at any point in time, a profile of the desired speed of the motion control system has to be either available or generated. This profile represents the value of the speed at which the load should travel at any point in time; it does not necessarily mean that that is the speed at which it is actually travelling. The actual value from the speed feedback device represents the speed at which the load is actually travelling. In an ideal situation, the two profiles should be identical.

When a motion control system attempts to move the load from one point to another, one important criterion is time: to move the load between the two points in the minimum possible time. Hypothetically speaking, the load would attain top speed, v_{max} in zero time, would travel to the destination at v_{max} and then would be brought to a halt in zero time. This case is the optimum for minimising travelling time, but requires theoretically infinite acceleration, which is impossible. This is shown graphically in Figure A.1,[2] where it takes time t_1 to move the load. Obviously, this is not possible as infinite acceleration is not possible for three reasons:

1   A high value of acceleration requires excessively high values of torque from the motor, which are not possible. This also necessitates high values of current that would trip the power supply protection.
2   High acceleration values lead to high shock forces on the equipment and could lead to mechanical failure.
3   Human safety and comfort: the human body cannot be subjected to high values of acceleration for safety and comfort reasons. For example, jet pilots who are subjected to high

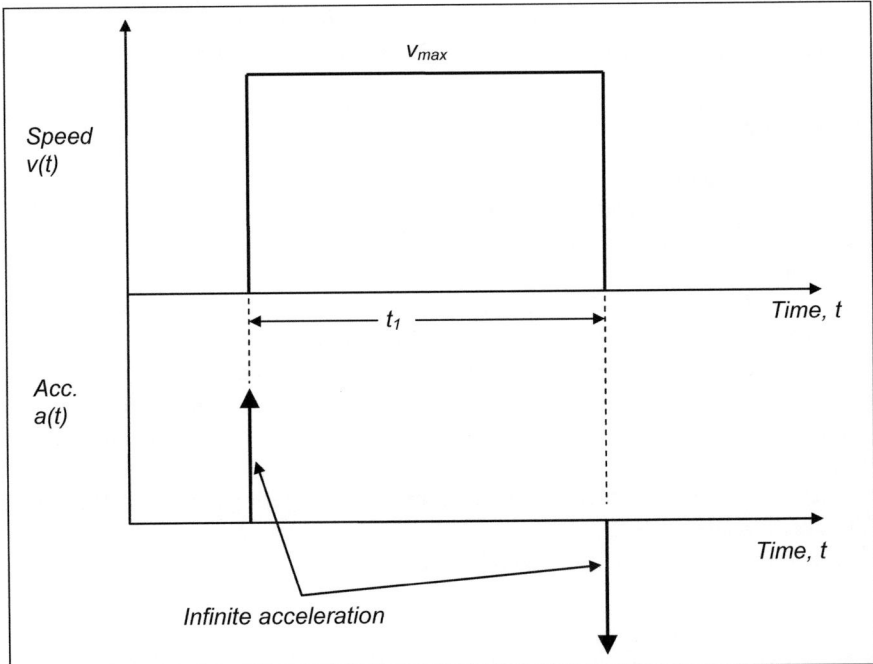

Figure A.1

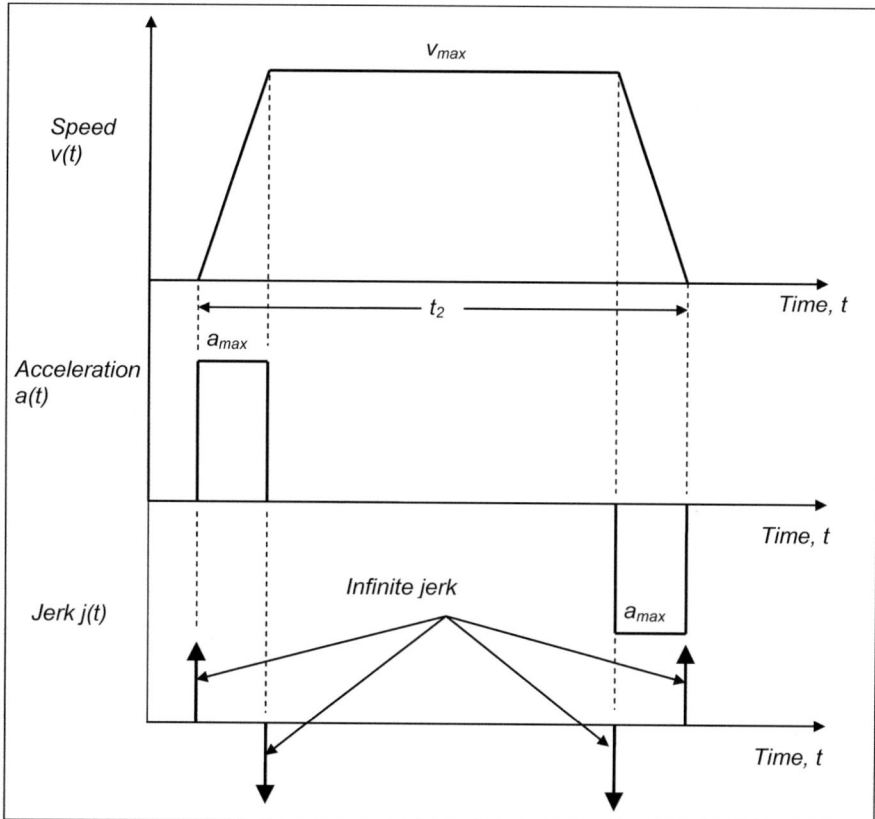

Figure A.2

values of g forces suffer blackouts. High acceleration forces could lead to miscarriages for pregnant women.

So, in order to avoid infinite (or extremely large) values of acceleration, it is necessary to allow the load to accelerate towards the top speed, v_{max}. It attains a maximum value of acceleration during the acceleration phase, which will be denoted as a_{max}. In this case, the top speed is attained via a finite value of acceleration, a_{max}.

This removes the problem of high acceleration values, and thus removes the requirement for high torque values, high starting currents and shock forces on the mechanical equipment. This makes it suitable for moving inanimate loads. The time taken in this case to move the load has increased to t_2 where $t_2 > t_1$. This case is shown graphically in Figure A.2.

However, it can be noticed that infinite values of jerk are experienced in the second case (see the jerk values in Figure A.2). When moving humans, this abrupt change from zero acceleration to maximum acceleration causes a noticeable level of discomfort. It is thus desirable to have a gradual change in the value of acceleration by limiting the top value of the jerk to j_{max}. This can be done by 'filleting' the speed-time profile at the 'corner' points. This is shown graphically in Figure A.3. This is sometimes referred to as the *S*-curve facility, and is an important feature of many of the modern variable speed drive systems, where it can be programmed with all the relevant parameters.

The price paid for the improved riding comfort is the increased journey time, t_3. Note that $t_3 > t_2 > t_1$. The selection of the values of the acceleration and jerk is a compromise between

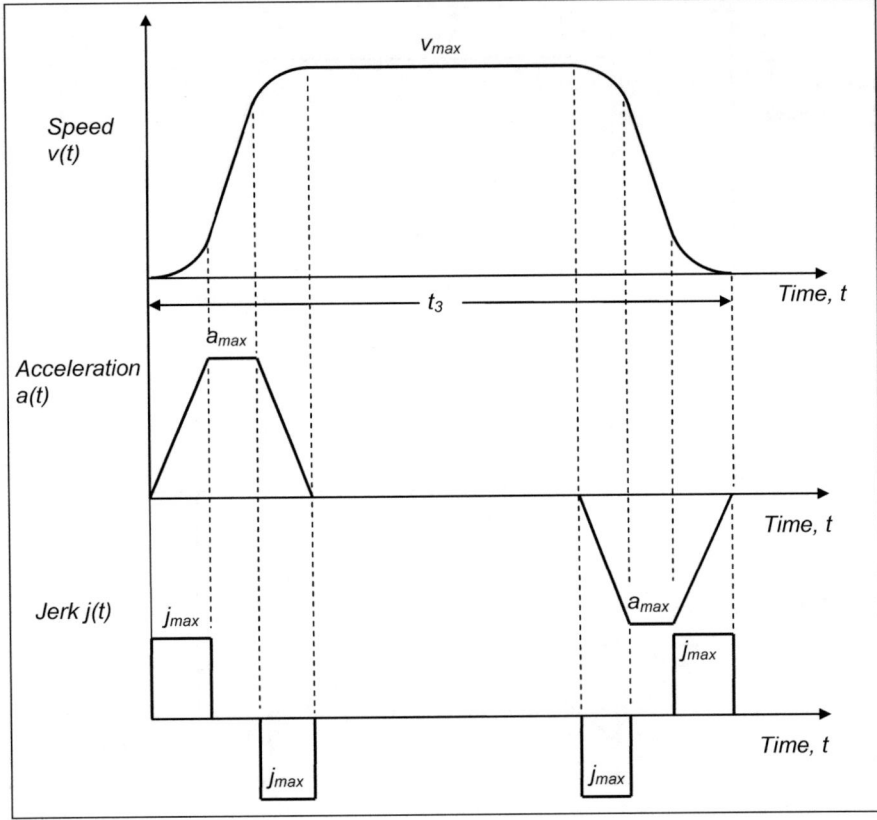

Figure A.3

minimising travel time and maximising comfort. Typical values for human comfort in lifts for example are 1 m·s⁻² for acceleration and 1 m·s⁻³ for jerk. Japanese manufacturers use lower values (around 0.5 m·s⁻² for acceleration) while higher values are used in North America (around 1.5 m·s⁻² for acceleration). Europeans use intermediate values.

The argument above can be continued to higher orders. For example, the rate of change of jerk against time is called jounce (or snap), and it is possible to try to limit its rate of change against time to a finite value. This is rare, however, although it has been reported that jounce is controlled in fairground rides. It is worth noting that there is no human limitation on the top speed attained.[3] This is effectively limited by the capability of the equipment and rate of change of pressure and its effect on the human eardrum.

A real-life trace of a speed curve taken from a lift with a variable voltage AC drive is shown in Figure A.4. It shows the reference speed-time curve, the actual speed-time curve measured using a tacho-generator coupled to the motor, as well as the drive and brake signals generated inside the drive.

It is worth noting that the curve in Figure A.4 has a low speed approach phase towards the end. This is when the load is gradually approaching the target position. The speed profiles shown later in this document have been plotted for profiles that do not have a levelling phase; however, the same principles also apply in these profiles.

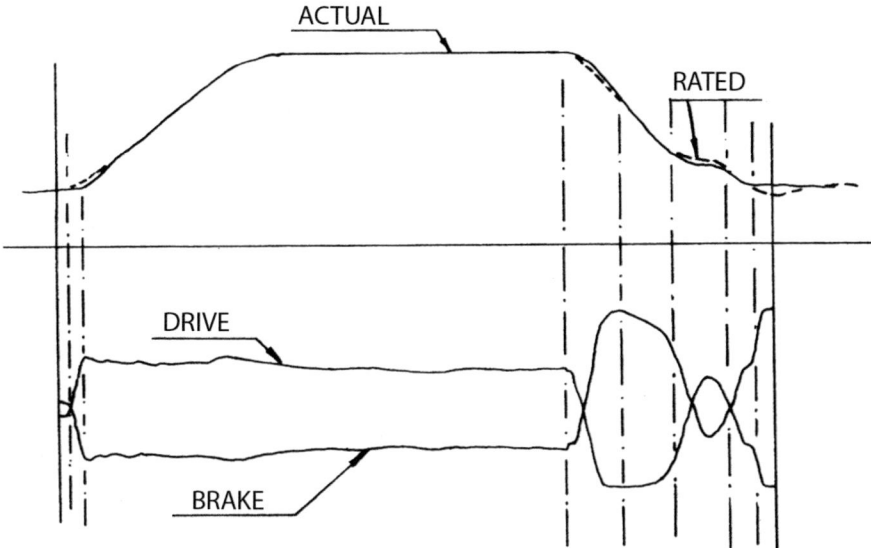

Figure A.4

A.4 The applications of kinematics in lift systems

There are four main applications for kinematics in lifts:

1 Precise control of the speed of the hoist motor: the drive system precisely controls the speed of the lift hoist motor. Such a feature enhances the levelling accuracy, and achieves riding comfort for the passengers.
2 Simulation: any lift traffic simulation package must have an accurate internal kinematic generator. The software must generate time slice by time slice values for the jerk, acceleration, speed and position for each of the lifts in the group. Full sets of equations for doing so have been documented (Peters, 1996c; Peters, 1995b).
3 Lift group control: being able to do the correct kinematic calculations is critical for lift group control. The landing call allocation decisions that the group controller makes must be based on the expected time it will take for the lifts to arrive at their desired destinations. The kinematic calculations can also be used to decide on the latest possible point in time (or distance) at which an allocation can be made to a moving lift.
4 Round trip time calculations: kinematic calculations are critical to the calculation of the round trip time. The evaluation of the round trip time is at the centre of the process of lift traffic design.

The main purpose of this article is the fourth point above (i.e., calculation of the round trip time). Most of the material from this point onwards concentrates on the use of kinematics in order to find the value of the round trip time.

A.5 The most basic case of constant acceleration

When considering kinematics during school physics, the following equations are developed:

$$s = ut + 0.5at^2 \qquad\qquad\qquad (A.1)$$

$$v = u + at \tag{A.2}$$

$$v^2 = u^2 + 2as \tag{A.3}$$

These formulae make the assumption (for simplicity) that the values of acceleration are instantaneously attained. Equations of motion describing real systems, such as the movement of a lift car in a shaft, cannot attain instantaneous values of acceleration, owing to such factors as the time delay for drive motor currents to reach working values, mechanical stiction, brake release delays, etc. termed 'start delays'. A body can only attain its maximum value of acceleration at a specific rate of change of acceleration. This physical effect is called jerk.

In lift kinematics, however, the value of acceleration changes continuously. This is explored in the next section.

A.6   The different cases of lift journeys

In doing the kinematic lifts calculations, there are four parameters to take into consideration:

1   The rated jerk of the lift denoted as j and having units of m/s^3. In the lift kinematics context, the value of the jerk at any point in time is either $+j$, 0 or $-j$. The instantaneous value of the jerk at any point in time is denoted as $j(t)$.

2   The rated acceleration of the lift denoted as a having units of m/s^2. The instantaneous value of the acceleration at any point in time is denoted as $a(t)$. The acceleration can be evaluated at any point in time by integrating the jerk, as shown in equation (A.4) below.

$$a(t_1) = \int_0^{t_1} j(t)\, dt \tag{A.4}$$

3   The rated speed of the lift, denoted as v having units of m/s. The instantaneous value of the velocity at any point in time is denoted as $v(t)$. The velocity can be evaluated at any point in time by integrating the acceleration, as shown in equation (A.5) below (in effect, the velocity is equal to the area under the acceleration-time curve).

$$v(t_1) = \int_0^{t_1} a(t)\, dt \tag{A.5}$$

4   The distance to be covered in the journey, denoted as d and having units of m. Of particular importance to the discussion in this section is the one floor distance, denoted as d_f. The instantaneous value of the displacement as a function of time is denoted as $d(t)$. The position/displacement can be evaluated at any point in time by integrating the velocity, as shown in equation (A.6) below (in effect, the displacement is equal to the area under the velocity-time curve).

$$d(t_1) = \int_0^{t_1} v(t)\, dt \tag{A.6}$$

Depending on the relative values of these four parameters, four cases arise. These four cases are discussed in the next four sub-sections.

A6.1 Case I: rated speed attained and rated acceleration attained in one floor journey

The first case is where the rated speed and rated acceleration are both attained in one floor journey. In the case shown below in Figure A.5 the rated speed is 1.6 m/s, and is attained for around 0.2 s. It can be seen that the rated jerk is 1 m/s^3, and attains values of 1, 0 or −1 m/s^3.

Each curve is the integration of the curve above it.

The equation/inequality that can be used to check if the rated speed will be attained in one floor journey is shown in equation (4) below (Peters, 1996c; Peters, 1995b):

$$d_f \geq \frac{a^2 v + v^2 j}{aj} \tag{A.7}$$

If such a condition is met, then the journey time can be calculated as shown in equation (A.8) shown below (Peters, 1996c; Peters, 1995b):

$$t_f = \frac{d_f}{v} + \frac{v}{a} + \frac{a}{j} \tag{A.8}$$

This case makes the round trip time calculations simple. In fact, the round trip time for the most basic of cases was based on the important assumption that *the rated speed was attained in one floor journey and that all floor heights were equal.* Under such conditions, it becomes very easy to calculate the difference in journey time for a one floor journey and a two floor journey, or between a two floor journey and a three floor journey. The difference can simply be found by adding the term t_v, which is the time required to traverse a floor while travelling at the rated speed, as shown below:

$$t\left(2 \cdot d_f\right) - t\left(d_f\right) = \left(\frac{2 \cdot d_f}{v} + \frac{v}{a} + \frac{a}{j}\right) - \left(\frac{d_f}{v} + \frac{v}{a} + \frac{a}{j}\right) = \frac{d_f}{v} = t_v \tag{A.9}$$

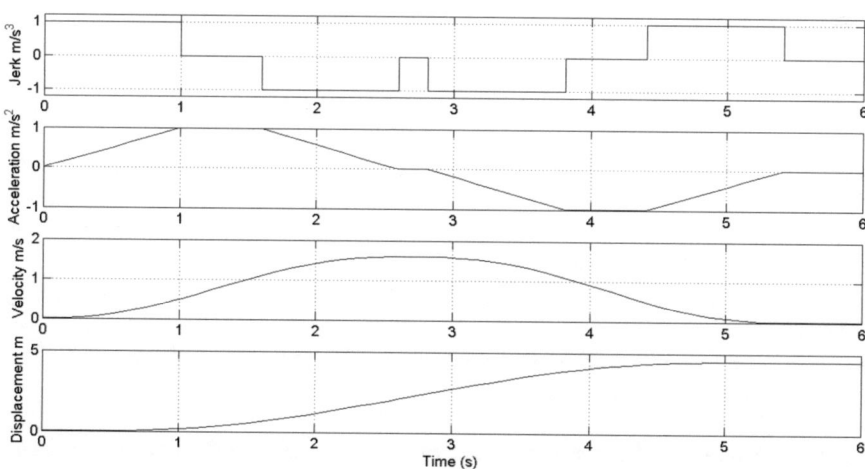

Figure A.5

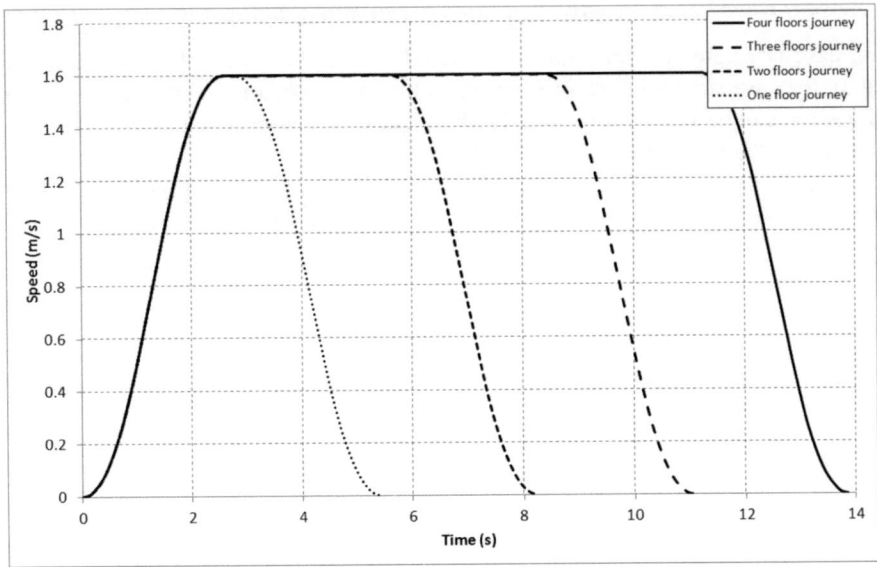

Figure A.6

In general, *linearity* is preserved for journeys of consecutive length:

$$t\left(2\cdot d_f\right)-t\left(d_f\right)=t\left(3\cdot d_f\right)-t\left(2\cdot d_f\right)=....=t_v=\frac{d_f}{v} \tag{A.10}$$

This linearity also offers another simplification in the derivation of the basic form of the round trip time equation (derived in chapters 4 to 9 earlier in the book). The difference between t_f and t_v is, in effect, the additional time (like a 'tax') incurred for every stop that takes place. Such a stopping 'tax' can be added to the stopping time and multiplied by the number of stops in a round trip, thus further simplifying the basic form of the round trip time equation.

$$t_{acc}+t_{dec}=t_f-t_v=\frac{v}{a}+\frac{a}{j} \tag{A.11}$$

The speed-time profile curves are shown in Figure A.6 for a lift that has equal floor heights (d_f=4.5 m) and where the rated speed is attained in one floor journey (v=1.6 m/s; a=1 m/s²; j=1 m/s³). By visual inspection, it can be seen that the difference in time between the successive journeys is equal to t_v (d_f/v=4.5/1.6=2.8 s, approximately).

In fact, the combination of these two conditions (i.e., top speed attained in one floor journey and rated speed attained in one floor journey) are very closely linked and render the derivation of a round trip time formula much easier, even where this is made as an intermediate assumption, later building up to a more complex formula.

A6.2   Case II: rated speed not attained and rated acceleration attained in one floor journey

As the rated speed increases, it becomes no longer possible for the lift to attain the rated speed in one floor journey. It can then require two, three or more floors for the lift to attain the rated

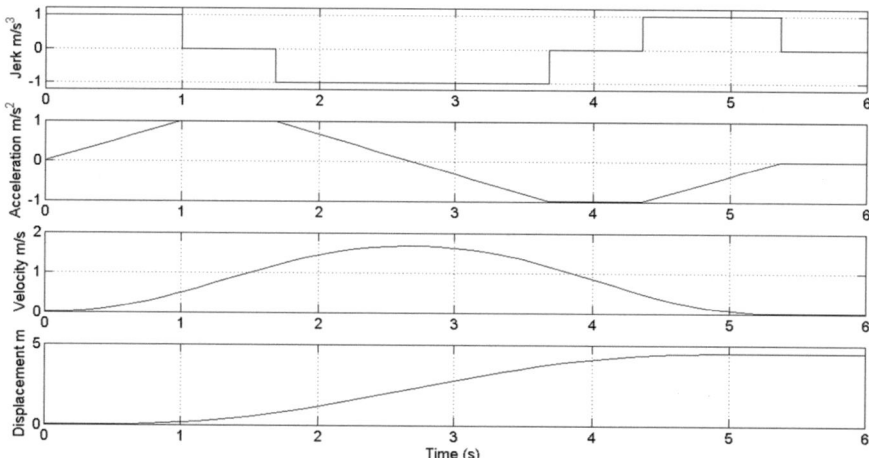

Figure A.7

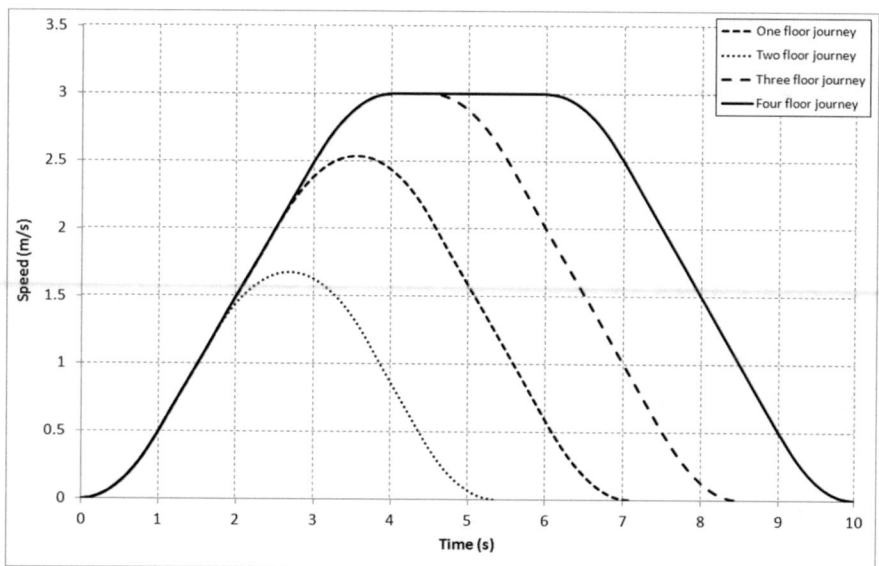

Figure A.8

speed. So, this case assumes that the rated acceleration is attained in one floor journey, but that the rated speed is not attained in one floor journey.

The plots of jerk, acceleration, velocity and displacement for the case of rated acceleration attained and rated speed not attained in one floor journey are shown in Figure A.7.

The following equation/inequality can be used to check to see if the conditions for this case exist (Peters, 1996c; Peters, 1995b).

$$\frac{2a^3}{j^2} \le d_f \le \left(\frac{a^2 v + v^2 j}{aj} \right)$$

(A.12)

If the equation/inequality is correct, then the time to traverse the one floor journey distance can be found as shown below (Peters, 1996c; Peters, 1995b):

$$t_f = \frac{a}{j} + \sqrt{\frac{4d_f}{a} + \left(\frac{a}{j}\right)^2} \tag{A.13}$$

As opposed to Case I (where both rated speed and rated acceleration were attained in a one floor journey), the time difference between a one floor journey and a two floor journey cannot simply be found by adding t_v. The relationship becomes non-linear.

The speed-time profile curves are shown in Figure A.8 for the same building that was used in the last sub-section. The floor to floor distances remain unchanged and the rated acceleration and rated jerk remain unchanged. The only difference is the rated speed, which has been raised from 1.6 m/s to 3 m/s. The difference can be clearly seen in the figure, where the rated speed is not attained in one or two floor journeys. It requires at least a journey of three floors or more to attain the rated speed.

A6.3   Case III: rated speed not attained and rated acceleration not attained in the journey

This third case occurs less frequently, and will take place where the rated jerk is relatively low and the rated acceleration is relatively high. A plot is shown for this case for the jerk, acceleration, velocity and displacement against time in Figure A.9. The rated jerk has been made lower than normal (at 0.5 m/s³) and the rated acceleration is set to 1 m/s². It can be seen that the rated acceleration value of 1 m/s² is not attained.

A check can be made to ascertain if the condition occurs for a certain floor height as in equation (A.14) below (Peters, 1996c; Peters, 1995b):

$$d_f \leq \frac{2a^3}{j^2} \tag{A.14}$$

If the condition is correct, then the time to traverse the floor can be found as shown in equation (A.15) below (Peters, 1996c; Peters, 1995b):

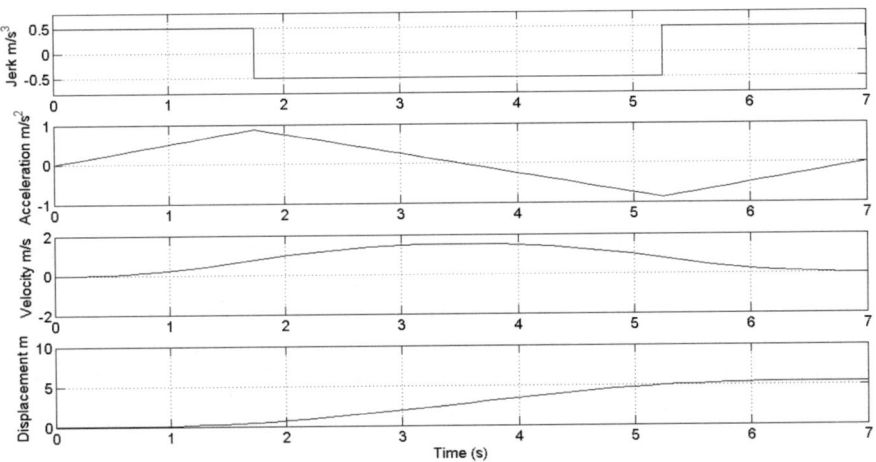

Figure A.9

$$t_f = \left(\frac{32d}{j} \right)^{\frac{1}{3}}$$
(A.15)

This case is unlikely to take place for practical building and lift values. For example, if the values of rated jerk and rated acceleration are set to 1 m/s^3 and 1 m/s^2, respectively, then the floor height would need to be at most 2 m for this condition to take place. Floor heights in buildings are rarely less than 2.5 m.

A6.4   *Case IV: rated speed attained and rated acceleration not attained in one floor journey*

Although this case is possible theoretically, it is very unlikely to take place in a lift context. It can take place when the rated speed is relatively low and the rated acceleration is relatively high. It is listed here for completeness and will not be discussed any further.

A.7   Derivation of the total travelling time for the case where the top speed is attained and top acceleration is attained

In this section, an equation for the total travelling time for Case I (i.e., the case where both the rated speed and the rated acceleration are attained in the journey) is derived. It is a geometrical method that is based on finding the area under the speed-time curve, equating it to the total distance and then finding the sum of the seven time parts.

It is assumed that the journey is split into seven parts: d_1, d_2, d_3, d_4, d_5, d_6 and d_7, which are all labelled in Figure A.10. It is also assumed that the time duration for each part is t_1, t_2, t_3, t_4, t_5, t_6 and t_7, respectively. The total distance of the journey is d, the rated speed is v, the rated acceleration is a and the rated jerk is at a value of j.

It is worth noting from the symmetry that the following applies:

$$t_1 = t_3 = t_5 = t_7$$
(A.16)

. . . and that:

$$t_2 = t_6$$
(A.17)

Also from the symmetry, the following distances are also equal as follows:

$$d_1 = d_7$$
(A.18)

$$d_3 = d_5$$
(A.19)

$$d_2 = d_6$$
(A.20)

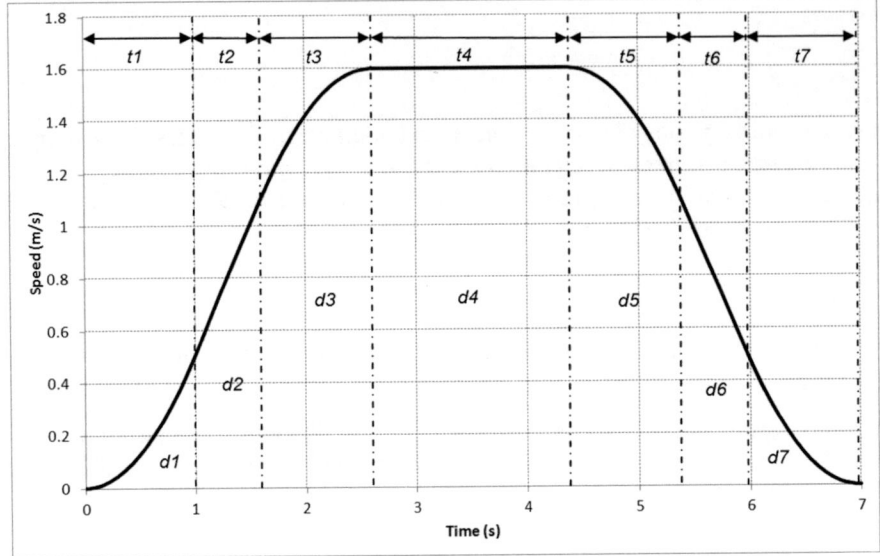

Figure A.10

At the end of the time t_1, the acceleration would have attained its rated value. Hence:

$$t_1 = \frac{a}{j} \tag{A.21}$$

The speed at the time t_1 is:

$$v(t_1) = \frac{1}{2} \cdot j \cdot t_1^2 = \frac{a^2}{2 \cdot j} \tag{A.22}$$

The speed at the time $t_1 + t_2$ attains the following value:

$$v(t_1 + t_2) = v - \frac{a^2}{2 \cdot j} \tag{A.23}$$

The distance covered at time t_1 is:

$$d_1 = \int_0^{t_1} \left(\left(\frac{1}{2} \cdot j \cdot t^2 \right) \cdot dt \right) = \frac{a^3}{6 \cdot j^2} \tag{A.24}$$

The distance d_3 is equal to the area of a rectangle less the area d_1 (from symmetry).

$$d_3 = (v \cdot t_1) - \frac{a^3}{6 \cdot j^2} = \left(\frac{v \cdot a}{j} \right) - \left(\frac{a^3}{6 \cdot j^2} \right) \tag{A.25}$$

The time t_2 can be calculated by dividing the change in speed by the rated acceleration, as follows:

$$t_2 = \frac{\left(\left(v - \dfrac{a^2}{2 \cdot j}\right) - \left(\dfrac{a^2}{2 \cdot j}\right)\right)}{a} = \frac{v}{a} - \frac{a}{j} \tag{A.26}$$

The area d_2 is the area of a trapezoid, calculated as follows:

$$d_2 = \left(\frac{\left(\left(v - \dfrac{a^2}{2 \cdot j}\right) + \left(\dfrac{a^2}{2 \cdot j}\right)\right)}{2}\right) \cdot t_2 = \left(\frac{v}{2}\right) \cdot \left(\frac{v}{a} - \frac{a}{j}\right) = \left(\frac{v^2}{2 \cdot a} - \frac{v \cdot a}{2 \cdot j}\right) \tag{A.27}$$

The total distance $d_1 + d_2 + d_3 + d_5 + d_6 + d_7$ can be calculated as follows:

$$d_1 + d_2 + d_3 + d_5 + d_6 + d_7 = 2 \cdot \left(\frac{v \cdot a}{j} + \left(\frac{v^2}{2 \cdot a} - \frac{v \cdot a}{2 \cdot j}\right)\right) = \frac{v \cdot a}{j} + \frac{v^2}{a} \tag{A.28}$$

The distance d_4 can thus be evaluated by subtracting the sum $d_1 + d_2 + d_3 + d_5 + d_6 + d_7$ from the total distance, d:

$$d_4 = d - \left(d_1 + d_2 + d_3 + d_5 + d_6 + d_7\right) = d - \frac{v \cdot a}{j} - \frac{v^2}{a} \tag{A.29}$$

During the time t_4, the elevator is travelling at constant speed. Thus, the time t_4 can be found by simply dividing the distance d_4 covered during t_4 by the rated speed, v:

$$t_4 = \frac{\left(d - \dfrac{v \cdot a}{j} - \dfrac{v^2}{a}\right)}{v} = \frac{d}{v} - \frac{v}{a} - \frac{a}{j} \tag{A.30}$$

Finally, the total time can now be found by adding t_4, twice the value of t_2 and four times t_1, QED:

$$t = 4 \cdot t_1 + 2 \cdot t_2 + t_4 = 4 \cdot \left(\frac{a}{j}\right) + 2 \cdot \left(\frac{v}{a} - \frac{a}{j}\right) + \left(\frac{d}{v} - \frac{v}{a} - \frac{a}{j}\right) = \left(\frac{d}{v} + \frac{v}{a} + \frac{a}{j}\right) \tag{A.31}$$

Notes (shown in text as[1])

1 This appendix originated in the 2003 edition of the book (Barney, 2003a) as a result of industry collaboration described in a footnote, the text of which is reproduced here.

> *This Appendix is not claimed as original, but sets down in one place information on lift dynamics: it arose as a result of a number of events. In November 1987 the Author sent Mr. G. Doek of Wolter & Dros-Evli, The Netherlands, a 'Traffic Design Card', which included suggested lift floor to floor flight times, etc. Doek queried some of the values as being too optimistic and sent the Author*

a copy of a paper by H. D. Motz of Solingen, West Germany. At the same time (November 1987) the late Joris Schroeder of Schindler, Switzerland, published a paper on drive dynamics. In 1989, while preparing the book 'Elevator Electric Drives' [Barney & Loher, 1996] with Messrs. Loher of Ruhstorf, the Motz paper was translated into English. This, together with the work of Roschier and Kaakinen of KONE, Finland, published in 1980, was the basis of the computer program published in 'Elevator Electric Drives'. During the writing of the CIBSE Guide D: 1993, Mr. P. Day, the Volume Chairman, consulted Mr. M. Kaakinen who confirmed the details and pointed out one error. The Motz paper has been translated into Italian and appears in Elevatori *(1/1991). Thus, an interesting international, industrial, consulting, estate management, academic collaboration has thrown light on a complex subject.*

2 The infinite value of acceleration and jerk has been shown as an impulse function in the figures in this appendix. The impulse function is a mathematical concept that represents a function that has zero duration, infinite amplitude and an area of 1 under the curve. It is sometimes denoted as $\delta(t)$. It is shown in the graphs as an arrow.

3 Velocity cannot be measured directly by non-contact methods. Acceleration can be measured by non-contact (inertial) methods (velocity can then be obtained by integrating the acceleration signal). Similarly, a blindfolded person in a car cannot tell at what speed he/she is travelling, but can feel the value of acceleration of deceleration. See Yang *et al.* (2013) for an interesting application of acceleration measurements in lifts.

References and principal citations

Section references shown after —, eg: — 18.3.1

Al-Adem, M. D., 2012, 'Discrete event simulation of multi-car elevator systems', Final Year Graduation Project, Mechatronics Engineering Department, University of Jordan, Amman, Jordan. – 18.3.1.

Alani, A. F., Mehta, P., Stonham, J. & Prowse, R., 1995 'Performance optimisation of knowledge-based lift group supervisory control system', in *Elevator Technology 6*, proceedings of Elevcon 1995, Hong Kong, March 1995, pp. 114–121. – 9.3.

Alani, A. F. & Mehta, P., 1996, 'A knowledge based group supervisory control system incorporating computer vision', *International Journal of Lift Engineering*, vol. 1, 1996, pp. 13–23. – 12.3.4.

Alexandris, N. & Barney, G. C., 1976, 'Three buildings surveyed', The design, evaluation and control of lift systems, Part 16, Control Systems Centre report no. 350, UMIST, August 1976. – 19.2.

Alexandris, N. A., 1976, 'Expected number of stops and highest reversal floor in lift performance', *Control Systems Centre report no. 314*, UMIST. – 7.7.3, 13.1.

Alexandris, N. A., 1977, Statistical models in lift systems, Ph.D. thesis, UMIST. – 1.5, 5.13.1, 13.3, Part D.1

Alexandris, N. A., Barney, G. C. & Harris, C. J., 1979a, 'Multi-car lift system analysis and design', *Applied Mathematical Modelling*, 1979, vol. 3, August, pp. 269–274. – 13.3, 18.6.1, 18.6.2.

Alexandris, N. A., Barney, G. C. & Harris, C. J., 1979b, 'Derivation of the mean highest reversal floor and expected number of stops in lift systems', *Applied Mathematical Modelling*, vol. 3, August 1979, pp. 275–279. – 7.9.1, 18.6.1, 18.6.2.

Alexandris, N. A., Harris, C. J. & Barney, G. C., 1981, 'Evaluation of the handling capacity of multi-car lift systems', *Applied Mathematical Modelling*, vol. 5, February 1981, pp. 49–52. – 18.6.1, 18.6.2.

Alexandris, N. A., 1986, 'Mean highest reversal floor and expected number of stops in lift-stairs service systems of multi-level buildings', *Applied Mathematical Modelling*, vol. 10, April 1986, pp. 139–143. – 18.6.1, 18.6.2.

Alexandris, N. A., 1988, 'LIFTES – An expert system for lift system design', in *Elevator Technology 2*, proceedings of Elevcon 1988, Karlsruhe, Germany, June 1988, pp. 1–9. – 9.3.

Al Masri, D. R. & Rishmawi, S. W., 2011, 'Design and verification of lift traffic analysis in buildings with multiple entrances', Final Year Graduation Project, Mechatronics Engineering, The University of Jordan, Amman, Jordan. Supervisor: Dr. Lutfi Al-Sharif, January 2011. – 19.2.

Al-Sharif, L., 1992a, 'Predictive Methods in Lift Traffic Analysis', Ph.D. Thesis, Oct 1992, UMIST, Manchester, United Kingdom, Supervisor: Dr. G. C. Barney. – 13.4, 19.2.

Al-Sharif, L. R., 1992b, 'New concepts in lift traffic analysis: the inverse S-P method', *Elevator Technology 4*, IAEE Publications. – 13.4.

Al-Sharif, L. R., 1993, 'Bunching in lift systems', *Elevator Technology 5*, IAEE Publications. – 13.5.

Al-Sharif, L. R., 1996, 'Escalator handling capacity', *Elevatori*, 4/96 and 5/96. – 2.8.

Al-Sharif, L., Dahyat, H. & Al-Kurdi, L., 2012, 'The use of Monte Carlo simulation in the calculation of the elevator round trip time under up-peak conditions', *Building Services Engineering Research and Technology*, vol. 33, no. 3 (2012) pp. 319–338, doi:10.1177/0143624411414837. – 6.13, 18.2.2.

Al-Sharif, L., Abdel Aal, O. F., Abu Alqumsan, A. M., 2013a, 'Evaluating the elevator passenger average travelling time under incoming traffic conditions using analytical formulae and the Monte Carlo method', *Elevator World*, June 2013, pp. 110–123. – 18.4.

Al-Sharif, L., Abu Alqumsan, A. M. & Abdel Aal, O. F., 2013b, 'Automated optimal design methodology of lift systems using rules and graphical methods (the *HARint* plane)', *Building Services Engineering Research and Technology*, vol. 34, no. 3 pp. 275–293, August 2013, published online before print, 12 April 2012, doi: 0143624412441615; 10.1177/0143624412441615. – 18.2.

Al-Sharif, L., Abu Alqumsan, A. M. & Khaleel, R., 2013c, 'Derivation of a universal elevator round trip time formula under incoming traffic', *Building Services Engineering Research and Technology*, doi: 0143624413481685, first published online 13 June, 2013 as doi: 10.1177/0143624413481685. – 18.2.1.

Al-Sharif, L. & Al-Adhem, M. D. 2013d, 'The current practice of lift traffic design using calculation and simulation', *Building Services Engineering Research and Technology*. Sage Publications, vol. 35, no. 4, pp. 438–445, published online before print, 26 September 2013 as doi: 10.1177/0143624413504422. – 18.4.

Al-Sharif, L., Abdel Aal, O. F., Abu Alqumsan, A. M. & Abuzayyad, M. A., 2014a, 'The *HARint* space: a methodology for compliant lift traffic designs', *Building Services Engineering Research and Technology*, vol. 36, no. 1, pp. 34–50 first published online 20 June, 2014, doi: 0143624414539968. – 6.13.

Al-Sharif, L., Abu Alqumsan, A. M., Ghanem, W., Jarrar, A. & Tayeh, I., 2014b, 'Modelling of elevator traffic systems using queuing theory', *4th Symposium on Lift and Escalator Technologies*, September 2014, Northampton, U.K. – 20.4.

Al-Sharif, L., Hamdan, J., Hussein, M., Jaber, Z., Malak, M., Riyal, A., AlShawabkeh, M. & Tuffaha, D., 2014c, 'Establishing the upper performance limit of destination elevator group control using idealised optimal benchmarks', *Building Services Engineering Research and Technology*, accepted for publication December 2014. – 20.6.

Al-Sharif, L. & Hammoudeh, A., 2014d, 'Evaluating the elevator round trip time for multiple entrances and incoming traffic conditions using Markov Chain Monte Carlo', *International Journal of Industrial and Systems Engineering*, Inderscience Publishers, 2014, vol. 18, no. 1, pp. 51–64. – 18.2.2, Appendix.

Al-Sharif, L., 2014e, 'Modern Elevator Traffic Engineering, Part III: Elevator Kinematics', *Lift Report*, September/October 2014. – Appendix.

Al-Sharif, Abu Alqumsan, A. M. & Khaleel, R., 2014f, 'Derivation of a universal elevator round trip time formula under incoming traffic with stepwise verification', *Building Services Engineering Research and Technology*, vol. 35, no. 2, 2014, pp. 198–213, doi: 0143624413481685. –Appendix.

Al-Zubi, H. Y., Al-Faouri, L. M., Al-Banna, H. F., Al-Abed Roby, M. & Hussien, A., 2011 'Vertical transportation survey for elevator design', Mechatronics Engineering Department, The University of Jordan, course project, Selected Topics in Mechatronics 0908589, Supervisor: Dr. Lutfi Al-Sharif, 5 October 2011. – 19.2.

ANSI/BOMA Z65.1, 1996, *Methods of measuring floor area in office buildings*, ANSI-USA. – 6.4.4.

Aoki, H., 2010, 'An advanced zoning system for a skyscraper', in *Elevator Technology*, 18, proceedings of Elevcon 2010, Lucerne, Switzerland, June 2010, pp. 28–37, IAEE. 8.3.2.3.

Aqel, N. S., 2011, 'Analysis of the Passenger Arrival Process in an Elevator Lobby', Final year graduation project, Mechatronics Engineering, The University of Jordan, Supervisor: Dr. Lutfi Al-Sharif, May 2011. – 19.2.

Bahjat, Z. S. & Bittar, J., 1992, 'Automated selection of high traffic intensity algorithms for up-peak period', *US patent no.: 5 168 133, 1 December 1992*. – 12.2.3, 12.3.1.

Bailey, N. T. J., 1954, 'On Queueing Processes with Bulk Service', *Journal of the Royal Statistical Society*, Series B (Methodological), vol. 16, no. 1, 1954, pp. 80–87.

Barney, G. C. & Dos Santos, S. M., 1975, 'Improved traffic design methods for lift systems', *Building Science*, vol. 10, no. 4, December 1975, pp 277–285, 6.3, 6.11.1,

Barney, G. C. & Dos Santos, S. M., 1977, *Lift traffic analysis design and control*, Peter Pereginus. – 4.5, 6.6.3, 13.3, 14.2, 14.2.2, 15.3.

Barney, G. C. & Dos Santos, S. M., 1985, *Elevator traffic analysis design and control*, Peter Pereginus. – 11.1.4, 11.5.1, 20.7.

Barney, G. C., 1988, 'Questions and answers', *Elevator World*, November. – 8.3.1.2.

Barney, G. C. & Loher, A. G., 1990, *Elevator Electric Drives*, Ellis Horwood. – A1.

Barney, G. C., 1992, 'Uppeak revisited', *Elevator Technology 4*, Proceedings of Elevcon 1992, Amsterdam, The Netherlands, published by the International Association of Elevator Engineers, pp. 39–47. – 8.5.2, Table 18.1, 20.6.

Barney, G. C., 1993, *Guide D: Transportation systems in buildings*, Section 3, CIBSE. – 8.3.3.

Barney, G. C., 1996, 'Load and capacity', *Elevatori*, no. 1, 1996. – 5.10.1.

Barney, G. C., 1998, 'Uppeak, down peak and interfloor performance revisited', *Elevator Technology*, 9, Proceedings of Elevcon 1998, Zurich, Switzerland, International Association of Elevator Engineers, pp. 31–40. – Table 18.1.

Barney, G. C., 2000a, 'Engineering: Load and capacity', *Elevator World*, March 2000. – 5.10.1, 11.8.

Barney, G. C., 2000b, 'Uppeak handling capacity improved, but what is lost?', *Elevatori*, no. 3, 2000. – 11.8.

Barney, G. C., 2000c, 'Uppeak handling capacity improved – what gives?', *Elevator World*, May 2000. – 8.5.1

Barney, G. C. & Pearce, C., 2000d, 'Calculating basement service', *Elevatori*, November/December 2000. – 8.3.3.

Barney, G. C. & Pearce, C., 2001a, 'Calculating basement service', *Elevator World*, January 2001. – 8.3.3.

Barney, G. C. & Imrak, E., 2001b, 'Application of neural networks to lift traffic control', *Elevator World*, May 2001. – 11.4.3, 11.10.

Barney, G. C., 2002, 'Behaviour of lifts and their use for evacuation', *Elevatori*, November/December 2002. – 20.5.

Barney, G. C., 2003a, *Elevator Traffic Handbook*, Taylor & Francis, 2003. – 1.6.2, 18.2.1, 18.5.

Barney, G. C., 2003b, 'Vertical transportation in tall buildings', *Elevator World*, May 2003, pp. 66–71. – 8.3.2.1

Barney, G. C., 2005a, 'Emergency Egress' (note to the editor), *Elevator World*, March 2005. – 20.5.

Barney, G. C., Peters R. D., Powell, B. & Siikonen, M. L., 2005b, 'Towards agreed traffic definitions', *Elevator World*, February 2005; *Elevatori*, 1/2005; *Elevation*, no. 42. – 1.7, 6.8.4, 18.4.

Barney, G. C., Cooper, D. & Inglis, J., 2009, *Elevator & Escalator Micropedia*, CIBSE, 5th edition. – 1.7, Table 6.9, 6.14.

Basagoiti, R., Beamurgia, M., Peters, R. & Kaczmarczyk, S., 2012, 'Origin destination matrix estimation and prediction in vertical transportation', *2nd Symposium on Lift and Escalator Technologies*, University of Northampton, Northampton, UK, 27 September 2012. – 12.3.3.

Basagoiti, R., Beamurgia, M., Peters, R. & Kaczmarczyk, S., 2013, 'Passenger flow pattern learning based on trip counting in lift systems combined with on-line information', *3rd Symposium on Lift and Escalator Technologies*, University of Northampton, Northampton, UK, 26–27 September 2013. – 12.3.3.

Bates, V. Q., 1993, 'What's in a word?', *Elevator Technology 5*, IAEE Publications. Preface.

BCO, 2009, *BCO Guide to Specification 2009*, British Council for Offices, 2009. – 18.5.

BCO, 2014, *Guide to Specification 2014*, British Council for Offices, 2014. – 18.5.

Bedford, R. J., 1966, 'Lift traffic recording and analysis', *GEC J.*, 33. – 15.1, 15.5.

Beebe, J. R., 1980, 'Lift Management', Ph.D. Thesis, Control Systems Centre, UMIST, 1980, pp. 114–116. – 11.1.2, 12.3.1.

Beebe, J., 2015, *Standard Elevator Information Schema*, http://www.jonathanbeebe.co.uk/. – 20.7.

Bittar, J. & Thangavelu, K., 1988, *Contiguous floor channeling with up hall call lift dispatching*, US patent no. 4 792 019, 20 December 1988. – 12.2.3.

Bittar, J. & Thangavelu, K., 1989, *Contiguous floor channeling lift dispatching*, US patent no. 4 804 069, 14 February 1989. – 12.2.3.

Bolat, B. & Imrak, C. E., 2006, 'Reducing the average journey time in the group of lifts by means of simulation', in *Elevator Technology 16*, proceedings of Elevcon, Helsinki 2006, Helsinki, Finland, June 2006, IAEE, pp. 26–33. – 12.2.2.

Brandin, B., Chen, T., Derventizis, C. & Pang, G., 1989, *The control and supervision of groups of elevators using the blackboard architecture approach*, IEEE International Conference on Systems, Man and Cybernetics, 1989, 14–17 Nov 1989, pp. 99–104, vol. 1. – 12.2.2.

Braun, R., 2003, *Need a Lift? An Elevator Queueing Problem*, United Technologies Research Centre, 14/8/2003, pp. 1–28. – 18.6.1.

Bril, J. & Marsh, A., 1984, 'Data logging and simulation', *Elevator World*, December 1984, pp. 36–43. – 19.3.

Browne, James J. & Kelly, James J., 1968, 'Simulation of elevator system for world's tallest buildings', *Transportation Science*, vol. 2, no. 1, Feb 1968, pp. 35–56. – 4.5, 8.2.1, 12.2.3.

BS, 2006, *BS ISO 4190–5 Lift (elevator) installation – Part 5: Control devices, signals and additional fittings*, British Standards. – 10.6.4.

BSI, 1972, BS CP407, 1972, *Electric, hydraulic and hand-powered lifts*, British Standards Institution. – CP9, CP13

BSI, 2003a, BS EN81–70, 2003a, *Safety rules for the construction and installation of lifts. Particular applications for passenger and goods passenger lifts. Accessibility to lifts for persons including persons with disability*, British Standards Institution. – 2.11, 4.1.

BSI, 2003b, BS EN81–72, 2003b, *Safety rules for the construction and installation of lifts. Particular applications for passenger and goods passenger lifts. Firefighters lifts*, British Standards Institution. – 4.1, 5.2.2.

BSI, 2008a, BS 9999, 2008a, *Code of practice for fire safety in the design, management and use of buildings*, British Standards Institution. 5.2.2

BSI, 2008b, BS EN115, 2008b, + A1, 2010, *Safety rules for the construction and installation of escalators and passenger conveyors*, British Standards Institution. – 2.8.

BSI, 2009, BS 8300, 2009, + A1, 2010, *Design of buildings and their approaches to meet the needs of disabled people*, British Standards Institution. – 2.11.

BSI, 2011a, BS 5655: Part 6, 2011a, *Code of practice for the selection and planning of new lifts*, British Standards Institution. – 5.2.2, 6.6.1, 10.6.4.

Building Regulations (the), 2004, *Approved document M (2004): Building Regulations: 2000, Access to and use of Building*, Approved Document M, The Stationery Office. – 2.11.

CEN (Comité Européen de Normalisation (European Committee for Standardization)), 2011, DD CEN/TS 81–76, 2011, *Safety rules for the construction and installation of lifts: particular applications for passengers and goods passenger lifts –Part 76: Evacuation of disabled persons using lifts.* – 2.11.

CEN, 2014a, PD CEN/TR 81–12, 2014a, *Safety rules for the construction and installation of lifts – Basics and interpretations, Part 12: Use of EN 81–20 and EN 81–50 in specific markets.* – 5.10.3.

CEN, 2014b, EN81–20, 2014, *Safety rules for the construction and installation of lifts. Lifts for the transport of persons and goods. Passenger and goods passenger lifts.* – 2.9, 5.10.3, Table 5.8.

Chan, W.L. & So, A.T.P., 1995, 'Dynamic zoning in lift traffic control', in *Elevator Technology 6*, International Association of Elevator Engineers, Proceedings of Elevcon 1995, Hong Kong, March 1995, pp. 133–140. 10.6.3

Chan, W.L. & So, A.T.P., 1996, 'Dynamic zoning for intelligent supervisory control', *International Journal of Elevator Engineering*, vol. 1, 1996, pp. 47–59. – 11.8.

Chenais, P. & Weinberger, K., 1992, 'New approach in the development of lift group control algorithms', in *Elevator Technology 4*, proceedings of Elevcon 1992, Amsterdam, The Netherlands, IAEE, Ed., G.C. Barney, pp. 48–57. – 12.2.1.

Cho, Y.C., Gagov, Z., Kwon, W.H., 1999a, 'Time Petri net based approach for elevator group controls', *Proceedings of the 1999 IEEE/RSJ International Conference on Intelligent Robots and Systems*, IEEE, pp. 1265–1270. – 12.2.2.

Cho, Y.C., Gagov, Z. & Kwon, W.H., 1999b, *Elevator group control with accurate estimation of hall call waiting time*, Proceedings of the 1999 IEEE International Conference on Robotics and Automation, Detroit, Michigan, May 1999, IEEE, pp. 447–452. – 12.3.2.

Cho, Y.C., Park, J.H., Chang, H.C., Gagov, Z., Kwon, W.H., Kim, C. S., Han, P.H. & Seo, J.H., 2000, 'A computer-aided design for the specification and performance evaluation of elevator systems', in *Elevator Technology 10*, proceedings of Elevcon 2000, Berlin Germany, May 2000, IAEE Publications, pp. 204–211. – 9.3.

Christensen, I., 1988, 'Why do users press call buttons unnecessarily?' *Elevator Technology II*, IAEE Publications. – 11.6.1.

Christy, T., 2012, 'The impact of traffic variations on elevator performance', in *Elevator Technology 19*, proceedings of Elevcon 2012, the 19th International Congress on Vertical Transportation Technologies, 22–24 May 2012, Miami Beach, USA, IAEE Publications. – 18.4.

Christy, T., 2014, 'An Evolution of Lift Passenger Grouping', in *Elevator Technology 20*, proceedings of Elevcon 2014, Paris, France, 8–10 July 2014, IAEE Publications, pp. 41–50. – 12.2.3.

CIBSE, 2000, *CIBSE Guide D: Transportation systems in buildings, Chartered Institution of Building Services Engineers*, Second Edition, 2000. – Part D.1.

CIBSE, 2010, *CIBSE Guide D: Transportation systems in buildings*, Fourth Edition, Chartered Institute of Building Services Engineers. – 1.6.2, 18.2.1, 18.4, 18.5.

CIBSE, 2015, *CIBSE Guide D: Transportation systems in buildings*, Fifth Edition, Chartered Institute of Building Services Engineers. – 2.12, 8.2.3.

Closs, G. D., 1970, The computer control of passenger traffic in large lift systems, Ph.D. thesis, UMIST. – 10.4, 11.6.2.2, 11.10.

Closs, G. D., 1972, Lift control conservation or progress, *Electron. Pwr.*, 18, Institute of Electrical Engineers. – 10.4.

Cook, H. B., 1920, 'Passenger elevator service', paper presented at US Elevator Manufacturers Association, May 1920. (Reported by Gray, L., 2011, *Elevator World*, September 2011.) – 1.8.

Cortes, P., Larraneta, J. & Onieva, L., 2004, 'Genetic algorithm for controllers in elevator groups: Analysis and simulation during lunch-peak traffic', *Applied Soft Computing*, vol. 4, no. 2, May 2004, pp. 159–174. – 12.2.2.

Courtney, R. G. & Davidson, P. J., 1974, 'A survey of passenger traffic in two office buildings' – 19.3.

Crites, R. H. & Barto, A. G., 1998, 'Elevator group control using multiple reinforcement learning agents', *Machine Learning*, vol. 33, pp. 235–262, 1998. – 12.2.2.

CTBUH, 2004, *Emergency evacuation elevator systems guideline*, Council on Tall Buildings and Urban Habitat, 47 pp. – 20.5.

Danapalasingam, K. A. & Marzuki, K., 2005, 'Design of an elevator group supervisory controller using ordinal structure fuzzy logic with context adaptation', in *proceedings of the 9th International Conference on Mechatronics Technology*, 5–8 December 2005, Kuala Lumpur, Malaysia. – 12.2.1.

Day, P & Barney, G. C., 1993, Section 11, *Transportation systems in buildings*, CIBSE. – 17.7.1.

Day, P., 2001a, 'Passenger comfort: are you travelling comfortably?' *Elevator World*, April 2001. – 5.10.1, 5.10.3.

Day, P., 2001b, 'Lift passenger comfort have we got it right?', *Elevatori*, September 2001. – 5.10.1, 5.10.3.

De Grout, P. J., 2014, 'Comment: further evaluation of "fastest elevator"', *Elevator World*, December 2014, pp. 14–16. – 8.3.2.1

Department of Health, 2008, HTM08–02, 2008, Specialist services, *Health Technical Memorandum 08–02: Lifts*, DH. – 8.4.5.

Ding, B., Zhang, Y-M., Yu, D-L. & Tang, H-Y., 2013, 'A hybrid approach for the analysis and prediction of elevator passenger flow in an office building', *Automation in Construction*, vol. 35, November 2013, pp. 69–78. – 18.7.

Dos Santos, S. M., 1972, 'Lift simulation', M.Sc. dissertation, UMIST. – 17.3.

Eguchi, T., Hirasawa, K., Hu, J. & Markon, S., 2004, 'Elevator group supervisory control systems using genetic network programming', *Congress on Evolutionary Computation*, 2004, IEEE, 19–23 June 2004, pp. 1661–1667 (vol. 2). – 12.2.2.

Equality Act, 2010, *Equality Act 2010*, The Stationery Office, London. – 2.11.

Finschi, L., 2010, 'State-of-the-art traffic analyses', in *Elevator Technology 18*, Proceedings of Elevcon 2010, Lucerne, Switzerland, IAEE Publications, pp. 106–115. – 18.4.

Fletcher, P. T., 1954, *The planning of lift installations in commercial buildings*, RIBA J., vol. 61, 1954. – 10.2, 10.5.2.

Fortune, J. W., 1986, 'Top/down sky lobby design', *Elevator Technology*, Ellis Horwood, 1986. – 8.2.1, 8.3.2.4.

Fortune, J. W., 1990, 'Top/down lift design –a case study', *Elevator Technology 3*, IAEE Publications. – 8.2.1.

Fortune, J. W., 1992, 'Elevatoring Frank Lloyd Wright's mile high building', *Elevator Technology 4*, IAEE Publications. – 8.3.2.5.

Fortune, J. W., 1995, 'Modern double deck elevator applications and theory', *Elevator Technology 6*, IAEE Publications. – 8.2.1, 8.3.5.

Fortune, J. W., 1996, Modern double deck elevator applications and theory, *Elevator World*, August 1996. – 8.2.1, 8.2.2.

Fortune, J. W., 1997, 'Mega-high rise elevatoring', *Elevator World*, December. – 8.3.2.2.

Fortune, J. W., 2002, 'Electronic up-peak booster options', in *Elevator Technology 12*, proceedings of Elevcon 2002, Milan, Italy, June 2002, The International Association of Elevator Engineers, pp. 84–90. – 12.2.3.

Fortune, J. W., 2010, 'Emergency building evacuations via elevators', *CTBUH 2010*, Remaking Sustainable Cities in the Vertical Age, 3–5 February 2010, Mumbai, India. – 20.5.

Forwood, B. & Gero, J. S., 1971, 'Computer simulated lift design analysis for office buildings', *Architectural Science Review*, vol. 14, no. 2, 1971. – 13.3.

Fruin, J. J., 1971, *Pedestrian planning and design*, Metropolitan Association of Urban Designers and Environmental Planners. – 2, 2.2, 3.1, 5.10.1.

Fujitec, 2014, 'FLEX-NK System: virtual passenger optimization method', www.fujitec.uk.com/#/other-products/4559919872. – 12.3.2.

Gaver, D. P. & Powell, B. A., 1971, 'Variability in round trip times for an elevator car during uppeak', *Transportation Research*, vol. 5, no. 4. – 7.4.1, 7.7.2.

Gerstenmeyer, S. & Peters, R., 2014, 'Reverse journeys and destination control', *4th Symposium on Lift and Escalator Technologies*, University of Northampton, UK, 25–26 September 2014. – 12.4.

GLC 25, date unknown, *Traffic generation: users guide and review of studies*, Greater London Council. – 6.5.

Godwin, A. M., 1993, 'Unique design for a high rise office building', *Elevator Technology 5*, IAEE Publications. –CS11.

Godwin, M., 1986, 'Bush House: lifts for the world', *Elevator Technology*, Ellis Horwood. – 11.5.3.

Gray, L., 2011, *Elevator World*, September 2011. –1.8.

Green, M. F. & Stafford Smith, B., 1977, 'A survey and analysis of lift performance in an office building', *Building and Environment*, vol. 12, 1977, pp. 65–72. – 19.2.

Gudwin, R., Gomide, F. & Netto, M. A., 1998, 'A linear fuzzy elevator group controller with linear context adaptation', *The 1998 IEEE International Conference Fuzzy Systems*, vol. 1, IEEE World Congress on Computational Intelligence, 4–9 May 1998, pp. 481–486. – 12.2.1.

Guifeng, C., 2001, 'Landing call allocation based on linear programming method', in *Elevator Technology 11*, proceedings of Elevcon 2001, Singapore, July 2001, The International Association of Elevator Engineers, pp. 1–10. – 12.2.2.

Hakonen, H. & Siikonen, M. L., 2008, 'Elevator Traffic Simulation Procedure', *Elevator Technology 17*, in Proceedings of Elevcon 2008, Thessaloniki, Greece, pp. 131–141. – 18.4.

Halpern, J. B., 1992, 'Variance analysis: a new way of evaluating elevator dispatching systems', *Elevator World*, September 1992. – 11.5.1.

Halpern, J. B., 1993, 'Variance analysis of hall call response time', in *Elevator Technology 5*, proceedings of Elevcon 1993, Vienna, Austria, November 1993, pp. 98–106. – 11.5.1, 12.2.2.

Halpern, J. B., 1995, 'Statistical analysis of call response times', *Elevator Technology 6*, IAEE Publications. – 11.5.1.

Hikita, S., Amano, M. & Ando, H., 2001, 'The latest lift group control system', in *Elevator Technology 11*, proceedings of Elevcon Asia 2001, Singapore, pp. 35–43. – 11.6.1, 12.2.1.

Hiller, B., Klug, T. & Tuchscherer, A., 2013a, 'Improving the performance of elevator systems using exact re-optimisation algorithms', *Models and Algorithms for Optimisation in Logistics*, Dagstuhl Seminar Proceedings 09261. – 12.2.2.

Hiller, B., Klug, T. & Tuchscherer, A., 2013b, 'An exact reoptimisation algorithm for the scheduling of elevator groups', *ZIB Report, Zuse Institute Berlin* 12–43 (November 2012, revised April 2013). – 12.2.2.

Hirbod, S., 1975, 'Simulation extensions for lift supervisory control', MSc, UMIST. – 11.1.3, 17.3.

Ho, M. & Robertson, B., 1994, *Elevator group supervisory control using fuzzy logic*, Canadian Conference on Elevator and Computer Engineering, 2. – 11.10.

Hunt, S., 1975, 'Control of high speed lifts – continuous pattern system', *GEC J.*, vol. 32, no. 3. – 7.4.1.

ISO, 1984, ISO 4190–6, 1984, *Planning and selection of passenger lifts to be installed in residential buildings*, International Organization for Standardization. – 5.2.2, 5.5.1.

ISO, 1996, ISO/TR 11071–2, 1996, *Comparison of worldwide lift safety standards – Part 2: hydraulic lifts, ISO.5.10.1*, International Organization for Standardization. – 5.10.1, 7.10.

ISO, 2001, ISO 4190–2, 2001, *Lift (US: elevator) installation – Part 2: Class IV lifts,* International Organization for Standardization. 7.10.

ISO, 2006, ISO/TR 11071-2, *Comparison of worldwide lift safety standards – Part 2: Hydraulic lifts (elevators)*, International Organization for Standardization. –5.10.1, 7.10, 10.6.4.

ISO, 2010, ISO 4190–1:2010, Lift (elevator) installation – Part 1: Class I, II, III and VI lifts, International Organization for Standardization. –Table 5.8, Table 5.9, 7.10.

ISO, 2012, ISO 18738–1, 2012, *Measurement of ride quality – Part 1: Lifts (elevators)*, International Organization for Standardization. –19.2.

Jackson, C., 1970, 'Analytical techniques: simulation case study', *Architects Journal Information Library*, 8. – 17.5.

Jappsen, H. M., 2000, 'Elevators for the Commerzbank building in Frankfurt', *Elevator Technology 10*, Elevcon Books. CS11.

Jappsen, H. & Rieke, O., 2012, 'Difficulties in comparing the results of lift-traffic simulations', in *Proceedings of Elevcon 2012, the 19th International Congress on Vertical Transportation Technologies*, 22–24 May 2012, Miami Beach, USA. – 18.4.

Jarrar, A., Tayeh, I. & Ghanem, W., 2013, 'The use of queuing theory to assess passenger waiting time and queue length', Final year graduation project report, Mechatronics Engineering Department, The University of Jordan, Supervisor: Lutfi Al-Sharif, April 2013. – 18.6.5.

Jones, B., 1923, 'The probable number of stops made by an elevator', *GE Rev.*, vol. 26, no. 8. – 1.8, 4.5, 6.3, 13.4.

Jones, B. W., 1971, 'Simple analytical method for evaluating lift performance during uppeak', *Transpn. Res.*, vol. 5, no. 4. – 5.5.1, 7.4.1, 8.3.6.

Kaakinen, M. & Roschier, N. R., 1991, 'Integrated Elevator Planning System', *Elevator World*, March 2000. – 19.3.

Kameli, N., 1994, *Predictor elevator for traffic during peak conditions*, US patent no. 5 276 295, 4 January 1994. – 12.2.3, 12.3.3.

Kameli, N., 1996, *Floor population detection for an elevator system*, US patent no. 5 511 635, 30 April 1996. – 12.2.3.

Kavounas, G. T., 1989, 'Elevatoring analysis with double deck elevators', *Elevator World*, November 1989. – 8.2.2.

Kavounas, G. T., 1992a, 'Lowest call express floor, a third order consideration for round trip calculations', *Elevator World*, January. – 13.2.

Kavounas, G. T., 1992b, 'Bunching' Private communication. – 13.5.

Kavounas, G. T., 1993a, 'Lowest call express floor: a full theoretical treatment, Part 1', *Elevatori*, January/February. – 13.2.

Kavounas, G. T., 1993b, 'Lowest call express floor: a full theoretical treatment, Part 2', *Elevatori*, March/April. – 13.2.

Kavounas, G. T., 1993c, 'Designing elevator systems also for the average waiting time', *Elevator World*, January. – 6.8.2.

Kavounas, G. T., 1993d, 'Pedestrian traffic calculations among linearly connected vertical modes', *Elevator Technology 5*, IAEE Publications. – 2.

Kim, C., Seong, K. A., Hyung, L-K. & Kim, J. O., 1998, 'Design and implementation of a fuzzy elevator group control system', *IEEE Transactions on Systems, Man and Cybernetics, Part A: Systems and Humans*, vol. 28, no. 3, May 1998, pp. 277–287. – 12.2.1.

King, S. S. & Bouketir, O., 2008, 'Simulation of a four-car elevator operation using MATLAB', *Modern Applied Science*, vol. 2, no. 6, November 2008. – 18.6.4.

Klote, J. H., 1993, 'A method for calculation of elevator evacuation time', *Journal of Fire Protection Engineering*, vol. 5, no. 3, 1993, pp. 83–95. – 20.5.

Koehler, J. & Ottiger, D., 2002, 'An AI-based approach to destination control in elevators', *AI Magazine*, vol. 23, no. 3, Fall 2002, pp. 59–78. – 12.2.2.

Krumke, S. O., 2001, 'Online optimization – competitive analysis and beyond', Thesis, Technical University of Berlin, 2001. – 20.6.

Kuusinen, J. M., Sorsa, J., Siikonen, M. L. & Ehtamo, H., 2012, 'A study on the arrival process of lift passengers in a multi-storey office building', Building Services Engineering Research and Technology published online before print, 10 November 2011, doi: 10.1177/0143624411427459, vol. 33 no. 4, November 2012, pp. 437–449. – 7.7.4, 19.2.

Kuusinen, J. M. & Malapert, A., 2014, 'The effect of randomisation on constraint based estimation of lift trip origin-destination matrices', *4th Symposium on Lift and Escalator Technologies*, University of Northampton, UK, 25–26 September 2014. – 12.3.3.

Kuusinen, J-M., Sorsa, J. & Siikonen, M.-L., 2014, 'The lift trip origin-destination matrix estimation problem', *Transportation Science*, articles in advance, pp. 1–18, 2014, the Institute for Operations Research and the Management Sciences (INFORMS). – 12.3.3.

Lauener, J., 2007, 'Traffic performance of elevators with destination control', *Elevator World*, September 2007. – 12.4.

Lauer, R. J., 1984, 'Group supervisory control', *Elevator World*, November. 11.5.3.

Lea, A., 1990, 'Lift traffic – actual vs. theoretical', Course Report, Supervisor: Dr. G. C. Barney, Department of Building, Session 1989/1990, UMIST. – 19.2.

Li, J-F., Leng, J-W. & Zhang, J-L., 2011, 'Chaos characteristics of peak elevator traffic flow', *Proceedings of the 30th Chinese Control Conference*, 22–24 July 2011, pp. 5552–5555, Yantai, China. – 18.7.

Li, Z., Tan & H-Z., Zhang, Y., 2007, 'Particle swarm optimization applied to vertical traffic scheduling in buildings', *Proceedings of the 11th international conference, KES 2007, and XVII Italian workshop on neural networks conference on knowledge-based intelligent information and engineering systems: Part I*, pp. 831–838. – 12.2.2.

Lim, S. H., 1983, 'A computer based lift control algorithm', Ph.D. thesis, UMIST. CS16.2, CS16.3. – 11.1.4, 11.5, 15.3, 15.6, 17.3.

Little, J.D.C., 1961, 'A proof for the queuing formula: L = λW'. *Operations Research*, vol. 9, no. 3, 1961, 383–387. doi:10.1287/opre.9.3.383. JSTOR 167570. – 18.6.2.

Malet, M. A., McKay, E. M. & Smith, J. T., 1980, 'Lifts in local authority high-rise flats: Proposals towards reducing tenant grievance', *Building Research Establishment Information Paper*, IP 26/80, November 1980. – 19.2.

Markon, S. & Nishikawa, Y., 1991, 'Threshold accepting optimisation for elevator group control', *Proceedings of the 24th Joint Conference of Automatic Control*, November 20–22 1991, pp. 351–354. – 12.2.2.

Markon, S., Kita, H. & Nishikawa, Y., 1994, 'Adaptive optimal elevator group control by use of neural networks', *Transactions of the Institute of Systems, Control and Information Engineers*, vol. 7, no. 12, 1994, pp. 487–497. – 12.2.2.

Markon, S., Kita, H., Kise, H. & Bartz-Beielstein, T., 2006, *Control of Traffic Systems in Buildings*, Springer. – 1.8, 18.6.1.

Matlab from Mathworks, http:www.mathworks.com/.

Mayo, A.J., 1966, 'A study of escalators and associated flow system', M.Sc., Imperial College. – 2.8.

McKay, E.M., 1980, 'Lift systems in high-rise flats: an exploratory study of their traffic performance', *Building and Environment*, Pergamon Press, 1980, vol 15, pp. 17–25. – 19.2.

Miravete, A., 1999, 'Genetics and intense vertical traffic', *Elevator World*, July 1999. – 11.10.

Mitric, S., 1975a, 'Elevator systems for tall buildings, part I: single mode lift systems', *Transportation Science*, vol. 9, no. 1, Feb 1975, pp. 54–73. – 12.2.3.

Mitric, S., 1975b, 'Elevator systems for tall buildings, part II: mixed-mode lift systems', *Transportation Science*, vol. 9, no. 1, Feb 1975, pp. 74–85. – 12.2.3.

Morley, C. G., 1962, 'Installations mechanical: lifts escalators paternosters', *Architects J.*, vol. 136. – 7.5.

Motz, H. D., 1976, 'On the kinematics of the ideal motion of lifts', (in German) *Foerden und Heben*, 26. – Appendix.

Motz, H. D., 1991a, 'On the ideal kinematics of lifts, part I' *Elevatori*, January/February 1991, pp. 41–46. – Appendix

Motz, H. D., 1991b, 'On the ideal kinematics of lifts, part II' *Elevatori*, March/April 1991, pp. 39–43. – Appendix.

Moussallati, M. Z., 1974, 'Lift performance under balanced interfloor traffic conditions', M.Sc. dissertation, UMIST. – 17.3.

Nagatani, T., 2003, 'Complex behavior of elevators in peak traffic', *Physica A*, vol. 236, 2003, pp. 556–566. – 18.7.

Nahon, J., 1990, 'Traffic analysis for small scale projects', *Elevator World*, February. – 8.3.3.

Negnevitsky, M., 2005, *Artificial intelligence: a guide to intelligent systems*, 2nd edition, Addison Wesley. – 12.2.2.

NEII-1, 2000, *Building transportation standards and guidelines*, NEII. – 6.14, 6.8.1.

Oldfield, P., Trabucco, D. & Wood, A. (eds.) (2014) *Roadmap on the Future Research Needs of Tall Buildings*, Council on Tall Buildings and Urban Habitat. – 20.5.

Pang, G., & Nandy, B., 1989, 'Intelligent scheduling of a group of elevators', *IEEE International Symposium on Intelligent Control*, 1989, 25–26 September 1989, Albany, New York, pp. 144–149. – 12.2.2.

Papoulis, A., 1965, *Probability random variables and stochastic process*, McGraw Hill. – 7.7.2.

Pearce, C., 1995, 'Effects of basements on lift traffic analysis – a review', *Elevator Technology 6*, IAEE Publications. – 8.3.3, 8.3.4.

Pearce, C., 1996, 'The lobby – a trivial element in lift design', *Elevator Technology 7*, IAEE Publications. – 8.3.5.

Pepyne, D. L., & Cassandras, C. G., 1997, 'Optimal dispatching control for elevator systems during uppeak traffic', *IEEE Transactions on Control System Technology*, vol. 5, no. 6, November 1997. – 12.2.2.

Peters, R., 1990, 'Lift traffic analysis: formulae for the general case', *Building Services Engineering Research & Technology*, vol. 11, no. 2, 1990, pp. 65–67. – Table 5.8, 5.10.3, 7.9.1, 7.9.2, 17.3, 18.4.1, 18.2.1.

Peters, R. D., 1995a, 'General analysis double decker lift calculation', *Elevator Technology 6*, IAEE Publications. – 8.2.2.

Peters, R. D., 1995b, 'Ideal lift kinematics: complete equations for plotting optimum motion', in *Elevator Technology 6*, proceedings of Elevcon '95, Hong Kong, March 1995, pp. 175–184.

Peters, R. D., Mehta, P. & Haddon, J., 1996a, 'Lift passenger traffic patterns: applications, current knowledge and measurement', in *Elevator Technology 7*, proceeding of Elevcon 96 in Barcelona, Spain, published by the IAEE. – 2.10.5, 19.3.

Peters R. D., Mehta P & Haddon J, 1996b, Lift traffic analysis: general formulae for double decker lifts, *Building Serv. Eng. Res. Technol.*, vol. 17, no. 4, 1996. – 7.9.3.

Peters, R. D., 1996c, 'Ideal lift kinematics: derivation of formulae for the equations of motion of a lift', *International Journal of Elevator Engineers*, vol. 1, no. 1, 1996, pp. 60–71. –Appendix.

Peters, R. D., 1997a, 'Vertical transportation planning in buildings', Eng. D. thesis, Brunel University. – 7.9.1, 8.3.3, 8.6.

Peters, R. D., 1997b, Private communication. – 7.4.3.

Peters, R. D. & Mehta, P., 1998, 'Green lift control strategies', *Int. Jour. of Elev. Eng.*, IAEE. – 11.4.2.

Peters, R. D., 2000, *Guide D: transportation systems in buildings, section 4*, CIBSE. – 8.2.2.

Peters, R.D. & Sung A.C.M., 2000, 'Beyond the uppeak', *Elevator Technology 10*, Elevcon Books. – 16.2.1, 16.2.2.

Peters, R., 2006, 'Understanding the benefits and limitations of destination dispatch', in *Elevator Technology 16*, proceedings of Elevcon Helsinki 2006, June 2006, IAEE, pp. 258–269. – 12.2.2.

Peters, R. D. & Evans, E., 2008, 'Measuring and simulating elevator passengers in existing buildings', in *Elevator Technology 17*, proceedings of Elevcon 2008, Thessaloniki, Greece, pp. 289–300. – 19.2.

Peters, R.D., Smith, R. & Evans, E., 2011, 'The appraisal of lift passenger demand in modern office buildings' *Building Serv. Eng. Res. Technol.*, published online before print, 9 November 2010, doi: 10.1177/0143624410385378, vol. 32, no. 2, May 2011,pp. 159–170. – 19.1.

Peters, R.D., 2012, 'Lift performance time', *2nd Symposium on Lift and Escalator Technologies*, The University of Northampton, UK, vol. 2, pp. 53–64. – 19.2.

Peters, R. D., 2013, 'The application of simulation to traffic design and dispatcher testing', *3rd Symposium for Lift and Escalator Technologies*, September 2013, Northampton, UK, pp. 128–139. – 20.4.

Peters Research Ltd., 2014, 'Elevate'. https://www.peters-research.com/index.php/elevate/about-elevate. – 1.7, 18.3.

Petigny, B., 1972, 'Le calcul des ascenceurs (Lift Calculations)', *Transportation Research*, vol. 6, no. 1, March 1972, 19–38. – 7.4.1, 7.7.3.

Phillips, R.S., 1973, *Electric lifts*, Pitman. – 5.5.1, 7.4.1, 7.4.3, 7.5, 7.6, 8.3.6.

Port, L.W., 1961, *Australian patent specification 255218*. – 11.6.1.

Port, L.W., 1968, *The Port elevator system*, University of Sydney, June, 1968. – 11.6.1.

Powell, B. A., 1971, 'Optimal elevator banking under heavy up-traffic', *Transportation Science*, vol. 5, no. 2, May 1971, pp. 109–121. – 12.2.3.

Powell, B. A., 1975, 'Elevator banking for high rise buildings', *Transportation Science*, vol. 9, no. 3, August 1975, pp. 200–210. – 12.2.3.

Powell, B. A., 1992, 'Important issues in up peak traffic handling', in *Elevator Technology 4*, proceedings of Elevcon '92, Amsterdam, the Netherlands, May 1992. – 12.2.3.

Powell, B. A., Sirag, D. J. & Whitehall, B. L., 2000, 'Artificial neural networks in elevator dispatching', in *Elevator Technology 10*, proceedings of Elevcon 2000, Berlin, Germany, IAEE, pp. 1–10. – 12.3.2.

Powell, B.A., 2001, 'Elevator planning and analysis on the web', *Elevator Technology 11*, Elevcon Books. – 11.8.

Prowse, R.W., Thomson, T. & Howells, D., 1992, 'Design and control of lifts systems using expert systems and traffic sensing', in *Elevator Technology 4*, proceedings of the Elevcon 1992, Amsterdam, The Netherlands, 1992, IAEE, pp. 218–226. – 11.10, 12.2.1.

Qun, Z., Ding, S., Yu, C. & Xiaofeng, L., 2001, 'Elevator group control system modelling based on object-oriented Petri net', *Elevator World*, August 2001, pp. 99–105. – 11.10, 12.2.2.

Roschier, N.R. & Kaakinen, M.J., 1980, 'New formulae for elevator round trip time calculation', *Elevator World*, August 1980. – 7.4.1, A1.

Ruokokoski, M., Ehtamo, H., Sorsa, J. & Siikonen, M-L., 2008, 'Passenger allocation to capacitated elevator problem', *Transportation Science*, manuscript 2.1 22.10.2008. – 12.2.2.

Russett, S., 2002, 'An open and shut case', *RIBA Journal*, August 2002. – 11.6.3.

Saadeh, H., Nofel, F. & Younes, M., 2012, 'Analysis of the passenger arrival process in an elevator lobby', course project, supervisor: Dr. Lutfi Al-Sharif, Mechatronics Engineering, The University of Jordan, 24 December 2012. – 19.2.

Sakai, Y. & Kurosawa, K., 1984, 'Development of elevator supervisory group control system with artificial intelligence', *Hitachi Review*, vol. 33, no. 1, 1984, pp. 25–30. – 12.2.3.

Sanjeev, S.R., 1991, 'Knowledge based system to generate lift software', in the *Proceedings of Interlift 1991 Convention*, Munich, 23–24 September 1991, pp. 90–98. – 9.3.

Schofield, A.J., Stonham, T.J. & Mehta, P.A., 1995, 'A machine vision system for counting people', *Proceedings of Intelligent Building Congress 1995*. – 19.3.

Schroeder, J., 1955, 'Personenaufzuege (passenger lifts)', *Foerden und Heben*, 1. (in German) – 4.5, 6.3., 9.8.

Schroeder, J., 1984, 'Down peak handling capacity of elevator groups', *Proc. Int. Lift Sym.*, Amsterdam. – 14.3.2.

Schroeder, J., 1989a, 'Sky lobby elevatoring', *Elevator World*, January 1989. – 8.2.1, 8.3.2.4.

Schroeder, J., 1989b, *Value elevatoring by computer*, Profi Press. – 8.2.1.

Schroeder, J., 1990a, 'Elevator traffic: elevatoring, simulation, data recording', etc, *Elevator World*, June. – 13.5.

Schroeder, J., 1990b, 'Elevatoring for modern supervisory systems', *Elevator Technology 3*, IAEE Publications. – 6.3.

Schroeder, J., 1990c, 'Advanced dispatching, destination hall calls', *Elevator Technology 3*, IAEE Publications. – 11.6.1.

Schroeder, J., 1990d, 'Elevatoring calculation: probable number of stops and reversal floor, M10 destination hall calls + instant call assignment', in *Elevator Technology 3*, proceedings of Elevcon 1990, Rome, Italy, March 1990, IAEE, pp. 199–204. – 12.4.

Scott, A., 2014, 'British Council for Offices (BCO) Guide to Specification 2014 – Vertical Transportation', *4th Symposium on Lift and Escalator Technologies*, The University of Northampton, 25–26 September 2014. – 6.4.

Siikonen, M-L., 1991, 'Simulation – a tool for enhanced elevator bank design', *Elevator World*, April 1991. – 19.3.

Siikonen, M-L., 1998, 'Double deck elevators: savings in time and space', *Elevator World*, July. – 8.2.2, 16.2.1.

Siikonen, M-L., 2000, 'On traffic planning methodology', in *Elevator Technology 10*, proceedings of Elevcon 2000, Berlin, Germany, The International Association of Elevator Engineers. – 12.2.2, 14.5, 18.3.1.

Siikonen, M-L., Susi, T. & Hakonen, H., 2001, 'Passenger traffic flow simulation in tall buildings', *Elevator World*, August. – 11.10.

Siikonen, M-L. & Hakonen, H., 2002, 'Efficient evacuation methods in tall buildings', in *Elevator Technology 12*, proceedings of Elevcon 2002, Milan, Italy, June 2002, IAEE, pp. 237–246. – 20.5.

Siikonen, M-L., 2002a, Private communication. – 15.5. 15.6, 16.2.2, Part D.1.

Siikonen, M-L., 2002b, 'Procedure for controlling an elevator group where virtual passenger traffic is generated', *US patent no. 6 345 697 B1*, 12 February 2002. – 12.3.2.

Siikonen, M-L. & Ylinen, J., 2006, 'Elevator group control method using destination floor call input', *US patent no. 7 083 027 B2*, 1 August 2006. – 12.2.3, 12.4.

Siikonen, M-L., 2008, 'Elevator group control method', *European patent no. EP 1 549 581 B1*, 10 September 2008. – 12.2.3, 12.4.

Smith, R. & Peters, R., 2002, 'ETD algorithm with destination dispatch and booster options', in *Elevator Technology 12*, proceedings of Elevcon 2002, Milan, Italy, June 2002, pp. 247–257. – 11.1.3, 11.6.1, 12.2.2.

Smith, R., 2003, 'The divine algorithm', *Elevator World*, March 2003. – 20.6.

Smith, R. & Peters, R., 2004, 'Enhancements to the ETD dispatcher algorithm', in *Elevator Technology 14*, proceedings of Elevcon 2004, Istanbul, Turkey, April 2004, IAEE, pp. 234–243. – 12.2.2, 20.2.

So, A.T.P., Chan, W. L., Kwok, H. S. & Liu, S. K., 1992, 'A computer vision based group supervisory control system', in *Elevator Technology 4*, proceedings of Elevcon '92, Amsterdam, The Netherlands, May 1992, pp. 249–258. – 12.3.4.

So, A.T.P., Chan, W. L., Kwok, H. S. & Liu, S. K., 1993a, 'A comprehensive solution to computer vision based group supervisory control', in *Elevator Technology 5*, proceedings of Elevcon 1993, Vienna, Austria, November 1993, pp. 203–211. – 12.3.4.

So, A.T.P., Kuok, H. S., Liu, S. K., Chan W. S. & Chow, T. Y., 1993b, 'New developments in elevator traffic analysis', *Elevator Technology 5* proceedings of Elevcon '93, Vienna, IAEE, pp. 212–221.

So, A.T.P., Beebe, J. R., Chan, W. L. & Liu, S. K., 1996, 'An artificial neural-network based traffic patterns recognition system', *International Journal of Elevator Engineering*, vol. 1, 1996, pp. 35–46. – 12.2.1, 12.3.1.

So, A.T.P. & Suen, W.S.M., 2002, 'New formula for estimating average travel time', *Elevatori*, vol. 31, no. 4, 2002, pp. 66–70. – 6.8.3, 11.4.2, 18.4, 18.5.

So, A.T.P., 2014, 'Fastest elevator – a competition in high technology', *Elevator World*, September 2014, pp. 152–160. – 8.3.2.1.

So, A.T.P., Al-Sharif, L. & Hammoudeh, A. T., 2014, 'Traffic analysis of a simplified two-dimensional elevator system', accepted for publication in *Building Services Engineering & Technology*, December 2014. – 20.2.

Sorsa, J., Hakonen, H. & Siikonen, M-L., 2005, 'Elevator selection with destination control system', in *Elevator Technology 15*, International Association of Elevator Engineers, proceedings of Elevcon 2005, Peking, China. – 12.2.3, 12.4.

Sorsa, J., Kuusinen, J. M. & Siikonen, M. L., 2012, 'Passenger batch arrivals at elevator lobbies', in *Elevator Technology 19*, proceedings of Elevcon 2012, Miami Beach, USA, IAEE, pp. 264–273. – 7.7.4.

Sorsa, J. & Siikonen, M. L., 2014, 'Up-peak roundtrip time in theoretical calculation, traffic simulation and reality', in *Elevator Technology 20*, proceedings of Elevcon 2014, Paris, France, pp. 347–357, published by the IAEE. – 19.2.

Sorsa, J. S., Ehtamo, H., Siikonen, M-L., Tyni, T., Ylinen, J., (year unknown) 'The elevator dispatching problem', submitted to *Transportation Research*, manuscript 1. sal.aalto.fi/publications/pdf-files/msor09.pdf. – 12.2.2.

Stanley, J., Williams, D., Simcik, P., Honma, H. & Mori, T., 2011, 'Elevator traffic control including destination grouping', *US patent no. 7 921 968 B2*, 12 April 2011. – 12.4.

Strakosch, G. R., 1967, *Elevators and escalators*, 1st edn., Wiley. – 4.3.2, 4.4, 5.5.1, 7.4.1, 7.5, 7.6, 8.3.3, 8.3.6, 10.2, 10.8.1, 13.3, 14.1, 14.4.1.

Strakosch, G. R., 1983, *Vertical transportation*, 2nd edn., Wiley. – 5.10.1, 5.10.2, 8.2.2.

Strakosch, G. R., 1988a, 'Questions and answers', *Elevator World*, November, 1988. – 8.3.1.2.

Strakosch, G. R., 1998b, *Vertical transportation handbook*, 3rd ed., Wiley. – 5.8, 5.10.1, 5.10.2, 6.7.1.

Swindells, W., 1975, 'Software for the computer control of lift systems', M.Phil. thesis, UMIST. – 17.3.

Taha, H. A., 2011, *Operations research: an introduction*, 9th edn. (international edn.), 2011, Pearson. – 18.6.4.

Tam, C. M. & Chan, A.P.C., 1996, 'Determining free lift parking policy using Monte Carlo simulation', *International. Journal of Elevator Engineering*, 1996, 1: pp. 24–34. – 12.3.5.

Thangavelu, K., 1989, 'Optimised "up-peak'" elevator channelling system with predicted traffic vol. equalised sector assignments', *US patent no. 4 846 311*, 11 June 1989. – 12.2.3.

Thumm, G., 2004, 'A breakthrough in lift handling capacity', in *Elevator Technology 14*, proceedings of Elevcon 2004, Istanbul, Turkey, April 2004, IAEE. – 20.2.

Thyssenkrupp Elevators, 2014, 'MULTI achieves the elevator industry's Holy Grail', *Elevation*, no. 81, April 2014, pp. 50–53. – 20.2.

Tolosana, N., Larrode, E., Cuartero, J., Miravete, A., Calvo, J. & Castejon, L., 1998, 'Application of artificial neural networks on vertical traffic simulation', in *Elevator Technology 9*, proceedings of Elevcon 1998, Zurich, Switzerland, October 1998, pp. 265–271. – 12.2.2.

Tregenza, P. R., 1971, 'The prediction of passenger lift performance in multi storey buildings', Ph.D. thesis, University of Nottingham. – 2., 15.1.

Tregenza, P. R., 1972, 'The prediction of passenger lift performance', *Archit. Sci. Rev.* – 2, 5.13.2, 6.11.1, 7.4.1, 7.5, 7.7.3, 13.1, 13.3, 15.1.

Tregenza, P. R., 1976, *The design of interior circulation: people and buildings*, Crosby Lockwood. – 2, 15.1.

Tsuji, S., Amano, M. & Hikita, S., 1989, 'Application of the expert system to elevator group supervisory control', IEEE, in *Proceedings of the Fifth Conference on Artificial Intelligence Applications*, 6–10 March 1989, Miami, Florida, pp. 287–294. – 12.2.1.

Tutt, D. & Adler, P., 1990, *New metric handbook, planning and design data*, Butterworth. – 3.5.3, 8.4.11., CS11.

Tyni, T. & Yilnen, J., 1999, 'Genetic procedure for allocating landing calls in an elevator group', 25 May 1999, *US Patent no.: 5907137*. – 12.2.2.

Umeda, Y., Uetani, K., Ujihara, H. & Tsuji, S., 1989, 'Fuzzy theory and intelligent options', *Elevator World*, July 1989, pp. 86–91. – 12.2.1.

Wit, J., 2010, 'Evacuating using elevators: enhancing the CTBUH approach for the Dutch high-rise covenant', in *Elevator Technology 18*, proceedings of Elevcon Lucerne 2010, June 2010, IAEE. – 20.5.

Wit, J., 2014, 'Sharing elevator capacity: exploring the unused potential of stacked mixed-use high-rise buildings', in *Elevator Technology 20*, proceedings of Elevcon 2014, Paris, France, July 2014, IAEE. – 20.2.

Yamashita, K., Fujino, A., Tobita, T., Okabe, R. & Kawano, K., 2000, 'Improvement of single elevator operation by using prior standby control', in *Elevator Technology 10*, proceedings of Elevcon 2000, Berlin, Germany, IAEE, pp. 252–259. – 12.3.5.

Yang, T., Kaji, K. & Kawaguchi, N., 2013, 'Elevator acceleration sensing: design and estimation recognition algorithm using crowdsourcing', *2013 IEEE 37th Annual Computer Software and Applications Conference Workshops*, pp. 534–539. –Appendix.

Yoshikawa, T., Toriyabe, S. & Hoshino, T., 2013, 'System and display for an elevator group supervisory and method for supervising a plurality of elevators', *European patent specification no. EP 1 719 727 B1*, 10 April 2013. – 12.4.

Zhang, J. & Zong, Q., 2014, 'Energy-saving-oriented group-elevator dispatching strategy for multi-traffic patterns', *Building Services Engineering Research & Technology*, appeared online before print: 17 March 2014, doi: 0143624414526723. – 12.2.2.

Zimmermann, F. O., 1973, 'Elevator traffic analysis', *Elevator World*, May. – 13.3, 14.1, 14.4.1.

Zong, Q., Yue, Y-j., Cao, Y-f. & Shang, X-g., 2000, 'Application of intelligence forecasting method in traffic analysis of EGCS', *Transactions of Tianjin University*, vol. 6, no. 1, May 2000. – 12.3.6.

Bibliography

The following publications have not been referenced in the text, but readers may find them useful.

Aoki, H., 2012, 'Average waiting time of destination control systems', in *Elevator Technology 19*, proceedings of Elevcon 2012, Miami Beach, FL, IAEE, pp. 48–55.

Fortune, J. W., 2008, 'Individual car loads – the myths and the reality', in *Elevator Technology 17*, proceedings of Elevcon 2008, Thessaloniki, Greece, June 2008, IAEE, pp. 79–86.

Gagov, Z., Cho, Y. C. & Kwon, W. H., 2001a, 'Implementation of velocity profiles in group control of elevator systems', in *Elevator Technology 11*, proceedings of the Elevcon Asia 2001, July 2001, Singapore, IAEE, pp. 25–34.

Gagov, Z., Cho, Y. C., & Kwon, W. H., 2001b, 'Improved concept for derivation of velocity profiles for elevator systems', *Proceedings of the 2001 IEEE International Conference on Robotics and Automation*, Seoul, Korea, May 21–26, 2001, pp. 2419–2423.

Gagov, Z., Cho, Y. C., Kwon, W. H., Han, B. H., 2000, 'Timing considerations and kinematics parameters of lift group control systems', in *Elevator Technology 10*, proceedings of Elevcon 2000, Berlin, Germany, May 2000, IAEE, pp. 245–251.

Hakonen, H. & Siikonen, M. L., 2006 'Generalised calculation of round trip time', in *Elevator Technology 16*, proceeding of Elevcon 2006, Helsinki, Finland, June 2006, IAEE, pp. 90–99.

Hall, E. T., 1966, *The hidden dimension*, Doubleday.

Halsey, L., 2015, 'Hybrid lift group control systems', *Elevator World*, January 2015.

Jappsen, H. & Rieke, O., 2010, 'Difficulties in comparing the efficiency of different destination control systems', in *Elevator Technology 18* proceedings of Elevcon 2010, IAEE, 435–437.

Jones, B., 1922, 'A method of determining resultant input from individual duty cycle and determining rating', *General Electric Review*, vol. XXV, no. 7, July 1922, pp. 405–417.

Kaakinen, M. & Roschier, N. R., 1991, 'Integrated elevator planning system', *Elevator World*, March 1991.

Li, Z., Mao, Z. & Wu, J., 2004, 'Research on dynamic zoning of elevator traffic based on artificial immune algorithm', *8th International Conference on Control, Automation, Robotics and Vision*, Kunming, China, 6–9 December 2004.

Maehara, T., Hatori, T., Tsuboi, A., Nishida, T., Aida, K. & Yoshikawa, T., 2014, 'Group management control that improves transport capability and enhances energy efficiency', in *Elevator Technology 20*, proceedings of Elevcon 2014, Paris, France, IAEE, pp. 186–194.

Markon, S., Aoki, K., Nakagawa, M. & Sudo, T., 2008, 'Recent trends in elevator group control systems', *23rd International Technical Conference on Circuits/Systems Computers and Communications*.

Mulvaney, D. & Hamdi, M., 2002, 'Real-time dynamic scheduling and its application to lift systems', in *Elevator Technology 12*, proceedings of Elevcon 2002, Milan, Italy, IAEE, pp. 206–216.

Peters, R. D. & Smith, R., 2010, 'Designing elevator installations using modern estimates of passenger demand', *Elevator Technology 18*, proceedings of Elevcon 2010, Lucerne, Switzerland, IAEE, pp. 284–295.

Peters, R. D., 2014, 'Elevator dispatching', in *Elevator Technology 20*, proceedings of Elevcon 2014, Paris, France, IAEE, pp. 125–138.

Powell, B. A., 2002, 'Elevator planning and analysis on the web', *Int. J. Elev. Eng.*, 4.

Powell, B., 2008, 'An alternate approach to traffic analysis for residential buildings', in *Elevator Technology 17*, proceedings of Elevcon 2008, Thessaloniki, Greece, June 2008, IAEE, pp. 313–323.

Schroeder, J., 1987, 'Elevator trip profiles', *Elevator World*, November 1987.

Shenquan, G. & Suzhou, F. S., 1991, 'Up-peak traffic simulation of elevators', *Elevator World*, February 1991.

Siikonen, M-L., 1993, 'Elevator traffic simulation', *Simulation*, vol. 61, no. 4, pp. 257–267, October 1993.

Siikonen, M-L., Susi, T. & Hakonen, H., 2000, 'Passenger traffic flow simulation in tall buildings', *IFHS, International Conference on Multi-Purpose High-Rise Towers and Tall Buildings*, November 2000.

So, A.T.P. & Yu, J.K.L., 2001, 'Intelligent supervisory control for lifts: dynamic zoning', *Building Services Engineering Research & Technology*, vol. 22, no. 1, 2001, pp. 14–33.

Sorsa, J. & Siikonen, M. L., 2006, 'Double deck destination control system', *Elevator Technology 16*, proceedings of Elevcon 2006, June 2006, Helsinki, Finland, IAEE, pp. 318–327.

Sudo, T. & Markon, S., 2001, 'The performance of multi-car linear motor elevators', in *Elevator Technology 11*, proceedings of the Elevcon Asia 2001, July 2001, Singapore, IAEE, pp. 141–149.

Suzuki, N., Kobori, S., Iwata, M. & Hikita, S., 'Elevator supervisory control system with cars cooperative method', *Elevator Technology 16*, proceedings of Elevcon 2006, June 2006, Helsinki, Finland, IAEE, pp. 338–346.

Yang, S., Tai, J. & Shao, C., 2009, 'Dynamic partition of lift group control system with destination floor guidance in up-peak traffic', *Journal of Computers*, vol. 4, no. 1, January 2009, pp. 45–52.

Yoshikawa, T., Muraoka, K., Hoshino, T., Toriyabe, S. & Aida, K., 'Elevator group control system with the advanced forecasting trajectory technique (AFTT) to achieve equal time interval operation', *Elevator Technology 16*, proceedings of Elevcon 2006, June 2006, Helsinki, Finland, IAEE, pp. 373–382.

Zubia, K., 1996, 'Time, distance, speed, acceleration and jerk in elevator starting and stopping', *Elevator Technology 7* proceedings of Elevcon 1996, Barcelona, Spain, IAEE, pp. 277–286.

Index